COMPENDIUM 0F ARTICLES ON EHV SUBSTATIONS & PROTECTIONS FOR BUDDING AND PRACTICING ENGINEERS OF TRANSMISSION UTILITIES

Er. K.K.MURTY

notionpress.com

INDIA • SINGAPORE • MALAYSIA

ISBN 979-8-89475-020-0

DEDICATION

I dedicate this book to my;

parents - Late Dr.K.Bhaskar Murty & Late Mrs.K.Sundari Murty.

beloved wife - Late Mrs. K.Vijaya Murty, who inspired and encouraged me to pen-down my experiences for future generation.

And I would like to thank my children, grandchildren and the people that are close to me for supporting me through thick and thin, without their undying love and support, this book would not have been possible today.

I would also like to thank my grandson, Kaustubh, for editing and proofreading my creation.

My grand-daughter has written a poem, dedicated to me

MY HERO-MY GRAND FATHER

Like a sailor
He sails across
the depths of wisdom,
Battling the tides that swept the surface

With his mighty sword,
He spreads his refined erudition
Each word he sings
Ignites a flame in
The hearts of Lord Indra's warriors
Like a gentle breeze,
He guides the pollens of
Love and care
To people he holds close

Like the man he is,
He holds his passion close
And his loved ones closer....
Aanvi

Contents

Section-1
Articles related to EHV Substation.

Section-2
Articles related to transformers.

Section-3
Articles on Relays & Protections

Gratitude and Acknowledgments

- I acknowledge with gratitude to my Gurus, guides, viz; Mr.P.Ramachandran, Dr. K. Rajamani, Mr. M.L. Jain, Mr. Virendra Lakhiani, Mr. Y.K. Pandhripande, Mr. Sanjay Patki, Mr.S.Ganesan, Mr. Pramod Rao and Dr. C.S. Narasimhan.
- I acknowledge with gratitude to my mentors;

  i. Mr. P.P. Wahi, Former Director, CBIP New Delhi, the platform of CBIP was given by him to me to grow and it is my proud privilege that he wrote the "Foreword" for this book.

  ii. Mr. B.V. Raghavaiah, Former Director CPRI Bhopal, gave me opportunities to carry out Third Party Inspections on behalf of CPRI, Bangaluru/Bhopal and also gave me the platform of CPRI to grow.

- I acknowledge with gratitude to all the MDs of MPPTCL including Mr.Sunil Tiwari and all the C.E.(T&C)s who had/have permitted me to use data of MPPTCL and also permitted the serving Engineers to get associated with me as co-authors.
- I am grateful to Mr. P.K.Pattanaik, Sr.GM, OPTCL and Dr. Rajesh Arora, Sr. Manager, DTL, who gave me opportunity to be a member of knowledge-sharing group"Spark Ignited to Share" and "Igniting Power Engineers" respectively and thus enhanced my knowledge.
- I am thankful to Dr. Mladen Banovic, Chief Editor "TRANSFORMERS MAGAZINE", published from Croatia (Europe) by giving me opportunities to write for this journal, which has global repute.
- I am thankful to the Editors/Sub-editors of ieema journal, and Electrical Mirror who publish my Articles in their respective technical journals.
- I feel honoured as following professionals gave me opportunity to speak in Webinar/presentation before a group of professionals;

  1. Dr. Rajesh Kumar Arora, Sr. Manager, Delhi Transmission Ltd.,
  2. Mr.Yugal Agrawal, Jt. Dir, CPRI, Bhopal,
  3. Dr. Mukesh Kumar Mishra, IEI, MP State Unit, Bhopal
  4. Mr. Nihar Raj, Sr.V.P(O&M). Transmission, ADANI Ltd. Ahmedabad.

- I am grateful to all those who helped me in writing the Articles and papers viz. Dr. Kulkarni, Prof. IIT, Mumbai, Dr. Monalisa Biswal, Asstt. Prof. NIT Raipur, Mr. Palande, MPPTCL, Mr Umesh Gupta of BSES, Rajdhani, New Delhi, Mr. Dwijen Roy of NHDC and Mr. Rakesh Patil of Siemens, Kalwa.
- I am thankful to Mr. J.J.L. Kapil, Mr. Santosh Dubey and Mr. S.K. Chaturvedi, for their immense help and support and to Mr.Santosh Bagade, who scrupulously scrutinizes my Articles/papers
- I am thankful to all my co-authors of Articles/Papers.
- I express my gratitude to all our family friends and also my friends from school level to Professional field for encouranging me.
- I express my sincere apologies to those, whose names have been inadvertently, omitted.

K.K. Murty

Foreword

I feel immense pleasure in introducing this prestigious book captioned as "COMPENDIUM OF TECHNICAL ARTICLES RELATING TO SUBSTATIONS AND RELAY PROTECTION"authored by Er.K.K.MURTY, Former Chief Engineer Testing and Commissioning of Madhya Pradesh Power Transmission Utility.

I would like to bring out that Mr. Murty had the bright academic and professional career spanning over long 45+ years.

He did his Bachelor's degree in Electrical Engineering in honours. Thereafter, he served MPSEB / MPPTCL with merit continuously for 33 years. After his superannuation from Madhya Pradesh Transmission Utility, he climbed to bigger glorious heights by serving various national and international organisations for 12 more years as an Expert Professional.

Besides above, he had the distinction of getting national award from the Institution of Engineers Calcutta in the year 2015 owing to his excellent contribution and eminence at national level.

During his active service in Madhya Pradesh for 33 years, he remained associated with operation and maintenance of substation equipments besides working as a specialist for testing and commissioning of the various Power Stations and EHV Sub stations up to and including 400kV.By virtue of his precious field experience acquired while handling several challenging assignments as above, he took the initiative of bringing out several technical papers from time to time.

Recently the author has taken extraordinary pains to collate his life-time experience in the shape of this book for the benefit of Engineering students as well as for the Practicing Power Engineers.

Here, I would like to focus on the two points which deserve the special mention;

A. That this useful document of its own kind incorporates the precious technical information which is generally not taught to the students in the Engineering colleges.

B. In the other part of this book, the author has excellently contributed and compiled several technical papers for dissemination of knowledge for the fresh and serving Power Engineers working in the Transmission segments .

As mentioned above, these papers are the outcome of his wide experience and expertise on substations and protection relays etc. while his serving in the power sector.

The specific mention of his works relating to Transmission domain can be seen by the readers on the back cover of this book. For further information of the readers, it is clarified that this book has been divided into three sections as per details given hereunder: The first section relates to dissemination of knowledge for the engineering students while the other three sections relate to the fresh and practising Engineers.

1. Articles related to EHV substation equipment and substation bus configuration etc.

2.Articles exclusively on Power transformers.

3.Articles related to protection relays and protection system etc.

Based on his meritorious work and valuable contribution, I, as the Director CBIP.had the privilege of inviting him as a learned speaker for participation in several national and international conferences organised by the Central Board of Irrigation and Power w.e.f. the year 2017 onwards.

I feel honoured to mention here that the presentations made by him in all these conferences were outstanding, illuminating and beneficial to the power Engineers and others. Besides his presentations in CBIP, he had presented several tech. papers in many other organisations as well, which brought him laurels and appreciation all around .Though he started writing the technical papers right from the peak of his professional career in Madhya Pradesh, yet he possess the same passion and zeal even today at the age of 79 years in bringing out more and more technical papers for the benefit of the power sector fraternity.

Lastly, I wish to make a very humble request to all the prospective readers that, the sincere efforts and pains with which the author has written this book will only be reciprocated, if his contributions are effectively utilised by them for accentuation of their technical knowledge and also for practical use in their day to day Engineering works.

I WISH MR. MURTY, ALL THE SUCCESS IN HIS ENDEAVOUR OF COMPILING THIS BOOK AS WELL AS FOR FUTURE ENDEAVOURS.

P.P.Wahi,
Former Director,
Central Board of Irrigation & Power, New Delhi.
Mobile: +91 98108 01555

Preface

After completing almost 45+ years in Electric Power Sector, I have gained enough experience in a prestigious State Transmission Utility, other fields and teaching the new entrants to the Power sector in India, this thought of compiling all my works came to my mind. My book contains as under;

A. EHV Substations.

B. Transformers

C. Protection & Relays.

Section wise salient points covered for and Transmission Engineers.;
A. EHV Sub Stations;

Bus-bar configuration, EHV equipment viz; LAs- selection criteria, Circuit breakers, GIS(Gas Insulated Substation) CVTs play a dual role in Grid network, Transient free back to back closing of capacitor banks using "Control Switching Device" Capacity assessment of Station battery-set, VRLA Batteries an over view, Innovative use of Auxiliary Bus (Transfers bus) to made available 220kV supply to a needy S/S, Synchronizing without Synchroscope by an innovative idea of drawing vector diagrams of incoming feeder & running bus Secondary voltages.

B. Transformers;

Transformers & reactor an over view, Pre-delivery Inspection of Power Transformer at manufacturers works, KYT (Know Your Transformer). In house extension of bushing turret for accommodating 300 mm long tail end bushing, In house repair of OLTC of 160 MVA TELK Transformer, Retrofitting of TELK 420kV draw-rod type bushing in place of a BHEL's draw-lead type bushing, Assessment of remnant life of old transformers on % solubility of moisture in oil methodology, OIP bushings-concept to condition monitoring, Life cycle of Transformers etc.

C. Protection & relays;

Concept of Time current Grading, Numerical O/C & E/F Relay, Need for Directional relay, Differential relay for Power transformers part 1(with conventional method & Part-2(With Numerical Differential Relay), REF protection- a complementary Protection to Differential Protection, Line Differential Protection, Concept & characteristics of Distance protection relays and setting philosophy. Carrier aided Auto-reclose operation, Bus-Bar protection, CTs- Concept, characteristic and constraints-A guide for budding Engineers, En masse tripping of 35 nos. EHV Transformers on Buchholz relay operation due to earth quake (seismic condition) and normalization thereof.

I am hopeful that this book shall prove to be a guide for budding Engineers as well as practicing Substation & Protection Engineers/Managers.

K.K.Murty.

SECTION-1

ARTICLES RELATED TO EHV SUBSTATION.

1. Concept of Bus-Systems in Substations/ Switch-Yards

Introduction: Bus bars are very vital arrangements in any substation/ switch yard. Bulk power is received from the Generators and grid-network at the bus and simultaneously the power is Transmitted/ distributed to the either to; inter connected Sub-stations at the same voltage level to other HV/EHV Sub-stations or at stepped down voltage level to Sub-Transmission Sub-stations.

1.0. Type of bus schemes:

1.1 Single bus scheme:

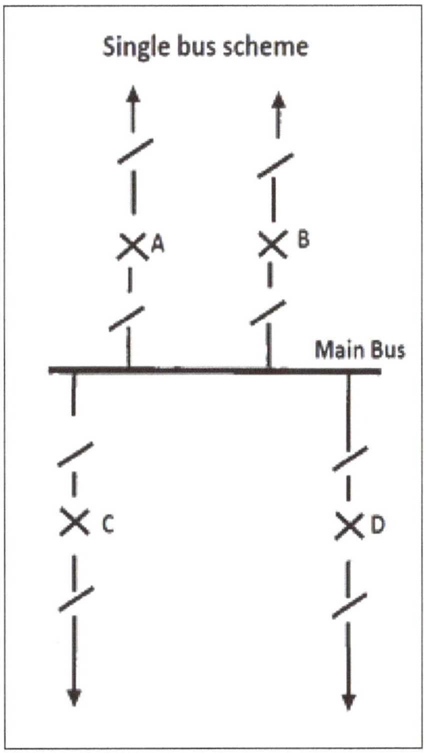

Figure 1

- This is simplest form of bus system arrangement.
- The disadvantage of this system is that the entire sub-station shall be taken out of service in case of any fault on the busbar or bus isolator.
- Also, in case of maintenance circuit breaker or of isolators, the said feeder shall be taken out of service.
- If there is some problem with the bus side isolator of any feeder the bus has to be made dead.

1.2 Single Bus scheme with bus section breaker:

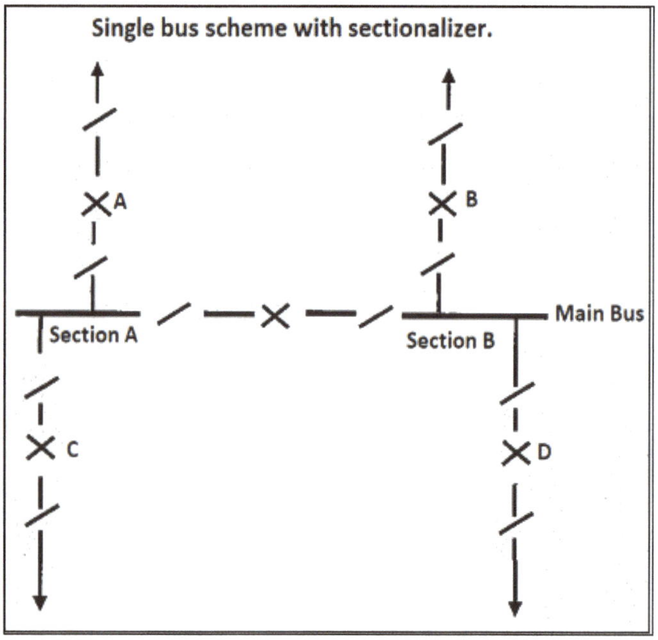

Figure 2

- This bus system has an edge-over the single bus system narrated in 1.1 as it has bus sectionalizing arrangement.
- The advantage of this bus system is;if there is some problem in either of the bus section, faulty bus section with any fault on that section or on the bus isolator of any feeder or transformer on that bus section can be isolated
- and the feeder or transformer on the healthy section of the bus can remain in service condition.

1.3 Double Main bus Scheme with bus coupler;

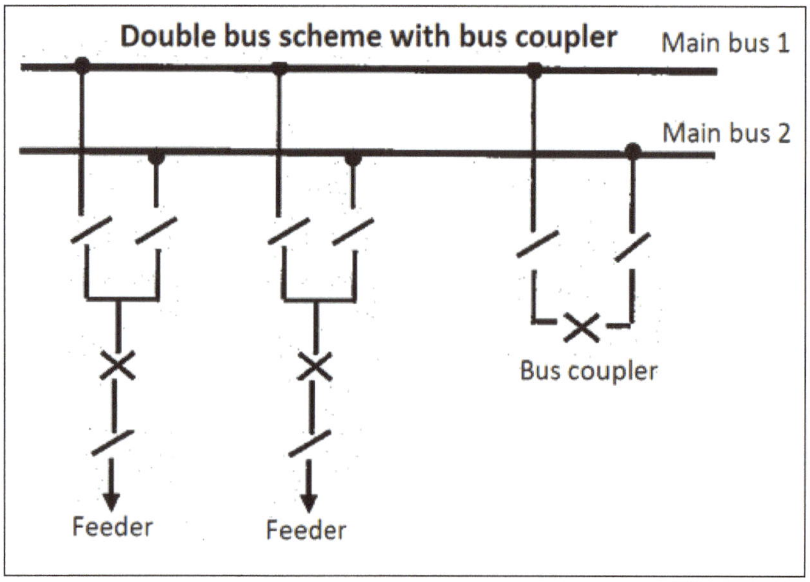

Figure 3

- This is a double bus system arrangement.
- Each circuit can be connected to either of these bus through an isolator.
- Bus coupler breaker is also provided so that the circuits can be switched on from one bus to another bus on load.
- The double bus arrangement is very suitable for large Generating stations as well as large grid stations

1.4 (a) 1-Main Bus & Transfer bus /Auxiliary bus scheme

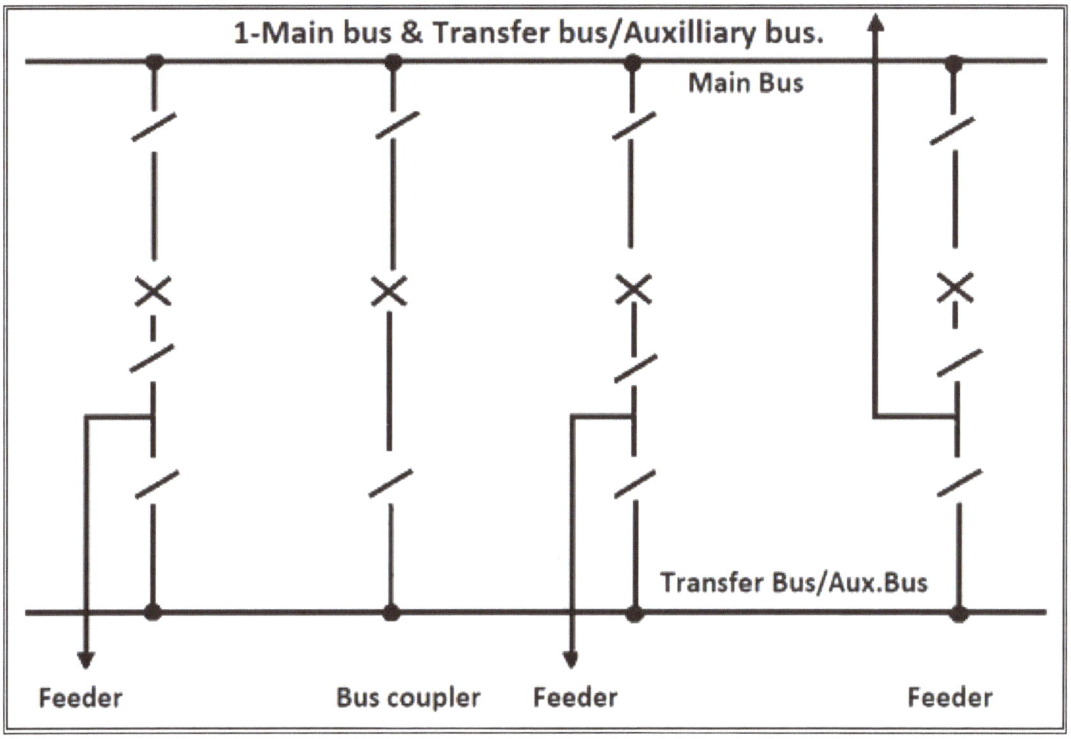

Figure 4

- In this arrangement, facilities are provided for BY-PASSING a Circuit breaker which is under maintenance.
- Each circuit is connected to the main bus-bar through a Circuit breaker and through an Isolator to transfer bus/Aux bus.
- Advantage: With-out interrupting load on any feeder/Transformer /capacitor bank, facilitates maintenance of their circuit breaker.

***Note:** Procedure for transferring a feeder on to the Auxiliary bus is narrated in sub-sequent para.2.0.

1.4 (b) 2-Main bus & Transfer bus/auxiliary bus scheme;

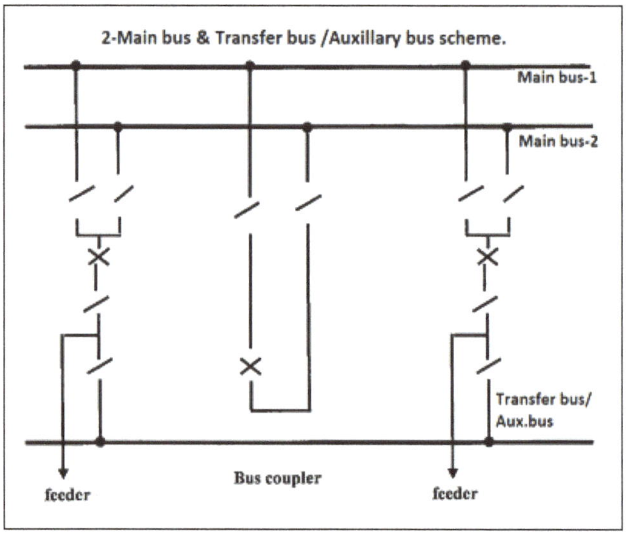

Figure 5

- This is combination of "Main and transfer bus scheme and Double Main bus scheme with a bus coupler".
- This scheme is similar to the scheme described in 2.3(a) only difference is it has 2- Main buses.
- Some Generating Stations adopt this scheme.
- In this scheme loads are distributed on the two buses each with its own transformer feed for normal operation, so that one bus fault will not cause a complete outage of the station.

1.5 (a) Breaker & half Scheme (conventional);

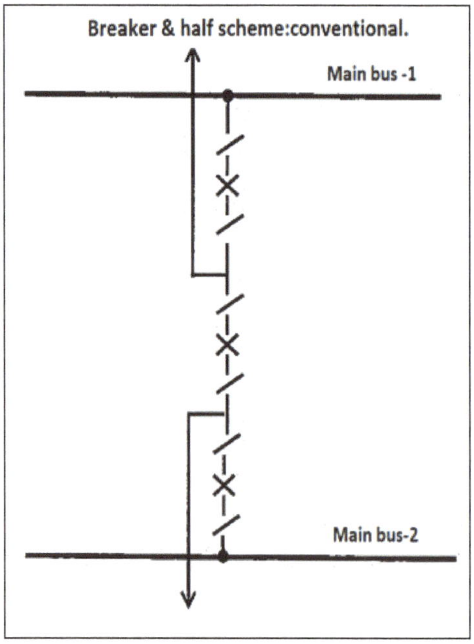

Figure 6

- It has been named as 1 & 1/2 Beaker scheme; it is derived from a simple mathematical equation; 3 Breakers / 2 Buses = 1&1/2.
- In this scheme 3-Circuit breakers are used for controlling two feeders.
- Normally both the buses shall be in service and in case of fault of any bus it is cleared by opening the affected bus through Bus bar protection, thereby ensuring continuity of supply.
- Similarly, any Circuit breaker can be taken out for maintenance without causing any interruption. This scheme is suitable for Station which handles large amounts of POWER in EHV/UHV substations.

1.5 (b) Breaker & half Scheme (D-Type–2bays on one side);

The concept of functioning and advantages are same as conventional breaker and half scheme.

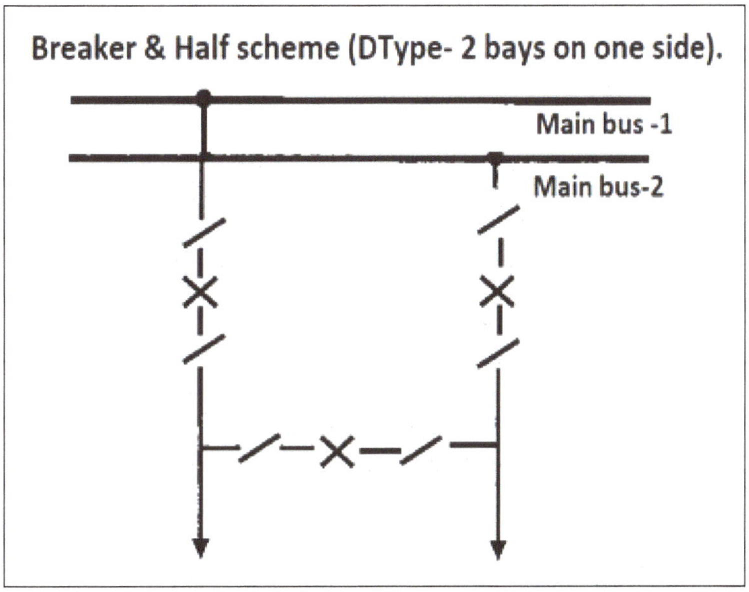

Figure 7

- Only difference in this scheme is; the 2-feeders are on the same direction.
- Requirement of space too is less as compared to the conventional breaker and half scheme.

2.0 Procedure for transferring a feeder on to the Auxiliary bus;

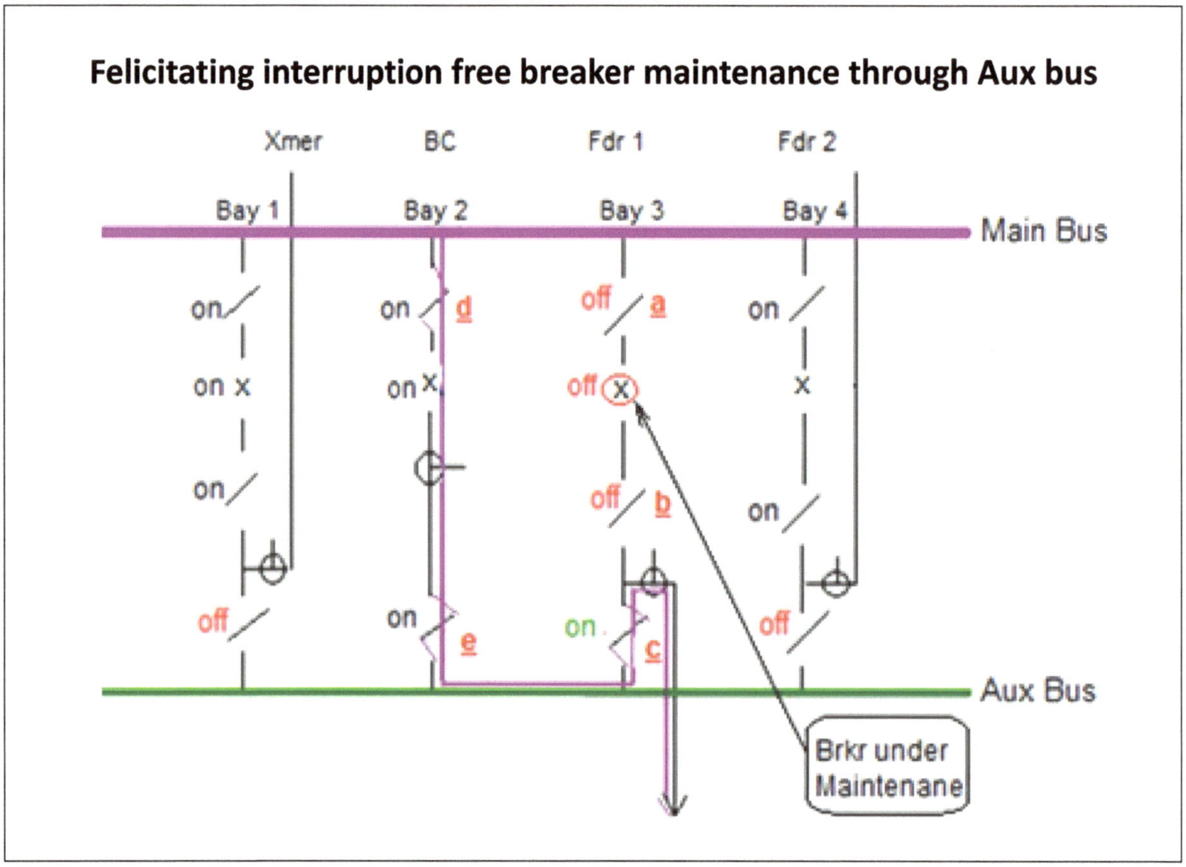

Figure 8

Operations;

i. Close Bus coupler Aux bus isolator "e".(bay2)
ii. Close Bus coupler Main bus isolator "d".(bay2)
iii. Close Aux bus isolator of feeder No1 "c".(bay3)
iv. Close (Switch on) Bus coupler breaker (bay2).

Note:

a. Now the load is being fed from feeder no 1 and the bus coupler through Aux bus, verify load sharing between both.
b. If there is trip-transfer scheme, take it into service.

v. Switch off feeder1. Beaker (bay no 3).
vi. Open Line side isolator "b" of feeder 1 (bay3).
vii. Open Main bus side isolator "a" of feeder 1 (bay3).
viii. Now feeder 1 breaker is free for maintenance. (bay3).

Note: Bus coupler bay C T ratio should be selected in such that it can take care for all the feeders.

2. Innovative Use of Auxiliary Bus in EHV Switch-Yard

Synopsis: - A Bus System in any Sub-station / Switch-yard in a power system are analogous to a bulk supply reservoir. Bulk Power supply is received on a bus from the generator units through GTs and /or from other nearby supply sources and is transmitted through different feeders to other adjoining sub stations. Bus arrangements in the Switch yards/ sub stations facilitate easy operations. There are two types of Bus arrangements in 220kV and above S/Ss generally followed;

- 1&1/2 Breaker scheme.
- 2 main buses and an auxiliary bus system.

In both the schemes, the main focus is on taking out Circuit breakers for maintenance/replacement without interruption of supply to the feeders/ transformers. In the state of MP right from the MPEB days the norms of 2 main buses and an auxiliary Bus system is followed for 220kV switch-yards / sub-stations.

Authors had inonvated an idea of extending supply to a 220kV S/s which was deprived of 220kV supply due to failure of its circuit breaker through the 220kV auxiliary bus and implemented this arrangement successfully.

1.0 Auxiliary bus Configuration: The Auxiliary bus facilitates interruption free breaker maintenance /breaker replacement,

- It is strung with the same sized conductors as the main bus.
- It has connection to the main bus 1&2 through main bus 1&2 main bus isolators (as the case may be) and an aux bus isolator.
- Each bay is equipped with auxiliary bus isolator.

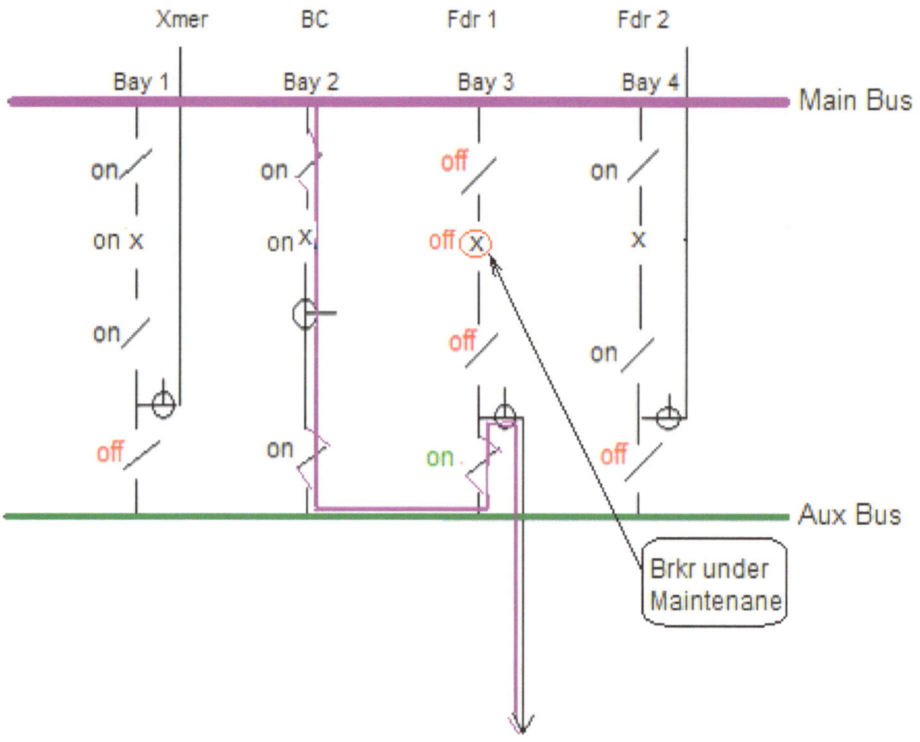

Fig: Fecilitating interruption free breaker maintenace through Aux bus

Figure 1: Use of Aux.bus in 1-main bus + an auxiliary bus configuration

3.0 Extending 220kV supply to needy 220kV S/S Sarni through innovative idea using Aux bus;

3.1 Case history:

The figure no.1depicts the concept of an Auxiliary bus and how it is utilised in by-passing the bay Circuit breaker for maintenance/ replacement without interruption.

An incidence had occurred when all the important Express/ mail and goods train stopped plying with electric locomotives between the Itarsi Rly Jn. to Nagpur Rly Jn. (Grand-Trunk route between New Delhi to Hyderabad-Chennai) and vice versa consequent to non-availability of traction supply from 132/25kV Rly Traction Sub stations. This incidence occurred as the source 220 kV mother Substation at Sarni (MP) was deprived of 220 kV Supply.

3.2 Supply arrangement:

The 220kV substation Sarni is 3 Kms away from SSTPS(Satpura Super Thermal Power Station)at Sarni in Betul Distt. It had an installed capacity of 2x100 MVA, 220/132kV Transformers. It gets 220kV supply from SSTPS's 220kV Bus through single 220 kV Transmission line. The other 220 kV feeders emanating from SSTPS Switch yard are 220kV Itarsi feeders- 1,2,3 & 4 & 220 kV Ambazari feeder(inter- state fdr).(Please refer fig. No2)

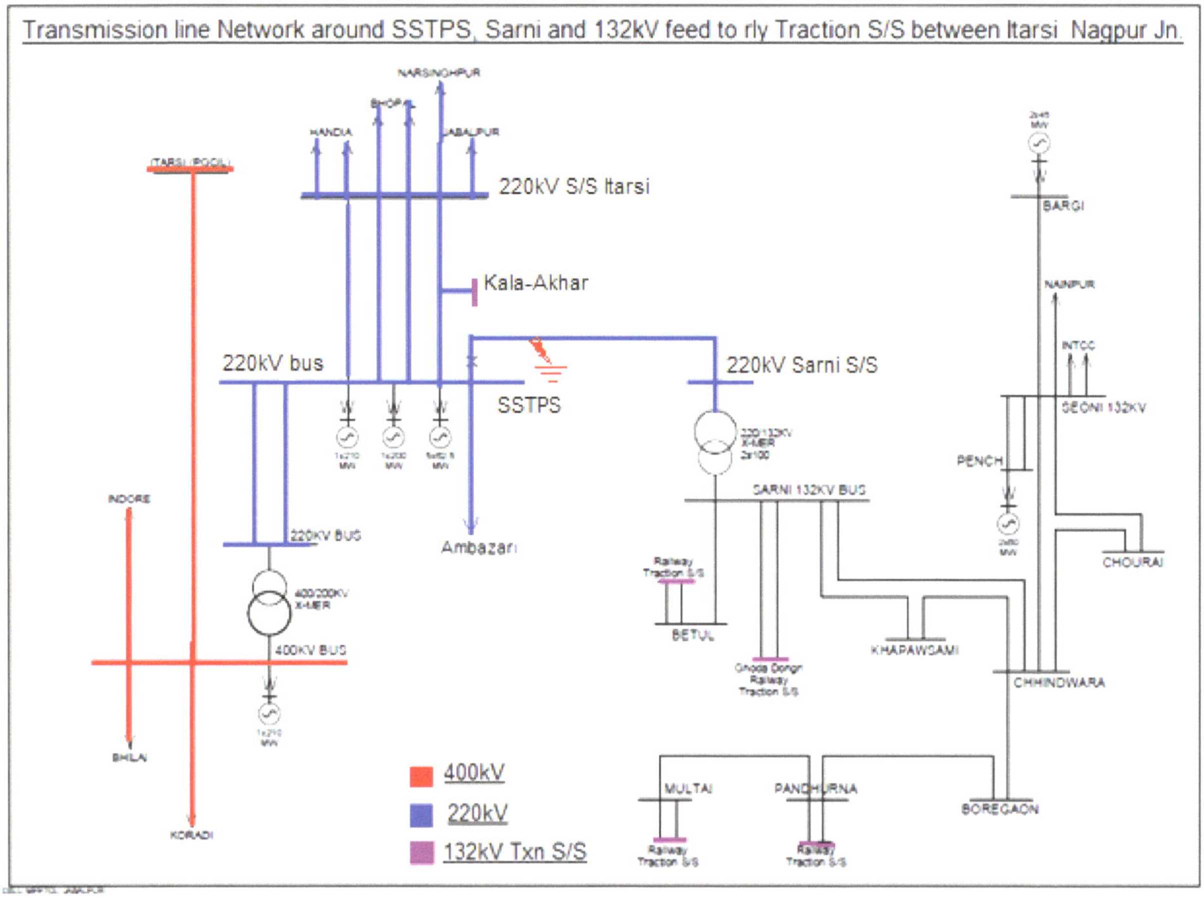

Figure 2: SLD Transmission System around Satpura Super Thermal Power Stn, Sarni, Fault on Sarni S/S Fdr & Traction S/Ss.

1.3 The 132kV feeders emanating from 220kV Substation Sarni are(Ref.fig.2);

i. Betul fdr,
ii. Ghora Dongri 1&2.
iii. Chhindwara
iv. Khapaswamy.

1.4 Traction S/S between Itarsi Rly Jn to Nagpur Rly Jn. in chronological order;

i. 220kV Txn S/S Kala-Akhar is tapped from 220kV Sarni-Itarsi Transmission line line.
ii. 132 KV Txn S/S, Betul fed through132kV Txn feeder no. I&II. From 132 kV s/s Betul S/S.
iii. 132KV Txn S/S, Ghoradongri is fed through132kV Ghoradonri feeder no. I &II from 220kV S/S Sarni.
iv. 132kV Txn S/S, Multai gets feed through Txn fdr no.I &II from 132kV S/S Multai which gets supply infeed from 132kV S/s Chhindwara via 132kV S/S Boregaon and 132kV S/S, Pandurna.

Note:-132kV S/S, Chhindwara gets supply from 220kV S/S Sarni and alternately from 132kV long lines from Pench Hydel Power stn, Bargi Hydel Power Stn or through Balaghat which gets supply from Bhilai (now in Chhatisgarh).

v. 132kV Txn S/S, Pandurna gets supply through Txn feeder no I&II from132kV S/S Pandurna.132kV Pandurna S/S gets supply from the upstream S/S- 132kV S/S Boregaon. Please refer to the note in (iv). (ref.fig.3)

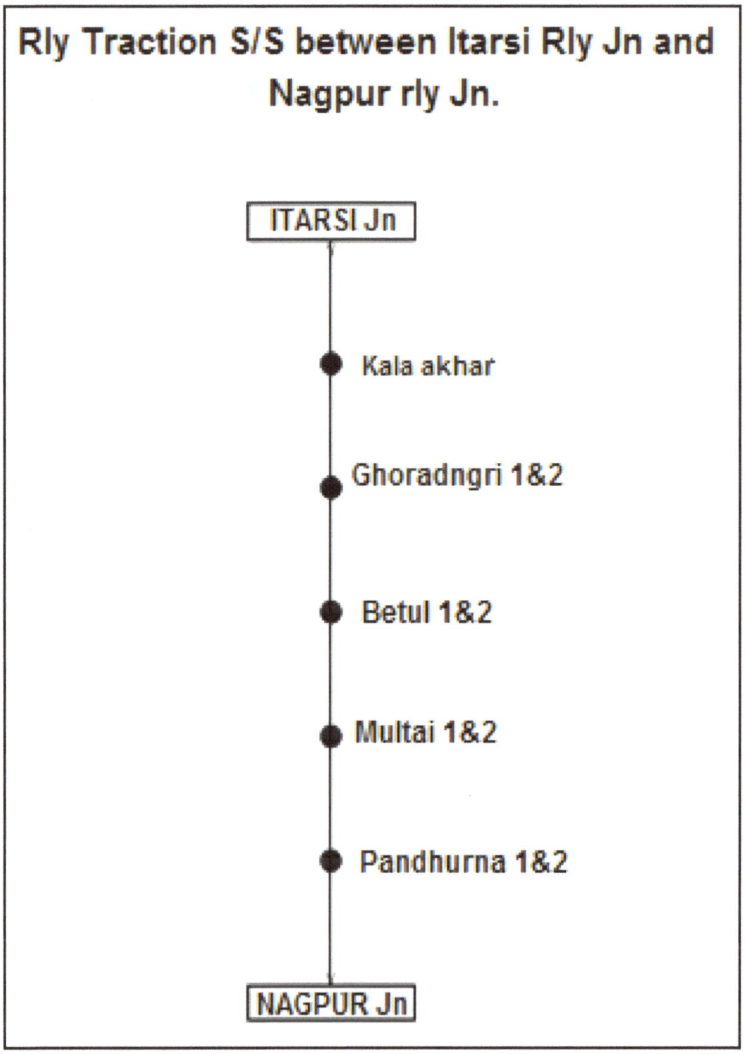

Figure 3: Rly Traction S/S between Itarsi Jn & Nagpur Jn.

4.0 Occurrence of incidence:

An incidence had occurred in SSTPS causing all the feeders and Gen Step-up Transformers to trip due to non-operation of 220kV, SF6 Circuit breaker (Westing House make) on a fault on 220kV SSTPS-220kV Sarni S/S feeder. This incidence had rendered;

1. System disturbance
2. Interruption of supply to all 132/25kV Traction Sub stations between Itarsi Rly Jn to Nagpur Rly Jn causing all the en-route trains to halt wherever they were in between Itarsi Rly Jn to Nagpur Rly Jn and vice versa(Grand Trunk route).

The System was normalized immediately by ousting the 220kV SSTPS-220kVSarni S/S feeder.

5.0 Reason for Non-operation of Circuit breaker of SSTPS-220kV Sarni S/S fdr. On thorough investigation it was found that the 220kV SF6 breaker of 220 SSTPS- 220kV Sarni S/S feeder did not trip though the DPR had sent the trip-command to the circuit breaker due to some mechanical problem (At that point of time there was no Stuck breaker Protection in operation.)

6.0 Reasons for non-operation of Electric Locos in this Trunk route; The reason for non-operation of Electric Locos was due to non-availability of 132KV supply to the Railway's 132/25kVTraction S/Ss consequent to non-availability of 220kV supply at 220kV Sarni S/S.

7.0 Steps/ measures taken by The Railways to combat the situation:

- The Railways had to switch-over to use diesel locos for pulling the important Mail, Express & Goods Trains between Itarsi Jn to Nagpur Jn and vice versa.
- Frantic FAX messages sent to the Chairman, MPSEB, requesting resumption of supply to their 132/25 KV Traction Substations immediately.

8.0 Extension of 220kV supply to 220kV Sarni S/s implementing Innovative idea:

- It was a challenge for the Protection and the EHV Maintenance Engineers as to how to extend 220kV supply to the needy 220KV Sarni S/S?
- As the saying goes in English language" Necessity is the Mother of Invention" keeping this in mind, an innovative idea struck to the mind of the principal author as mentioned here under;

➢ "Since, the old 220kV corridor (through 220kV Itarsi-4nos feeders) for evacuation of Power of 300 MW STPS has almost become redundant, why not use Aux. Bus available in the 220kV Switch-yard of SSTPS as tie link to connect 220kV Sarni S/S feeder with one of the 4 Nos. -220kV feeders from SSTPS to 220kV S/S Itarsi and convert this feeder as 220kV Itarsi-220kV Sarni S/S feeder with revising the DPR settings accordingly at Itarsi end."

Note:-

❖ Before Introduction of 400kV System in the WREB grid 220kV feeders namely 220kV STPS – Itarsi feeders-4 Nos were the only corridor for power evacuation from 300 MW STPS.

❖ With the introduction of 400kV system and with enhanced the installed capacity of 1130 MW, turning this Thermal Power Station as Super thermal Power Station, the 4 Nos - 220kV feeders from SSTPS to Itarsi had lost their importance.

❖ To transfer load from SSTPS to 220kV S/S Itarsi was adequate to be catered with 3- no. 220kV SSTPA-220kV Itarsi feeders.

8.1 Execution of Innovative idea;

The following single line diagram depicts the use of Aux bus at 220kV SSTPS Switch yard.(ref.fig.4)

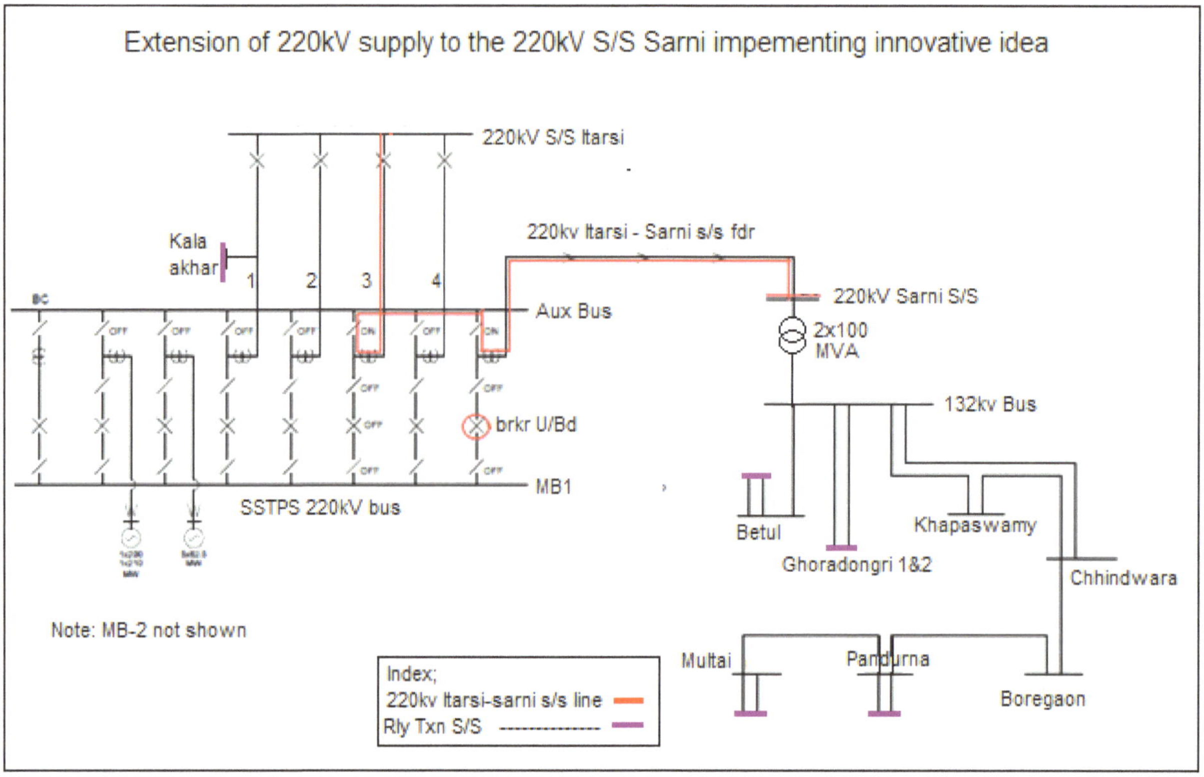

Figure 4: Innovative use of 220kV Aux bus at SSTPS Switch yard for tying 220kV SSTPS-220kv Sarni s/s feeder with 220kV SSTPS-220kV Sarni S/S feeder and to convert it as 220kV Itarsi-220kV Sarni S/s feeder.

220kV SSTPS-Itarsi fdr no 3 had been converted into 220kV Itarsi-220kV Sarni feeder by tying them together with closing the Aux Bus Isolator of 220kV SSTPS -220kV Sarni S/s and the Aux.bus isolator of 220kV SSTPS-220kV Itarsi fdr-3 keeping their Main bus isolators and Line side Isolators and the circuit breaker of 220kV Itarsi feeder no3 open, with revised DPR settings at Itarsi end.

With execution of this operation the 220kV supply could be extended to the 220kV S/s Sarni, and thus 132kV feeders could be energised and supply to 132/25 kV Rly Traction S/s could be resumed.

9.0 Conclusion:

9.1 Advantages;

By implementing this innovative method, following advantages could be achieved;

- Curbing dissatisfaction of the Indian Railway Authorities towards MPPTCL.
- Saving on time,
- Saving on investment on arranging any other method for tying any 220kV feeder with 220kV Sarni S/S feeder outside the Switch yard.
- Saving on manpower.
- This system is foolproof and can last for any length of time without any system constraints.

9.2 Caution: This episode is an eye opener episode/incidence calling for attention to keep the Stuck breaker protection (Now known as 52BFP) always in healthy condition in all the EHV sub stations/ Switch yards of the Power Stations.

3. Synchronizing Two Supplies without Synchroscope

1.0 Synopsis: Synchronising is a process of paralleling two supplies without hampering/disturbing the stability of the power system. The authors had synchronised two supplies without synchroscope. It was established successfully using innovative idea of drawing vector diagram of secondary voltages, taking the Mother-Earth as reference at 220kV S/s Katni (under-construction).

2.0 Key Words: Mother-Earth, Synchroscope, Synchronizing CVT, Vector diagram.

3.0 Theory of Synchronizing;

This is a Process/ Methodology of paralleling two Power supplies in the Power system (Grid-system) without losing stability of the Power System. In the process of paralleling, when closing a circuit breaker between the two live energized parts of the power system, it is essential to match three parameters namely voltage, frequency and phase angle on both sides of the circuit breaker. If this matching or «synchronizing» process is not done correctly, power system disturbance will occur resulting into probability of equipment (including generators) failure / damage. The three parameters of power supplies are called the synchronizing parameter variables and are:

i. **Voltage magnitudes:** The allowable voltage **difference** is ± 5% with reference to running bus voltage.

ii. **Frequency of the voltages**: Prior to closing the circuit breaker the frequencies should be within 0.19% of 50 Hz (ref: Protective Relays Application Guide by GEC Measurements.).

iii. **Phase angle difference between the voltages**: The third synchronizing parameter is the Voltage phase angle difference. The voltage phase angle difference is the difference between the zero crossings of the voltages on either side of the open circuit breaker. Ideally, the voltage phase angle should be as close to zero degrees as possible before closing the circuit breaker. However, allowable difference is $\pm 35^0$ (ref: Protective Relays Application Guide by GEC Measurements.).

4.0 Synchroscope

A synchroscope is an equipment that is used to monitor the three synchronizing parameters A basic synchroscope is shown in the (Fig:1).

Replica of Primary voltages in terms of the secondary voltages of the incoming Supply and that of running bus are the input parameters to the synchroscope If the voltage waveforms are at the same frequency, the synchroscope does not rotate. If the voltage waveforms are at different frequencies, the synchroscope rotates in proportion to the frequency difference. The synchronoscope needle always points to the voltage phase angle difference.

The synchroscope is normally mounted on the Control-Panel/Synch panel". The synchroscope synch panel also contains two voltmeters so that the voltage magnitudes can be simultaneously compared with before closing the circuit breaker manually at the right moment.

The synchroscope in figure:1 below reflects a slight voltage magnitude mismatch, and a stationary synchroscope with a phase angle of approximately 30°.

The fact that the synchroscope needle is not rotating indicates frequency is the same of the running bus and the incoming feeder.

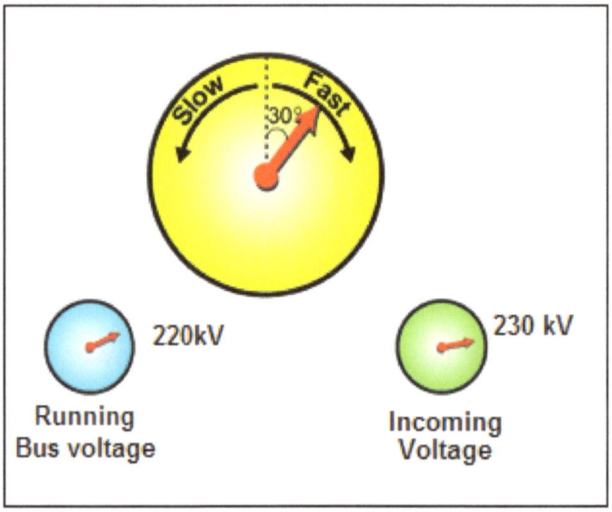

Figure 1: Synchronoscope with non-rotating pointer indicates same frequency of Running bus & Incoming Supply

Figure 2: Conventional Synchronoscope

5.0 Brief description of the incidence:

The construction, erection Commissioning work of 220kV S/S at Katni was under progress in the year 1998-1999. Following works were already completed and commissioned.

i. 220/132kV Auto Transformer TELK Made on Dtd.11.02.1998
ii. 132kV Kymore feeders No 1 on Dtd.11.02.1998.
iii. 132kVKatni Interconnetor on dtd.11.02.1998
iv. 220kV Traction Feeder no.1 on Dtd.24.02.1998.
v. 220kV Traction Feeder no.2 on Dtd.21.01.1999.
vi. 220kV PGCIL, Satna feeder on Dtd. 05.01.1999.

Supply to 220kV S/S Katni was being fed through 220kV PGCIL feeder from 400kV S/S of POWERGRID at Satna, which was the only 220 kV source. On Dtd. 04.02.1999, the 220kV PGCL feeder, went into breakdown was declared faulty. Thus emergency had been created as 220kV S/S Katni, it was was deprived of 220kV supply.

Owing to this break down of 220 kV PGCIL feeder, acute emergency had emerged. It became essential to avail supply from the Sanjay Gandhi Thermal Power Station at Birsinghpur. The 400 kV Transmission line No2 from Birsinghpur Switch Yard (118 kms away) was to be charged with 220kV supply for the 1st time from the Birsinghpur end and supply to be availed there from. It was pre-decided to charge the said Transmission line with 220kV supply as 400kV S/S at Katni was not in existence at that time.

As a special case of emergency the 220kV Birsinghpur feeder No.2 was to be check synchronosed at 220kV S/S Katni. It is to mention here that no phasing out of 220kV Birsighpur feeder No.2 was ever done with 220kV Katni Bus before. Therefore following works were to be carried out at 220kV S/S at Katni at that time on Dtd 05.02.1999.;

i. Phasing out of the 220kV Birsighpur feeder No2.
ii. Synchronization of 220kV Birsighpur feeder No.2 with 220kV bus at 220 kV S/S at Katni.

The single line diagram of 220kV S/S at Katni (partially constructed) the time of this incidence (on Dtd.05.02.1999) is as follows;

220 KV SUBSTATION KATNI (AS ON 05.02.1999)

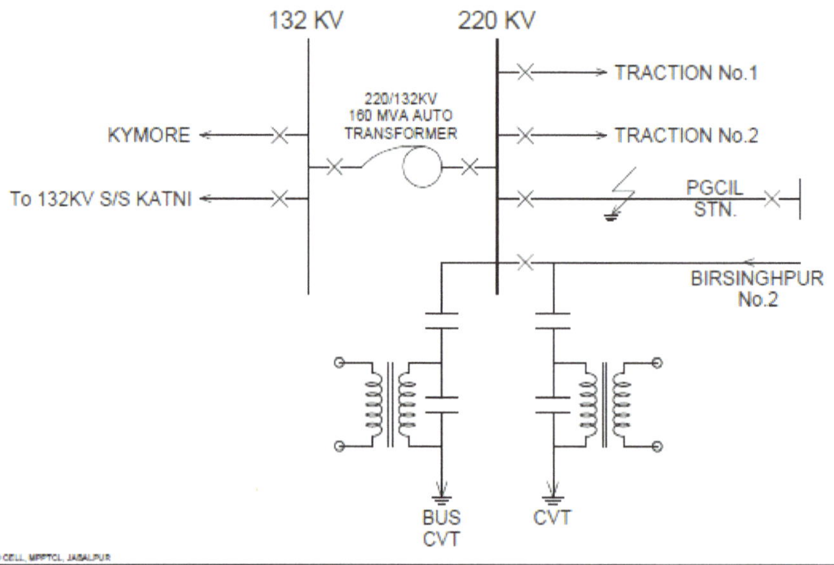

It was a challenging task to check synchronize live 220kV SGTPS Birsinghpur feeder No.2 at the 220kV S/S Katni on to the live 220kV bus without a Synhroscope.

6.0 Innovative Idea:

The Authors had no option but to ponder different ways and means to Synchronize the 220kV Birsinghpur fdr No2 with the running Bus of 220kV Katni S/s(Construction under progress)

As per saying in English language goes "NECESSITY IS THE MOTHER OF INVENTION" An idea struck to the mind of the principal author considering the following points;

- Earth is a conductor.
- The 245kV CVTs of the 220kV Katni bus as well as those of 220kV Birsingpur Feeder are single phase units.
- The earthing flats from CVTs of each phase of the BUS CVTs is brought to the earth -pit exclusively dug tor CVT Earthing and are connected to the earth -pit forming the Neutral (star-point) and the same is deemed to have been earthed.
- The earthing flats from CVTs of R & B phases of the 220kV Birsinghpur –Katni fdr no.2 are brought to the earth -pit exclusively dug tor CVT Earthing of 220kV Birsingpur fdr no.2 and are connected to the earth-pit forming the Neutral (star-point) and the same is deemed to have been earthed.

- **Note**: Since the 220kV supply is balanced 3 phase supply wherein all the phase voltages vectors are120 degrees apart in the Phase sequence of RYB thus the R & B phase vectors at the incoming Voltage shall also follow the rule and shall be 120 degrees apart.
- The Earth-pit of CVTs are connected to common Earth Mat of the 220kV S/S of Katni.
- Whole Earth Mat is at eqvi-potential.

6.1 Drawing of Vector diagram of Secondary Voltages of CVTs;

Considering all the above reasons as mentioned in para 6.0, an innovative idea struck to the mind of the principal Author as under;

a. Since primary side of EHV CVTs are forming Star/neutral point at Earth-potential, considering their secondary replica of primary voltages as measured in the relay panel, Vector diagram could be drawn with Neutral point of the secondary voltage vectors superimposed one-over the-other.
b. Following vector diagram of measured secondary voltages was drawn.
c. With following parameters.;

Suffix" bus" stands for 220kV bus voltage in terms of secondary volts.

Suffix "brp" stands for 220kV SGTPS Feeder Birsinghpur CVT voltages in terms of secondary volts.

$R_{bus} - Y_{bus}$= 118 Volts, $Y_{bus} - B_{bus}$=118Volts,

$B_{bus} - R_{bus}$=117 volts, $R_{brp} - B_{brp}$=120Volts.

$R_{brp} - Y_{bus}$=124volts, R_{bus}-B_{brp}= 125volts. Type equation here.

$R_{brp} - B_{bus}$=106Volts, Y_{bus}- B_{brp}=105 volts.

$R_{bus} - R_{brp}$=20 volts, R_{bus}- N = 68 volts, R_{brp} - N = 71 volts

The mf (multiplying factor) = 220000/√3 ÷ 110/√3=2000.

R_{bus}- N voltage vector was very close to 90° and hence,

$$\tan\theta \approx \frac{\text{Rbus} - \text{Rbrp}}{\text{Rbus} - \text{N}} \approx \frac{20}{68}$$

$$\therefore \textbf{Phase difference } \angle\theta \approx \tan^{-1}\frac{20}{68} \approx {}^{*}16.39^{\circ}$$

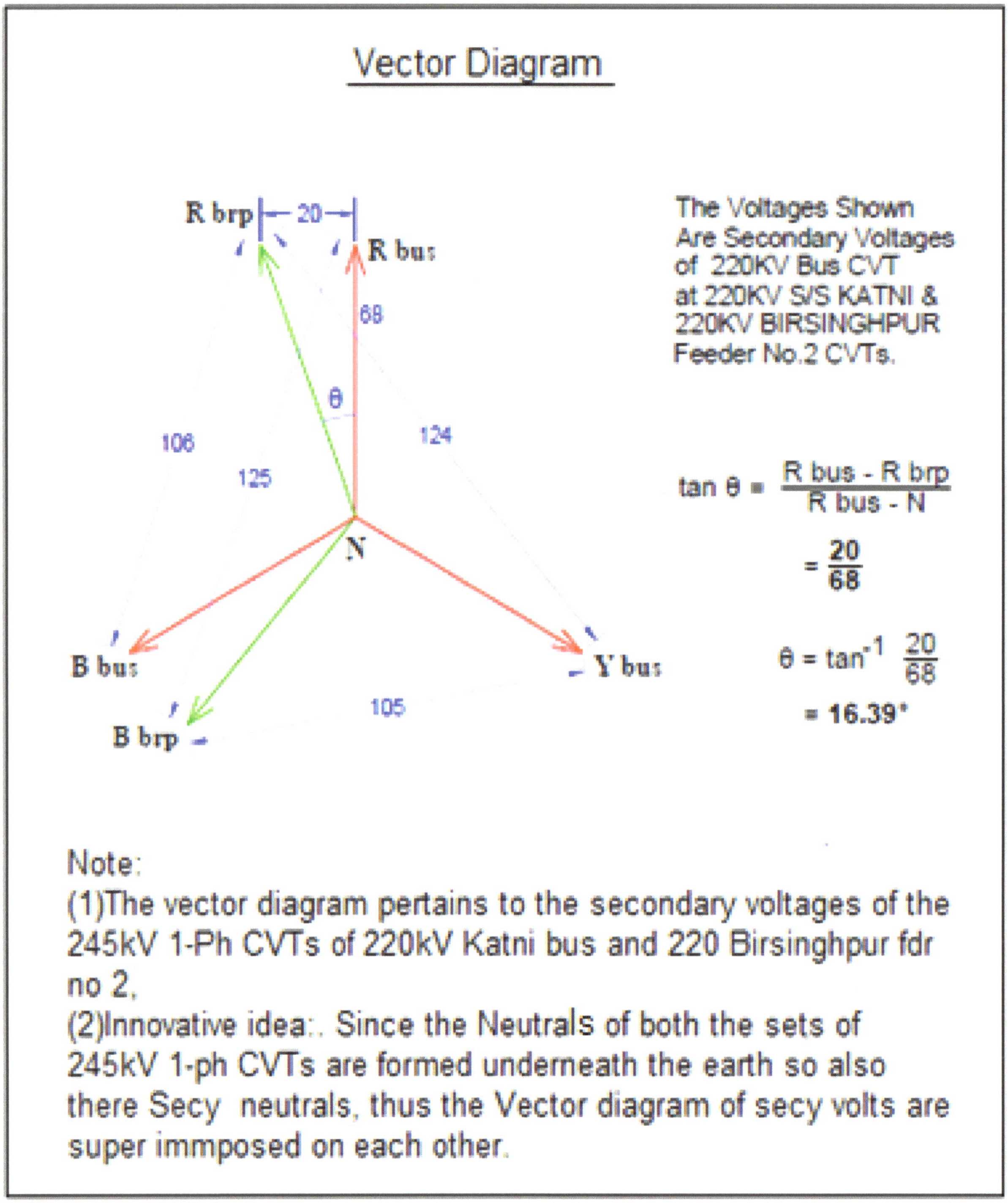

Figure 4: Vector diagram of the secondary voltages of 245 CVTs of 220kV bus and 220kV Birsinghpur feederNo.2

$$\mathbf{tan\ \theta} = \frac{\text{Rbus} - \text{Rbrp}}{\text{Rbus} - \text{N}} \approx \frac{20}{68},$$

$$\therefore \textbf{Phase difference } \angle\theta \approx \mathbf{tan}^{-1}\frac{20}{68} \approx {}^{*}\mathbf{16.39^{o}}$$

On confirming that the phase difference was only *16.39^{0} approx which is well within permissible limit of ± 35° [ref: Protective Relays Application Guide by GEC Measurements.).

(*Note:

The reasons for phase difference: As could be seen from the vector diagram that the incoming voltage vector is on the leading side of bus voltage vector. It is due to the capacitance effect of the 118km line at the 220kV S/S Katni end).

Since the 220kV Katni S/S was connected to the main Grid-network through the 132kV system, so also the Sanjay Gandhi Thermal Power Station, Birsinghpur, they were operating on the same frequency, the feeder was then check synchronized successfully by the Authors at 13.30 Hrs on 05.02.1999 and immediately it started feeding the load of around 45 MW to 132 kV feeders and 220 kV supply extended to 220 kV Railway Traction sub-station.

7.0 Conclusion:

In the present scenario of Numerical relying with Synchronism check relay (ANSI-device no 25), synchronizing of two supplies without losing System stability takes place. However, in such exigencies as mentioned in this paper, this method can be used by the field Engineers without hitch or hesitation as it has been also scientifically proved. This is an exemplary true case history which was accomplished with the help of the "MOTHER EARTH"

8.0 References:

1. Protective Relays Application Guide by GEC Measurements.
2. Network protection & Automation guide by ALSTOM.
3. IEEE 80-2000 "Guide for Safety in AC Substation Grounding".

9.0 Author

K.K.Murty

4. Valve Regulated Lead Acid Batteries-An Over View

Introduction

The conventional lead acid batteries used in signaling circuits as a source of

D.C supply suffers from number of maladies which enclosed regular topping- up a separate storage arrangement which amount to increases maintenance and reduced reliability. The new improved VALVE REGULATED LEAD ACID (VRLA) battery overcomes much of these earlier deficiencies.

As the name suggests, these are sealed valve regulated batteries with absorptive glass mat (AGM) separator to permit the oxygen evolved from the positive plate to the fuse to the negative plate. The micro-porous glass separator as used in VRLA is not completely saturated with electrolyte and the wide space thus available allows an unimpeded access of oxygen to the negative plate. The oxygen gas gets reduced at the negative plate surface, thereby effectively suppressing the evolution of hydrogen. Consequently, Power stacks cells do not lose any water under normal operation and therefore, no topping- up is required.

Valve regulated lead acid battery (VRLA) use a patented MFX alloy for the positive grid, which exhibits excellent deep discharge performance. In addition it has a universally acclaimed maintenance free lead calcium alloy for the negative grid. The power stack cells so formed require very little maintenance and have longer life duration.

1.0 Salient features and advantage

VRLA batteries are based on "Starved electrolyte design and Oxygen recombination principle." The important functional features and advantages are as under.

- No periodic topping up of electrolyte is required.
- No spillage of electrolyte as it is sealed.
- These cells are supplied in fully charged condition; therefore, no initial charging is required.
- These have very low self-discharge the order of 20% to 25% of conventional lead acid batteries.
- These can be used as either vertically or horizontally. They are available in stackable steel trays capable of being mounted one over the other. Therefore, they occupy very less space.
- They do not emit any toxic, corrosive acid fumes; hence can be installed in the equipment room itself, saving the battery room.
- They are lighter in weight by about 25% when compared to conventional lead acid batteries.
- They have 30% to 40% longer life than the conventional lead acid batteries.

2.0 Constructional features

The main components of lead acid cell are as under.

2.1 Positive plates

Positive plate consists of a number of flat tubes, containing patented MFX alloy

2.2 Negative plate

It is made of lead-calcium alloy.

2.3 Container

Container of battery is made of suitable flame retardant Polypropylene Copolymer plastic (PPCP) with high insulating strength to resist acids and low permeability to water vapors. The container is enclosed in epoxy coated steel trays. The steel tray is so designed as to permit both vertical and horizontal stacking of cells.

2.4 Electrolyte.

The sulfuric acid and water used as electrolyte for the battery confirming to IS 266-77 and IS 1069-64 respectively.

2.5 Separators

Separators consist of either of micro-porous glass mat or synthetic material having high acid absorption capabilities and also resistant to sulfuric acid.

2.6 Safety Vent Plug

The plug is explosion resistant and is of pressure regulating type. Safety Vent Plug does not permit ingress of air into the unit.

3.0 `Recombination Principles

Under typical charging conditions, oxygen at the positive plate occurs before the evolution of hydrogen at the negative plate. This feature is utilized in the design of VRLA batteries.

The oxygen gas evolved at the positive plate instead of bubbling upwards is transported in the gaseous phase through the separator medium to the negative plate. The separator which is of highly absorbent glass mat type has very high porosity designed to have pore volume in excess of the electrolyte volume (starved electrolyte design), due to which the Oxygen gas finds an unimpeded path to the negative plate. Reaction reduces the Oxygen gas with the spongy lead at the negative plate turning a part of it into a partially discharged condition, there by effectively suppressing the hydrogen gas evolution at the negative plate. This is what is known as the Oxygen recombination principles.

3.1 Principle of VRLA Battery

The VRLA battery is designed so that the negative plate does not have to be fully charged even when the positive plate is charged fully. Here no H_2 gas is generated from the negative plate though O_2 gas is generated from the over charged positive plate. The O_2 gas so generated from the positive plate reacts with charged spongy lead (Pb) of the negative plate and forms Lead Monoxide (PbO). The Lead Monoxide in turns reacts with Sulfuric Acid (H_2SO_4) in the electrolyte and forms Lead Sulphate ($PbSO_4$), thus allowing the negative plate to be discharged. In other words, O_2 from the positive plate is absorbed by the negative plate without being expelled to the outside. Since the negative plate develops discharging with the help of O_2 evolved at positive plate during charging, the negative plate as a result never generates H_2. This completely prevents the loss of water. The overall reactions may be expressed as follows.

$$Pb + \tfrac{1}{2} O_2 \longrightarrow PbO \text{ (-ve plate)}$$

(-ve charged)

O_2 generated from positive plate

$$H_2O + PbSO_4 \longleftarrow H_2SO_4 + PbO \downarrow$$

$$PbSO_4 + 2e^- + 2H^+ \longrightarrow H_2SO_4 + Pb$$

(All above reactions are exothermic reactions)

- The batteries come in pre-charged condition and thus no initial charging is required. These are ready to use.
- Battery requires no topping up and maintenance except for periodic cleaning of the terminals.
- The battery has self-sealing vent plug, which normally does not open out in service.
- Under normal conditions they emit no corrosive fumes or gases.

4.0 VRLA batteries- Effect thereon;

4.1 Effect of temperature on VRLA batteries

The lead acid batteries (both conventional as well as VRLA) are normally designed to give the full rated capacity and the expected life at the operating temperature of 27°C. When the operating temperature is higher than 27°C, the corrosion rate at the positive grid increases. This is due to Arrhenius principle, which states that, the rate of chemical reaction doubles for every 10°C rise in the ambient temperature.

Note: The effect of increased temperature is more pronounced on VRLA batteries as less volume of electrolyte is available and also because there is no addition of electrolyte during the use of the battery. It is said that for every 10°C rise, the reduction in life of VRLA batteries is up to 50%.

1.2 Effect on capacity when stored for long duration

A fully charged battery loses certain capacity when stored. This process of capacity loss is called as "self-discharge". The capacity loss for VRLA batteries is around 1% per week. The various parameters that affect the self- discharge include.

- Alloys used for positive and negative grids.
- Purity of raw materials used in the manufacturing process for grids.
- Processing techniques employed in manufacturing of grids.
- Also storage conditions of batteries at the site.

1.3 Storage interval

- Batteries in storage lose capacity, due to its self-discharge. The rate of self- discharge increases with temperature. The loss in capacity becomes permanent due to sulphation of battery.
- Batteries may be stored for a period up to six months before installation. (i.e. from the date of shipment and the date of installation).
- In case the batteries are to be stored for longer duration, then the batteries must be unpacked at the intervals of six months and they must be given charged afresh.
- The storage interval at the various elevated temperatures before which the battery is due for freshening charge is shown in the table below.

Sr.No	Temperature	Storage interval.
1	32°C	6.0 Months
2	37°C	4.5 Months
3	42°C	3.0 Months
4	47°C	2.25 Months
5	52°C	1.5 Month

5.0 Installation,

For installation of VRLA batteries, the following instructions should be followed.

5.1 Location

- Install the battery in a clean, cool and dry location, an ideal temperature being 27° C.
- Free space of one meter should be provided on all sides of the battery for carrying out periodic checks.

5.2 Ventilation

Under normal operating conditions, natural ventilation is adequate.

6.1 Battery charging

6.2 General Instruction

For a good performance, it is better to follow the procedure defined by the manufacturer for charging. The procedure is as under.

6.2.1 Ensure that the entire cell in the battery-set is of the same type and capacity. Charging individual cell separately is not advisable.

6.2.2 Correctly connect the battery charger output terminals to the battery set i.e. positive to positive and negative to negative.

6.2.3 Ensure that all electrical connections between the battery charger and the battery are tight and offer good conductivity.

6.2.4 Voltage settings of power plant

In the float mode, the float Voltage is 2.25 V per cell. In normal cases, the float voltages work out to be as under.

- 56 Cells Battery Bank -126.0 +1.0,- 0.5 Volts.
- 56 Cells Battery Bank -126.0 +1.0,- 05 Volts.

6.3 Freshening Charge

In case the batteries are to be stored for longer duration, the battery must be charged for the freshening charge.

- Charging must be carried out at constant potential with the current limited to the maximum 0.2 C10 Amperes (where C10 refers to the numerical value of the nominal capacity of the battery).
- In charge mode, at a charge voltage of 2.3 Volt per cell for a period of 16 hours.
- In the float mode, at a float voltage of 2.25 Volts per cell for a period of 72 hours. Say, for 48 Volts, 400 AH battery consisting of 24 cells, charging must be carried out in the charge mode at a charge voltage of 55.2 Volts ± 0.1 Volts for a period of 16 hours. Or charge in the float mode at a float voltage of 54 Volts ± 0.1 Volts for a period of 72 hours. The current should be limited to maximum of 80 Amps. (400X0.2 = 80).

Note: The above charging procedure is to be used for giving a freshening charge to the battery for extending the storage period.

- This charging procedure is also valid for charging the battery to determine the rated capacity of the battery after installation.
- Battery located in areas having very low average ambient temperatures (below 15° C) will require the charging time recommended above to be doubled in order to fully charge the battery.

6.3 Float charging method.

Under normal condition the batteries are charged with float charge method. This type of charging sustains the battery in a fully charged condition and also makes it available to resume the emergency power requirements in the event of an AC power interruption or charger failure.

- Float charging of battery is carried out as under.
- The battery is connected in parallel with a constant voltage charger and the critical load circuits. The charger should be capable of maintaining the required constant voltage at battery terminals and also supply normal connected load.
- After the batteries have been given its freshening charge the charger should be adjusted to provide the recommended float voltage at the battery terminals.
- Adjust the charger voltage to 2.25V per cell. (i.e. in a 24V system, 12 Cells are floated 27 Volts).
- During charging as the cells approach full charge, the battery voltage rises to approach the charger output voltage.(i.e. 27 Volts in the above case).
- The charging current decrease to the float current value of around 50 mA per 100 AH.
- When the charging current has stabilized at the float current for three consecutive hours or the voltage across the battery bank terminals is constant for six consecutive hours, the battery bank can be considered as having reached the fully state of charge.
- Recommended float charge voltage for state of charge is given the table below.

Sr No.	Item	Nominal required value
1	Float Voltage setting at battery terminals.	2.25 X No. of cells
	(i) 24 V system	27 Volts
	(ii) 110 V system	123.75 Volts
2	Normal battery charging currentwhen battery put into charge after discharge.	10% of AH capacity i.e. for 100 AH battery to be charged at 10 A current.
3	Maximum battery charging current when battery put in to charge after discharge.	20% of AH capacity.
4	Ripple content in DC voltage.	Should be less than 2% rms
	(i) 24 V system	0.48 V rms (Max.)
	(ii) 110 V system	2.20 V rms (Max.)
5	Battery charging voltage (High protection).	Charger should trip at 2.37 Volts per cell
	(i) For 24 Volts system	Charger should trip at 28.44 Volts
	(ii) For 110 V system	Charger should trip at 130.35 Volts.
6	Battery under voltage (Trip on discharge).	Charger should trip 1.75 Volts per cell
	(i) For 24 Volt system	Disconnect the load at 21 Volt
	(ii) For 110 Volt system	Disconnect the load at 96.25 Volt

6.4 Equalize charging

Under normal operating conditions an equalizing charge is not required. An equalizing charge is a special charge given to battery when non- uniformity in voltage has developed between cell. An equalizing charge should be given when the following conditions exist.

6.4.1 The float voltage of the pilot cell is at lest 0.05 V below the average floats voltage per cell in the bank.

6.4.2 A recharge of battery is required in a minimum time amount of following an emergency discharge.

6.4.3 Accurate periodic records of individual cell voltage show a sudden variation from the previous readings.

6.5 Equalizing charging method.

i. Equalizing charging is made using constant voltage method.
ii. Determine the maximum voltage that may be applied to the system (i.e. 2.30 X No. of Cell)
iii. Establish the maximum Voltage apply to the system for equalizing charge in shortest time duration as per the table below.

SN	Temperature	Volts/Cell	Time
1	Below 15°C	2.30 V	30 Hours
2	15-32° C	2.30 V	20 Hours
3	Above 32°C	2.30 V	12 Hours

6.6 Discharging battery

- The discharge duration for the determined rated capacity of the battery is 10 hours. Discharged the battery for 10 hours so as to get an end voltage of 1.75 Volts per cell at an average ambient temperature of 27° C.
- It is important to immediately charge the battery in the charge mode after every
- discharge and maintain the battery in the charged condition by keeping it connected to the battery charger in the float mode.
- Ensure that the battery is charged optimally and no overcharging, over gassing or heating takes place.

7.0 Difference between conventional flooded batteries and VRLA batteries;

Sr. No	Conventional flooded batteries.	VRLA batteries
1	Constant current charging method is normally used for charging these batteries.	Constant voltage with current limiting method is the recommended for charging these batteries.

2	As excess of free acid is available. Hence electrolyte stratification takes place. To maintain uniform specific gravity of electrolyte equalizing/extended charging is required.	As there is no stratification of electrolyte, no equalizing charging is required.
3	Frequent topping-up of distilled water and acid is required since equalizing/extended charge method is used which results in loss of water.	Under normal recommended operating conditions there will no loss of water. Hence the periodic topping- up is not required if the voltage, current, ripple and regulation are in specified limits.

Sr. No	Conventional flooded batteries.	VRLA batteries
4	Post corrosion is observed due to acid mist.	No post corrosions in normal operation since there is no acid mist.
5	These batteries are needed to be placed only vertically since they required frequent topping-up with distilled water and can lead to spillage's as free and floating acid is present inside the battery.	The batteries can be mounted in any position (normally horizontal position is recommended) since they do not require any periodic topping and are both spill and leak proof.

8.0 General maintenance

- Care must be taken to ensure that the working circuits do not get affected during general cleaning of the battery.
- In all cases which require disconnecting as re-charging or replacement of the battery, disconnection notice must be issued to the ASM on duty and work commenced after the same is allowed and necessary precautions have been taken as laid down in SEM.
- After the maintenance and works are completed, test must the immediately be carried out to ensure that connected gears are functioning properly.
- Check each cell voltage of the battery set once every week, or as per instructions issued by the Railways.
- Avoid sources of heat or cooling directed on to the battery.
- Check that physical condition of the battery is good, i.e. there are no cracks, bulges and heating marks on it.
- The battery rooms and location box should be kept well ventilated and also free from water, oil and dirt.
- Since the voltage to be measured require an accuracy of 0.05 Volts, only digital multimeters are to be used for this purpose.

8.1 General Cleaning

- Remove dust or dirt from the batteries. Clean the entire battery box, terminals and connectors etc.
- Gently clean the terminal post sulphation by using fine emery paper. After cleaning the terminal-post with dry cotton cloth.

8.2 Record maintenance

- Record individual cell voltage after 30 minutes discharge on load.
- Maintain the record of the battery terminal voltage and the individual cell voltage using the service record formats as given below.

1. Nominal capacity of battery. ---------------
2. Nominal voltage of the battery. ------------
3. Date of installation. --------------------------

Date	Cell No.	Voltage	Total voltage of the battery	Charger voltage in float / charge mode	Charger current	Average ambient temperature	Remarks	Signature of supervisor
1	2	3	4	5	6	7	8	9

Note: The end cell from which the positive terminal has been connected to the power plant is designated as cell number one.

9.0 Battery charger Voltage / Current settings.

1.1 Float mode

i. Check the setting of battery charger. The battery voltage is 2.25± 0.05 voltsper cell in float mode.
ii. Special care must be taken to ensure that the voltage of the power plant hasbeen set exactly to a value within the specified range is given below.

Voltage Setting	
2V,56 Cells	126V
2V,54 Cells	122V
Charging voltage	2.25 V per cell
Voltage ripple	5% rpm
Ripple current	10% rpm
Charging current	Max 20% of AH rated current

Note: Ensure that the battery charger is so adjusted that charger trip at 2.35 V per C ell to avoid over voltage of the battery.

9.2 Boost mode

Check for cell voltage, if it is falling below 2.0 Volts, boost charge should be given with the following settings.

i. Set the power plant voltage to 2.3 Volts per cell.
ii. Set the maximum charging current limit to 20% of AH rating and charge the battery for 12 hours.
iii. Recommended voltage setting during boost charging is as under.

Voltage setting	
2V,56 Cells	128.8 V
2V,54 Cells	124.2 V
Charging Voltagele	2.3 V per cell
Vottage ripple	Less than 5% rpm
Ripple current	Less than 10% rpm
Charginmg curret	Max 20 % of AH rating, current limit to be 20 % be set between 10 to 20 %based on the time for recharge.

10.0 Safety precautions.

Before proceeding with the unpacking, handling, installation and operation of the valve regulated lead acid storage battery, the following information and the recommended safety precautions are to be followed.

i. Wear a rubber apron, rubber gloves and safety goggles or other eye protection to prevent injury if contact is made with the acid.
ii. Under abnormal usage batteries can generate explosive gases. Under such conditions, the safety vent opens while the gases are being generated so keep sparks, flame and smoking materials away from the battery area.
iii. Ensure that tools are adequately covered with insulation tape or suitable non- conducting material to minimise possibility of shorting across connections.

11.0 Trouble shooting

S N	Trouble	Cause	Remedy
1	Over - charging	Continuous charging of batteries at a higher voltage.(i.e more than specified value)	Adjust the float voltage to 2.25V per cell at 20%(max.) current limit of the rated capacity.
		The higher voltage can be due to. • Either set point in the charger it self is higher. • Or it is connecting more or less number of cells than required.	Adjust the power plant or connect the required number of cell.
		Over voltage protection in charger is not provided or is defective.	Attend charger or replace with the proper charger.
2	Under - charging	Continuous charging of batteries at a lower voltage (i.e. Less than the specified value.)	Charge the cells at 2.25 V until all the cells reach the normal condition.
		Terminal bolts are not properly tightened.	Tighten the terminal bolts properly.
		Malfunctioning of charger or sudden failure of charging voltage controlling circuit or higher ripple current.	Attend the charger and rectify it.
3	Corroded terminals.	Sulphation formed on the terminal post or both positive and negative plates got sulphated (i.e. forming of $PbSO_4$)	Remove the connectors and clean the terminals with brush. After making connections apply a thin film of petroleum jelly.
4	Total voltage of battery suddenly drops	Showing negative voltage when measured across terminal of a cell.	Charge the cell separately at 2.30V with maximum current of 20% of its rated capacity.

Do's and Don'ts

12.1 Do's

a. Unload the batteries carefully and place them upright on the floor in the single tier.
b. Store the batteries in cool and dry location.
c. Charge the batteries every six months for the storage period.
d. Unpack the batteries as per the unpacking instructions.
e. Install the batteries in cool and dry location.
f. Monitor the charge and float voltages of the power plant at monthly intervals and adjust, if required.

g. Check the tightness of all electrical connections at monthly intervals.
h. Always use a spring washer wherever bolt connections are provided
i. After a discharge, re-charge the batteries immediately.
j. Check the compatibility of power plant and replace with the modified one if required.
k. Maintain service record as per the instructions.

12.2 Don'ts

a. Do not exceed the charging voltage above 2.30V maximum per cell.
b. Do not open the safety valve for adding water or acid.
c. Do not attempt to dismantle the battery.
d. Do not boost charge the batteries for more than 12 hours.
e. Do not mix ordinary conventional, low maintenance batteries with maintenance free VRLA batteries.
f. Do not expose the packed batteries to rain and sunlight.
g. Do not install the batteries in rooms with varying temperature pockets due to sun light or ventilation ducts.
h. Do not keep the cells in idle condition for more than the specified storage period without charging the batteries.
i. Do not short-circuit the battery or cells during assembly.
j. Do not over tighten the terminals.
k. Do not charge the batteries in sealed cubicles.
l. Do not mix batteries of different types or makes.
m. Do not make tap connections.
n. Do not tamper with the cell vents.

5. D.C. Earth Leakage- Identification & Rectification

1.0 Introduction: D.C. System in a Sub -Station or Power Station is the life line of the Protection system. DC supply takes over the responsibility of faithfully and successfully operating the protective gear for operation of control circuits of relays, control and indication circuits of the C&R panels, operating relays for finally Tripping of Circuit breakers and also for control room emergency lighting in case of total failure of AC supply.

Thus maintaining healthiness of DC supply in an EHV S/S is of utmost importance.

2.0 Arrangement of DC supply: In our EHV S/Ss in MPPTCL, we have 110V DC as well as 220V. DC. having individual battery set (Lead-Acid) of 2Volts each connected in series so as to form the set of required Voltage.

The Battery set has a Charger of adequate capacity connected to the terminals of the Battery set and also feeding the DC load of the S/S parallel.

The Function of the Battery charger is 2 folds, ie it keeps on charging the Batteries in float mode as well as feeds the DC loads of the S/S simultaneously and constantly.

The Battery set is always kept ready to take over the load of the system at the time of failure of AC system.

The DC system of S/S is usually ungrounded.

3.0 Phenomenon of DC Leakage:

DC Leakage is likely to take place if either the (-) or the (+) of the DC supply gets grounded due to some reason such as

- loss of insulation of the control cable due to ageing
- Biting of the control cable insulation by the rodents in the S/S /Power Station switch yard.
- Improper terminal connections by the Electrical technicians which may foul with the metallic kiosks, C&R panels etc.
- In some cases, while cabling and making terminal connections, the technicians are in a habit of twisting a few of the strands of the cable and connect to the DC supply at one end but due to constraints of terminals being of smaller size, where the said cable is to be terminated, a few strands thereof are left unconnected and hanging, such that they may touch the metallic kiosk body etc.

4.0 Hazards of D.C. leakage:

DC leakage can lead to

- Spurious trippings leading to system disturbance.
- Spurious indications which may make the operator take faulty decisions.
- Such leakages also lead to running down of battery set in case of AC failure.

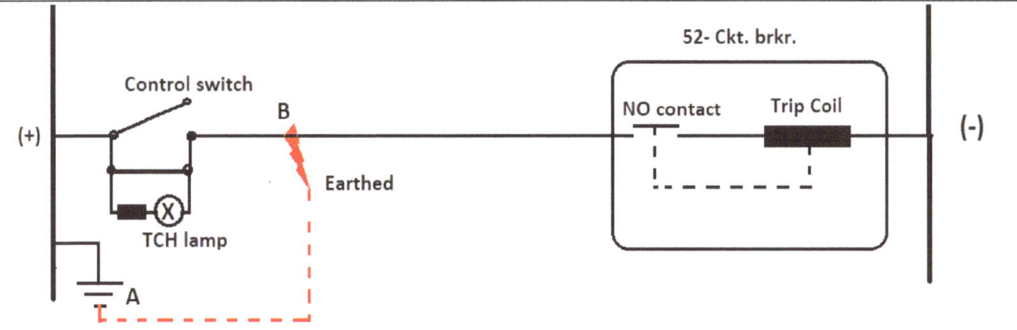

Fig. Showing A point is already earthed (+) the lead from control cable to the Circuit breaker Trip Coil gets earthed at point B.

Figure 1.

Example of (+) getting Earthed: In The Figure 1 the (+) was already getting earthed at point (A). With (+) or (-) is getting earthed at one point it is not going to effect in any way. In this case, subsequent earthing of (+) has taken at place at location (B) after the relay/switch contact. With this kind of E/L tripping shall take place without any indication putting the concerned personnel in confusion/doubt.

4.1 Location of DC E/L:

- Detection of D C leakage is a tedious job.
- Following procedure may be followed;
- Have a DC source of similar rating.
- Make -before -break switching arrangement should be there.
- Connect the new set of batteries with a CZA and change over each feeder one by one by one.
- If deflection is observed on any feeder that means that particular feeder is suffering from DC leakage.

4.2 Method/Circuit to identify Earth Leakage:

1. The set of Batteries say 220VDC (2 Voltsx110 No's) is in use in the S/s.
    - Identify the midpoint of the set.
    - Connect a CZA (Central Zero Ammeter) at the midpoint and earth it through a very high resistance.
    - The Operators of the S/s should record the deflection of the CZA in particular intervals as per the standing instructions of the Authority.
    - The Circuit pertaining to the same is shown here under.

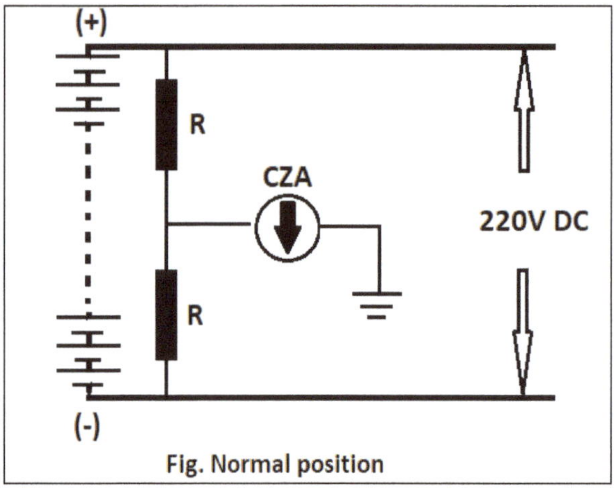

Figure 2: normal position.

(+) ve Earthed condition;

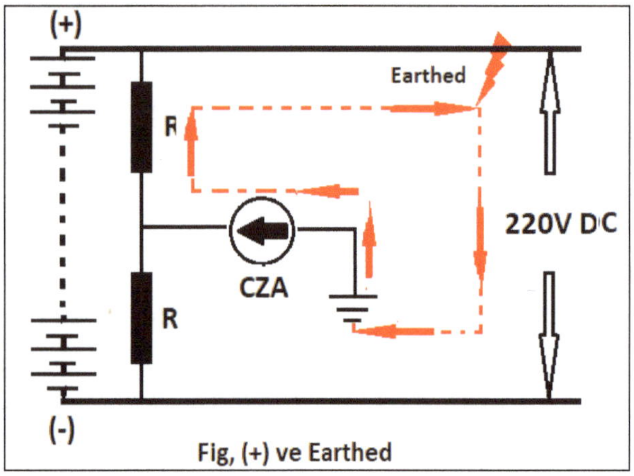

Figure 3: (+) ve Earthed.

(-) ve Earthed contition;

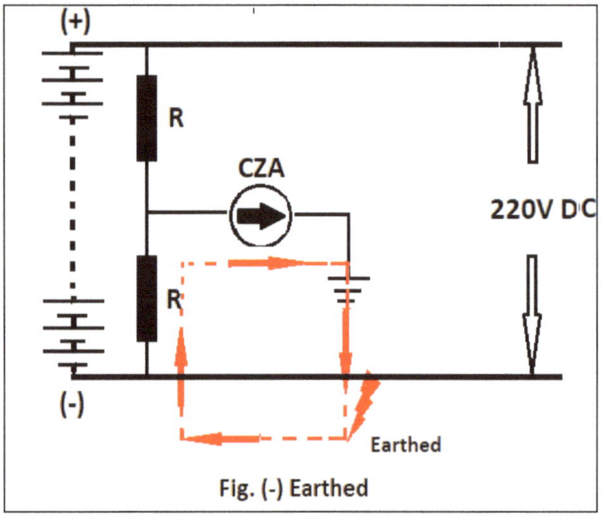

Figure 4: (-) ve Earthed.

2. A separate circuit is also in vogue for identification of the DC earth leakage. which is shown hereunder refer fig.5;

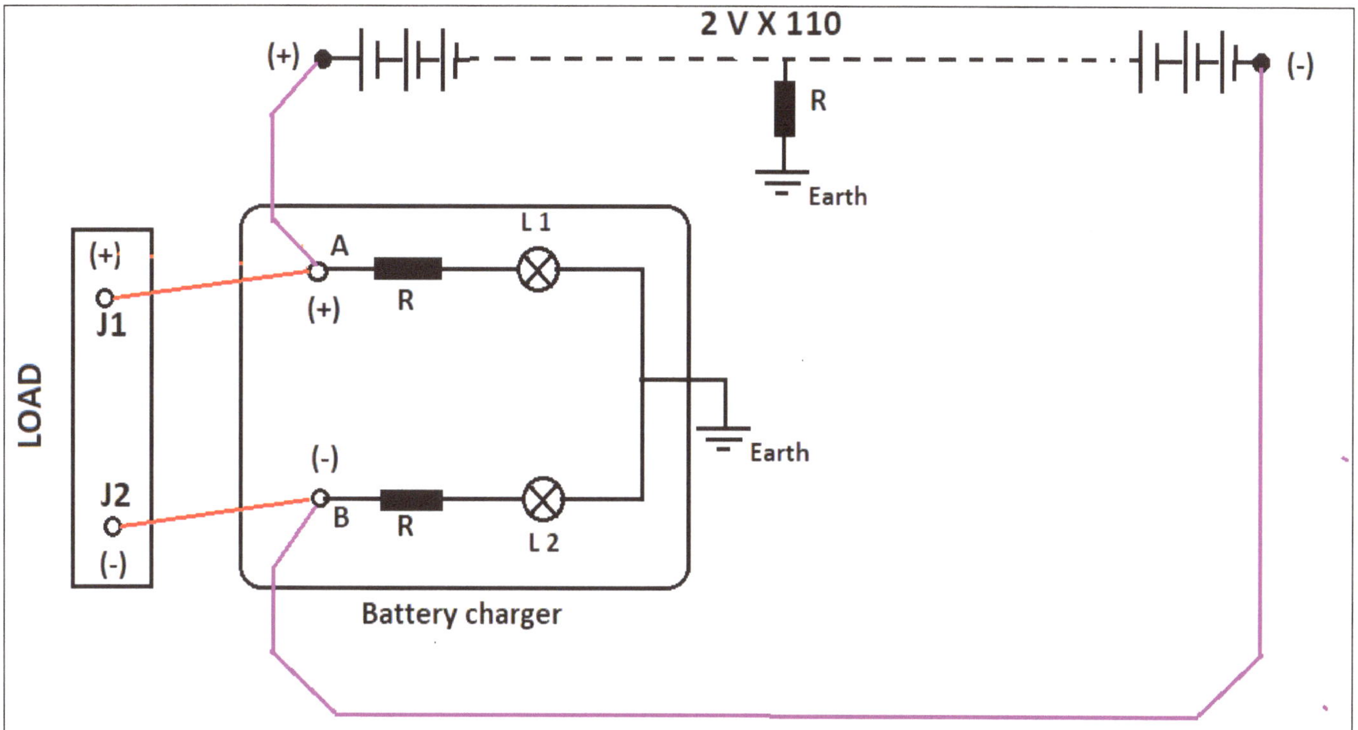

Note: (1)Battery set (+) terminal is connected to the Battery charger (+) terminal (in pink font line) & battery set (-) terminal is connected to the (-) terminal of the battery charger(shown in pink font line).
(2) Battery charger (+) terminal is connected to the (+) terminal J1 of the load(shown in red font line) & Battery (-) ternial is connected to the (-) terminal J2 of the Load(shown inred font line)

Figure:5.

- Identify the midpoint of the Battery set.
- Connect a very high resistance to the identified midpoint and earth the same.
- Connect the battery set + and – ve terminals to the Battery Charger + and – ve terminals through N/C of the contactor. This contractor shall de-energize when AC shall fail and make the connections thru to the terminals of the Charger, from where the load is also being fed.
- Across the Battery charger terminals a closed circuit is to be introduced having a very high resistance and an indication lamps on either side of the midpoint thereof which should be earthed through a very high resistance.
- If the – ve is being earthed somewhere, the indication lamp connected in between the – ve terminal of the Charger and the midpoint which has been earthed shall become dim and at the same time the indication lamp connected to the + ve terminal shall glow brighter.
- If there is + ve E/L, similar phenomena shall take place with the indication bulb connected in between the + ve terminal and the midpoint earth.

6. Effective earthing system in EHV S/S

We, in MPPTCL (Erstwhile MPEB) are very conservative and we follow the practice of earth pit having 4 x 40 mm x 3000 mm GI pipes connected to each other with MS flats at Right angles to form a quadrilateral shape of 1000 mmx1000 mm and in addition, their diagonals are also connected to opposite GI pipes Pl see the attached drawing). This practice is still being followed in about 320 Nos. EHV S/S and about 780 Nos. EHV Transformers installed therein. I have never heard of any mishap occurring due to poor earthing or Earth potential rise or Step and Touch Potential beyond limits, affecting the life of men and equipments. Such earth pits are many in Number and in-turn they are connected to each other and also connected to main earth mat through appropriate sized MS flats. The bodies of all other equipments except Transformer body are connected to the earth mat through MS flat riser. The Transformer body too is connected to the earth mat in 2-places. Further, all structures, shielding Earth wire, cable armours / cable screens, cable trays, C&R panels, battery chargers water-pipings passing though the substation yard/premises and the S/s peripheral barbed boundary fencing etc and MOST IMPORTANT IS THAT THE OTHER PARTS OF THE SUB-STATION SUCH AS 132KV AND 33KV SWITCHYARDS/ SUB STATIONS (in some cases at different levels)ARE JOINED TOGETHER FIRMLY TO HAVE EQUI- POTENTIAL BONDING.

Generally, as a standard practice in vogue in MPPTCL following salient features, parameters/ methodology are mentioned here under;

a. Forming of Earth mat; a). MS flats:
    - 400 & 220kV S/s- 75x8 mm.
    - 132 kV S/S - 65x8 mm.
    - 33kV S/S – 50x6 mm.
    - Risers – 50x 6 mm (to be connected to equipment body at 2-places).

b. Size and no. of spikes;
    - 25 mm dia x 3000 mm.
    - No. Standard assumption as 500A of E/F current/per spike.

(Note: If a S/S has 40kA as fault current then no. of spikes =40000/500=80 nos.)

- Distance of burial-7 to7.5 meters.

c. Making of Mesh thereof through lap type welding joints etc.
d. Depth at which the mat is to be buried from the ground level - 500mm.

{Note: The main Earth mat should be extended (Length x breadth- wise) 2000 mm beyond the Peripheral boundary.}

a. Details of Earth pit construction: Pl. ref. drg. no. SK/ED-15.
b. Details of Transformer earthing: Pl. ref. drg. no. SK/ED- 6.
c. Location of Earth pits in the switch-yards;

   i. Two earth pits with four pipes each for HV & LV Neutrals of the power transformers of ratio 220/132KV or 132/33KV.
   ii. One earth pit each exclusively for lighting arresters of the power transformer on high & low voltage side.
   iii. One earth pit each for a set of three PTs in 220/132KV & 33KV Class.
   iv. One earth pit for each set of 3 LAs in the 220KV and 132KV Class, with three individual leads, to be connected, one from each phase unit directly to the grounding.
   v. One earth pit for each of the 220/132KV feeder bays for connection of LAs and other equipments..
   vi. One earth pit between two feeder bays in the 33KV System.
   vii. On earth pit each for main buses.

Perhaps the above salient features are sufficient to project the practice of Earthing System in an EHV S/S in vogue in MPPTCL.

Enclosed: Drg of Earthing;

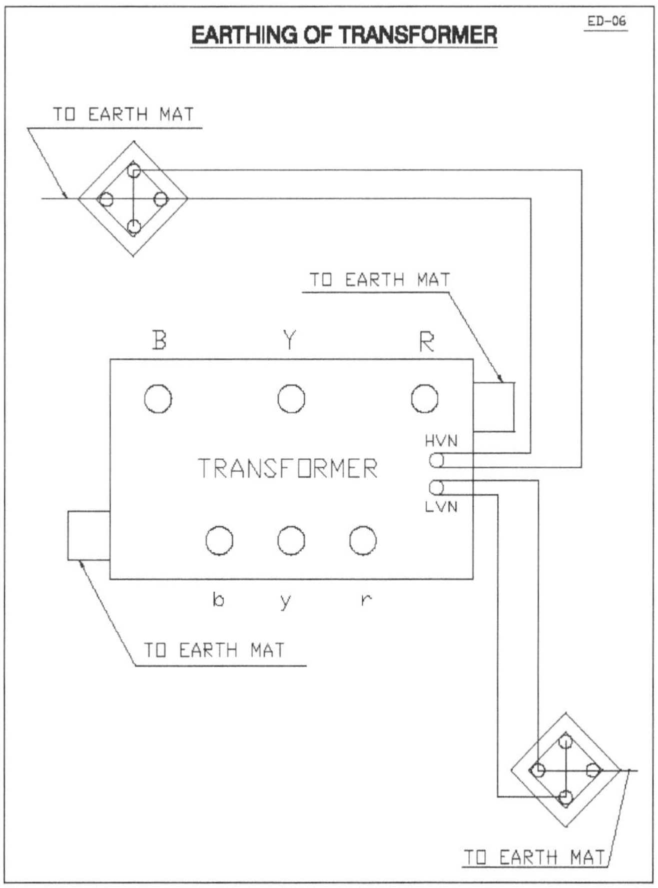

ED-15

DETAILS OF EARTH PIT

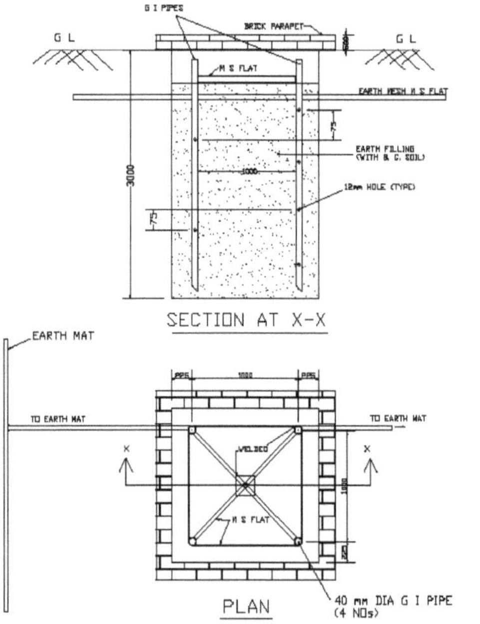

7. Achieving Tremedous Reduction in Earth-Resistance Through Deep Driven Ground Rods and Drilled Ground Wells in a Stubborn Rocky S/S Site in MPPTCL

1.0 Abstract: To ensure safety to the Operating Personnel and to the equipments in any EHV Sub-Staion in a Power system or in any switch-yard of a Power Station is of utmost importance. Such safety measures can only be implemented through appropriate earthing /Grounding design within the available size of the site, such that hazardous voltages are contained within the permissible limits.

The Authors are narrating a true case-history of challenging task in a stubborn rocky 132kV sub-station site wherein the earth resistance has been reduced from 1780 Ω(max) to 0.72 Ω(max){permissible limit of <1.0 Ω) through the use of deep driven ground rods and drilled ground wells. {Ref IEEE-80-2013 clause 9.5(b)}

2.0 Introduction:

Designing of Earthing system is of utmost importace following are the reasons to have a good earthing system;

- to safe-gaurd against non-fatal & fatal accidents of human beings under normal and during fault conditions.
- to safe-gaurd against equipment failure under normal and during fault conditions.
- for forming of star point of primaries of 3x1ph PTs/CVTs in the earth and the neutral gets grounded in the earth itself for their use in 3ph system.
- providing a conductive path to a part of Zero sequence Earth fault currents($3\ I_0$), return to their source in case of Star connected Tranformers and the Generators.
- to safely discharge the High frquency Lightening Surge and switching surge Transient Voltages to the Earth/ground.

3.0 Adequacy of earthing design: The earthing design is said to be adequate if;

- Earth/Grounds potential rise is within permissible limits
- Step and touch potential are to be within permissible limits as defined in the IEEE 80-2000.
- Earth resistance should be <1Ω for EHV Sub Stations and,<0.5 Ω for Powerstations,

4.0 Soil models with repect to resistivity;

- A uniform soil model is recomended for use where there is moderate variation in resistivity.
- Two-layersoil model consists of a upper layer of finite depth with different resistivity than a lower layer of finite thickness.

5.0 Prervialing practice of Earthing-System in MPPTCL;

The Sub station design Section headed by an E.E under the control of C.E (Plg&Desgn), MPPTCL is responsible for designing the respective proposed S/s projects right from layout drawing vis'a vis availablity of land configration etc. including earthing-system design etc. As a general rule Earthing design as per 2-layer soil model is considered suitable for almost all the green field sites.

The general practice for eathing as followed for the 132kV S/Ss in MPPTCL is mentioned here under;

5.1 Size of MS Flats:-

M.S. Flat to be utilized for formation of earth mat shall be of following sizes;

i. For 132KV Switchyards – 65x8mm
ii. For 33KV Switchyards – 50x6mm

5.2 Earth Mat: Earth-Mat is a mesh of M.S. Flats buried under the earth at a depth of 500 mm in the entire substation area. The **earthing rods or spikes (size-25mm dia X 3000 mm)** are driven into the ground/earth at each of the cross-sectional points of the Earth-mat and are welded to the earth-mat establishing firm contact to the earth-mat

i. Number of earth rods depends on the fault level in the substation and 500 Amps per rod is an empirical consideration. Therefore for a fault level of say 40KA, the requirement would be 80 nos. vertical Earth spikes/rods.
ii. All the joints of the earth-mat are welded using lap joints and "risers" are used for connecting the Earth-Mat to the structures, equipment body, surge arresters etc with 2-nos risers each and also the switch-yard fencing is connected to the earth mat..
iii. The "Power Engineers Hand book "of year 1985 of TNEB Engineers Association indicates that the Earth resistance values as under:-

    Large power / generating station – less than 0.5 Ω
    Major sub-station – Less than 1.0.Ω

Considering above practices followed by various utilities of different countries, the design of Earth Mat for 220KV and 132KV substations in the erstwhile MPEB/MPPTCL has been based on limiting value of the Earth -Resistance to < 1.0 Ω

6.0 Location of earth pits as per practice in MPPTCL:

i. Two earth-pits with four pipes each for HV & LV Neutrals of the power transformers of ratio 220/132KV or 132/33KV.
ii. One earth-pit exclusively for lighting arresters of the power transformer on high /low voltage side.
iii. One earth-pit each for a set of three PTs in 220/132KV Class. In case of 33KV PTs, it is adequate to provide two lead connections from the tank earth of the PT/CT units to the station earth mat.
iv. One earth pit for each set of 3 LAs in the 220KV and 132KV Class, with three individual leads, to be connected, one from each phase unit directly to the grounding.
v. One earth-pit for each of the 220/132KV feeder bays for connection of LAs and other apparatus.
vi. One earth-pit between two feeder bays in the 33KV System.

7.0 Prima facie activities;The piece of land for any proposed Electrical S/S (AIS)in M.P, alloted by the State Govt. is generally away from the inhabited area of the town, irrespective of its staus.

As a first and foremost acivity after the site is idenified by the revenue authorities, earth resistivity/ resistances at different locations in the area/piece of the ear-maked land for the proposed substation are measured and tabulated. In majority of cases, the conventional earthing of 2- layered soil -eartning design has proved to be adequate.

8.0 Case history of Earthing of 132kV Saori S/S in a rocky terain;

8.1 Site alloted for 132kV Saori sub-station in Chhindwarea Distt;

Figure 1: Site allotted for132kV Saori S/s in the Chhindwara Distt.s

8.2 Layout drawing of 132kV Saori sub-station;

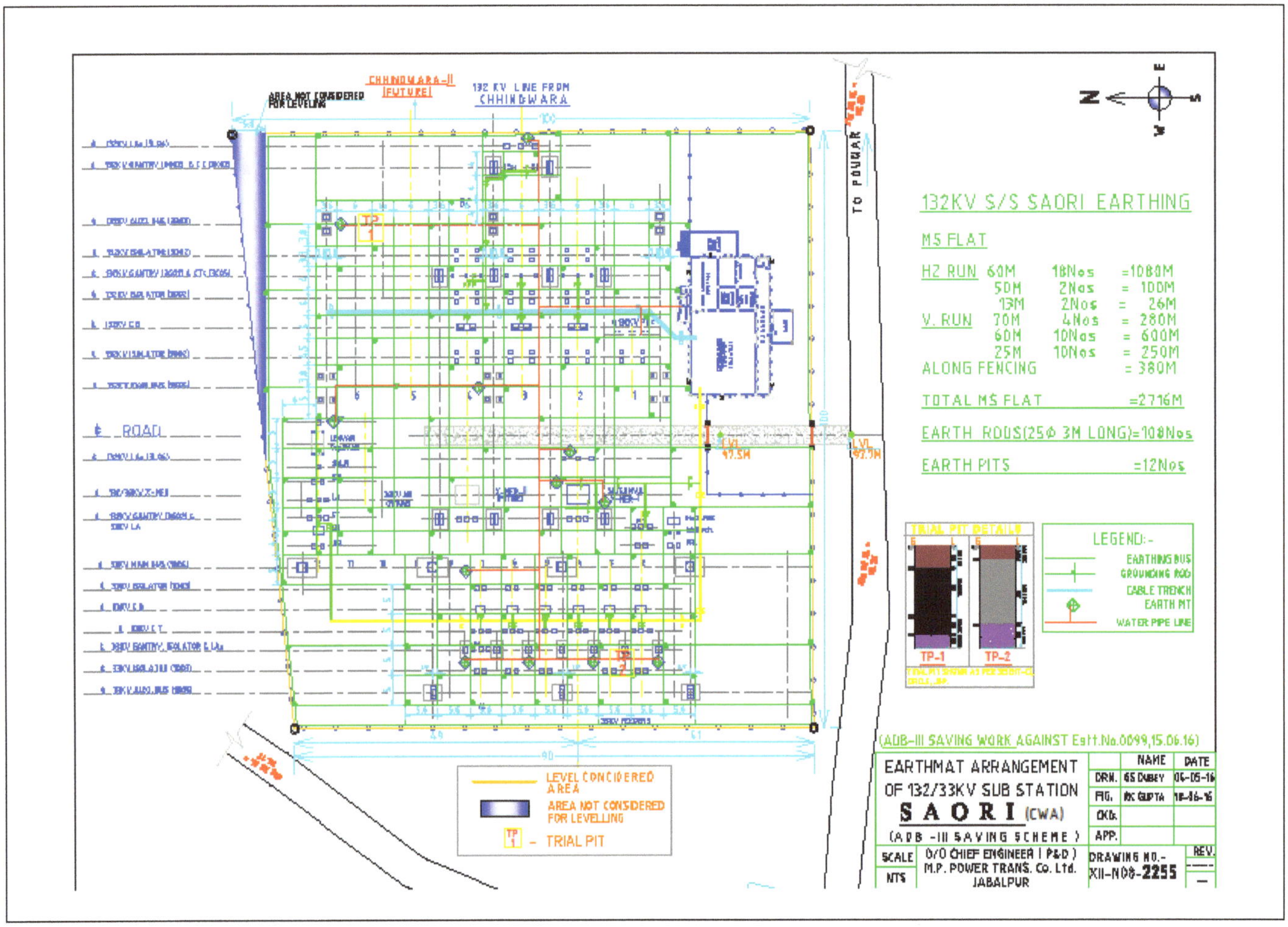

Figure 2: Earth mat Lay out Drg of 132kV Saori s/s

8.3 Measurement of Soil resistance at different locations: As per the conventional practice earth resitances at different locations as per the Lay out drawing was taken to assess the type of the soil of the site. To the surprise of the field officers, the resitance were foundvery high beyond imagination in the band of 550 Ω to 1780 Ω. (Pl.Refer figure No.3 & table No.1).

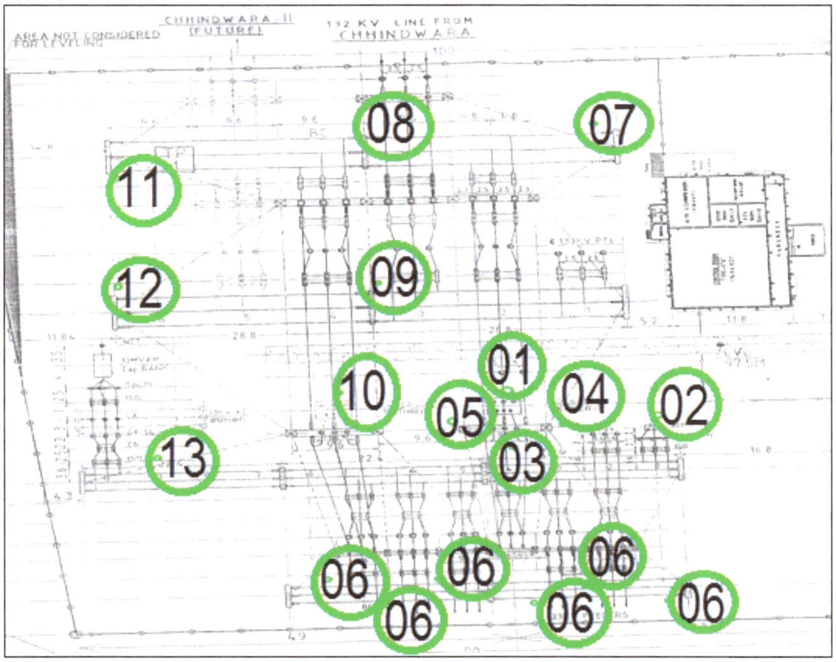

Figure 3:Locations of measurement of earth resistance

S.N.	Loc.	Location where E/R measured	E/R measured
1	2	Near Station Transformer	900.00 Ω
2	3	Near Transformer LV Side	670.00 Ω
3	4	Near Transformer LV side (tap changer side)	950.00 Ω
4	5	Near Transformer LV side (Conservator side)	870.00 Ω
5	6	33kV Aux. Bus side at various location(6 loc.)	812.00 Ω
6	7	132kV Aux. Bus structure side	595.00 Ω
7	8	132kV Aux. Bus structure (middle)	560.00 Ω
8	9	132kV Main Bus structure (middle)	550.00 Ω
9	10	Near Future Transformer structure	560.00 Ω
10	11	132kV Aux. Bus structure (near Cap.Bank Bay)	1780.00 Ω
11	12	132kV Main Bus structure (near Cap.Bank Bay)	1730.00 Ω
12	13	33kV Main Bus structure (near Cap.Bank CB side)	1750.00 Ω

Table 1: Earth resistance initially measured.

Thus the piece of land alloted for 132kV Saori S/s in Chhindwara Distt was found to be having a stubborn Rocky strata.

9.0 Challenge- Conquering very high earth resistance and slashing down to stipulated level of <1 Ω.

It was a real challenging task for the field EHT Const. Engineers for reducing such high earth resistance values to the required earth resistance of <1 Ω.

9.1 Trial earth pit & its effect on Earth resistance;

A trial bore of 6”x100 feet was drilled. There after 2 MS flats welded together in order to extend the length of the MS Flat so as to reach up to the bottom of the drilled bore. The extended MS flat was welded on to the 6ft long Casing pipe and was lowered down deep into the drilled bore. There after the bore was filled with black cotton soil

Earth resistances of all the 12 locations with reference to the trial earth-pit were measured. Appreciable reduction in the earth resistances was observed. This observation was very encouraging. The earth resistances measured initially (Pl. ref. table 1) and thereafter w.r.t. the trial earth pit (pl. ref. Table 2) as stated below may please be compared to;

S.n.	Earth Pit no.1, reference for E/R measurement	E/R measured
1	Near 50 MVA transformer Tap changer side	5.61 Ωb
2	Near 50 MVA transformer HV side	10.12 Ωa
3	Near 50 MVA transformer CB side	3.94 Ω
4	Near 132kV PT structure side	7.85 Ω
5	Near Control room side	5.89 Ω
6	In front of Station Transformer	12.77 Ω
7	Beside of Station Transformer	29.00 Ω
8	Near Main road side	6.99 Ω
9	Near Yard fencing	7.53 Ω
10	Near 33kV Aux. Bus side	14.62 Ω
11	Near 33kV Aux. Bus side	8.97 Ω
12	Near Incomer 33kV CB side	3.37 Ω

Table 2:Earth resitances w.r.t. Trial –pit.

Bold and challenging decision was taken by the field EHT Engineers to adopt deep driven ground rods earthing methodology in addition to the conventional earth mat and earth pit concepts. This proposal could also get a nod to go ahead by the top Brasses of EHT Const. set-up of MPPTCL.

This methodology is in line with the clause 9.5(b) as stipulated in the IEEE 80-2013, so called the bible of earthing system for safety in AC sub -stations which reads as follows;

”Use of deep driven ground rods and drilled ground wells.”.

Though, it is referred to as per the aforesaid clause but further, innovative idea had also to be implemented to reduce the resistance of the stubborn rocky site to the required level of < 1 Ω.

This was a challenging and herculean task which needed; extra efforts, extra finances, extra men & materials, extra skilled work-forces, additional heavy machinary over and above all, strict monitoring & scrupulous supervion of works on 24x7 basis maintaining utmost patients.

10.0 Civil works:

Breaking of rocks and boulders and levelling of the Sub-station site was caried out with JCB Excavator etc.

Figure 4:Breaking of rocks /boulders and leveling in progress.

10.1 Civil foundation works: Since the sub -station site had hard rocks, it was not possible to have conventional foundation for the Transformer, other out-door eqipments and structures, it was decided to conduct the pullout test of hard rock strata. On the basis of results obtained of Pullout test the Civil authorities designed Pile foundations for each equipment in including Transformer, other out-door equipments and strucures. The designed foundations were constructed with the machine-drilled bores by filling machine mixed M-25 concrete with proportionate addion of non-shrinkable compound with grouting of the foundation bolts a per respective templates thereof.

Figure 5: 50MVA 132/33kV Transformer installed on the pile foundation

Figure 6: Construction of foundation for 33kV Ckt Brkr. under progrss

11.0 Earth Pits-with deep drilled earthing bores.

The number of earth pits as per convention and as depicted in the Lay out drawing are 14Nos. The experience of drastic reduction in earth resitance (Pl.ref.table 2)with deep driven eartning conductor had given a vision for modifying the conventional earth pit with an additional earth coductor to be driven 100 feet down into the earth and to be filled them with blak-cotton soil and the Bentonite comound.

11.1 Revised drawing for Earth pits: Since decision was taken to revise the conentional earth pit in vouge in MPPTCL, a revised drawing was developed as ED 15A. Which is depicted here under;

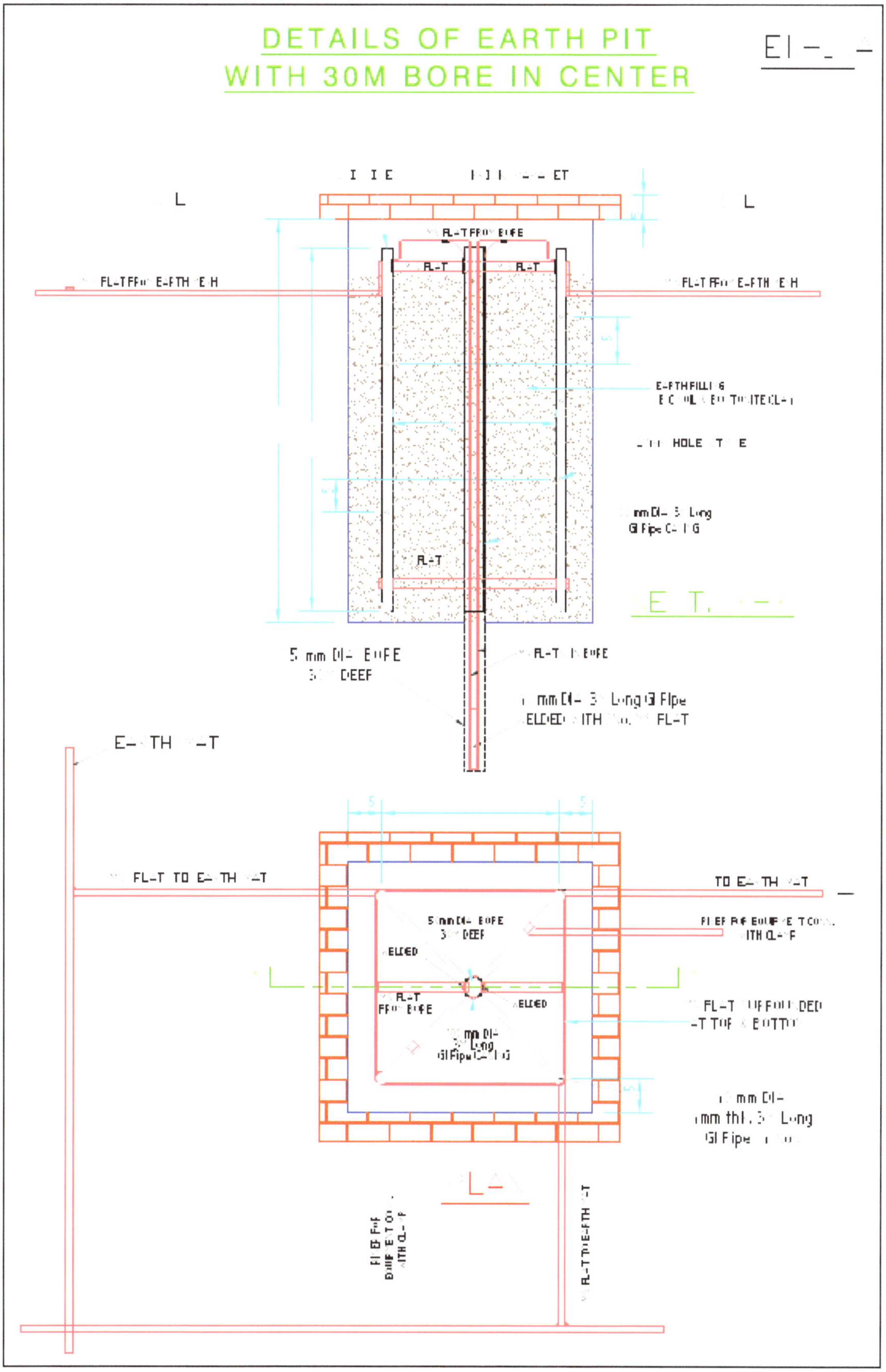

Figure 7: Revised drg of Earth-pit with additional bore of 30 meter depth.

Hard stubborn rocks were found in some of the locations of Earth pits, wherein even blasting of the rocks was also carried out. Then boring of 30 meters was carried out.

Figure 8: Drilling of borewell 30 mtrs deep in addition to conventionl earth pit.

As could be seen from the revised drg(Pl. ref.fig.7);a casing of GI pipe(100mm dia X 3 mter)was lowered in th the drilled bore of 150 mm dia carrying of 2no. extended flats(30 metrs)inserted into the GI Casings and welded-on. In each of the 30mtrs bores, the black cotton soil was also filled up completely. The preparation of 100mmX 3 meters GI pipe casing is depicted in the fig.9.

Figure 9:Casing pipe containig 2 MS flats welded together and inturn welded to the casing pipe.

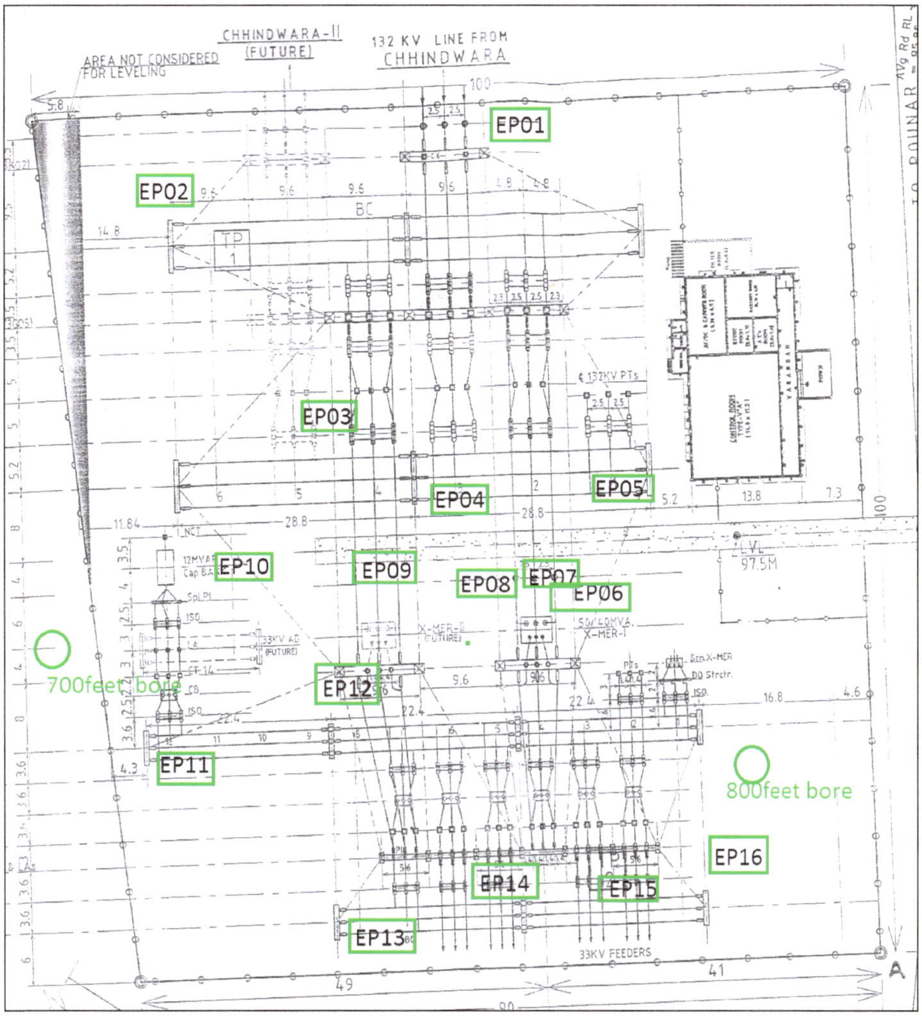

Figure 10: Marked earth-pits(EP)and Ground water bores 700ft & 800ft.

Fig11: Earth pit for neutral Grounding

Fig12: Earth pit in a Bay of the Transformer

As could be seen from the fig 11 &12, the MS flats from the deep driven bore are welded onto the opposite MS flats connecting the 4 nos. 3 meter pipes of the earth pit in a square formation.

1. All such earh pits are inturn connected to the laid earth-mat 0.5 mter below the ground level.
2. The Extended Earth pit is filled with Black cottonsoil and bentonite compound.
3. As a convention each earth pit is provided with water taps for watering the earth pits in the dry weather seasons.

13.0 Earth Mat and earth spikes:

Fig13 MS Flat laying workin progress

Laying the Earth mat with the MS Flats and driving MS spikes were another hurdlels to cross due to the Rocky strata of the site.

12.1 Earth mat:

As per practice MS flat of size (65x8mm)is to be laid 500mm below the ground level for formation of an earth-mat in transverse and longitudinal directions i.e. along and across the bays. The straight run joints and cross joints of the MS flat shall be "arc welded". All MS flat connections below the ground shall be made of electric arc welding. To prevent erosion of welded joint under the ground, Bitumen paint (2mm thick) shall be applied. This shall ensure good contact with the earth. The earthing strips shall also run around the control room building and earthing of all metallic but non-current carrying parts shall be done with the strip, which shall be finally connected with main earth mat. Switchyard fencing shall be at least 2mtrs inside the earth-mat and all steel work of the fence shall be connected to it at a regular spacing not exceeding 10 mtrs. Fence gate shall be separately grounded with flexible connections to permit movements.

The trenches for laying the earth mat too needed cutting of rocks with concrete cutting Machine. Adequate layer of Black cotton soil was spread all over the trench/groove made for laying the earth mat. After the earth mat was laid thereon the whole mat had also been covered with Black cotton soil.

12.2 Driving the Spikes into the earth: Drilling of the rocky-earth was done for almost all the 90 nos. of MS spikes(25mmx3000mm)to the of size of bore of 40mm X 3mtrs. At each of the cross joints of the MS flats of the earth mat, the MS spikes were firmly welded onto those cross joints. In each and evey pocket for the Earth spikes, Bentonite compound was filled in.

13.0 Earthing/Grounding of each out door equipment:

i. At the equipment end, 2 nos each of MS flat (6X50mm) known as Risers are connected to 2 locations of the metallic portion of each equipment installed in the AIS.
ii. The other ends of the risers from each out-door equipment is connected to the earth mat.

13.1 Earthing of Structures: Each structure have also been connected to the earth mat through 50x6mm. MS flats

13.2 Shielding wires and OPGW:

1. All the shielding wires over the gantry structures have been also connected to the earth mat with Risers.
2. Incoming 132kV Transmission line's OPGW has also been connected to the S/s shielding wire.

14.0 Wells for meeting the water requirement and assisting in improving earthing System;

2 nos. water well bores 700ft & 800ft respetively are drilled at the locations maked on the fig 10 for day to day use of the sub station including watering of Earth pits. It is also to mention that the casing of these borewells are connected to the Earth mat which inturn add their contribution to the earth mat in improving the earthing system.

15.0 Final Earth resistance measurements:

After completing the earthing system improvement works as narrated in the foregoing paras the earth resistances measured in the sub station are as tabulated hereunder;

S.No.	Earth Pit No.	Location	E/R measured
1	E/P-1	132KV LA of 132KV feeder.	0.65 Ω
2	E/P-2	132KV Aux. Bus near Cap. Bank.	0.26 Ω
3	E/P-3	132KV Main Bus.	0.15Ω
4	E/P-4	132KV Main Bus Structure (middle).	0.40Ω
5	E/P-5	132KV Main Bus.	0.25Ω
6	E/P-6	132/33KV Transformer OLTC DM.	0.65Ω
7	E/P-7	132/33 KV Transformer HV side.	0.28Ω
8	E/P-8	132/33 KV Transformer HV side.	0.65Ω
9	E/P-9	132/33 KV Future Transformer Structure.	0.29Ω
10	E/P-10	33KV Cap. Bank CB side.	0.73Ω
11	E/P-11	33KV Cap. Bank bay side.	0.35Ω

12	E/P-12	33KV Future X-mer Incommer side.	0.28Ω
13	E/P-13	33KV Aux. Bus near B/C bay.	0.28Ω
14	E/P-14	33KV Mujawar Bay	0.74Ω
15	E/P-15	33KV near Main Bus Lawaghughri feeder.	0.73Ω
16	E/P-16	33KV Main Bus near 33KV PT.	0.60Ω

Table 3: Earth resistances measured after earthing improvement measures.

16.0 Observation: It could be observed from the table 1&3 that the earth resistances have been slashed down drastically from the initial range of 550 Ω - 1780 Ω to the required range of <1Ω.

17.0 Completion and commissioning of the sub station: Other works of the sub station were completed and the sub station was commissioned on 23rd August 2018. A few of the photographs are depicted hereunder;

Figure 14: 33kV bus under const.

Figure 15:33kV PTs & StationTransformer

Figure 16: Completed 132kV Gantry Structure of Transformer bay and work-men seen working on the on-looking Gantry structure of 132kV Chhindwara Transmission line.

Figure 17:The 132kV Saori sub station was commissioned on the 23rd August 2018.

18.0 References:

1. IEEE 80-2000 & IEEE 80-2013 Guide for A.C.Sub station Grounding.
2. CBIP Earthing manual-339 of 2017.

19.0 Authors: K.K.Murty, Pramod Behere & Arun Vaidya

8. Integrity of Earthing System-Its importance in the EHV Switch yards/sub stations.

Preamble: Earthing system plays a vital role in Power system protection, safety to men and equipment in the EHV Switch yards/sub stations. The earthing system in an EHV Switch yards/sub stations as a whole consists of;

i. an Earth grid which is formed by a mesh of a horizontal strip electrodes and vertical ground electrodes of adequate sizes buried in the earth, providing a common ground for electrical devices or metallic structures, usually in one specific location.
ii. Earth pits.
iii. Risers from the earth mat to the equipment.

Integrity of this system means rendering service with reliability and loyalty in the power System. Any disconnection in continuity in the earthing grid is synonymous to breach in loyalty.

The authors have described some of the basic definitions like Earth potential rise, Equipotential bonding and concept/methodology of verifying integrity of Earthing system highlighting some case studies as well.

Key words; Grounding, Equipotential bonding, GPR (Ground Potential Rise). Primary injection kit.

1.0 Introduction;

1.1 Brief description of earthing grid;

iv. **Choice between Copper and M.S.**: Prior to about half a century copper strips /rods were used in the earthing grids of HV and EHV switchyards. However, since a few decades the use of copper has been phased out by the Transmission utilities and in the present scenario MS flats/ MS rods are being used on following counts;

- Cost-benefit analysis favours use of MS over Copper, despite being prone to corrosion.
- Copper is theft prone being a costly metal.

1.2 Earthing-grid: The Earth grid consists of a mesh of equally placed horizontal electrodes (MS Flats / solid rods) covering the whole switchyard area and some numbers of vertical rod-electrodes are dug underneath the ground from the point of crossing of horizontal conductors as per design. The whole earth grid is embedded below the surface of ground to the depth of 500 mm. Generally, a site of low resistivity is preferably selected for construction of the proposed sub-station, if available, however, it is the responsibility of the utility to construct the substation on any site made available to them, designing the earthing-grid it in such a way that all the parameters towards safety to men, equipment, material and the living beings around the vicinity of the substation are fulfilled.

a. **Some of the recommended sizes of MS flat are shown in the table 1.**

kV class	Size of MS flat for Earth mat	Size of MS flat for Risers
33kV	65x8 mm	50x6 mm
132kV	65x8 mm	50x6 mm
220kV	75x8 mm	65x8 mm
400kV	40 mm dia. solid rods	75x8 mm

Table1.

Note: MS flat sizes/solid rods shown in the table 1 are as per MPPTCL's practice, however, it may vary from utility to utility as per their approved designs.

b. **Size of spikes and placement thereof**; 25 mm dia. x 3000 mm and each spike shall be driven vertically at an interval of not less than 2 times the length of the rod (spike

If the rods are too close, resistance area of one electrode will interfere with that of other resulting in no effective gain. Therefore, to avoid such a situation, the distance between two spikes(rods) is kept as 2 times the length of the rod.

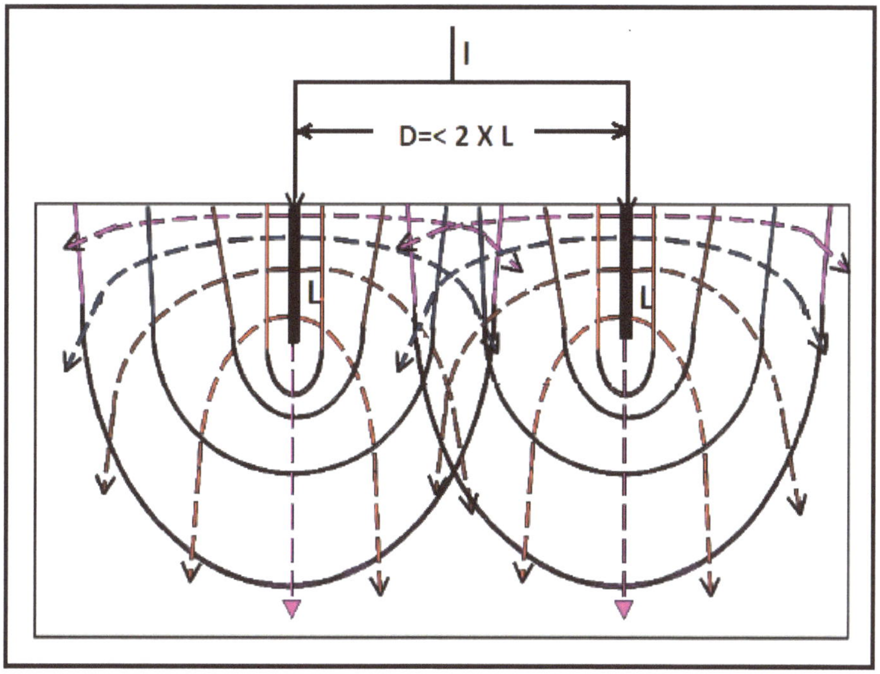

Figure1: Overlapping resistance area of two earth rod.[1]

c. **Concept of Earthing Grid;** Conceptual Drawing showing an Earthing Grid is in fig.2.

Note: Spreading of Gravel 0.08 to 0.15 m reduces the Step and Touch potential [1]and also restricts the movements of snakes and other reptiles.

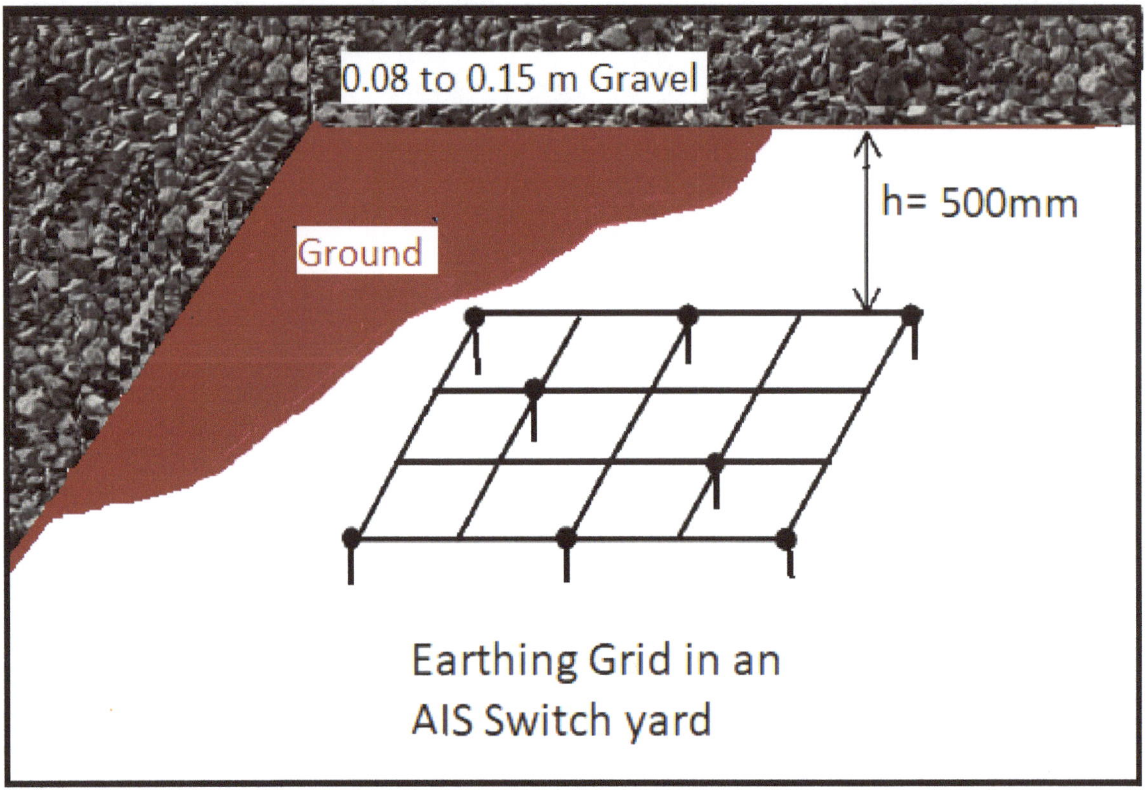

Figure 2: Earthing-Grid.

2.0 Function of earthing-grid; The earthing-system in the substations is vital. The main functions of earthing systems are as follows;

i. To provide earth connection to the neutrals of generators, transformers, capacitors banks and reactors.
ii. To earth the single phase EHV equipment viz; CVTs, CCs, PTs and LAs.
iii. Offer a low impedance path through the earth for the currents coming from ground faults, lightning strokes etc.
iv. * To contain the Ground Potential Rise/Earth Potential Rise to safe limits.
v. To limit the touch and step potential within tolerable limits.
vi. To earth the metallic body of equipment of EHV S/s, structures, metallic water- pipe lines, metallic gas-pipe lines and metallic fencing etc. for establishing equipotential bonding.
vii. providing safety of the living beings around the vicinity of the substation
viii. Discharging induction /static voltages to the earth of de-energized equipment prior to maintenance or other any work to be conducted.

***Note:** Please see the figure 3 and the definition of Earth Potential Rise/Ground Potential Rise.

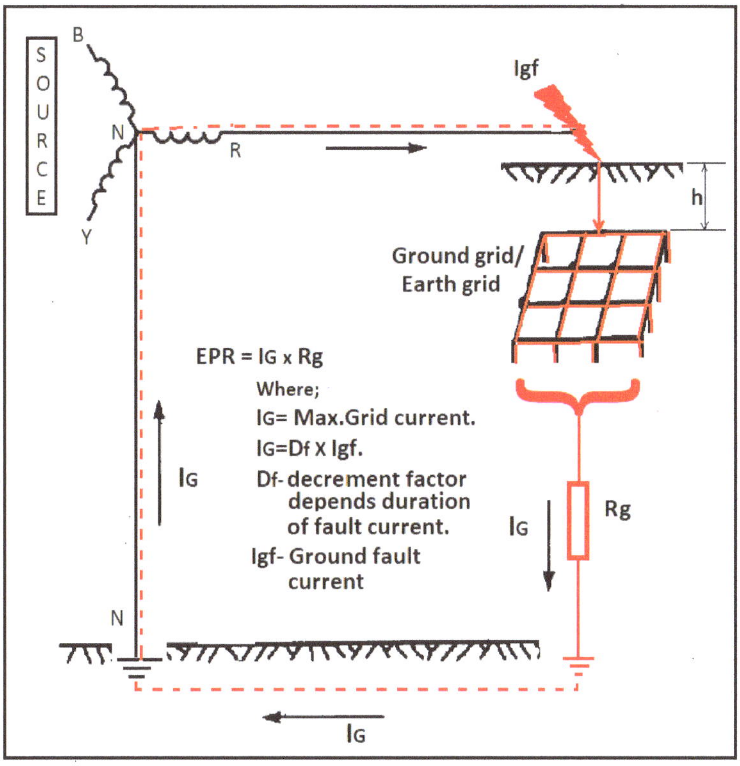

Figure 3: Earth/Ground Potential Rise.

3.0 Earth Potential Rise (EPR): The maximum electrical potential that a substation earthing grid may attain relative to a distant grounding point assumed to be at the potential of remote earth.

EPR/ GPR is the product of the magnitude of the grid current as a portion of the fault current, conducted to earth by the grounding system and the ground grid resistance.

EPR/GPR= IG X Rg,[2]

Where; IG =Df x Igf

Igf- Ground fault current,

Df- Decrement factor dependent on the duration of the ground fault current in secs.

Rg-Resistance of the earthing grid.

4.0 Maintenance of Earthing System, General practice:

Following activities are in vogue in most of the Transmission Utilities towards maintenance of the earthing system preferably in summer months.

- Cleaning the earth pits of any weeds or vegetation.
- The condition of tap for watering the earth pits.
- Cleaning of blackened spots/pitting-globules if any and tightening of nut- bolts of leads/flats firmly to the earth mat.

- Taking earth resistance of the earth pits.
- Welding is carried out, if required for strengthening the contacts of risers and other connections to the earth mat.

4.1 Figures;

Figure 4: Earth Pit.

Figure 6: WTI & OTI earthing of Marshaling-Box through riser.

Figure 5: Body earthing of transformer through riser.

Figure7: Body earthing of 220kV CT.

Figure 8: LA earthing.

Figure 9: Earthing of neutral through Neutral CT.

5.0 Equipotential bonding; Equipotential bonding is essentially an electrical connection maintaining various exposed conductive parts at substantially the same potential. An earthed equipotential zone is the one within which exposed conductive parts and extraneous conductive parts are maintained at substantially same potential by bonding. Thus, under fault conditions, the difference in potential between simultaneously accessible exposed and extraneous parts will not cause electric shock.

Bonding is the practice of connecting all accessible metal work-whether associated with electrical installation (known as exposed metalworks) or not (extraneous metalworks) to the system earth.

In each EHV Switchyard, equipotential bonding is established connecting following to the earthing-grid of;

i. each equipment in the switch yard. Such connection of bonding as sample cases have been shown with risers painted in green colour in the figures 5,6 & 7.
ii. all the conductive water-pipe lines and conductive compressed air pipe- lines.
iii. body of air-compressors and conductive portion of compressor room.
iv. conductive portion of Control room building.
v. all the structures in the switch yard.
vi. conductive fencing and gates.

5.1 Importance of equipotential bonding; As already mentioned that, "under fault conditions, the difference in potential between simultaneously accessible exposed and extraneous parts will not cause electric shock". In other words, there is every possibility of experiencing shocks or equipment getting damaged/failed, if the equipotential bond of the of equipment through the risers is lost with the earthing grid while attaining Earth Potential Rise consequent to heavy fault currents or lightening strokes.

6.0 Reasons for deterioration of Earthing system;

A general practice of maintenance of earthing system which includes earth resistance test as well is a mere voltage drop across surrounding soil and does not give the physical condition of the electrodes. It is also an irony that the most of the suitable soil are low resistivity soil and most corrosive.

While in service the earth mat is damaged during clearing the faults and discharging lightning currents, resulting into interrupting continuity and introducing high resistance across the bonds. Thus, hampering its capability to render successful and faithful service thereafter.

7.0 What is Earth-Integrity test and when to perform this test?

i. The integrity test, as the name defines, this test verifies the healthiness of the Earthing System so that it should render its service with loyalty.

This test verifies the continuity between two different ground points of the grid. In this way, healthiness of the earth grid is verified confirming that it can carry fault currents successfully. This test should be essentially performed in the old switch yards.

Note: The situation mentioned in clause 6.0 can be detected with the Earth-integrity test only.

ii. This test should also be essentially performed after completion of the job of earthing grid in a new substation prior to energizing it, for verifying that none of the joints /connections are inadvertently left unconnected or welding/ bracing job is not properly done.

8.0 Earth integrity test procedure;

i. **Requirement of Items and supply arrangement;**
   a. 1 ph. LT-AC supply,
   b. Primary injection kit; 230 Vac,50 Hz, 2 kVA, capable of supplying 200A for 1-3 minutes.
   c. High-current carrying copper leads/cables of 20 meters length preferably welding machine leads with proper clamping arrangement - 02Nos.

ii. **Test connections;** Please refer to the connection diagram shown in the figure10.

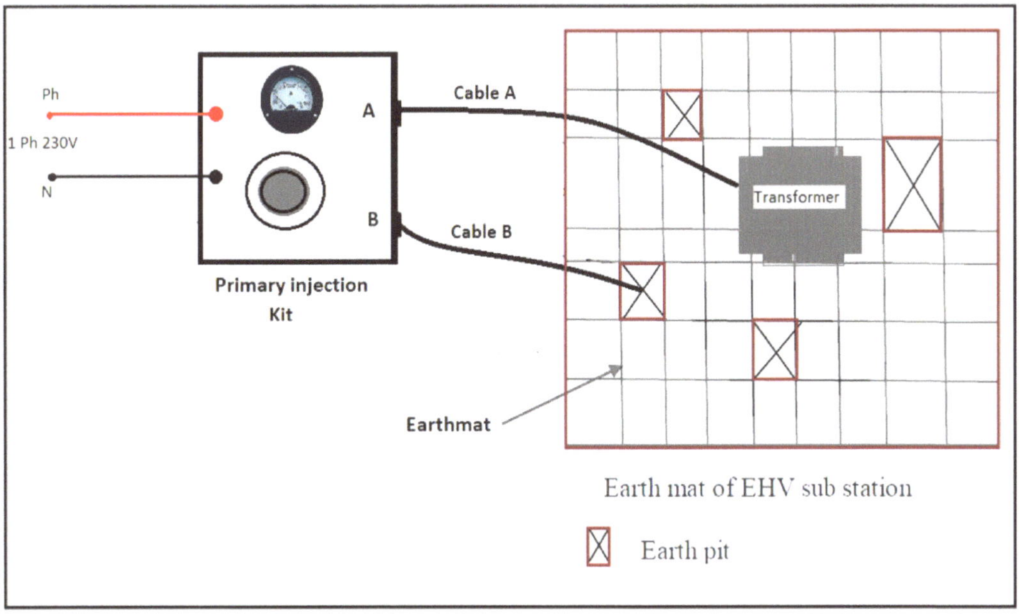

Figure 10: Schematic diagram of Integrity test setup.

a. Extend 1 ph.LT- AC supply to the Primary injection kit.
b. Connect 02 Nos. leads to the output terminals of the primary Injection kit.
c. Switch on the kit.
d. Short the loose end of the cables.
e. Inject 100A current and simultaneously measure the voltage drop at the Primary Injection kit output terminals.
   Calculate the resistance as per ohm's law (ie R=V/A). This calculated resistance is the burden taken by the leads.
f. Now, clamp one cable to an earth-pit and the other cable to the riser/or to the point of section of earth mat to be tested.
g. Inject the current to the extent of 100A and simultaneously measure the voltage drop at the PI output terminals.
h. Test all the equipment in the vicinity of the first one in line with the procedure described in g).
i. The voltage drop shall be almost equal in the vicinity of all the points of test..
j. At some points the injection of current does not rise up to 100 A then it is a clear indication of that the Joint is deteriorated and needs improvement.
k. If at some location, the current is not being injected at all, confirms that there is discontinuity of connection with the earthing grid.
l. **Method of calculation**: Voltage drop observed while testing object (VT) – voltage drop when loose ends of the both the cables were shorted together (Vsrt) and 100A current was injected in both the cases = difference in voltage (Vrg).

Therefore, Resistance of the object including Earth grid (Reg),

$$\text{Reg} = \frac{\text{VT} - \text{Vsrt}}{100}\Omega$$

iii. **Note**:

a. It is a practice, followed in MPPTCL to verify the healthiness of the earthing-grid by performing Earthing-grid integrity test in the Sub stations more than 20 yrs old at least one in a year.
b. It is also before charging of any new EHV substation.

8.1 Some of the pictures showing actual Earthing-Integrity Test being performed at site.

Figure 11; Primary Injection Kit.

Figure 12: One lead connected to earth pit

Figure 13: Measuring current with clamp-on Amp. meter while injecting 100A from the Primary Injjection kit.

Figure 14: Measurement of 100A shown in the clamp-on Amp meter display window.

Figure 15: Current injection process between the earth pit and the test point with the Primary Injection kit.

Figure 16: Measuring voltage at the terminals of Primary Injection Kit.

Fig17: Test results being noted down by the testing technicians.

8.1 Some true data of Earthing integrity test performed in an EHV sub-station in MPPTCL, which is > 50 years old and the results thereof are tabulated in the table 2;

Voltage drop on injecting 100 A, making both the loose ends shorted, voltage measured at the PI kit 2.66 V. therefore the leads resistance = 0.0266 Ω.

Sno.	33 KV Bay No.	Current Injecton in different point between adjacent structure of gantry /CT/ Isolator with Earth pit	Voltage measured across the Primary Injection kit terminals.	Voltage drop across earth mat = voltage drop across the terminals of PI kit -Voltage drop across the terminals of PI kit with shorted the leads & injecting 100A	Resistance in Ω as per calculation (V/I)	Remark
01	Bay-1	100 A	4.3 V	4.3 - 2.66=1.64	0.0164	
02	Bay-2	100 A	4.2 V	4.2 - 2.66=1.54	0.0154	
03	Bay-3	100 A	4.2 V	4.2 - 2.66=1.54	0.0154	
04	Bay-4	0 A				Current is not injected between bus isolator to earth pit
05	Bay-5	100 A	4.2 V	4.2 - 2.66=1.54	0.0154	
06	Bay-6	100 A	4.1 V	4.1 - 2.66=1.44	0.0144	
07	Bay-7	77 A	5.0V	5.0 - 2.66=2.34	0.0303	It seems that in these locations earth-strengthening by providing additional earth mat is required
08	Bay-8	77 A	5.0 V	5.0 - 2.66=2.34	0.0303	
09	Bay-9	80	5.10V	5.10 - 2.66=2.44	0.0305	
10	Bay-10	76 A	5.08 V	5.08 - 2.66=2.42	0.0318	

Table 2

8.2 Rectification job:

i. Afresh earthing was done in the Bay No.4 by laying MS flats from the 33kv bus isolator and connecting the same to the old earth mat.
ii. Additional earthing was done by wherever the resistance was found in the range of 0.0303 to 0.0318 Ω.
iii. After carrying out rectification /revamping jobs as stated in (i) & (ii), healthiness thereof was verified by performing earth integrity test and have been found in order.

9.0 Recommendations:

i. In view of above, it is recommended that the earth integrity test be carried out in yearly basis in the EHV substations (more than 20 years old), which could identify the status of earth connectivity-bonding with structure & its equipment. It is also recommended to be performed in new substations (before commissioning thereof).
ii. In view of importance narrated about testing/verification of integrity of grounding system in IEEE 80- 2013 and IEEE 81-2012, test kits for use in induction-prone EHV switch-yards have been developed by some of the leading manufacturers of global repute. Such test-kits can be procured by the interested Utilities as the same are available in the Indian market too.

10.0 References:

i. Dr. K. Rajamani - Application guide for power engineers Part-1 Earthing and Grounding of Electrical Systems.
ii. IEEE-80,2013; IEEE Guide for Safety in AC Sub Station Grounding.
iii. IS:3043-2018; Code of Practice for Earthing. (Second Revision)

11.0 Authors: K.K.Murty, Deepak Kumar

9. Methodology for Monitoring and Enhancing Life Expectancy of EHV Current Transformers

1.0 Introduction

In the erstwhile MPSEB (Madhya Pradesh State Electricity Board) 400kV system was introduced in early 80s. Prior to this the MPSEB Transmission system had 132kV and 220kV Transformers and Transmission lines only. The EHV current transformer like other equipments used, were mostly imported ones like ASEA, Brown Bowery and CDC France makes etc. The Transmission system of Madhya Pradesh was small due to less industrialization in the State in that era. The Transmission system of MP assumed greater importance with the introduction of 400 kV transmission system with interstate and Intra state lines in addition to rapid growth of 220kV and 132 kV S/S and Transmission lines. 400kV lines passing through Indore was the corridor for bulk power transfer to Gujarat through 400kV Indore Asoj Lines.

Since, the Transmission system was small and the Generating units too were less, the fault level too was less. The failure of CTs in the system at that time was less and was unnoticed too, as such failures were not affecting a large area of the state. With the growth of the Power system, the population of CTs has increased many folds and any failure of CT irrespective of the level of Transmission voltage affects a large area of the state. Catastrophic failure of 400kV and important 220kV CTs may even lead to System disturbance in the grid.

2.0 Need for monitoring and Life assessment of EHV CTs:

Though the investment on CTs is very meager as compared to the other equipment in a Power project / system, but are very important in the light of above and also, in the field of Power measurement and Protection. Thus the need for monitoring the health of the CTs and their Life assessment thereof is the need of the day. The State Transmission Utilities, Central Power Transmission Utility and also Private Entrepreneur's owning Transmission lines may resort to this.

3.0 Modus Operandi: The concerned Transmission Utilities should first identify the CTs age wise, manufacturer wise and importance of the feeder/equipment wise. They should plan shutdowns for condition monitoring accordingly.

4.0 Condition Monitoring of CTs:

Over the period of time of 15-18 years the aging of the CTs start. The CTs gradually deteriorate and lose their credibility.

The paper impregnated with oil/oil starts deteriorating/degrading with time and this process depends upon the thermal, electrical and other chemical complex reactions within the CTs. Due to ageing of insulation the dielectric loses increase which reflect as change in Dissipation factor (Tan δ). Other factors like Capacitance value, moisture content (paper/oil) also indicate the extent of aging of insulation. Sometimes due to voids/cavities in the primary insulation, carbonization takes place, due to the partial discharge, which may also lead to pre mature failure of CT.

The rate of insulation Ageing is directly proportional to the moisture content in the oil/paper and presence of oxygen still accelerates the deterioration process by 2.5 to 10 times the normal aging rate.

5.0 Life assessment methodology: The technique such as measurement of insulation resistance, Dielectric dissipation factor/loss factor (Tan δ) and capacitance measurement are generally used by the power utilities since long. In addition to these, dissolved Gas analysis of oil and thermo vision scanning of CTs would prove as diagnostic tools in identifying development of faults in the initial stages itself and call for remedial measures by the maintenance staff. Thus catastrophic failure could be avoided.

5.1 Analysis of Degradation process of paper/oil insulation is generally based on the following criteria;

i. Increase in Tan δ as compared to its original value. An increase of 0.1% per year is considered normal.
ii. Analysis of DGA results with rising trend of key gases is monitored. If there is sudden rise in any key gas is indicative of development of fault inside the CT. The evolution of gases in the CT may be based to IEC-60599.
iii. Monitoring of moisture content in the oil and paper. Higher moisture content in paper may accelerate aging.
iv. Increase in value of capacitance is due to shortening of insulation layers due to low intensity discharges, presence of moisture due to sealing/gasket failure and ageing of insulation etc. An increase in capacitance more than 1% per year is considered high and is a matter of concern.

6.0 Tan δ and Capacitance Values at site: The active tan δ and capacitance values at site require compensation for following factors;

i. Effect of CT temp.
ii. Effect of ambient humidity.
iii. Effect of electrostatic influence.

Normally for new CTs the Tan δ value ranges from 0.2 to 0.3%. Aging of CT under normal operating conditions will cause gradual increase in Tan δ value. Tan δ value up to 0.7% is generally considered within permissible limit during service, however CT having Tan δ value beyond this value, needs to be kept under close monitoring.

The major hurdles in measurements at site conditions should be taken care by use of proper testing equipments/instruments which shall measure with specific accuracy at site conditions.

Caution: The transmission lines should be properly earthed/discharged and discharge rods be kept in place till such time all the measurement of Tan δ and Capacitances are completed.

7.0 Interpretation of Tan δ Values; An increase in Tan δ value indicates excessive moisture in the insulation and other deterioration or contamination other than water.

In order to eliminate the effect of leakage current in case of CTs provided with Test Tab, the Tan δ should be measured in UST mode (Ungrounded Specimen Test mode).

8.0 Insulation Resistance Measurement:

(i) Dialectic Absorption Current: On application of Direct Voltage on any insulation, in add ion to charging current, dielectric absorption current also flows. This current is due to presence of polar molecules in the insulation material.

(ii)The ratio of Insulation resistance value taken after 60 seconds to 15 seconds is termed as Polarization Index (PI). The PI is indicative of health of the CT.

9.0 On line Partial discharge test; This Test has been recently adopted in the M.P. Power Transmission Co.Ltd. in case of 220kV & 400kV class CTs.

10.0 Probable Causes of CT failure:

i. Primary insulation puncture/ failure.
ii. Non uniform electric stress due to improper wrapping of insulation paper during manufacturing process.
iii. N2 gas leakage from sealing gaskets leading to entry of moisture inside the CTs

11.0 Implementation of Remedial measures: Monitoring of CTs so as avoid catastrophic break downs were discussed in the fore-going paragraphs, however in addition to the same following measures if implemented shall definitely reduce the failure rate of CTs.

i. Insisting for Partial discharge measurement at the manufacturer's works as well as at 220 & 400kV Class CTs while in service.

Note: This Test has been recently been adopted in the M.P. Power Transmission Co.Ltd. in case of 220kV & 400kV class CTs.

ii. Monitoring of change in Tan δ and Capacitance values at 0.1 to 1.1 pu of the rated voltage.
iii. Monitoring of change in capacitance and tan δ before and after I minute HV test.
iv. Monitoring of N2 Pressure for all CTs having N2 cushion at the top for taking care of Oil expansion/contraction.
v. The IR values should be monitored during each maintenance program. The values and megger to be used recommended are as follows;

a. Pry to Secy-by 5kV – 250 Meg ohms.min
b. Pry to Earth- by 5kV- 250 Meg ohms. min
c. Pry to Tan δ Test Tap –by 5kV - 250 Meg ohms. min

d. Secy to Earth –by 500V— 1 Meg ohm. min
e. Between each Secys—by 500V--1 Meg ohm. min

12.0 Conclusion: The importance of the CTs in EHV grid network has been brought forward in this paper. The Transmission Utilities, if follow the same may improve the quality of power supply, with negligible break downs on account of CTs and thus resulting into minimal down time in the system.

13 Author: K.K Murty

10. Selection and Application of Surge Arresters

INTRODUCTION;

Working Principle of Surge Arrester; When the voltage surge traveling along the conductor reaches the point at which a lightning arrester is installed it breaks down the insulation of the arrester momentarily and allows the voltage surge to discharge to the ground. As soon as the voltage surge is discharged to the ground, the Arrester regains its original condition ie it becomes an insulator.

In the present era gapless ZnO or Zinc (Metal) Oxide Surge Arrester (MOSA) arresters are mainly used for surge protection.

1.0 Working Principle of Surge Arrester; When the voltage surge traveling along the conductor reaches the point at which a lightning arrester is installed it breaks down the insulation of the arrester momentarily and allows the voltage surge to discharge to the ground. As soon as the voltage surge is discharged to the ground, the Arrester regains its original condition ie it becomes an insulator.

The Surge Arresters, shall be proved to be performing satisfactorily if they possess following qualities;

- ➢ Prevent the current to flow to the ground as long as the system voltage remains normal.
- ➢ Provide a path to ground, when the system voltage rises to a predetermined valve above normal and dissipate the energy from the surge without raising the voltage at which the circuit is operating.
- ➢ Arrest the flow of current to ground, as soon as the system voltage drops to normalcy and restores the insulating qualities between the conductor and ground.
- ➢ It should not get damaged during discharging process and shoud be capable of automatically restoring its property of insulation between the Transmission line conductors.
- ➢ It should be ready for performing further discharging duty.

Note:

i. The performance of any arrester is dependent on a good connection to ground. Arresters will not function without a proper ground.

ii. The arrester should be placed as close as possible to the equipment, that is to be protected and leads connecting arresters to ground should be kept as short as possible & without any loop.

2.0 Characteristic of Zinc oxide (ZnO)blocks: The zinc oxide is a semiconducting material of N-type. It is pulverised and finely grained. More than ten doping materials are added in the form of fine powders of insulating oxides such as Bismuth (Bi2O3), Antimony Trioxide (Sb2O3), Cobalt

Oxide (CoO), Manganese Oxide (MnO2), Chromium oxide (Cr2O3). The powder is treated with some processes, and the mixture is spray dried to obtain a dry powder.

The dry powder is compressed into disc-shaped blocks. The blocks are sintered to obtain a dense poly- crystalline ceramic. The metal oxide resistor disc is coated with a conducting compound to protect the disc from undesirable environmental effect.

Figure 1: ZnO blocks.

2.1 Electrical representation of a ZnO element; A ZnO element basically represents a capacitive leakage circuit. In its leakage current range, it may be electrically represented as shown in Figure where I_{ZnO} is the leakage current, capacitive in nature, and Ic and Ir its capacitive and loss components, respectively. The success of an element will depend upon its low loss-component, which would mean a lower loss during continuous operation, on the one hand, and a lower temperature rise of the element on the other.

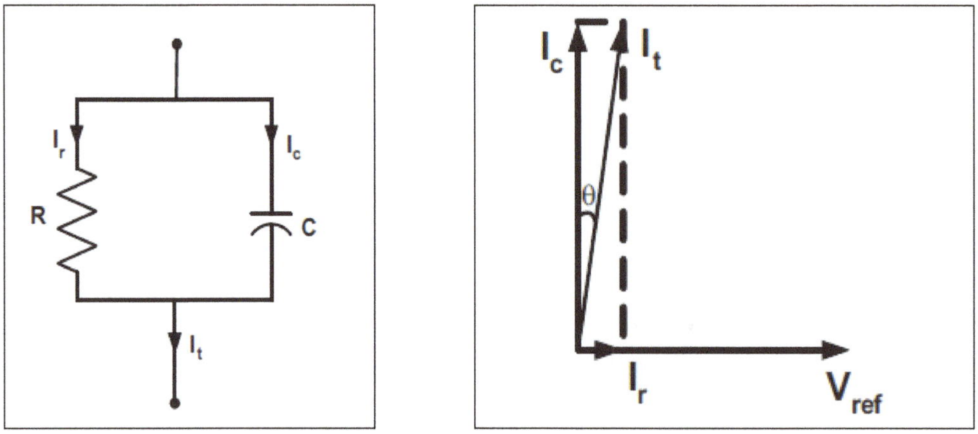

Figure 2(a): Electrical Representation of ZnO element. (b)Vector diagram

3.0 ZnO Surge Arrester:

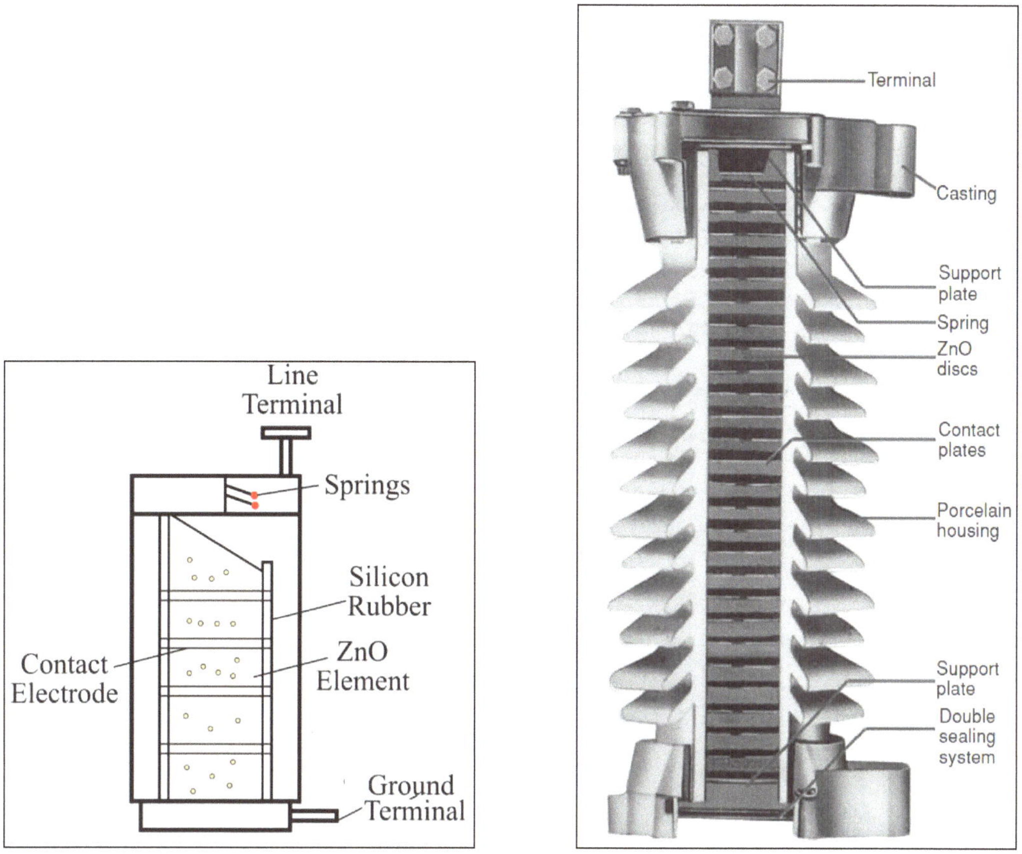

Figure 3: Internal arrangement of ZnO Surge Arrester.

Electrical equipment connected in power system are exposed to many stresses, and one of major concern is protecting them from transient over voltages. Transients over voltages are caused by lightning discharges and switching operations.

A surge arrester is a protective device connected in parallel with system equipment to be protected. Over -voltages at the protected equipment are limited by the arrester that conducts energy associated with surge to ground and protect the equipment. The highly non-linear characteristics of an arrester allow the arrester to limit the voltage across its terminal nearly a constant value over a wide range of arrester current. The voltage across the equipment to be protected is almost same as voltage across arrester (Unless there is large separation distance between surge Arrester and equipment to be protected & large lead lengths).

During conduction of surge current, arrester exhibits very low impedance & forms a voltage divider to applied surge voltage in conjunction with line surge impedance. During the time arrester is in conduction, a large percentage of surge voltage appears across line surge impedance and not across equipment to be protected.

By properly applying the arrester, the equipment insulation will not be exposed to damaging voltages, thus eliminating the threat of insulation failure. It is important to correctly select arrester parameters so that it can do the desired protection function without causing any nuisance in the system.

4.0 Over Voltages in Power Networks;

A. Temporary overvoltage

These occurs due to earth faults, load rejections, resonance and ferro-resonance or combinations of above. For insulation coordination purpose the representative temporary over voltage is considered to have the shape of the standard short duration (1 min) power frequency voltage. Generally, their amplitude does not significantly exceed 2.5p.u and duration vary from few cycles to several hours depending on system configuration.

(1 p.u = $\sqrt{2}$ Us / $\sqrt{3}$)

Where Us = maximum system voltage

B. Switching overvoltage

They generally arise from;

- Line energization & re-energization
- Faults & fault clearing
- Load rejections
- Switching of capacitive and inductive currents
- Distant lightning strokes to the conductor of overhead line

The representative voltage shape is standard switching impulse (250 / 2500 μs). Generally, the amplitude can go up to 3 p.u.

Steeper impulses with very high du/dt in the range from 0.1 to 10 μs & magnitude up to 4.0 p.u are possible in switching operations in inductive power circuits.

C. Lightning over voltages

They are caused by direct strokes to the phase conductor or back flashover or are induced by lightning strokes to earth close to the line.

Induced Lightning surge generally cause over voltage below 400 kV on the overhead line and are, therefore of most importance for medium voltage networks.

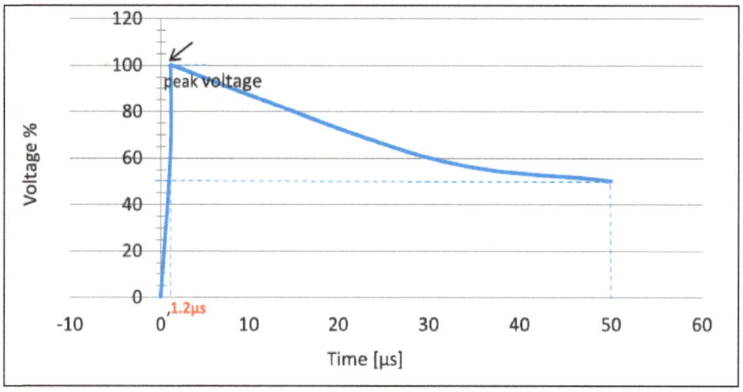

Figure 4: Representative shape of lightening wave

The representative shape of lightning wave form is 1.2/50 μs (figure 4). In a medium voltage network, the amplitude of lightning voltage can go up to 10 p.u.

4.0 SELECTION OF SURGE ARRESTER

The user, application engineers should be aware of important surge arrester parameters and how to select it with reference to system parameters.

A. Continuous operating voltage,

Uc – is the maximum permissible value of a sinusoidal power frequency voltage, which may be continuously applied between the arrester terminals.

Uc is selected with reference to the highest actual system voltage **Us** or if this voltage is not known or it changes in the course of time, the highest voltage for the equipment **Um** should be taken as reference.

Normally, arresters are connected phase to ground so Uc should be equal to or greater than Um/√3. Additionally, a factor of 1.05 can be taken for harmonic distortion.

B. Rated voltage, Ur –

It has no particular practical significance from the user point of view. It is defined as maximum permissible 10 second power frequency rms over voltage that can be applied between the arrester, as verified in the TOV and the operating duty test.

Ur is selected with reference to temporary power frequency over voltages expected on system.

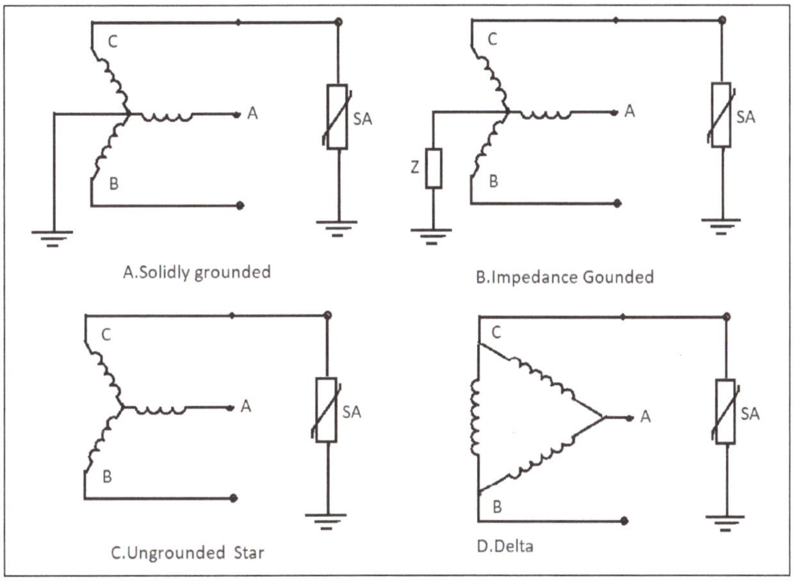

Figure 5: SA Grounding Systems.

In three phase system temporary over voltages can occur due to earth faults. Single line to ground earth fault is considered to be severe condition where other two phases get over voltage & its magnitude depends on earthing of networks.

The magnitude of the expected temporary over voltage is often defined using the earth fault factor, Kd.

The typical range of factor Kd for various network configurations is listed in table-1

Table 1

System configuration	Factor Kd
Solidly grounded 4 wire (Multi grounded	1.25
Solidly grounded 3 wire system	1.4
Impedance grounded system	1.73
Ungrounded or isolated neutral system	1.73 (Up to 2.3)

It is also important to note that the grounding of the neutral at the source transformer is the configuration referred to in determining the voltage rise of the system.

C. **Selection of continuous operating voltage Uc, rated voltage, Ur & TOV of surge arrester;**
1a) Solidly grounded three wire system – Kd 1.4 (Refer figure 5. A)

Provided a sufficient number of transformers have low-impedance earthed neutrals, the Kd factor will not exceed the value 1.4 for this network; clearance in such networks is very rapid so here also lower values of Uc & Ur can be selected giving better protection margin.

Ur = Um * Kd / √3

Uc = 0.8 * Ur

e.g. 11kV solidly grounded system

Ur = 12*1.4/√3 = 9.6 kV

Nearest rating can be selected as 9kV or 10kV

Ur = 10kV

Uc = 8 kV

D. **High Impedance grounded system (with earth fault clearing) – Kd 1.73**(Refer figure 5.B)

Here the magnitude of temporary over voltage is the same as in isolated neutral system but as there is earth fault clearing, lower values of Uc & Ur can be selected giving better protection margins.

Ur = Um * Kd / √3

Uc = 0.8 * Ur

e.g 11kV impedance grounded system

Ur = 12 * 1.73 / √3 = 12 kV

Uc = 12*0.8 = 9.6 kV.

E. **Ungrounded or isolated neutral systems–Kd 1.73 (up to 2.3)** (Refer Figure 5.C & D)

Generally, in these networks the phase to ground voltage of healthy phase would not exceed Um. There is no earth fault clearing so this voltage can continuous operating voltage.

Uc ≥ Um (considering Kd = 1.73) remain there till the time fault gets cleared manually.

It must be noted; however, the Kd factor can reach higher values under certain circumstances as a result of resonance phenomena. In such cases the Uc value should be increased accordingly.

F. **Rated voltage** – Once the Uc value selected as above there is no special attention needed to select Rated Voltage, Ur. Generally, there is margin of 20% between Uc & Ur. So, Ur can be selected as;

Ur = 1.25* Uc

e.g 11kV isolated neutral system

Uc ≥ 12 kV.

Ur = 15 kV.

Generally, the arrester manufacturer defines the curve temporary over voltage vs time. It means arrester can withstand specified over voltages for specified duration. The over voltage is defined per unit of rated voltage Ur and also with & without prior energy. e.g. TOV withstand with prior energy

Time (Sec)	TOV
1	1.15 * Ur
10	1.1 * Ur
100	1.05 * Ur

Table 2

Once the Ur & Uc values selected as per network configuration user should check the values defined in TOV curve of arrester datasheet with the expected over voltages & its duration at arrester location. User should keep in mind that at all time the arrester guaranteed TOV values should be greater than the expected TOV at arrester location, if not, arrester Ur & Uc to be increased accordingly.

In fact, Uc, Ur & TOV of arrester are all linked parameters to be derived from system highest voltage, system temporary over voltages and fault clearing time.

5.0 NOMINAL DISCHARGE CURRENT, In

The nominal discharge current is the peak value of lighting current impulse with wave shape as 8/20µs, which is used to classify an arrester, It is also a basis for calculating lightning impulse protection level, LIPL of a surge arrester

Standard 'In' values are 2.5 kA, 5kA, 10kA & 20kA but the value of the nominal discharge current alone does not give enough information about the performance of the arrester. Additional information about the application either Distribution or Station class and duty low, medium or high along with repetitive charge transfer 'Qrs' rating is required to be specified.

6.0 SPECIAL APPLICATIONS

A. Surge Arrester for transformer neutral;

Each unearthed neutral brought out through a bushing should be protected against lightning and switching over voltages by an arrester. The neutral insulation may be overstressed in case of incoming multiphase lightning over voltages, or in case of switching over voltages due to asymmetrical faults in the power systems

The Uc of surge arrester for transformer neutral should be selected as given hereafter

i. **Isolated neutral**

$Uc \geq Um/\sqrt{3}$ and the energy capability should be same as line to earth surge arrester.

ii. **High Impedance grounded system (with earth fault clearing)**

$Uc \geq Um/(\sqrt{3} * T)$, where T = 1.25 considering fault clearing within 10 seconds and margin of 20% between Uc& Ur.

iii. **Low Impedance grounded system**

$Uc \geq 0.4 * Um / T$, where T = 1.25 considering fault clearing within 10 seconds and margin of 20% between Uc& Ur.

B. Protection of Rotating machines –

If a generator under load condition disconnected from network, the generator voltage will rise sharply till the time regulator acts and readjust it. If surge arrester is connected at generator side, then special care should be taken during selection as this temporary over voltage in tune of 1.5-time normal voltage will appear across arrester for up to 10 seconds. The arrester Ur & Uc should be selected accordingly for proper functioning of surge arrester.

High voltage motors connected through VCB can experience high voltage surge during switching operations especially over voltage on account of multiple re-ignition.

Surge due to current chopping can have low magnitude but very steep. Surges due to current chopping can have very high magnitude but with lower steepness that can be handled by surge arrester.

The Ur & Uc of the arrester can be selected as mentioned in 'C'.

C. Surge Arrester for capacitor switching;

Arresters are installed at capacitor banks due to a variety of reasons.

i. To prevent capacitor failures at a breaker, restrike.

ii. To limit the risk of repeated breaker restrikes.
iii. To prolong the service life of the capacitors by limiting high overvoltage.
iv. For overall limitation of transients related to capacitor bank switching, which can be transferred further in the system and cause disturbances in sensitive equipment.
v. To serve as protection against lightning for capacitors banks connected to lines.

The possible arrester discharge energy is the most important parameter to be considered.

This energy depends on capacitor bank design, grounded – ungrounded, arrester installation phase-ground or phase neutral, breaker performance.

Arrester energy W can be roughly estimated as below;

$$W = \frac{1}{2} x\ C\ x\ [\ (3x\ U_0)^2 - (\sqrt{2} x Ur)^2]$$

Where,

C is the single-phase capacitance of the bank Uo is the phase-to-earth operating voltage; (peak voltage)

Ur is the rated voltage of the arrester (r.m.s. value).

The factor “3” comes from the assumption of a breaker restrike with full voltage of opposite polarity on the capacitor due to a previous break.

Furthermore, the operating voltage at the capacitor bank may be 5 to 10% higher than at other locations due to series reactors which may be used either to limit current when switching-in with parallel banks or to form a filter with the capacitors. This enhanced voltage must be considered when selecting the continuous operating voltage of the arrester.

C. Protection of the cable sheath;

Surge arresters for cable sheath protection are sometimes called Sheath Voltage Limiters (SVL). Due to thermal reasons (power losses in the cable sheath) cable sheaths of power cables in high voltage systems are earthed on one side only (majority of the cases depending on circuit length). The open cable sheath has to be protected against overvoltage.

The continuous operating voltage Uc of surge arrester to be selected based upon induced voltage in the sheath during short circuit conditions. Normally the short circuit current is defined for 3 seconds so TOV capability of selected surge arrester for 10 seconds should be greater than calculated sheath voltage during short circuit.

D. Surge arresters between phases;

Considerable overvoltage between the phase terminals of transformers or reactors may occur when a reactor or a reactive loaded transformer is switched off. The withstand voltage of the reactor or the transformer between phases may be exceeded without operation of the phase-to-earth arresters. If such switching operations are expected, surge arresters should be applied between phases in addition to those applied phase-to-earth. The phase-to-phase arresters should have a continuous operating voltage equal to or higher than 1.05 times the highest system voltage

F. Transmission line arrester, TLA – non gapped

TLA are in most cases directly suspended from the line conductor close to an insulator. The ground connection is connected to the tower steel structure. It's important to understand the objective of TLA installation. There shall be one or more targets possible as listed below, hence the selection should be appropriate to address the needs.

- Reduce the total number of trips for the line to a target level.
- Reduce or eliminate double circuit tripping.
- Reduce number of shielding failures.

Employment of disconnector in series with TLA is essential. The electrical characteristics of the disconnector are in general different from those of disconnectors for distribution arresters, because the operating duties, where disconnection shall not occur, are harder. It must be ensured that after disconnection no part (swinging in the wind) will be able to produce a flashover to ground.

i. **Selection of Ur –**

The Ur shall be selected so that the lightning and switching surge protective levels are coordinated below the LIWV, and SIWV of the line insulation respectively. Ur of TLA is normally higher than substation arrester Ur as this will reduce the risk of TLA unnecessary getting stressed by system power frequency high voltage, here there is no benefit of extra protection margin as the purpose is to avoid flashover of line insulators by shielding failure or back flashover.

ii. **Selection of energy –**

On shielded lines TLA typically have a nominal discharge current of 5 kA or 10 kA according to IEC with energy equivalent to classes 1 to 3, depending on their application.

iii. **Fault clearing & disconnector for TLA –non gapped**

Disconnectors are used to facilitate fast reclosing as TLA are connected directly across the line insulators, which are self-restoring. Disconnectors are usually not permitted to disconnect high voltage substation arresters automatically in the event of an arrester failure since the insulation of the substation equipment is generally not self-restoring and should not be switched in without protection.

These disconnectors in series with the TLA also serve as indicators making it simple to find overloaded TLA with visual inspection. The TLA disconnector must be capable of withstanding both higher impulse currents as well as longer duration impulses compared to disconnectors for distribution arresters; in fact, the disconnectors must pass all the type tests that the TLA is capable of and must know when disconnector shall operate and when it shall not operate. Below is the requirement to be complied for satisfactory TLA fail safe operation.

7.0 LOCATION OF ARRESTERS;

The ideal location for lightning arresters, from the point of view of protection, is directly at the terminals of the equipment to be protected.(Figure)

Figure 6: Bushing of transformer is connected very near the transformer bushing.

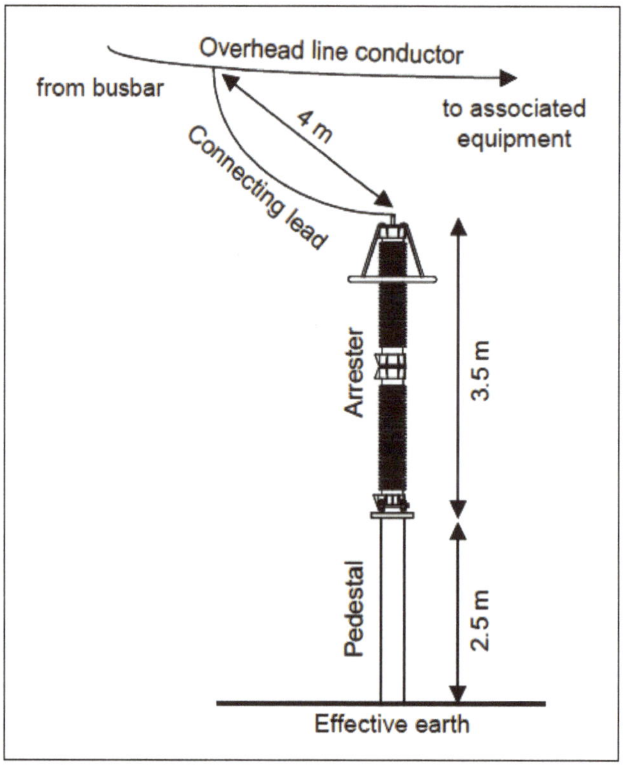

Figure 7: Arrangement of jumpering from Overhead line conductor to the LA.

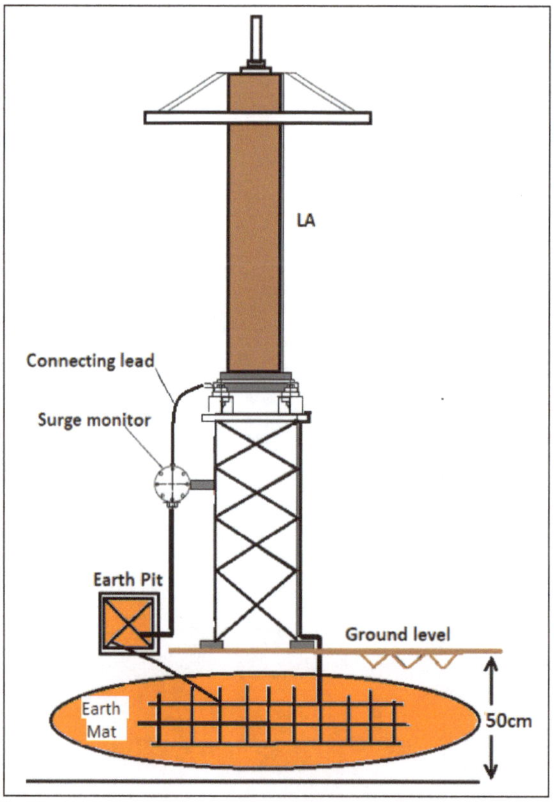

Figure 8: Earthing (Grounding) of Surge Arrester.

8.0 Protective Characteristic of gapless station class Surge arresters as per IEC 60099-4,2009:

The chart shown in Table 3 is only for academic interest only.

Table 18.9 Protective characteristics of gapless station class surge arresters for a nominal discharge current of 10 kA[a]

1	System voltage (50 Hz)	kV	420	245	245	*145*	123	72	36	36	24	12	12
2	Rating of surge arrester (at the reference current)	kV	390	216	198	*120*	96	60	36	30	18	12	9
3	Discharge class		III	III	III	*III*	III	II	II	II	I	I	I
4	Energy dissipation capability cumulative operation	(kJ/kV_r)	10	10	10	6.5	6.5	4.5	4.5	4.5	2.5	2.5	2.5
5[b]	High current impulse withstand of 4/10 μs wave shape	kA	100	100	100	100	100	100	100	100	100	100	100
6	Reference current of the arrester at ambient temperature	mA	5	5	5	3.25	3.25	2.25	2.25	2.25	1.5	1.5	1.5
7	Components of the continuous leakage current at COV												
	Resistive	μA peak	400	400	400	400	400	400	400	400	400	400	400
	Capacitive	μA peak	1500	1500	1500	1500	1500	1500	1500	1500	1500	1500	1500
8	Watt loss at MCOV per kV of rated voltage	W/kA	0.25	0.25	0.25	0.25	0.25	0.25	0.25	0.25	0.25	0.25	0.25
9	Maximum continuous operating voltage	kV r.m.s.	303	178	168	102	81	52	30.5	25.5	15	10	7.65
10	Max. residual voltage at lightning impulse of 8/20 μs wave kV peak												
		5 kA	860	567	518	333	249	162	103	85	60	38	33
		10 kA	900	602	550	*355*	264	175	110	90	65	40	35
		20 kA	957	668	610	390	304	202	128	104	75	46	41
11	Maximum residual voltage at switching current impulse of 30/60 μs.	1 kA	780	496	457	*294*	221	138	87	71	52	32	28
12	Maximum residual voltage at steep-fronted impulse (1/20 μs. IEC 60099-4) at	10 kA	1050	654	600	*386*	288	202	124	101	73	45	40
13	Temporary power frequency overvoltage capability												
	(a) 0.1 second kV peak		705	397	364	220	156	97	58	48	29	19	14
	(b) 1.0 second kV peak		580	382	350	212	149	93	56	46	28	18	14
	(c) 10.0 seconds kV peak		565	367	336	203	142	89	53	44	26	17	13
	(d) 100.0 seconds kV peak		550	305	280	169	135	84	51	42	25	16	12

Table 3

9.0.0 Conclusion

To safe guard electrical equipment in an electrical system against voltage surges, Surge Arresters are very essential. Following points are to be kept in mind;

- SA/LA divert the Lightning to the Ground.
- They clamp (limit) the voltage produced by the Lightning or switching.
- They protect only the equipment electrically in parallel with them.
- The arrester should be placed as close as possible to the equipment,
- The performance of any arrester is dependent on a good connection to ground. to be protected and leads connecting arresters to ground should be kept as short as possible of a conducting material.

10.0 Authors: K.K.Murty & Dr.Rajesh arora

11. HV/EHV Circuit Breakers; Concept and Brief Description Thereof

Preamble;Any failure of HV/EHV circuit breakers shall affect a large area in the DISCOMS and the Power Grid too. This paper contains basics of circuit breakers, type of circuit breakers and the Vacuum and SF6 Gas Circuit breakers. In the present scenario vacuum circuit breakers are Used up to 33kV and SF6 circuit breakers from 66kV and above voltage class.

The most important parameters like Arcing phenomena, Arc Voltage, TOV (Temporary Over Voltages), RRRV (Rate of Raise of Recovery voltage etc.) have also been narrated. Basic concept of how the main contact is saved from the effect of arcing through **Abstract:** HV/EHV Circuit breakers are the, most vital equipment in the Power arcing contacts.

Key words: Circuit breakers, SF6 Circuit breakers, Arc Voltage, TOV (Temporary Over Voltages), RRRV (Rate of rise of recovery voltage)

Introduction:

1.1 Circuit breakers: The circuit breaker is "a mechanical switching device in power-systems capable of;

- making, carrying and breaking currents under normal circuit conditions.
- breaking currents under abnormal/faulty conditions.

1.2 Application:

Switching and controlling of;

- ➢ Transmission overhead lines and Power Cables.
- ➢ Transformers and shunt reactors.
- ➢ Capacitor banks and harmonic filters,
- ➢ Generator switching.
- ➢ Motor switching.
- ➢ Bus tie & transfer switching.

2.0 Protection of;

- ➢ By clearing the fault of all the items mentioned above.
- ➢ Protection of the utility assets.
- ➢ Protection of personnel and public.

3.0 Classification of circuit breakers

i. Based on voltage;

- Low - less than 1kV
- Medium - 1kV to 52kV
- High/Extra High - 66kV to 765kV
- Ultra-High - above 765kV

ii. **Based on location;**

- Indoor
- Outdoor

iii. **Based on external design;**

- dead tank
- Live tank

iv. **Based on interrupting media;**

- Air-break
- Air-blast
- Bulk-oil
- Minimum-oil
- SF6 gas insulated
- Vacuum

v. **Based on design of operating mechanism;**

- Compressed air system for pneumatic mechanism
- Hydraulic Mechanism
- Spring Mechanism
- ND circuit breaker = gang operated circuit breaker
- Single-Pole Circuit breakers.

4.0 Status of usage in present scenario;

Usage of Vacuum and SF6 gas: All other interrupting media have been phased-out. Vacuum is used up to 33kV and above this voltage level SF6 gas is used as interrupting /insulation media.

5.0 Vacuum Circuit breakers; The dielectric strength of vacuum is considerably higher than other interrupting mediums. Hence, a contact separation of around 1 cm is enough to withstand high voltages. In addition, the rate of dielectric recovery of vacuum is much faster than that of air.

The interrupting technique of vacuum circuit-breakers is different from other types of circuit-breakers. The arc is not extinguished by an interrupting medium but by the metal vapor.

Nowadays, vacuum circuit-breakers are predominant used in medium voltage levels. They are considered as maintenance-free circuit-breakers due to their simple and reliable design.

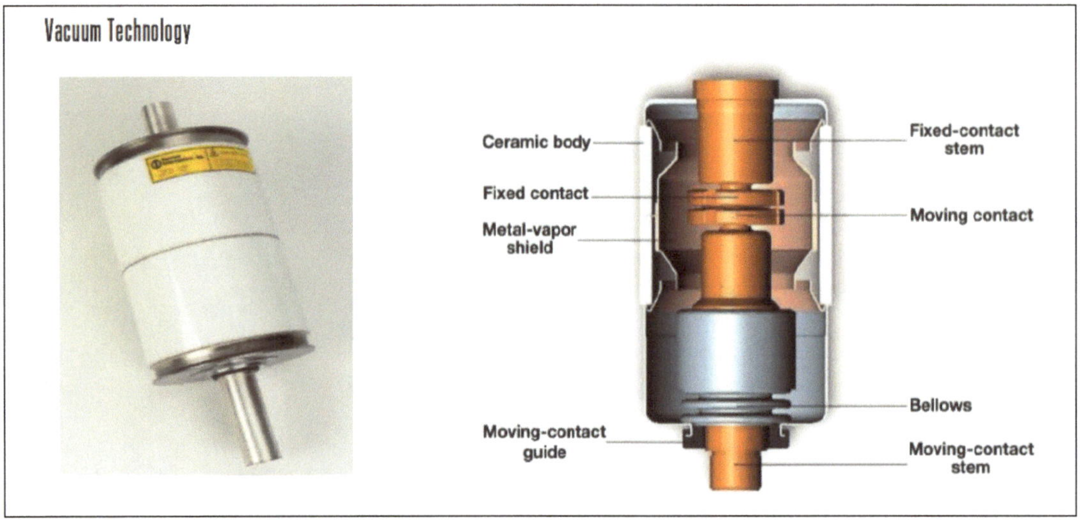

Figure 1A: Vacuum Circuit breaker's metal vapour arc extinguishing camber.

Figure 1B:33kV Vacuum interrupter assembly ready for erection.

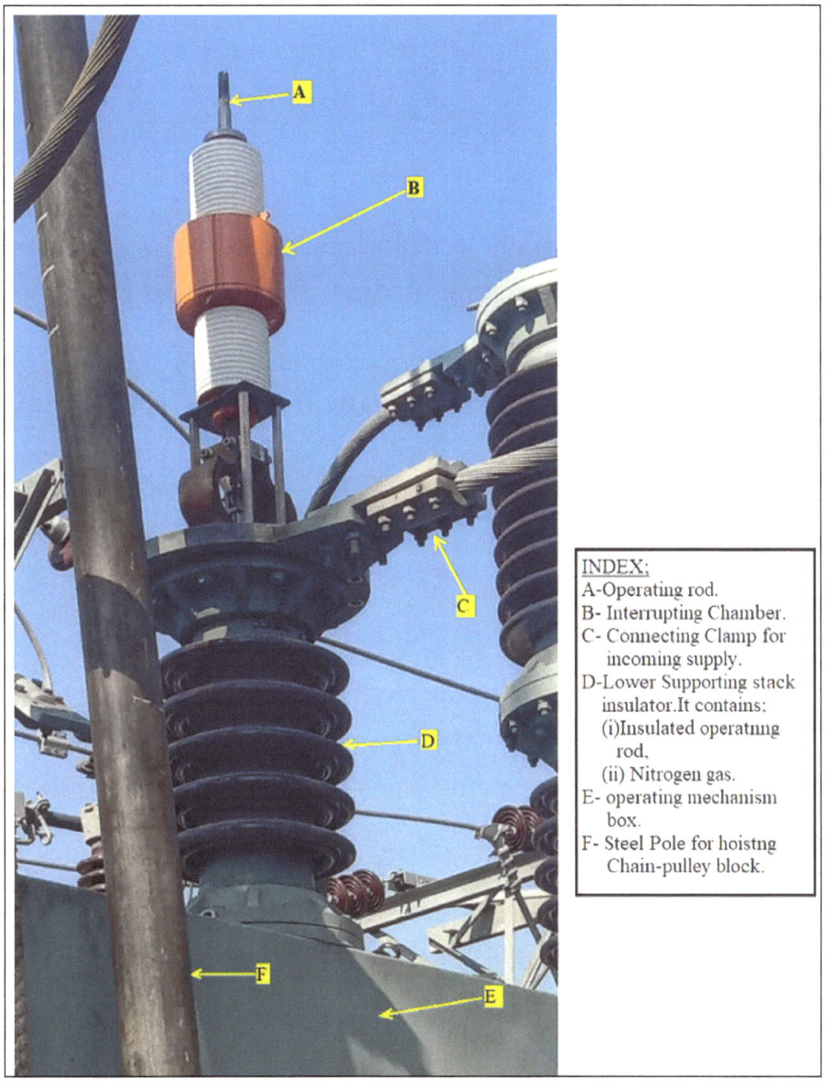

Figure 1C: 33kV Vacuum interrupter assembly shown erected on the lower insulator stack.

5.1 Process of repairing under progress: There had been a failure of 33kV Vacuum Circuit breaker damaging the vacuum interrupter in one of the stations in M.P. Repairing work in progress by replacing the Interrupting Chamber assembly with a new one,

- The photograph in the fig.1B is depicting the 33kV vacuum interrupter mounted on the bottom flange of the Interrupter Chamber.
- The photograph in fig.1C depicts 33kV Vacuum interrupter assembly shown erected on the lower insulator stack.

6.0 SF6 gas Circuit breakers;

6.1 Characteristics of SF6 gas;

- Highest dielectric strength of any known gas (e.g. 10 times higher than N2)
- Excellent arc extinguishing medium
- Approximately five times heavier than air

- Thermally stable
- Regeneration characteristics after arcing
- Excellent heat transfer properties
- Superior arc quenching capabilities
- Excellent sound attenuation properties
- Non-corrosive, non-toxic, non-flammable, colorless and odorless
- Currently the most predominant insulating medium for Live and Dead Tank circuit breakers at ratings ≥60 kV.

6.2 Schematic diagram of LT (Live tank) & DT (Dead tank) Circuit breakers.

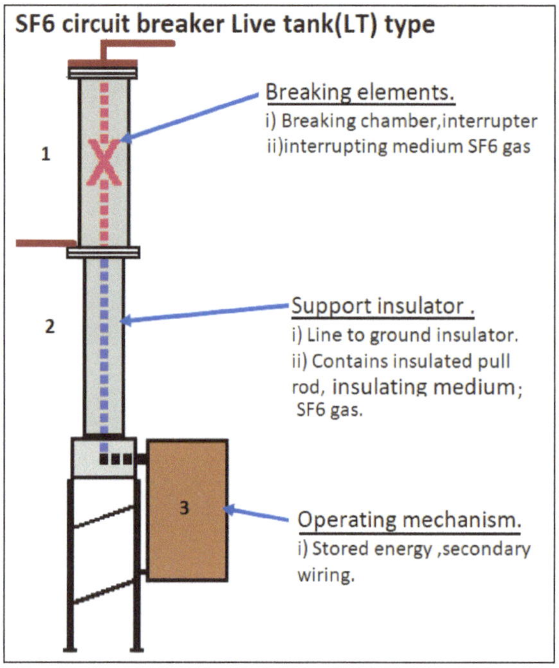

Figure2: Schematic Diagram of Live talk SF6 Circuit breaker.

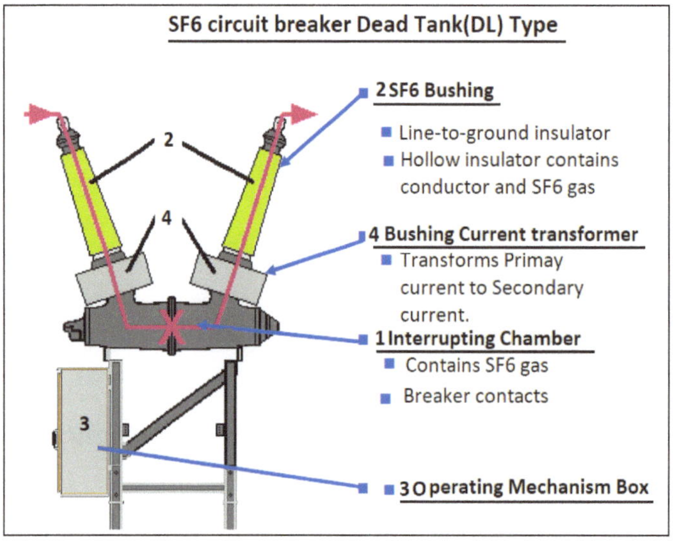

Figure3: Schematic Diagram of Dead tank SF6 Circuit breaker.

6.3 Comparison of Dead tank & Live tank SF6 circuit Breakers:

(a) (b) (c)

(a) Dead tank(DT) Circuit breaker.

(b) & (c) are Live tank(LT) circuit breakers.

Figure 4: Comparison between Dead tank and live tank SF6 circuit Breaker.

a. **Dead tank (DT) circuit breaker;**

- Grounding of Breaker chamber is essential.
- Bushing mounted current transformers.
- Used in US, Russia and Latin America.

b. & c. Live tank (LT) Circuit breaker;

- Have Breaking chamber at line potential.
- Current transformers are separately mounted in the Sub Station yard.
- Extensively used in UK, Europe, Asia and India.

7.0 Definition of Arc voltage, Restriking Voltage &RRRV;

i. **Arc voltage.:** The voltage across the contacts of Circuit breaker is the arc voltage when the arc persists. (Refer figure 5)

ii. **Restriking Voltage**: Interruption will take place at an instant when the alternating current reaches zero and the arc has quenched out. Interruption

current reaches zero and the arc has quenched. After the arc extinction, a voltage shall appear which is transient in nature. This transient component of voltage contains multi- frequencies. It is also known as **Transient Recovery Voltage (TRV).**

Requirement: After the arc extinction, the insulating medium between the contacts must withstand the rapidly increasing recovery voltage.

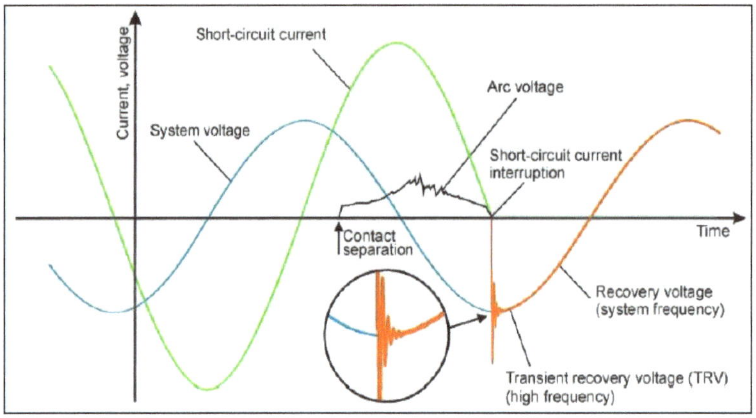

Figure 5: Graphical representation of System voltage, Short-circuit current,

Arc-voltage, High -frequency Transient Recovery Voltage (TRV) and

Recovery-Voltage of System Frequency.

iii. **Recovery Voltage:** The power frequency rms voltage which appears across the breaker contacts at the instant of arc being extinguished and the transient oscillations die out is called the recovery voltage. (Refer figure 5)

iv. **RRRV (Rate of rise of Restriking voltage); (Refer figure 6)**

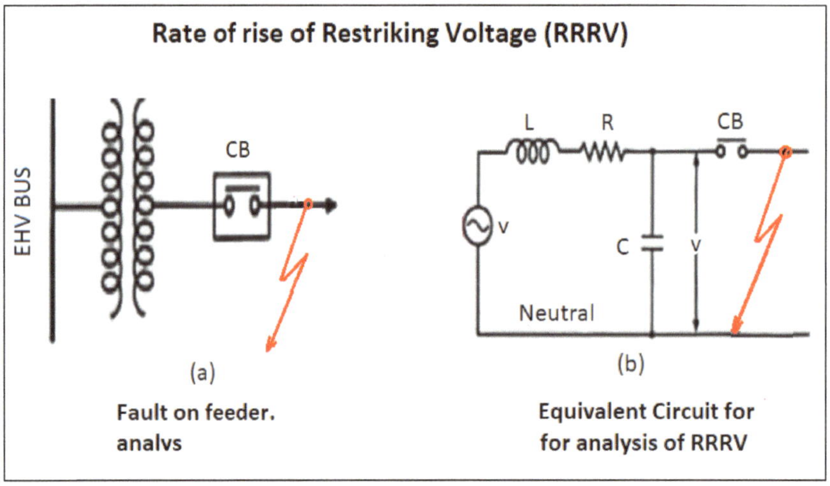

Figure 6: Equivalent Circuit for analysis of RRRV.

When the fault occurs {Refer figure 6(a)}, the circuit breaker opens and capacitance C comes in series with L. {Refer figure 6(b)}

This L-C series combination oscillates at the frequency $fn=\frac{1}{2\pi\sqrt{LC}}$ *resulting into development of high voltage across the contacts of the circuit breaker, which appears across the capacitor and hence across the contacts of the circuit breaker.*

The RRRV (Rate of Rise of Re-striking voltage) is in kV and time is in microsecond so RRRV is expressed in kV/μ sec.

Condition: It is the R.R.R.V, which decides whether the arc will re-strike or not. If R.R.R.V is greater than the rate of rise of dielectric-strength between the contacts, the arc will re-strike. The arc will fail to re-strike if R.R.R.V is less than the rate of increase of dielectric strength between the contacts of the breaker.

Point to note: This phenomenon is taken care-off in SF6 circuit breakers with

- SF6 puffer interrupters build up pressure through mechanical compression.
- Higher pressures lead to greater voltage withstand capability.
- Higher gas flow velocities cool the arc more efficiently.
- Efficient gas flow clears the contact region of hot gases, contaminants due to arcing and unstable post-arc conductivity.

8.0 Conceptual figure of SF6 Circuit breaker Operating chamber. (Refer figure7)

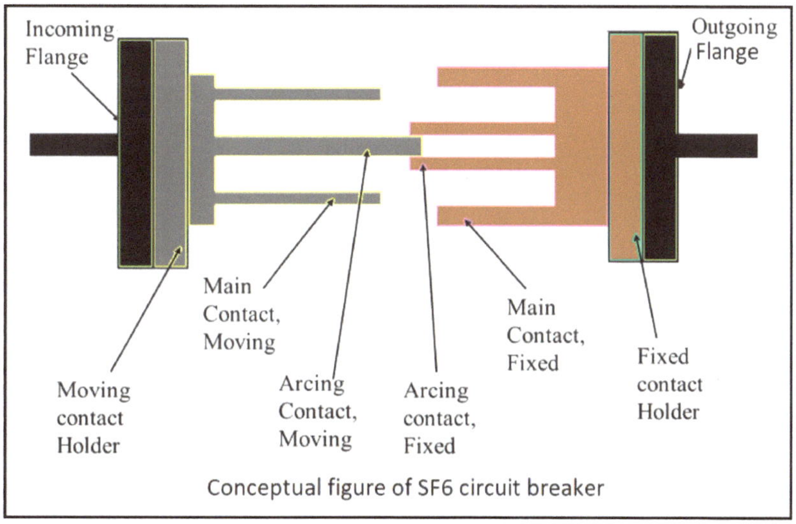

Figure7: Conceptual figure of SF6 Circuit breaker operating Chamber.

Note:

- It can be noticed that arcing contacts are longer than the main contacts.
- During closing or opening arc is produced.
- Closing operation: Arcing contacts carry the arc and get closed prior to the main-contacts. Thus, arc is quenched and the main contacts are not subjected to hazards of arcing.
- Opening operation: Arcing contacts carry the arc and get opened prior to the main contacts. Thus, quenching the arc and the main contacts are not subjected to hazards of arcing.

9.0 Main components of SF 6 circuit breaker;

9.1 Construction and Main Components

A complete assembly of a single pressure puffer type SF6 circuit breaker consists of the following parts:

9.1.1 Interrupter Unit:

The separation of circuit breaker contacts, quenching of the resulting arc and interruption of current takes place in the interrupter unit. It houses two sets of contacts which are commonly called the 'main' or 'normal current carrying contacts' and 'arcing contacts.' Both the types of these contact sets have one stationary contact whilst the other contact is moving contact. Current carriers (providing connection to the external circuit breaker terminals) are connected to the stationary and moving main contacts. The tips of all circuit breaker contacts are coated with a copper-tungsten arc-resistant material.

The main body of the interrupter (which is filled with SF6 gas) contains a moving puffer cylinder that can axially slide upward and downward along the contacts. There is one stationary piston inside the cylinder which is fixed with other stationary parts of the SF6 circuit breaker, in such a way that it cannot change its position during the movement of the cylinder. A nozzle is located at the opening of the cylinder.

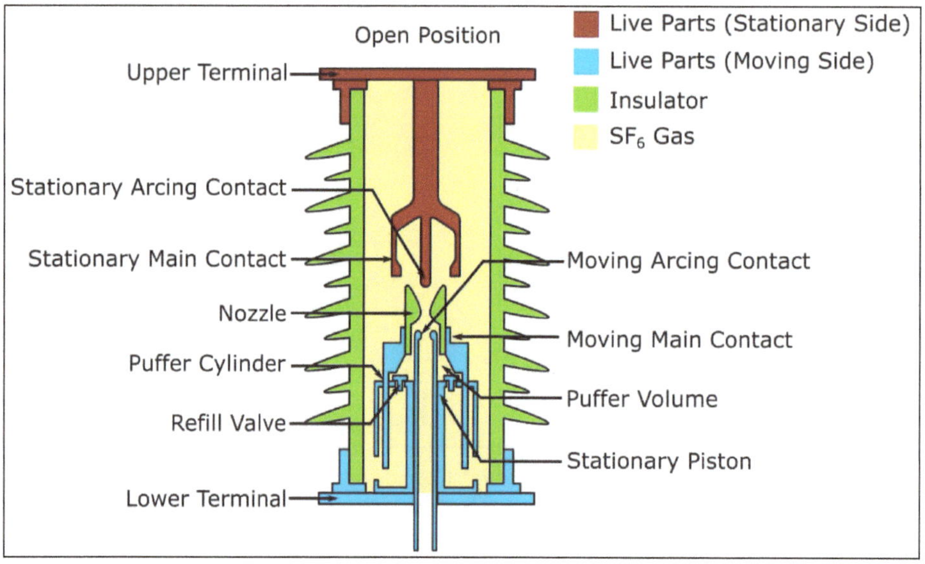

Figure 8. Main Components of SF6 Circuit Breaker Interrupter.

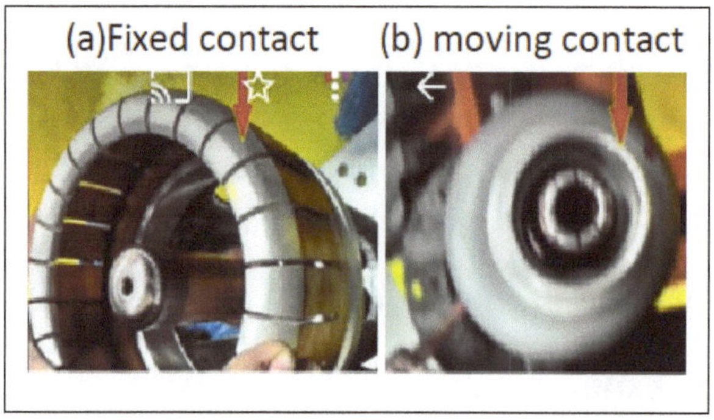

Figure 9: (a) Fixed contact (b) Moving contact.

In-house major maintenance of 132kV Circuit breaker: True photographs of Fixed contact assembly & Moving contact assembly respectively of a 132 kV SF6 Circuit Breaker;

In order to enlighten the reader about how the fixed contact assembly and moving contact assembly look-like, the Authors are posting 2 pictures. {Refer figure 9(a) & (b)}.

It may be noticed that

a. **Fixed contact assembly**-It contains a flexible MAIN CONTACT and a rod like contact placed in the middle of the Flexible contact is the ARCING CONTACT. The length of the Arcing contact is a little longer than the main contact.
b. **Moving contact assembly**-It is mounted in the lower portion of the SF6 gas Circuit breaker and is installed over the lower insulated stack. The Arcing contact is found to be little shorter than the main contact for clarity please refer fig.8.

Figure; 9 (a)Fixed contact - side view& (b) is moving contact of

132kV SF6 Circuit breaker.

Figure10: Top view of fixed contact assembly of SF6,132kV circuit breaker.

{Refer figure 9(a)}

9.1.2 Interrupting Chamber:

The interrupter unit is vertically mounted on top of an insulating stack which is made up of a hollow insulator encapsulating the drive rod which connects the mechanical operating mechanism of the circuit breaker to the moving contacts housed within the interrupter. Depending on the voltage rating of the system, the insulating stack may be a single piece, or multiple segments mechanically coupled in series. Just like any other insulator, it provides adequate line-to-ground dry arcing and creepage distance to prevent flashovers associated with transient over voltages and the ambient pollution. The complete circuit breaker unit is usually fixed on a steel structure which erected on the concrete foundation.

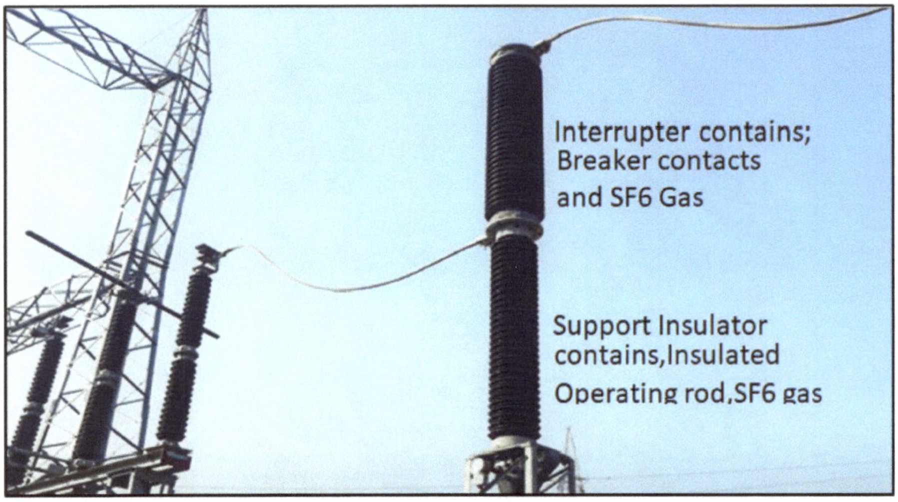

Figure11A: 220kV, SF6 circuit breaker installed in a 220 kV Switch-yard.

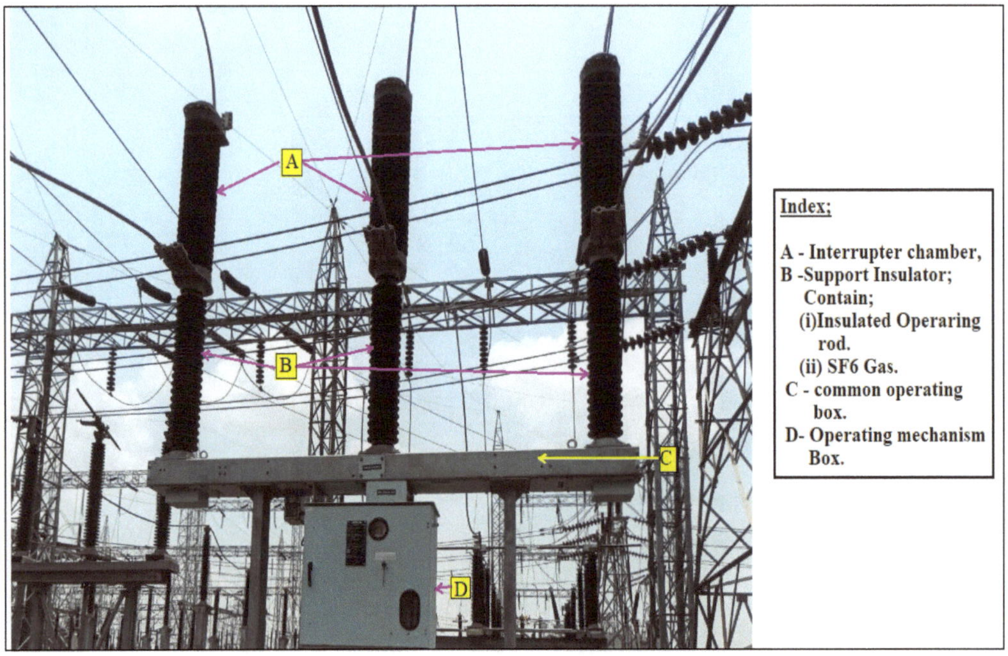

Figure11B: 132 kV, SF6 circuit breaker installed in a 132 kV Switch-yard.

9.1.3 Insulated operating rods;

Figure 11C: Insulated Operating rods of 132kV SF6 circuit breaker.

10.0 Operating Mechanism;

10.1 Spring Mechanism: -

For Circuit Breaker up to 245 KV, spring operating mechanism is used.

a. The closing spring gets charged by geared motor. During closing stroke, the closing spring is discharged. It closes the breaker and also charges the opening springs.
b. The Circuit breaker mechanism is basically very simple. It comprises of a motor-reducer and two springs (one for closing and one for opening), plus their solenoid operated latching devices.
c. The closing spring after having been charged by the motor-reducer is kept ready for closing by its latching device;

i. As soon as energized, the closing control coil triggers the closing latch thereby causing the closing spring to operate the breaker to the closed position and simultaneously charge the opening spring.
ii. Since then, trip latch keeps the breaker in the closed position and the opening springs is ready for opening.

The compact design of the mechanism avoids the need for the opening springs to be located close to each pole and permits the opening operation to be performed by only one spring accommodated in the control cabinet.

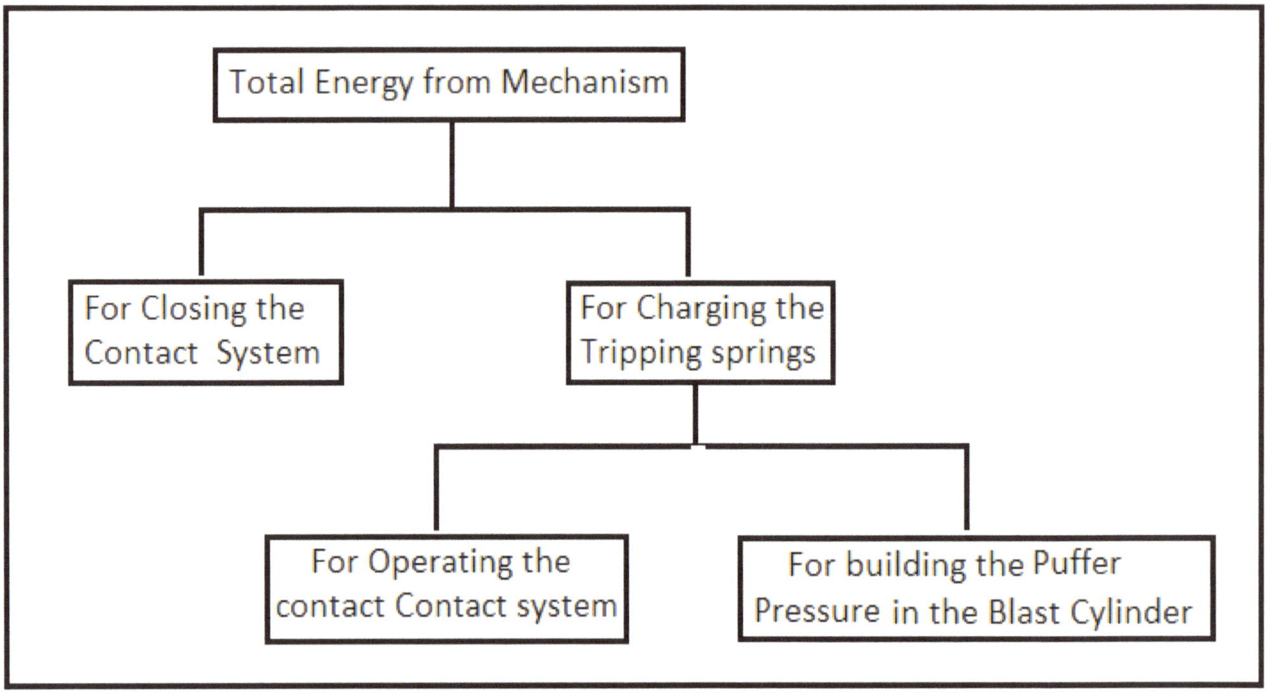

Figure12A: Mechanical Operating Mechanism;

Figure 12 B: Operating control box of 132kV SF6 Circuit breaker mechanism.

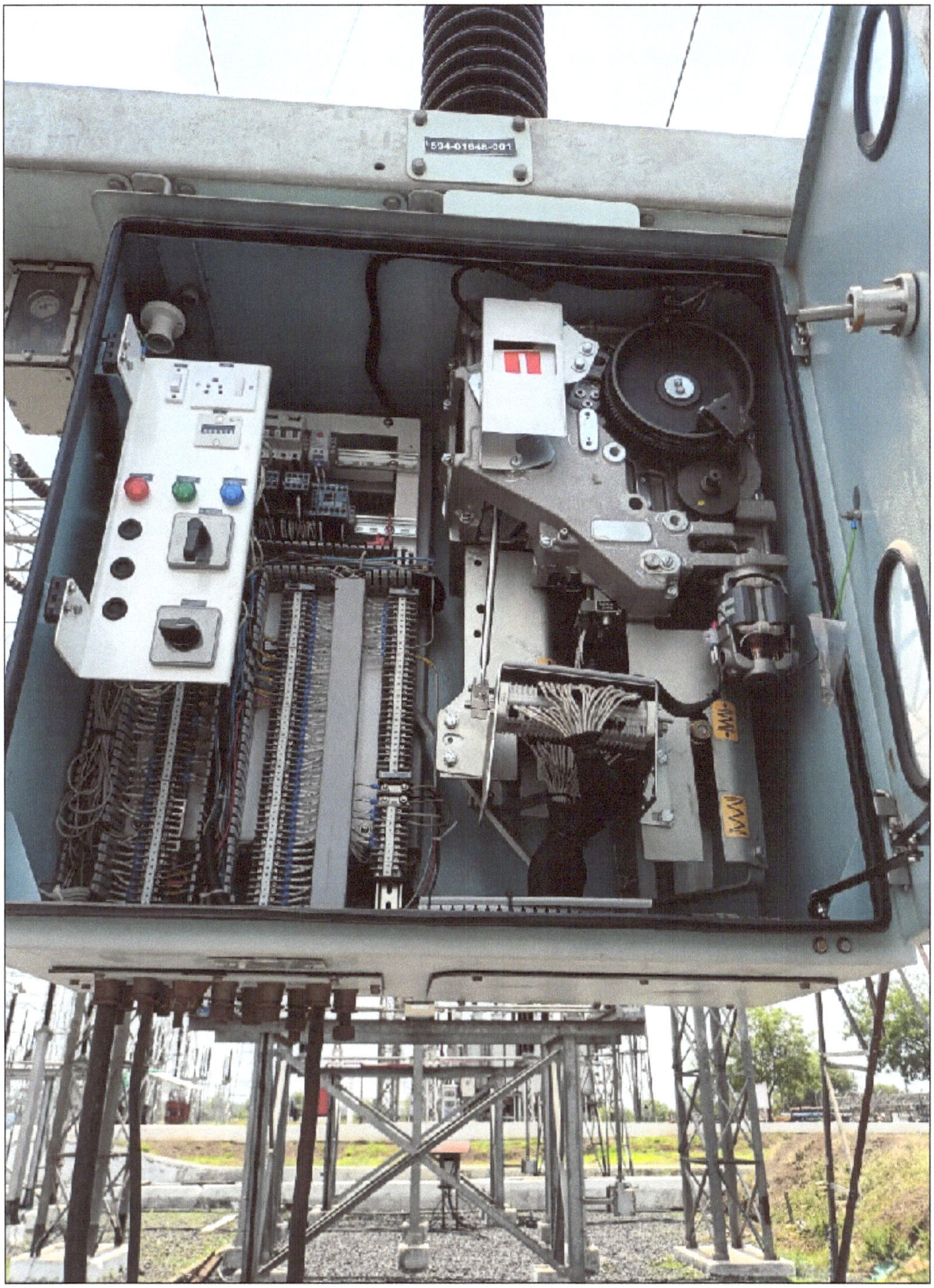

Figure 12 C. Operating mechanism box of 132kV SF6 circuit breaker.

10.3 Pneumatic Mechanism.

The pneumatic-operated mechanism uses compressed air as energy storage,

and pneumatic cylinders for operation. Solenoid valves allow the compressed air into the actuating cylinder for closing or for opening.

The compressed-air tank is replenished by a compressor unit. The use of pneumatic operating mechanisms is decreasing. Due to the high operating pressure, there is always a risk of leakage of air, particularly at low temperatures.

10.4 Hybrid Pneumatic spring Mechanism: In this mechanism there is no need of motor to energize closing spring, it is charged by pneumatic operated tripping mechanism.

Figure 12D.Operating mechanism box of 220kV Hybrid SF6 Circuit breaker.

Note: This circuit breaker has

- Closing operation: Motor is not needed. It is charged by pneumatic operated tripping mechanism.
- Tripping operation: Pneumatic operation.

Figure 12E. Pressure gauges of SF6 gas and Air pressure gauge for pneumatic charging and tripping operation in a Hybrid 220kV Circuit breaker.

The driving apparatus provides the kinetic energy required to open and close the circuit breaker contacts. It consists of a set of opening and closing springs that are charged manually, or with the help of a small electric motor.

10.5 Control Panel;

The control & Relay panels are used in the control-rooms for accommodating supervisory devices, remote Closing & opening operations and through system protective relays between the circuit breaker mechanical operating mechanism.

11.0 Working of SF6 Circuit Breakers:

Figure 8 is reproduced as figure 12 for ready reference indicating the parts of the SF6 Circuit breaker in the interrupter chamber;

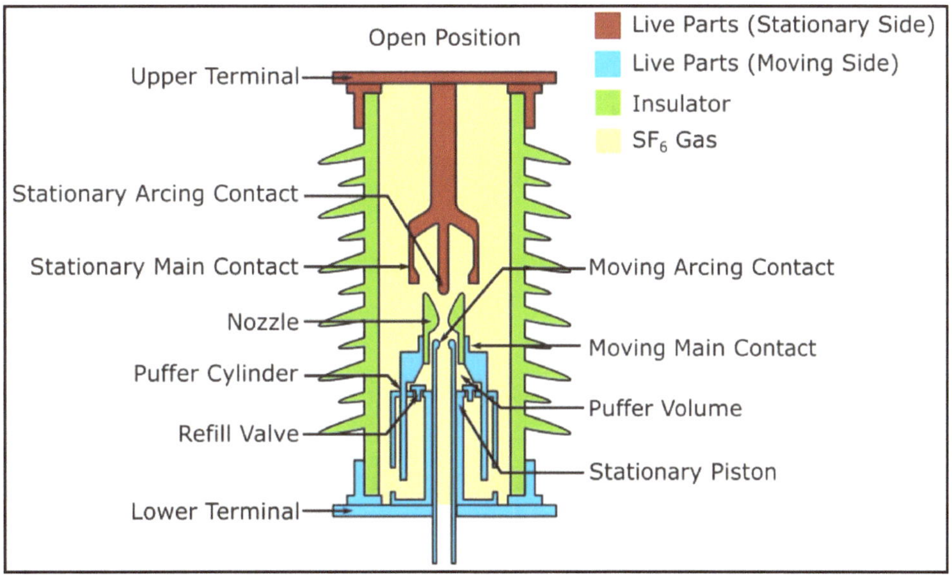

Figure 12. Indicates the parts in the interrupter chamber.

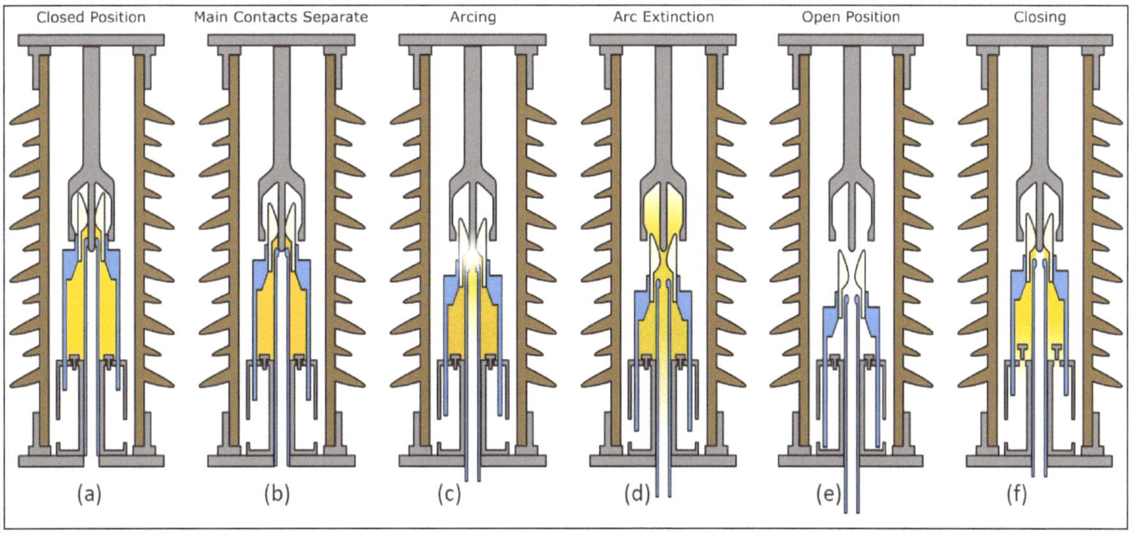

Figure 12 A:Operation of a Puffer Type SF6 Circuit Breaker.

In the normal condition, the circuit breaker contacts are closed and current flows from one contact carrier to the other via the main contacts. {Refer figure 12A(a)}.

11.1 Circuit Breaker Opening Operation:

- When the circuit breaker control panel receives an opening command (to clear a fault or disconnect part of a network), it sends a signal to the trip coil of the mechanical operating mechanism, which in turn releases the latch holding the charged opening spring.
- As the opening spring discharges, it pulls the drive rod (connected to the interrupter) in a linear direction, which causes the moving contacts and puffer cylinder to move downwards. {Refer figure12A (b)}
- The movement of the puffer cylinder against the stationary piston leads to a decrease in the puffer cylinder's internal volume, which causes compression of the SF6 gas inside the cylinder. Due to contact overlap, gas compression starts before any contacts open.
- As the downward movement continues, the main contacts separate and the current commutates to the arcing contacts which are still in the closed position (due to their physically longer in construction){Refer figure12A(b)}.
- During the course of further opening, the arcing contacts start to separate and an arc is established between them. {Refer figure12A(c)}
- As the arc flows, it blocks the flow of SF6 gas through the nozzle to some extent. Thus, the gas pressure in the puffer cylinder continues to increase.
- When the sinusoidal current waveform approaches zero, the arc becomes relatively weak and the pressurized SF6 gas inside the puffer cylinder flows axially (through nozzle) over the arc length.
- This blast of SF6 gas removes the thermal energy in the contact gap and reduces the degree of ionization (electrical conductivity) such that the arc is extinguished. {Refer figure12A (d)}
- When the arc is interrupted, transient recovery voltage (TRV) starts to appear across the contacts. {Refer figure12A (e)}.

Constraint:

- ➢ The opening speed of the circuit breaker contacts should be fast enough to create an adequate contact separation distance to withstand this voltage stress.
- ➢ **Re-ignition or Re-strike phenomenon**: In case the contact gap's dielectric strength is lower than TRV stress, the arc will be re-established, this phenomenon is commonly called circuit breaker re-ignition or re-strike.

11.2 Circuit Breaker Closing Operation;

- ➢ During the circuit breaker closing sequence, the closing coil releases the energy of the closing spring which causes the contacts to move towards each other, {Refer figure 12A(f)
- ➢ Ultimately bringing them to their normal closed position. {Refer figure 12(a)} At the same time, SF6 gas is redrawn into the puffer cylinder making the circuit breaker ready for the next operation.

Constraint:

- **Pre-Strike:** Whilst closing, a circuit breaker can sometimes experience an event known as pre-strike.
- As the contacts move towards each other during closing, the contact gap's dielectric strength decreases.
- At some point, the voltage stress across the contact gap exceeds its dielectric strength, thus producing a 'pre-strike 'arc which bridges the contacts.

12.0 OPERATING DUTIES:

It is mandatory to mention Operating sequence on the name plate of the Circuit breakers. This is requiredforAuto-reclosure applications on Transmission lines

The circuit breaker should be able to perform the operating sequence as follows:

O-T-CO-T-CO

Where,

- **O=** opening operation,
- **t** = time required for circuit breaker to be ready to receive closing command. from auto-reclosure relay (0.3 s to be used for rapid reclosure).
- **CO** = close operation followed by open operations.
- **T** = time required by the circuit breaker, insulating media for regeneration and operating mechanism (3 min).
- **CO** = close operation followed by open operations.

The operating duty cycle in accordance with IEC Std. 62271-100 is:

O - 0.3 s - CO - 3 min – CO

- Normally the time 3 minutes between the two close-open operations is valid,
- **Note:**An alternate timing as specified by the IEC is 15 Sec or 1 min.
- The dead time of 0.3 s is based on the recovery time of the air surrounding an external arc in the system (i.e. a short-circuit).
- The time of 3 min is the time needed for the operating mechanism to restore its power after a O - 0.3 s - CO.

Note: Modern spring and hydraulic operating mechanisms of SF6 Circuit Breakers do not need 3 min to restore their power.

12.1 OPERATING TIMINGS OF CB

Description	400kV	220kV	132kV
Close (max)	150ms	150ms	200ms
Mechanical operating time	25ms	35ms	45 ms

Trip (Max)	40ms	60ms	70ms
CO time	85ms	90ms	120ms
Dead time adopted for Auto-reclose application.	1 sec.	1 sec	-
Pole to pole discrepancy -trip	3.3ms	3.3ms	3.3ms
Pole to pole discrepancy - close	5.0ms	5.0ms	5.0ms

Table1

13.0 Conclusion and recommendation:

i. It is recommended to go-in for SF6 gas breakers for EHV level Circuit breakers and Vacuum circuit breakers for and up to 33kV level Circuit breakers.
ii. For Transformer & Reactor application it is essential to use the CSD
iii. (Control Switching Device) in the closing and opening operations through control panels.
iv. For transmission line applications in-conjunction with Auto-reclose facility, please verify, whether they are equipped with Single pole reclosing facility?
v. In view of fool-proof operations, please ensure that the Circuit-breakers are equipped with two Trip-coils.
vi. Please verify that the breakers are equipped with Pole-discrepancy facility.
vii. The operating mechanism, preferable is mechanical ones which can be readily got petty repairs carried out by the local mechanics in the remote locations of the sub-stations.

14.0 References;

1. saVree's narration about SF6 circuit Breakers.
2. IEC std. 62271-100
3. Tutorials by Er R.K. Tyagi, Director operations, POWERGRID.
4. ABB's Application guide.

15.0 Authors; K.K.Murty, JJLKapil, Santosh Dubey & S.K.Chaturvedi

12. Isolators for HV & EHV Systems

Preamble:

Isolators, also known as disconnector switch, are used in EHV, HV and MV installations to isolate a circuit or parts of a circuit and to ensure, for safety of work-men and for safe operation & maintenance of equipment. Isolators are the switches "to put-off or put-on" any feeder, Bus-bars and Equipment. They are installed in series with a load break switch e.g. circuit breakers. They have the same current carrying capacity and withstand capability that of the feeders, bus bars or equipment has been connected for.

Isolators also have provision of earthing switches which ground the equipment temporarily during maintenance. There is a mechanical interlock between the main isolator. It cannot be operated if the isolator is in on condition.

Isolators are operated either automatically (motorized) or manually.

Caution;

a. the isolators should be operated when any equipment or feeder is in de-energized condition only.
b. it should never be opened while in it is live condition (ie. current flowing through it.)

This article discusses an overview of isolator used in HV & EHV System., their types and applications.

General description;

Working of "off load Isolator-disconnector switch";

The operation of an Off-Load Isolator Disconnector is operated by manually through handle and with motorised operation. Isolator handle assembly is always lock in position when it is not in use. Isolator operation is done by only authorised person of switchyard.

When the isolator is in the closed position, the isolating blade connects the electrical contacts, allowing the normal flow of current. This energizes the connected equipment or circuit. However, before any maintenance or repairs can take place, the isolator must be opened through the operation of the isolating blade using the operating mechanism manually using Hand-gloves or through motorised operation.

Before opening the isolator make sure that the equipment/ feeder is in de-energized position and the circuit breakers thereof are in off position. This isolation provides a safe working environment for personnel and prevents unintended energization of the equipment/feeder.

33kV Isolators;

System parameters;

S.No.	Particulars	33 kV
a)	System Voltage (kV rms)	33
b)	Maximum Voltage (kV rms)	36
c)	Lightning impulse voltage (kVp)	170
d)	Power frequency withstand voltage (wet) (kV rms)	70
e)	Short circuit level (kA)	26.2 kA for two seconds

Note: Any 33kV isolator must comply with system parameters as well as weather conditions

Constructional Features for 33kV Isolators

The current density: The current density to be adopted for all the part s of isolator shall not exceed the following limits:

Hollow Tube Section:

Copper:2.00Amps/sqmm

Aluminum:1.25 Amps/sq mm

Other sections:

i. Copper:1.60Amps/sq mm
ii. Aluminum:1.00 Amps/sq mm

Please refer fig.1 & fig.2A & 2B

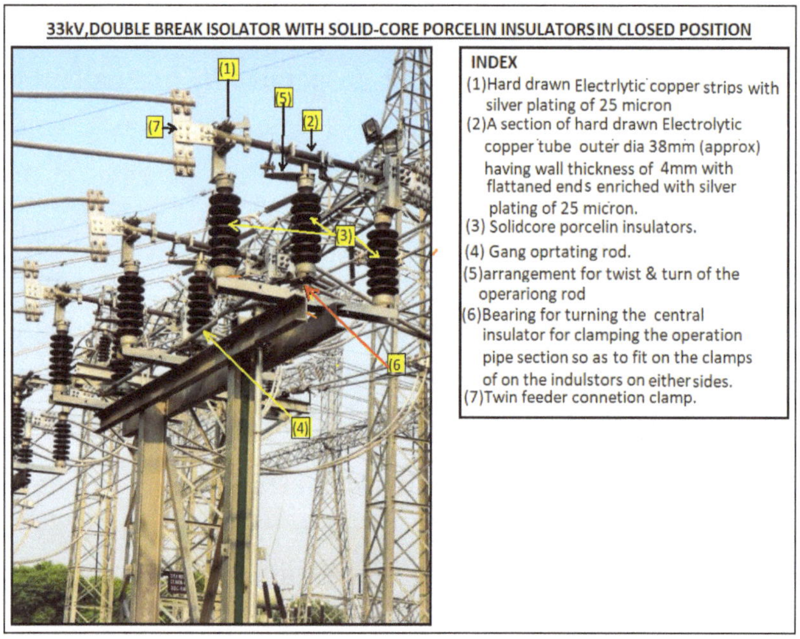

Figure 1:33kV double break isolator in closed position.

33kV,DOUBLE BREAK TYPE ISOLATOR WITH SOLID CORE PORCELIN INSULATOR-OPEN POSITION

Figure 2 A: 33kV, double break isolator in closed position.

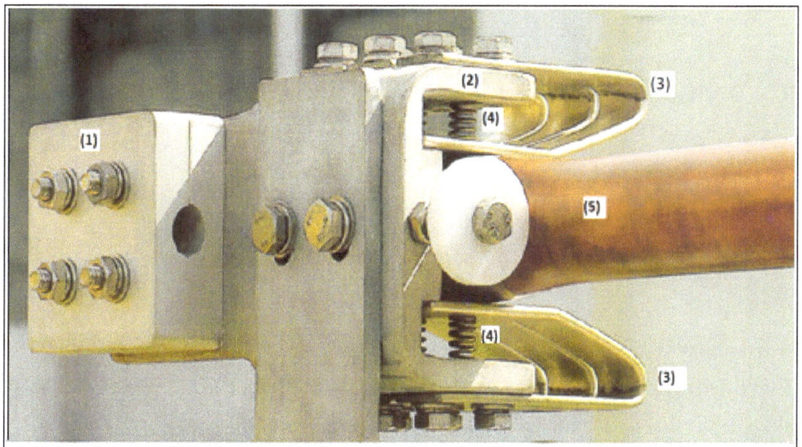

Index:

(1) clamp for tightening the conductor.
(2) C- channel made of HDG Mild steel.
(3)Hard drawn Electrolytic copper strips with silver plating of 25 micron. 6nos.
(4)Stainless steel springs 6 nos.
(5)Tubular section of hard drawn Electrlytic copper with outer dia 38mm & inner dia 34mm silver plating of 25 micron on flattened at both ends.

Figure 2B: fixed Clamp.

Fixed contacts & moving contact; The position and the labelling of each part shown in the fig.1 & fig.2A & 2B may be referred to;

i. Centre post top pin for holding the turn &twist assembly {Refer item(5) of fig.1} should be adequate in size to avoid any bending of turn twist mechanism. The U-fork holding the moving copper rod should be made out of GI strip of not less than 6 mm thickness.

ii. The switch blades forming the moving contacts shall be made from tubular section of hard drawn electrolytic copper having outer diameter not less than 38 mm and thickness of 4 mm. The surfaces of contacts shall be rendered smooth and silver plated. The thickness of silver plating shall not be less than 25 microns.

iii. The male and female contact assemblies shall be of robust construction and design so as to ensure conditions contained in the IS:9921-4.-1985 and other physical conditions as mentioned here under;

   a. Electro-dynamic withstand ability during short circuit without any risk of repulsion of contacts.
   b. The maximum current density shall be as under:-
      For hollow tube sections: - 2.0 Amp. /sq.mm
      For other sections and terminal connectors:- 1.6 Amp/ sq.mm (for Copper) & 1.0 Amp / sq.mm (for Aluminium).
   c. Thermal withstand ability during short circuit.
   d. No wiping action during closing and opening.
   e. Self-alignment assuring closing of the switch without minute adjustment.

INTER-LOCKS AND EARTH SWITCHES:

Earth switch shall be mechanically interlocked with the main switch so that is possible to close or open the earth switch only when the main switch is in the open position & its closing operation shall not be possible

Type Tests: Prior to procurement of the 33 kV Isolator it is essential to confirm that the offered 33kV isolators have with-stood the following type tests as per IS: 9921.4.1985;

a. Test to verify the insulation, level, including withstand tests at power-frequency voltages on auxiliary equipment. This clause is applicable wherever 33kV isolators are supplied with auxiliary equipment.
b. Tests to prove that the temperature rise of any part does not exceed the limits.
c. Test to prove the capability of the disconnector or earthing switch to carry the rated peak withstand current and the rated short-time withstand current.
d. Test to prove the short-circuit- making capacity of earthing switches.
e. Test to prove satisfactory operation and mechanical endurance.

Routine Tests;Each and every Isolator must withstand all the Routine tests as stipulated in the IS:9921-4-1985

BEARING: It plays a very vital role in smooth and reliable operation of the isolator. The bearings shall be of reputed make.{refer item no.(6) in Figure 1}

General technical Particulars as per IS: As per IS:9921.4.1985

132 KV ISOLATORS.

STANDARDS:

Performance and testing of the isolators shall conform to the latest revisions and amendments from time to time;

a)	**Nominal system voltage (kV rms)**	**132**
b)	**Highest system voltage (kV rms)**	**145**
c)	**Lightning impulse voltage (kVp)**	**650**
d)	**Power frequency withstand voltage (wet) (kV rms)**	**275**
e)	**Short circuit level (kA) for one second.**	**40**

Figure 3: 132 kV Double break Isolator without Earth Switch.

132kV Double break Isolator with earth switch

Index;
(1) Gang operating rod.
(2) Isolator manually operating handle.
(3) Solid core porcelin Insulator.
(4) Hard drawn Electrolytic copper strip with silver plating of 25 micon
(5) Earth switch.(Earthing rod)
(6)Eletrolytic copper pipe section with silver coated ends 25 micron.

Figure 4A: 132 kV Double break Isolator with earth Switch.

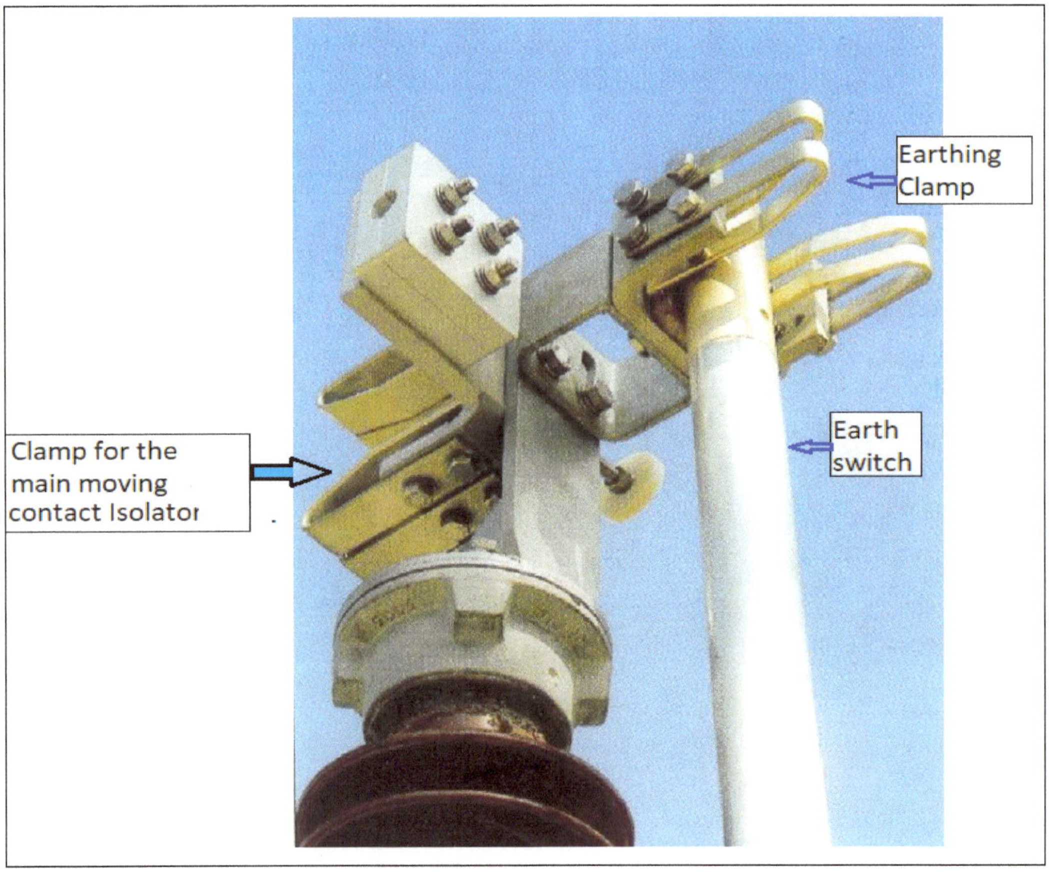

Figure 4B: Depicting Earth switch in closed position.

STANDARDS:

Performance and testing of the isolators shall conform to the latest revisions and amendments from time to time;

Sl. No.	Standard No.	Title
1	IS: 9921	Alternating current isolators (disconnectors) & earthing switches
2	IEC:60649 & 62271	Alternating current isolators (disconnectors) & earthing switches

4.0 SYSTEM PARTICULARS:

The Isolators shall be installed in the system with following parameters:-

a)	System Voltage (kV rms)	220
b)	Maximum Voltage (kV rms)	245
c)	Lightning impulse voltage (kVp)	1050
d)	Power frequency withstand voltage (wet) (kV rms)	460
e)	Minimum corona extinction voltage at 50 Hz AC System Dry Condition (kV rms)	176
f)	Short circuit level (kA) for one second.	40

Index;
A-Moving contact.
B-Turn & twist arrangement.
C-Sationary Solid Insulator.
D- Rotary solid insulator.
E-Earth switch.
F- Earth switch operating rod,
G-Earth switch operating Mechanism.
H-Handle for manually operating handle.
J-Isolators operating mechanism box.
K- Isolators operating rod.

Figure 5: 220kV double break isolator in open condition & earth Switch in closed position.

Figure 7: 220kV Centre break Isolator in open condition.

Figure 8: 220 kV Centre break Isolator in closed position.

400kV Pentagon Isolator;

Standards: The isolator & its components must confirm to the following Standards;

SYSTEM PARTICULARS:

The isolator shall be installed in the system with following parameters:

a)	System voltage (kV rms)	400
b)	Max. voltage (kV rms)	420
c)	Lightning impulse voltage (kVp)	1425
d)	Power frequency withstand voltage (wet) (kV rms)	520
e)	Switching surge withstand voltage (wet) (kVp)	1050
h)	Short circuit level (kA) for one second.	40

5 OPERATING MECHANISM AND CONTROL:

5.1 Phase AC motor operated type operating mechanism shall be provided for each pole of 400 kV Pantograph Isolator main switch for individual as well as simultaneous operation of all the three poles i.e. it should be possible to operate all the poles either simultaneously through electrical operation. Individually (master–follower mode) or The operating mechanisms of three phases should be well synchronized and interlocked. There shall also be manual cranking in the event of power failure.

400kV Pentagn Isolator;

Standards: The isolator & its components must confirm to the Std, IS: 9921

The isolator shall be installed in the system with following parameters:-

a)	System voltage (kV rms)	400
b)	Max. voltage (kV rms)	420
c)	Lightning impulse voltage (kVp)	1425
d)	Power frequency withstand voltage (wet) (kV rms)	520
e)	Switching surge withstand voltage (wet) (kVp)	1050
f)	Min. corona extinction voltage at 50 Hz A.C. system Dry condition (kV rms)	320
g)	Maximum radio interference voltage at one MHz for phase to earth voltage of 266 kV dry conditions (micro-volts)	500
h)	Short circuit level (kA) for one second.	40

GENERAL TECHNICAL PARTICULARS FOR 400 kV PENTOPGRAPH ISOLATORS WITH / WITHOUT EARTH SWITCH:

1. Unlike in lower kV rating isolators which are generally gang operated, the 400 kV Isolators are a set of three individual poles having individual drive for each pole isolator.
2. The pantograph isolators have one vertical break earth switch.
3. Each three phase pantograph isolator 3 No. solid core post insulators and 3 No. solid core operating rod insulators which shall conform to IS: 2544 and IEC:60168. The porcelain of solid core post insulator and operating rod insulator shall conform to minimum cantilever strength of 8 kN & 2.4 kN respectively. The pressure due to contacts shall not be transferred to the insulators after the main blades are fully closed.

4. Disconnects and earthing switches, including their operating mechanism are so designed and constructed that these do not come out of their open or closed positions by gravity, wind pressure vibrations, reasonable shocks or accidental touching of the connecting rods or their operating mechanisms

5 **OPERATING MECHANISM AND CONTROL:**

5.1 Phase AC motor operated type operating mechanism shall be provided for each pole of 400 kV Pantograph Isolator main switch for individual as well as simultaneous operation of all the three poles i.e. it should be possible to operate all the poles either simultaneously through electrical operation. Individually (master–follower mode) or The operating mechanisms of three phases should be well synchronized and interlocked. There shall also be manual cranking in the event of power failure.

SYSTEM PARTICULARS:

The isolator shall be installed in the system with following parameters:-

a)	System voltage (kV rms)	400
b)	Max. voltage (kV rms)	420
c)	Lightning impulse voltage (kVp)	1425
d)	Power frequency withstand voltage (wet) (kV rms)	520
e)	Switching surge withstand voltage (wet) (kVp)	1050
f)	Min. corona extinction voltage at 50 Hz A.C. system Dry condition (kV rms)	320
g)	Maximum radio interference voltage at one MHz for phase to earth voltage of 266 kV dry conditions (micro-volts)	500
h)	Short circuit level (kA) for one second.	40

5 **OPERATING MECHANISM AND CONTROL:**

5.1 Phase AC motor operated type operating mechanism shall be provided for each pole of 400 kV Pantograph Isolator main switch for individual as well as simultaneous operation of all the three poles i.e. it should be possible to operate all the poles either simultaneously through electrical operation. Individually (master–follower mode) or The operating mechanisms of three phases should be well synchronized and interlocked. There shall also be manual cranking in the event of power failure.

Figure 9: Operating box in 400kV Pantograph bus coupler bay with Master –follower mechanism.

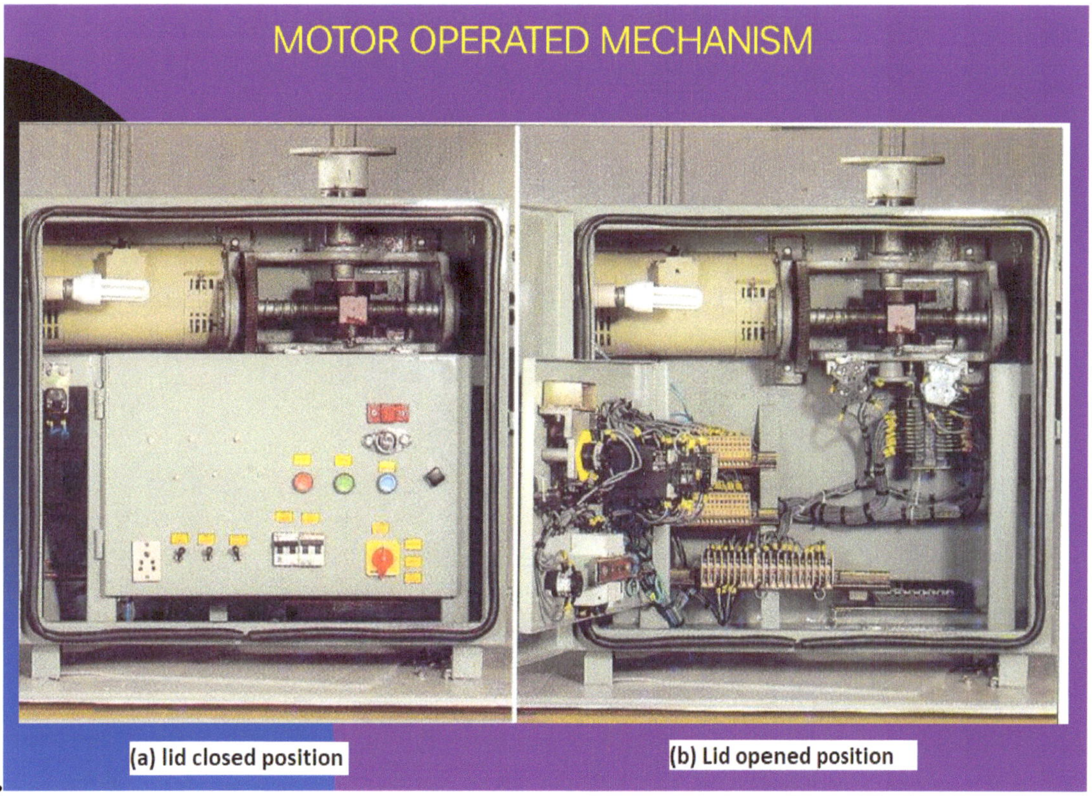

Figure 10: Motor operating Mechanism

Figure 11: Operating system for motor operation on the Bus coupler Isolator having the option for gang operation or individual operation

Figure 12: Operating system for motor operation for individual isolator.

7.1 400kV pantograph Isolator figures;

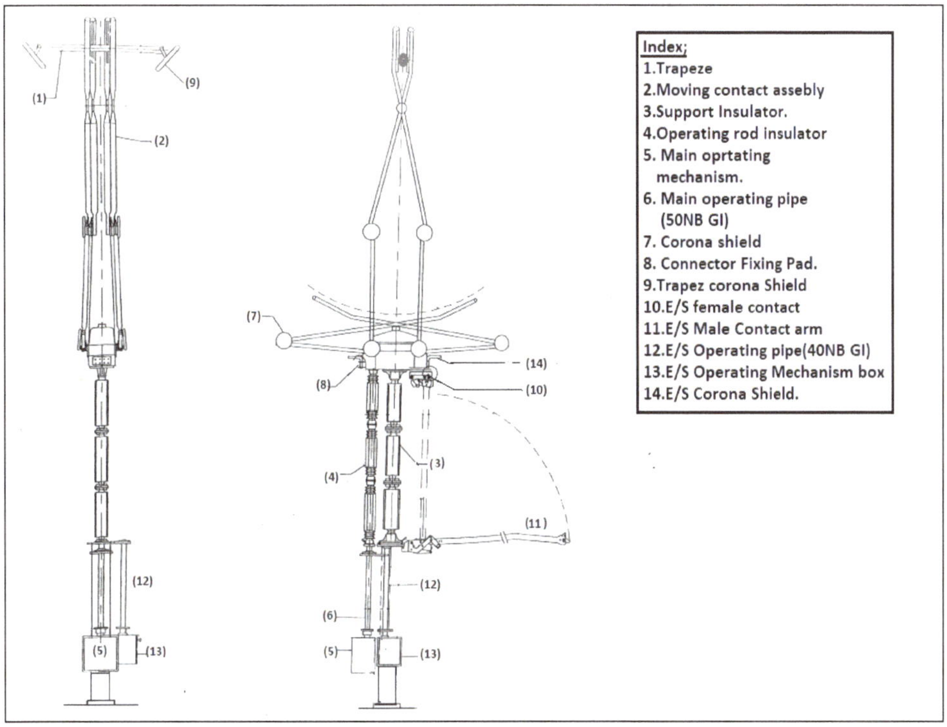

Figure 13: Diagram of 400 kV Pantograph Isolator with part name.

Figure 14: 400kV Pentograph isolator in Open condition.

Figure 15:A.400kV Pantograph isolator in closed condition.

8.0 Authors: K.K.Murty, JJL Kapil, SantoshDubey, S.K.Chaturvedi & Santosh Bagade

13. Dual Role of CVT in Grid Network

SYNOPSIS: *The CVT this acronym expanded as "CAPACITOR VOLTAGE TRANSFORMER" only. An EHV CVT is a versatile out – door equipment which is used in lieu of a (PT) Potential Transformer and a CC (Coupling Capacitor). Thus, it is cost effective too.*

In this paper, the author shall not only describe about its usage for Power - System Protections, Metering, Synchronizing, Tele-protection (Carrier Aided Protections), Tele-metering and controls including SCADA but shall also explains the concept of Voltage transformation from the primary to secondary voltages through Capacitor-Voltage Divider action and derivation of frequency for usage in PLCC (Power Line Carrier Communication)System. Using PLCC link is cost effective in the remote places wherein no other media of communication exist as compared to any other mode of communications including OPGW (Optical Fiber Ground Wire) etc.

Key words: CVT(Capacitor Voltage Transformer), Capacitor voltage divider [1] Ferro- Resonance Suppression Device (FSD) PLCC(Power Line Carrier Communication)

1.0 Introduction The Grid-network of Power in India and abroad consists of 220kV, 400kV, 765kV & 1200kV transmission lines covering practically all the states of our country or respective countries. The Instrument Transformers play vital roles in measurement of EHV/UHV level currents and voltages of the circuits in terms of their secondary quantities which in turn are used for measurement, protection and synchronizing, it also facilitates system control, system operation and SCADA through the PLCC link etc.

2.0 EHV/UHV CVTs;

2.1 Connection of CVT on the Transmission;

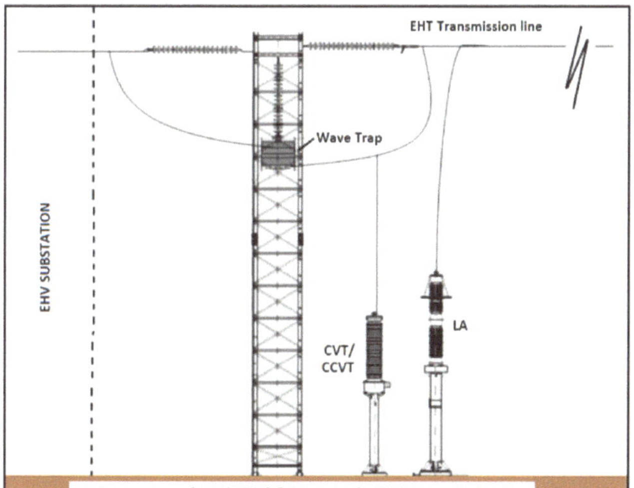

Fig;Connection of CVT(CCVT) & LA on to the EHT Transmission Line

The CVT is basically an Instrument Transformer. It is a voltage Transformer comprising of a capacitor divider unit, an electromagnetic unit so designed and interconnected, that the secondary voltage of the Electromagnetic unit is substantially proportional and in phase with the primary voltage applied to the capacitor divider unit. It has a high voltage terminal connected to the transmission line and the low voltage terminal connected to the carrier frequency transmission circuit or to the earth terminal as the case may be, this terminal too is a primary terminal.

2.2 Capacitors;

There is a high voltage capacitor (C1) and the intermediate voltage capacitor (C2) and for PLCC capacitor is C_N (sum of C1& C2 in series) They are made of aluminum foil, Poly propylene film and Kraft paper impregnated in mineral oil. The Capacitor C1 may be formed with series of capacitors and connected between the high voltage terminal and the intermediate voltage terminal, whereas the Capacitor C2 is connected between intermediate terminal and the low voltage terminal to the earth. Both C1 and C2 include stray capacitances of negligible values.

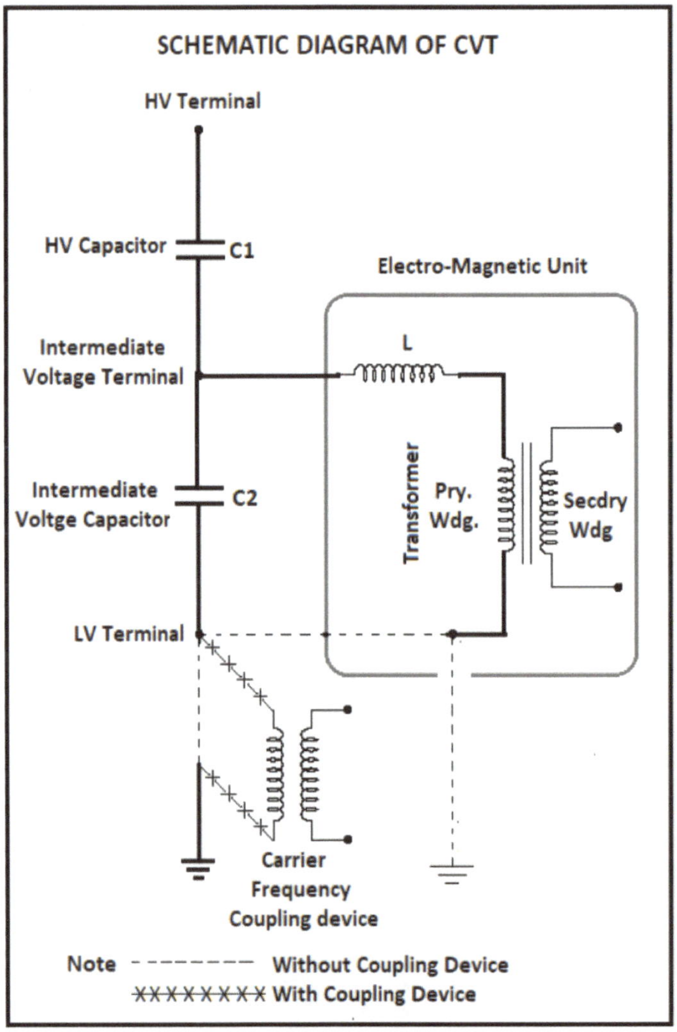

Figure 1:Schematic diagram of a CVT.[2]

1.3 Open Circuit Intermediate Voltage;

The ratio of $\frac{C1+C2}{C1}$ capacitor divider [1] corresponds to the ratio of primary voltage to the open Circuit intermediate voltage ie;

$\frac{C1+C2}{C1} = \frac{Vpry}{Vint}$, Where, "pry" sands for primary and "int" stands for intermediate

1.4 Electromagnetic Unit;

There is an electromagnetic unit too, which is connected across the intermediate terminal and the earth terminal. The electromagnetic unit has an inductive reactance

and a transformer. The value of inductance L is chosen such that at normal frequency L and C are in resonance and therefore cancel each other and the circuit behaves similar to a conventional transformer. Depending upon the intermediate voltage, the VT ratio of the windings is decided so as to get the secondary voltages as 110/√3 volts, etc as the case may be for protection and metering circuits depending on the requirement.

As an example please refers to Figure 2 wherein the CVT is of 245 kV is shown. The value of C1 is 4848 pf and C2.

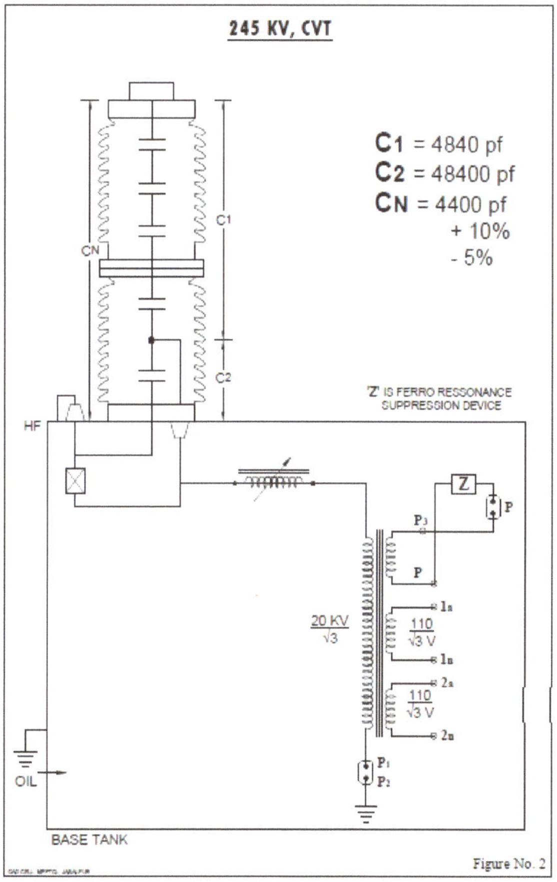

Figure No. 2

Figure 2

CN =is 48400 pf. and C_N is 4400 pf +10%, – 5%.

Therefore $\frac{Vpry}{Vint} = \frac{C1+C2}{C1}$

$$\frac{220/\sqrt{3}}{Vint} = \frac{4840+48400}{4840} = \frac{53240}{4840} = 11$$

$$\therefore Vint = \frac{220/\sqrt{3}}{11} = \frac{20}{\sqrt{3}} \text{ kV}$$

Therefore the VT should have Pry. Voltage of 20/√3 kV and the secondary voltages be 110/√3.volts, 110/ √3.volts and No. of secondary be as per the requirement of accuracy class depending upon metering and protection as per IS:3156 part 2 and part 3[2].similar to Potential Transformers. 2,5 Accuracy limits;

i. For easurement.

Class	%Voltage (ratio) Error	Phase Displacement minutes
0.1	**± 0.1**	**± 5**
0.2	**± 0.2**	**± 10**
0.5	**± 0.5**	**± 20**
1.0	**± 1.0**	**± 40**
3	**± 3.0**	**-**

ii. **For protection**:

Accuracy class	% voltage Error	Phase displacement minutes
3P	**± 3**	**± 120**
6P	**± 6**	**± 240**

Ferro-Resonance Suppression;

To overcome the phenomenon of Ferro-Resonance in CVTs, Ferro-Resonance Suppression Device (FSD) is provided in the VT secondary. It contains a suitable reactor which acts like a switch presenting a very high impedance under normal conditions and switches on a damping resistance across the secondary at a prescribed

voltage and switches off the damping load when voltage has normalized, Therefore significantly improves transient response performance of the CVT. Please refer figure No.2 wherein Z indicates the Ferro-Resonance Suppression Device provided in 245kV CVT).

2.6 Standard reference range of frequency:

As per IS: 3156 (part-6) and IEC: 60186 the deviation in frequency ranges for Metering is from 99% to 101% and for Protection from 96% to 102%. However the frequency range followed in India for security of the Grid network is from 49.9 Hz to 50.05 Hz that is 99.8% to 100.1%, which is well within requirement for the metering as well as protection purposes.

4.0 Why to use CVTs?;

a. **The CVTs are preferred over the conventional PTs for usage on the transmission lines of voltage class 132kV and above, on the following reasons;**

   i. The CVTs are manufactured to the highest system voltages unlike PTs, for example for use on 220kV and 400kV Transmission lines the CVTs are generally manufactured as 245 kV and 420kV class respectively. This takes care for the increase of voltage at the receiving end of the unloaded transmission lines or idly charged Transmission lines due to Ferranti effect.

   ii. Above 132 kV voltage level the PTs become very bulky, unmanageable and also uneconomical. Their turn ratios become very high such as for 220 kV level, the PT shall have 220000/√3 √ 110/√3=2000 and so on

   iii. Unlike PTs the CVTs are dual purpose equipment i.e. the carrier signals can also be transmitted / received for tele-protection, tele-metering, tele-control (in UHV system the manual tripping and closing the Circuit Breakers of 400kV and 765kV Transmission lines are required to be done simultaneously at both the ends through transmitting carrier signals to the remote end Circuit breaker control system), Auto- reclosing schemes of EHV and UHV Transmission lines and SCADA etc.

   iv. If the CVT is to be used for Carrier coupling, the line matching unit is to be connected to the HF bushing; otherwise this terminal is required to be earthed.

e. **Synchronization of Supplies:**

   i. The secondary voltages the replica of primary voltages of the incoming Transmission line CVTs and the bus CVTs are used for synchronization of that particular

Transmission line with the energized Bus of the running Sub - Station or the Power Station switch-yard through an instrument called Synchroscope. This instrument forms a part of the relay-control panel schemes in the S/S or PS Control room or a separate movable Synchronizing Desk is provided. The voltages received from the CVTs of the incoming Transmission line charged from the remote end to be synchronized with the running Bus and the voltages received from the running Bus CVTs are matched for magnitudes of voltages, Phase angle(it should be within $\pm 20^{o}$) and the frequency of both the systems. If the same are matched properly the closing switch of the Transmission line is closed and is synchronized with the running bus.

In the present scenario, Auto- synchronizing is carried out with the help of Sychronizm Check relay (ANSI-Device No.25) which compares (i) Phase angle difference of $+ 20^{o}$ to $- 20^{o}$(ii) voltage, < ±5 % of the rated voltage of the running bus,(iii)Rate of change of Slip frequency of 0.11% of 50Hzs, over a period of 2 secs.

5.0 Concept of PLCC;

Prior to using CVTs, CCs (Coupling Capacitors) were being used. Even now for 66 kV and 132 kV class transmission lines, the CCs are being used for PLCC links and PTs too are used in 66 kV and 132 kV Sub-stations.

5.1 The value of capacitor for Carrier communication -C_N is derived from the value of C1 and C2.

As an example in case of 245kV CVT (please ref. Figure 3),

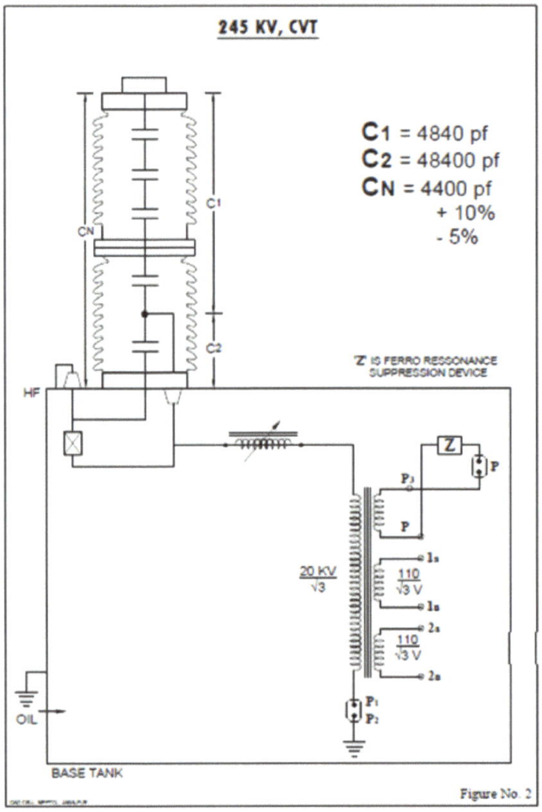

Figure 3 (Figure 2 reproduced)

For the Carrier Communication the value of Capacitance C_N is derived from the values of C1 and C2.

$$\frac{1}{CN} = \frac{1}{c1} + \frac{1}{C2}$$

$$\frac{1}{4840} + \frac{1}{48400} = \frac{10+1}{48400} = \frac{11}{48400} = \frac{1}{4400}$$

Therefore**;** CN = 4400pF +10% -5%

i. **CVT**; As already mentioned above, that CVT is a dual purpose outdoor equipment, This is used in lieu of PT as well as CC. This is connected in shunt condition to the Power conductor towards the line side.

The CVT acts like a high pass filter to pass Carrier frequency to carrier cabinet. C_N is the Capacitance of the Capacitor used for communication purpose. It is derived from the combination of C1 and C2 as shown in fig: 2.

Phase To Phase Coupling; Pl ref. the fig: 4, it pertains to Ph to Ph coupling. In this type of coupling, 2 Nos.- CVTs, 2 No.s -WTs and 1 No.-of LMU and 1 No- DLMU are required for each end of the transmission line.

a. The advantage with this type of coupling is that, the communication does not get disrupted due to the earthling of the line conductors for maintenance purposes.

Typical schematic diagram for Ph to Ph coupling is shown in the fig: 4.

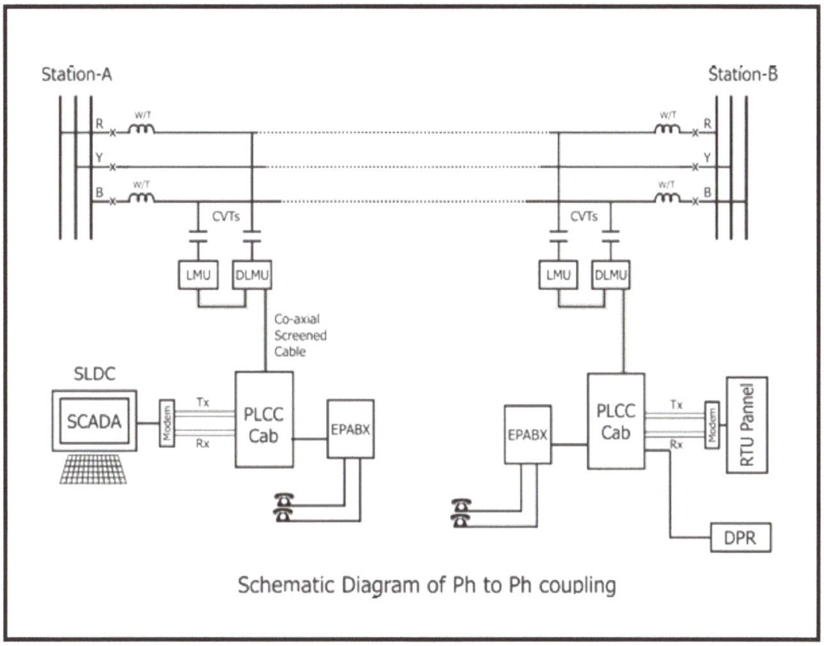

Figure 4: Phase to phase coupling

6.0 Advantages of using PLCC;

The PLCC system has an edge over other communication systems for use in Power system networks on following accounts;

1. It is the most economical and reliable system particularly where the sub-stations are situated in far-off remote places where no other mode of communication is available.
2. No clearance is required from external agencies like DOT/ Deptt of Forest and Environment/ Right of way etc. As far as clearance from Deptt of Forest and Environment is concerned, the same is not at all needed separately for PLCC as the Transmission line, as it is laid after obtaining due clearances from the Deptt. Of Forest and Environment. PLCC does not need laying of any separate conductor with separate supports etc. for carrying HF signals.
3. HF and Power frequency currents can simultaneously be transferred on the same Transmission lines.
4. Quick and easy installation and commissioning without site selection, as no radio path surveys/ Mast erection and stringing of OPGW etc are involved,
5. The carrier frequencies used in PLCC system is radio frequency 50kH to 500 kHz
6. **Author:** K.Murty & P.K.Devangan.

14. Gas Insulated Substation Vs Air Insulated Substations

ABSTRACT

This article presents a comparative analysis of gas insulated substation and air insulated substation. A brief study of both the substations is carried out. As the electricity demand is increasing day by day, the power companies are giving their active participation in fulfilling the consumer demand. The design of power substation is classified as air insulated substation and gas insulated substation where the air and gas have a key role in providing insulation. An AIS unit is basically a conventional switchgear technology where air is used as a primary dielectric for insulation. Key property of AIS unit is step-up systems in connection with power Stations and Transformer substations in the extra high voltage transmission system. Whereas the gas insulated substation works on the principle operation of completely enclosure of all the energized parts in metallic encapsulation compressed within sulphur hexafluoride (SF6) gas. Both the substations have equal merits and demerits in different aspects. Taking into account a comparative analysis is made regarding design aspects, cost and reliability.

Key words: Substations, AIS, GIS, Sulphur Hexafluoride Gas.

1.0 INTRODUCTION;

For the past three decades, SF-6 Sulphur hexa fluoride gas is being extensively used as a dielectric and extinguishing arc media in the area of high voltage electrical switchgear. SF-6 gas does not exist in the natural state and is a synthetic product, finds itself progressively adopted by all concerned users due to its improved performance characteristics and service quality which has reduced size and weight of the equipment. The reason for this choice lies in the fairly rare combination of dielectric and arc-extinction properties.

The dielectric strength is several times greater than that of other known media, at the same pressure. The arc-extinction properties are also unique could not discover a comparable media so far. The GIS are manufactured over a wide range of voltages from 132kV to 765 kV and capacity upto 4,000 amperes depending upon requirements in various designs suited to the particular requirements. In places where the cost of land or the cost of earthwork is high this can greatly influence the overall investment needed.

In case of GIS each individual item of switchgear is metal enclosed which is at earth potential. This metal encloser not only gives enhanced safety but is inherently reliable due its component being placed within a protective gas environment. Its major advantage is the small space, occupied by the equipment compared to air insulated equipment.When new sub-stations are planned, there is the choice of different designs.Conventional outdoor sub-stations are a viable option when there is a plenty of space, land prices are low, environmental conditions are normal and maintenance labour costs are low. GIS are advantageous when the space available is limited, land prices are high, environmental conditions are more extreme, and reliability of power supply is of more importance.

2.1 What is a Gas Insulated Substation?

A gas insulated substation (GIS) is a high voltage substation in which the major conducting structures are contained within a sealed environment with a dielectric gas known as SF6 or sulphur Hexafluoride gas as the insulating medium.

Figure 1: GIS substation housed in a room.

2.2 What is Air insulated Substation: A conventional (AIS) or Air-Insulated Substation, atmospheric air is the dielectric medium, as these types of substations are almost always located at outdoor locations.

Figure 2: AIS (Air Insulated Substation)

The selection criteria for AIS and GIS, make it more difficult for a user to reach to a conclusion. Therefore, once the functional needs of a substation have been specified, it is important to examine the requirement vis'a vis importance the user will be using the substation for.

3.0 Detailed evaluation of both GIS and AIS designs.

It is crucial to identify and organize the benefits and drawbacks of AIS and GIS in order to help with the decision-making process. After that, it is necessary to examine the aspects according to the requirements of the user.

1.1 Benefits & Draw backs of AIS & GIS;

1.2 Requirement of space;

The most notable benefit of GIS solutions is their ability to save space; their substation area is 15 to 25% smaller than identical AIS Substation. However, users are making use of additional advantages of GIS, such as enhancing the resilience of key infrastructure, avoiding hurdles in the way of obtaining permits, and shortening the duration of scheduled outages.

4.0 Regarding the evaluation of AIS and GIS substation solutions, the following are some definitions of factors that are commonly used:

4.1 Substation Altitude and dielectric strength,

- **Choice of Substations**: Equipment that uses air as an insulating and cooling media will experience a greater increase in temperature and a decrease in dielectric strength at elevations greater than sea level.
- However, if case of GIF is proposed Substation altitude does not matter.

4.2 Atmospheric Contamination;

- **AIS substations**: Exposed electrical insulation is vulnerable to air airborne pollution such as salt, dust, debris and industrial pollution. If they are not removed, the pollutants that have accumulated on the high-voltage insulators in your transmission, substation and distribution substation have the potential to wreak horrific devastation.
- This pollution can frequently result in Leakage current, arcing and ultimately a flash over, all of which can cause interruptions, outages etc. The following are examples of common forms of contamination that can build up:

- ➢ Salt fog.
- ➢ Desert sand,
- ➢ Snow accumulation,
- ➢ Coal dust,
- ➢ Cement dust,
- ➢ Sugar cane Carbon,
- ➢ Bird's excreta,
- ➢ Fly ash,
- ➢ Industrial vapour.

Figure 3: deposit of sand dust on the insulator string.

Figure 4: Snow/ice deposit on the insulator and structure.

- **GIS substation:** Indoor placement is the norm for GIS systems. In addition, most GIS parts are enclosed and hermetically sealed, making GIS the best option even when the air quality is low.
- Unlike in AIS following shall not affect the GIS substations;

- Desert sand
- Salt deposit in sea shore area.
- Snow accumulation,
- Coal dust,
- Cement dust,
- Bird's excreta,
- Fly ash,
- Seismic condition.
- Vandalism etc.

4.0 GIS cost VS AIS Cost;

GIS equipment is more expensive than AIS equipment. However, when life cycle costs are taken into account, it is frequently discovered that GIS is not only more cost effective but also performance wise superior.

Comparisons of costs of GIS & AIS substations should be based on the complete LCC(life cycle costs), which should include not only the price of equipment and land and development but also the expenses of routine operation and maintenance and the costs of forced outages based on reliability.

- Gas-insulated switchgear installed in a hall,(Figure 5)

Figure 5: GIS substation

Figure 6: AIS substation.

- **Availability of service;** Percentage of time that the service is available and the (steady-state) the probability that power will be available in AIS System is less.

5.0 Audible Noise and Inconvenience to public and remedy thereof;

Audible substation noise, particularly continuously radiated discrete tones (e.g., from power transformers), is the type of noise that the community may find unacceptable. Community guidelines to ensure that acceptable substation noise levels are maintained can take the form of governmental regulations or individual/community reaction (permit denial, threat of complaint to utility regulators, etc.).

Remedy: Instead a GIS may be considered though it will be a costly affair.

6.0 Control and Monitoring system;

Provisions for controlling and monitoring substation equipment can either be local, which means they are housed within the substation, or remote, which means they are situated in a different place, most commonly an operation centre. Because most switches are often motor operated, GIS may offer more options and features for control than other systems.

Figure 8: GIS substation control and monitoring through local control.

7.0 Substation capacity;

The capacity of a substation to carry a load, which is often described in relation to the MVA rating of a power transformer and is determined by a system load study. The AIS and GIS switching equipment needs to be coordinated with the capacity of the substation.

Additionally, there is a good chance that enhanced capacity projects will benefit further from the use of GIS.

8.0 Substation Expandability;

The AIS can be enlarged with some difficulty. On the other hand, GIS is an excellent alternative for expanding current AIS when space is restricted and

scheduled outages are difficult to achieve. This is a situation that typically exists when there is a combination of both of these factors.

When expanding an existing GIS, the initial design must incorporate provisions for future plans, and there may be instances in which the future infrastructure

must be deployed ahead of schedule. However, that same can be said for AIS substations.

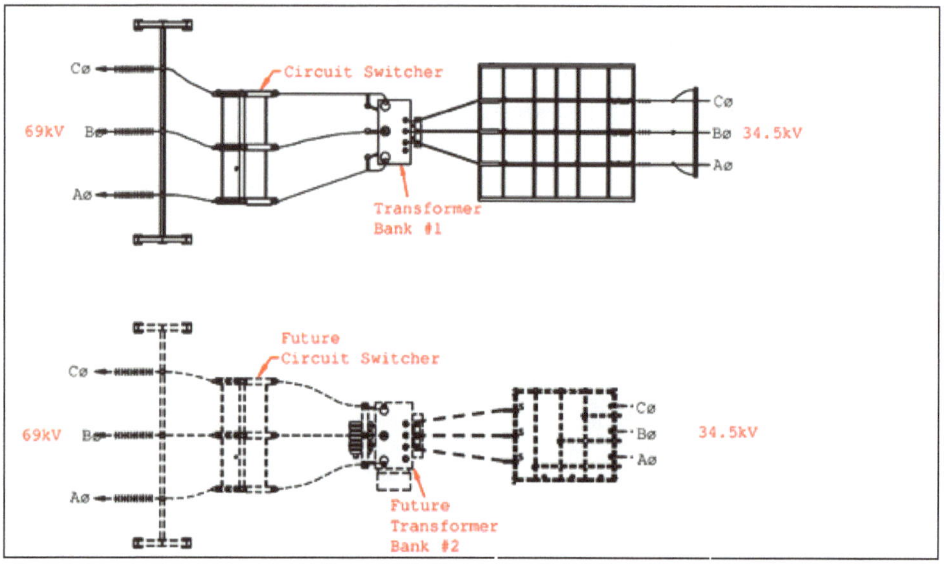

Figure 9: Proposed expansion of AIS.

Figure 10: Expansion of GIS in progress.

9.0 GIS Earthing:

Figure 11(a):Earthing of GIS.

The copper strips seen in the fig.11(a) are grounded/Earthed to the ground and connected to the earthing grid fig. 11(b) which is layed 500mm below the ground level of the substation.

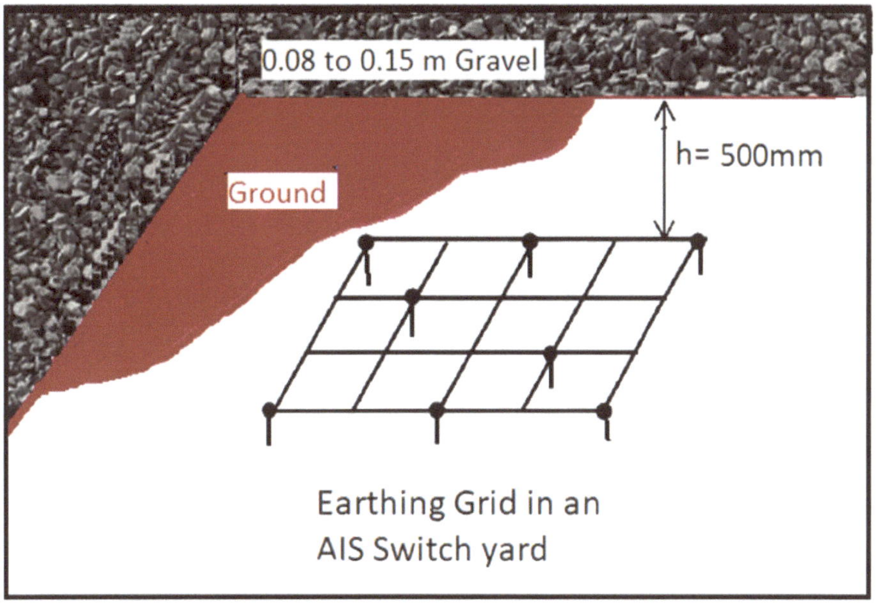

Figure 11(b) Earthing grid

10.0 Substation Security;

The substation could be at risk from things like vandalism, terrorism, or intruders. Since most substations are housed indoors and their electrical components are shielded**, GIS is often used to lessen security concern.**

11.0 Seismic withstand capability;

GIS substation equipment has ability to withstand forces generated by the earth quakes. GIS typically has better seismic withstand capability than the AIS.

Figure 12: GIS withstands earth Quakes effectively.

12.0 SF6 Emissions;

Sulphur hexafluoride (SF6) is a synthetic fluorinated compound with an extremely stable molecular structure. Because of its unique dielectric properties, electric utilities rely heavily on SF6 in electric power systems for voltage electrical insulation, current interruption, and arc quenching in the transmission and distribution of electricity.

Yet, it is also the most potent **"greenhouse gas" as known to-date". Since SF6 is a greenhouse gas, it is essential that it should be handled in the appropriate manner.**

As the greenhouse gas is emitted, it accumulates in the atmosphere in an essentially un-degraded state for many centuries. Thus, a relatively small amount of SF6 can have a significant impact on global climate change.

15. Back to Back Switching of Capacitor Banks Through Control Switching Device

1.0 Synopsis: In a large EHV Grid system, the circuit breakers are the most important equipment. Failure of any EHV circuit breaker may cause disastrous disturbance in the grid system leading to even grid failure. The failure of Circuit breaker may take place due to the Transient conditions developed while switching-in and/or switching-out conditions of the circuit breakers used for following equipment/Transmission Lines.

- Shunt capacitor banks.
- Shunt Reactors.
- No-Load transformers.
- No-Load Over head transmission lines.
- Shunt and series compensated Transmission lines, etc.

Use of "Controlled Switching Device", an intelligent and smart IED is the only answer to obviate such conditions and to safe guard the circuit breakers against failures due to switching Transients.

In this article a basic duty of a "Controlled Switching Device" in back to back switching of an EHV Shunt capacitor banks at PoW has been explained. The Transients developed are very high in nature and are very injurious to the circuit breaker used for this duty.

A case study, highlighting the need for implementing such device for back to back switching of shunt capacitor banks is also described.

2.0 Key Words: Controlled switching Device, IED (Intelligent Electronic Device), RDDS(Rate of Decrease Dielectric Strengh).

3.0 Smart "Controlled Switching Device"- An introduction: In order to safeguard the catastrophic failure of EHV circuit breakers and ill-effects on the associated EHV equipments while performing switching-in/off duties particularly in case of Back to back switching, a modern smart device named as "Controlled Switching Device" has been innovated which has an IED (intelligent Electronic Device) through which the programmed command is given for switching-in/off of the circuit breaker contacts at the point "0" position on the voltage wave. This shall safe guard the Circuit Breakers from ill-effects of Switching Transients.

4.0 Phenomenon of In-rush transients in Back to Back Switching;

The fig.1 shows the basic schematic single line diagram of a Power system feeding EHV bus, on which Capacitor banks C1 & C2, each having individual circuit breakers, CB1 & CB 2 are connected through bus-section circuit breaker BSCB.

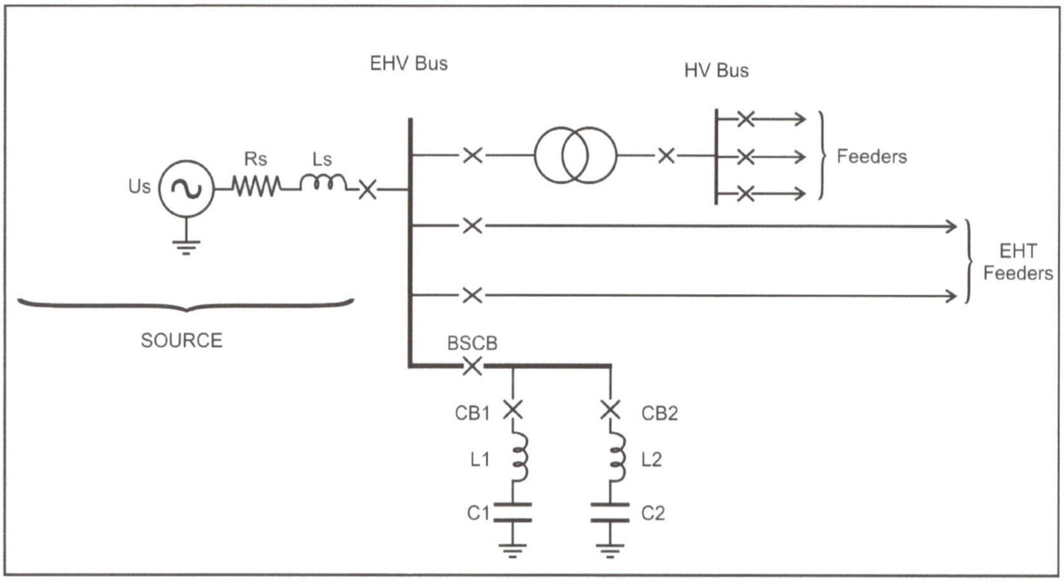

Fig: 1 SLD showing Shunt Capacitor banks, C1 & C2 connected to the common EHV Bus.

While Capacitor bank C1 is already connected to the energized EHV bus, the second capacitor bank C2 is now switched-on to the same bus in parallel to the existing Shunt capacitor bank C1. The inrush high frequency transient Currents shall rush to the 2nd Capacitor bank C2 through the circuit breaker CB2. indicated by the following formula;

$$i(t) = \frac{Us\ max(peak)}{Zo}.\mathrm{Sin}(\omega o.t)$$

Worst case occurs when the switching is takes place at U max in the $(+)$ half of the cycle or at Umax in the $(-)$ half of the cycle.

The peak value of inrush current will be, $I_o = \dfrac{Us\ max(peak)}{Zo}$

Where, Zo **is the Surge Impedance** $= \sqrt{\dfrac{Leq}{Ceq}}$

Natural frequency of un-damped oscillation $\omega o = \dfrac{1}{\sqrt{\mathbf{Leq.Ceq}}}$,

$$\textbf{Where Ceq} \neq \frac{\mathbf{C1.C2}}{(\mathbf{C1+C2})}$$

The inrush current to the tune of 40 to 100 times of the rated current at frequency 20 to 2000 kHz shall rush into C2 (ref.fig.2 showing wave shape) and also the bus shall experience voltage dips and transient over voltages. The transient phenomena of the inrush currents and the voltages may produce high mechanical and dialectical stresses on the Circuit breaker which may result into its break-down, the Capacitor banks, and the other Power equipments in the vicinity thereof shall also experience the ill-effects of Back to back switching (Ref. fig. 2 & 3 depicting wave shapes of the Transient Currents and the Transient Voltages).

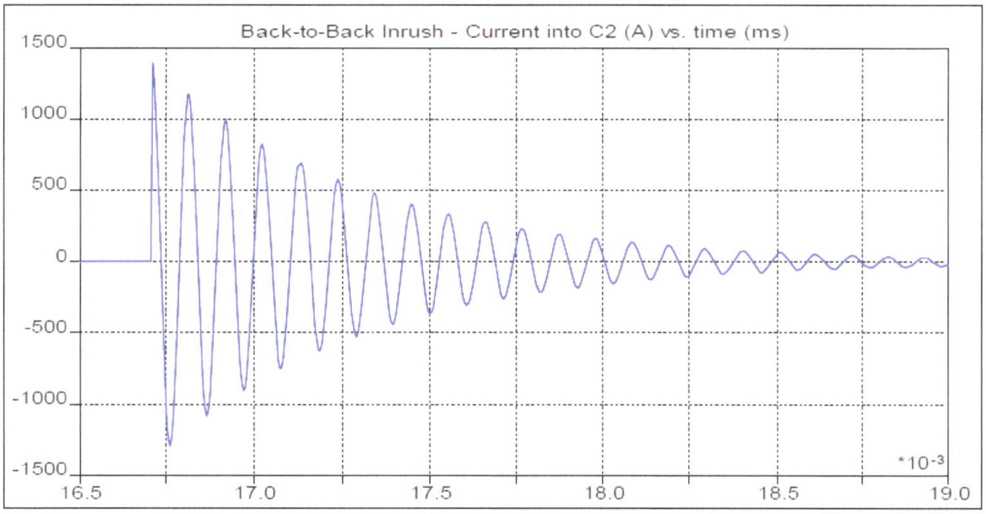

Figure: 2 Curve showing transient inrush current.

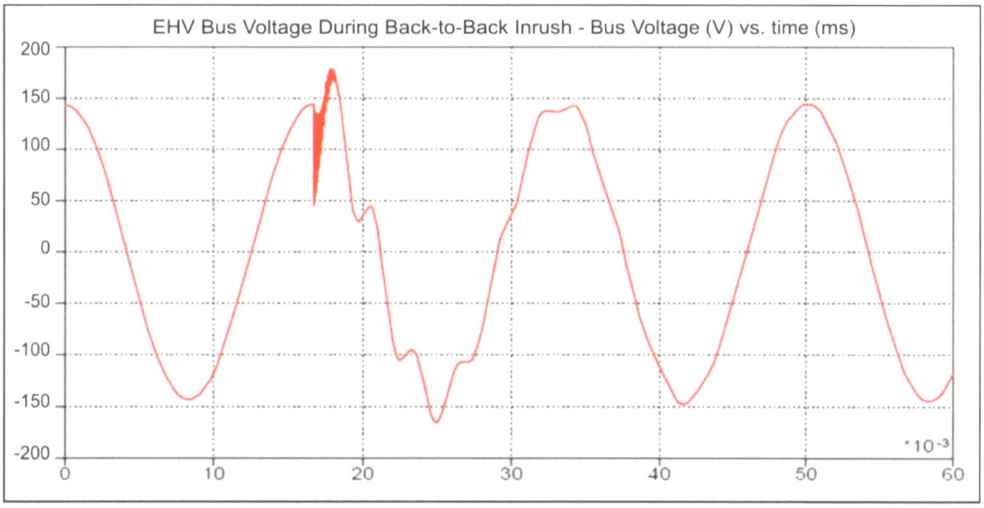

Fig;3.: Curve showing Transient voltage dips and over voltages on the EHV Bus.

5.0 Eliminating Switching Transients through smart "Controlled Switching Device:

This equipment has been developed consisting of **an Inteligent Electronic Device** (based on the digital technology) which receives the input voltage signal from the bus and is programmed to delay the switching command such that the actual contact making of the circuit breaker takes place at the voltage '0' position on the sinosidal voltage wave, thus eliminating the switching Transients.

5.1 Principle of operation of Controlled Switching Device/scheme:

The principle of operations of the Controlled Switching System, has been explained through a block diagram showing fuctions/operations involved therein, as depicted hereunder;(ref.4).

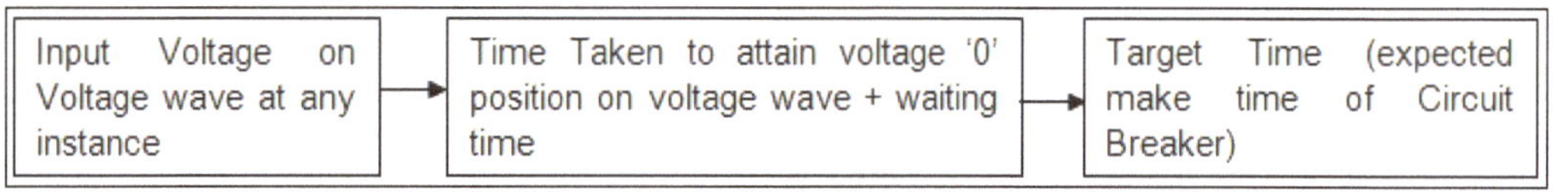

Fig: 4 Functions of Controlled Switching System for effectively eliminating switching transients.

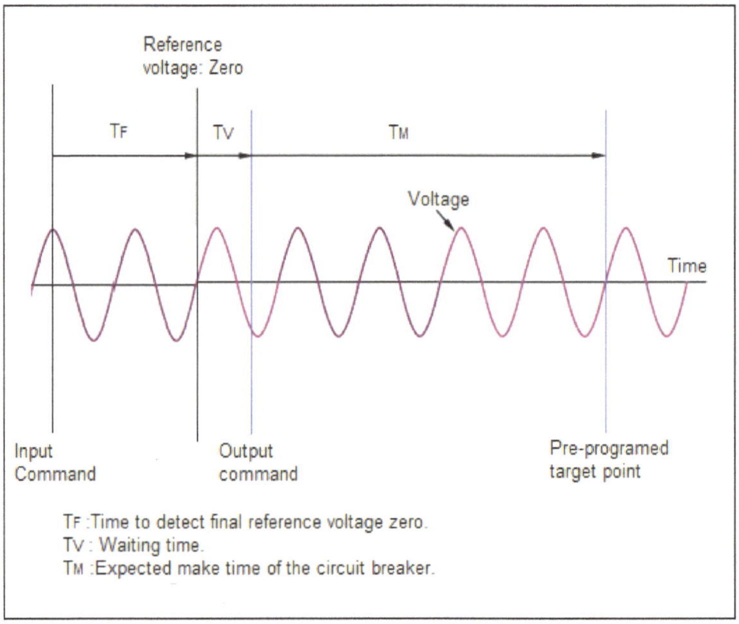

Figure 5: Ref. fig.4: Voltage vs Time for effecive elimination of switching transients

It is observed from fig.4 & 5 that;

- T_F – time to attain input voltage to '0', from the instant of giving switching command (input Command) at some point on the voltage wave.
- T_V – the waiting time with reference to voltage '0' to that of, out-put command.
- T_M – pre-programmed make time of the Circuit breaker to make contacts at voltage'0'.

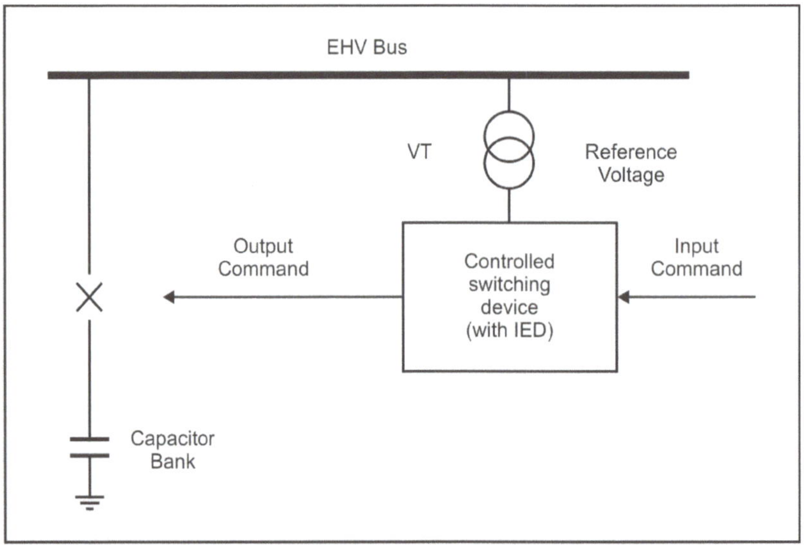

Fig: 6 Basic schematic diagram of Capacitor Bank switching in-conjunction with Controlled Switching System.

By means of the Controlled Switching System, the circuit breaker closing command is controlled with regard to point "0" position" on the voltage wave at the instance when the Breaker contacts make the contact (touch each other), followed by time taken for complete Closing action, therefore no harmful transients will be generated due to such switching operations.

Schematic diagram of Indivudual operated Circuit breaker.Such circuit breakers are known SPR breakers(Single Pole Reclosure Breakers).(Ref fig.7.)

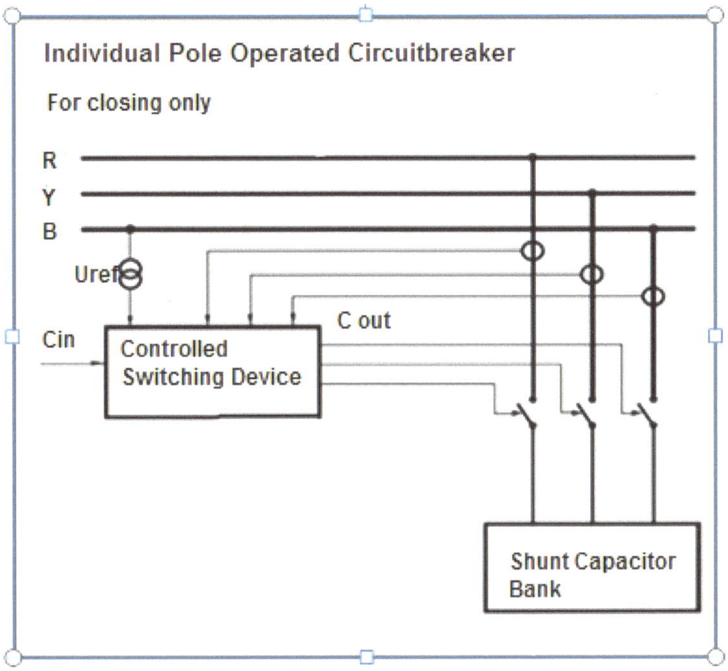

Figure 7: Schematic diagram of Individual pole operated circuit breaker.(closing operation).

6.0 In case of 3-pole operated Circuit breakers:

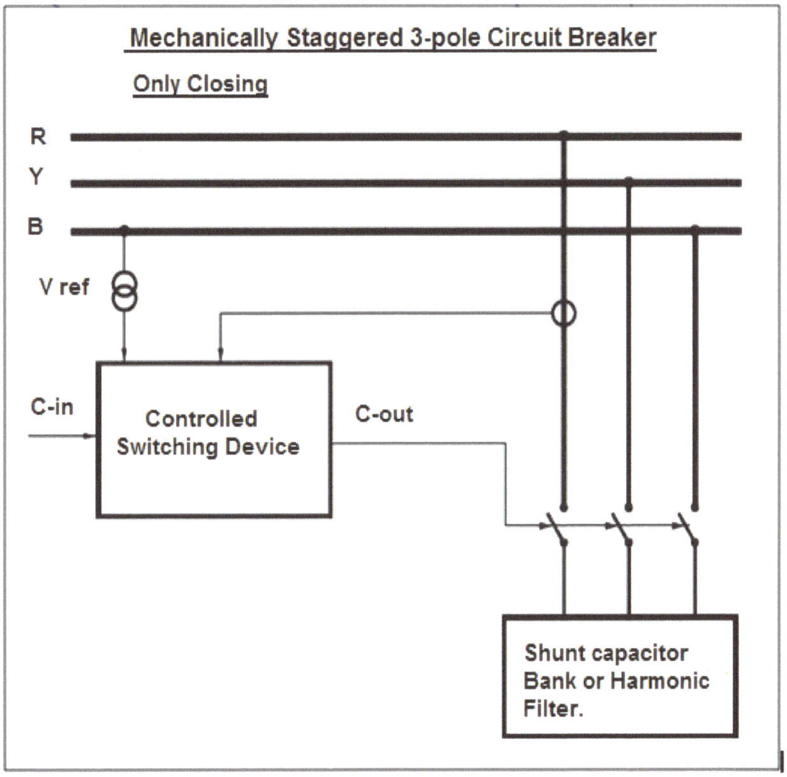

Figure 8 Schematic diagram of Machanically staggered 3-pole Circuit breaker(closing only).

It is essential to stagger the poles (mechnically to be adjusted) of the Circuit Breaker such that if one the poles closes at voltage '0' the other 2-poles should close simultaneously at voltage '0', instant. This is effected through mechanically staggering the circuit breaker poles (Ref.fig 8) w.r.t the different phases of the bus voltage such that they are adjusted to electrical degrees as shown below;

Phases as per cofiguration in the Bus	Circuit breaker Poles	Staggering of closing instant, Eletrical degrees.
R	C1	+60°
Y	B1	0°
B	A1	+120°

Table:1 Connection of Breaker poles to the Bus as per phase-wise configuration and phase difference in electrical degrees.

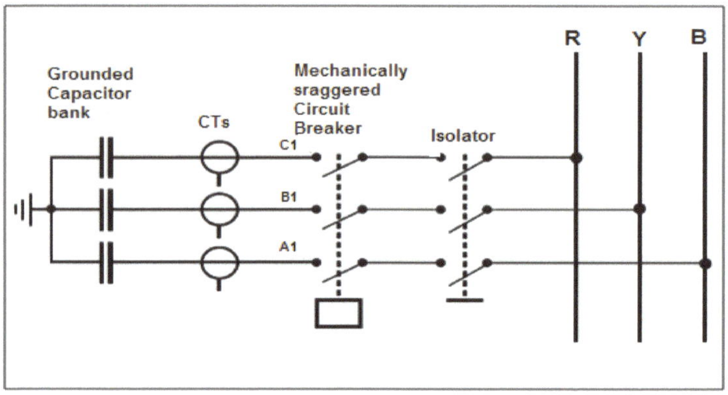

Figure 9: Connection of Breaker poles to the Bus as per phase-wise configuration.

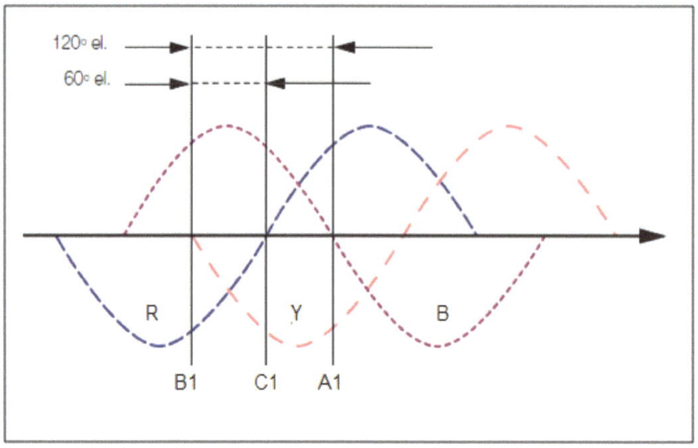

Fig:.10.The effect of staggering of breaker poles w.r.t.bus voltage-wave with eferene to table no.1, above.

Care should be taken to connect the circuit breaker poles such that R-ph, Y-ph & B- ph of the bus-voltage should be connected to C1, B1 A1 poles respectively of the circuit breaker.

It is observed from the fig.10,

- when the Y-ph voltage wave passes through '0', at this instant, B1 pole of the Circuit breaker contacts make.
- At this instant the R –ph voltage wave shall be lagging behind by 240 $^{\circ}$ ele and B- ph Voltage wave shall be lagging behind by 120° el to the Y-ph voltage wave.
- Thus the R-Ph voltage wave shall cross the point "0" position at 60° ele while traversing from +ve to –ve cycle.
- The B-ph voltage wave shall cross the point "0" position at 120° ele.while travesing from +ve to – ve cycle.
- Had the poles of C1 & A1, not stagerred and adjusted, the voltage waves of R ph would have crossed the '0' crossing after *3.33 ms and the B-Ph voltage wave would have crossed the '0' crossing after after *6.66 ms as per prevailing system frequency of 50 Hzs in INDIA.
- Since, stagerring of the Breaker poles is done, it resulted into closing of all the 3-Ph pole-contacts of the circuit breaker at the same instant.ie at voltage '0' position for all the 3 Phases.

Note:

1. The constraint of connecting the Circuit breaker poles to individual phase as tabulated in the Table. No1 is essential in case of grounded Shunt capacitor banks.
2. In case of un-grounded shunt capacitor banks such constraint is not there.
3. It is recommended that for Shunt Capacitor bank applications, 3-Pole operated Circuit breakers should be preferably used.

***Calculations to arrive at 3.33 ms and 6.66 ms.**

Since 360° ele = 1 cycle,

∴ 1 cycle = 20 ms,(As there are 50 cycles in 1 sec ie in 1000 ms).

∴ 60° ele -------- [(20 x 60)/360] = 3.33 ms.

∴ 120° ele --------------- 3.33 x 2 = 6.66 ms.

7. 0 Essential conditions for successful operations:

- The System is programmed in such a manner that the closing command to the individual poles of the Circuit breaker is targetted to make the contacts of the poles at voltage '0'.The Closing command to individual closing coils of the individual poles is given accordingly as per difference of electrical degrees between them, by the Controlled Switching Device.
- Thus, it is essential that all the three poles of the Circuit breaker should have individual operating features (Closing & Opening).
- This system to function satisfactorily, it is essential that the accuracy limit at the instant of '0' crossing of the system voltage (phase-wise) and the making of individual Circuit breaker poles should have an accuracy limit of ± 1ms (approx.).
- The Circuit breaker should have rapid RDDS(Rate of Decrease Dielectric Strengh) ie on closing operation, the dielectric strength of the Circuit Breaker should rapidly decrease from high starting value to zero value.

- This system has an inter-lock for preventing the switching close-command to pass to the closing coil for 10 minutes.This feature takes care for the dis-charging time of the Capacitor banks.

8.0 Specimen wave shapes of switching transients with and without CSD.

Following are the specimen Wave shapes of the switching operations.

a. **Uncontrolled Back to back Switching;**

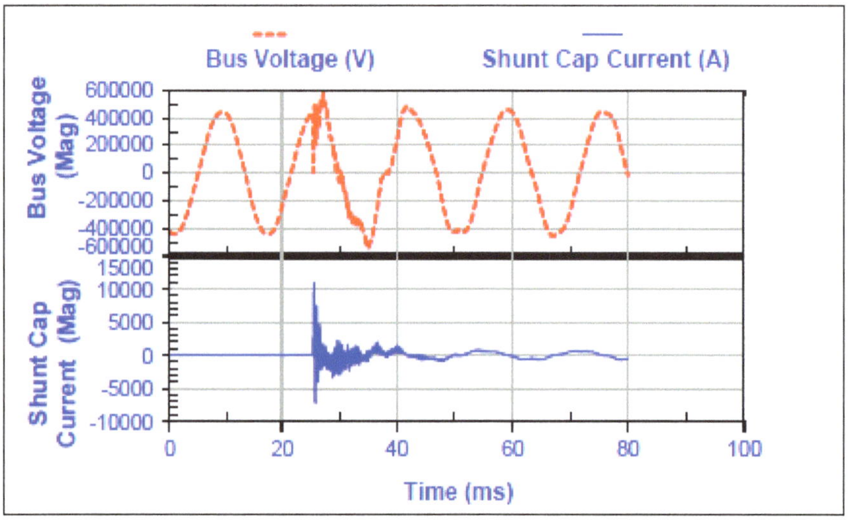

Figure 11A:Uncontrolled back to back switching.

Please refer to figure 11A, It is observed that the the transient variation in voltage is from 0 volts to 600 kV (500kV system) and that of current transients 0 to +12 kA to - 7kA.then decays.

b. **Controlled Back to Back Switching**;

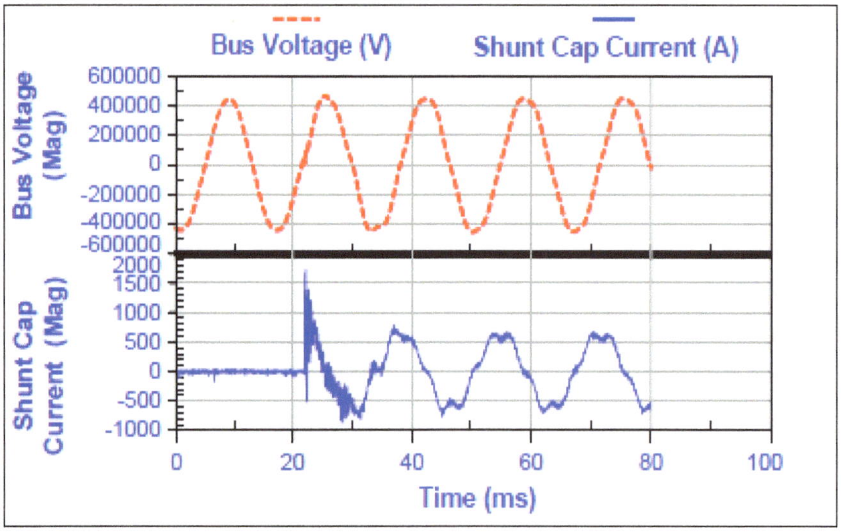

Figure 11B: Contolled Back to back switching.

Please refer fig 11B.the voltage transient is almost eliminated and the votlage wave peak is 500kV. Where as Current transient is also reduced to + 1.8 kA to -500A then decays.

9.0 A Case-Study in M.P.:

A case-study of 2x33 MVAR, 145kV, Shunt Capacitor Banks at a 220kV S/S Gwalior, with reference to installation of 145kV SF6 Circuit breakers with Controlled Switching Device for each 33 MVAR Shunt capacitor bank.

Figure 11A: Installed view of 2x33 MVAR, 145kV Shunt Capacitor Bank.

2x33 MVAR, 145kV, Shunt Capacitor Banks were connected onto 132kV bus at 220kV S/S Gwalior in M.P.

Please refer to the following figures;

Figure 11A – Installed view of 2x33MVAR,145kV Shunt Capacitor Banks

Figure 11B - Single line diagram of 2x33MVAR Capacitor bank with 1No. 145kV, SF 6 Circuit Breaker.

Figure 12.- Connection of 2x33 MVAR,145kV, Shunt Capacitor Bank on to the 132kV bus at 220kV S/S Gwalior in M.P.).

The 2x33 MVAR 145kV Capacitor banks are un-grounded banks having double star configuration.

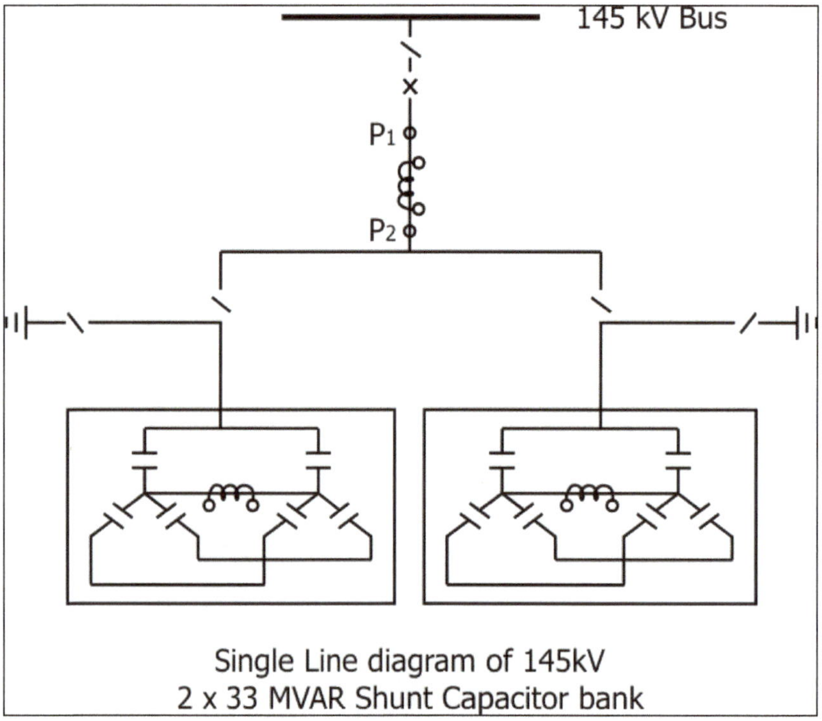

Figure 11B: Single line diagram showing 2x33 MVAR, 145kV.Shunt Cap. Banks, Controlled by 145kV, SF6 CB-1No. only.

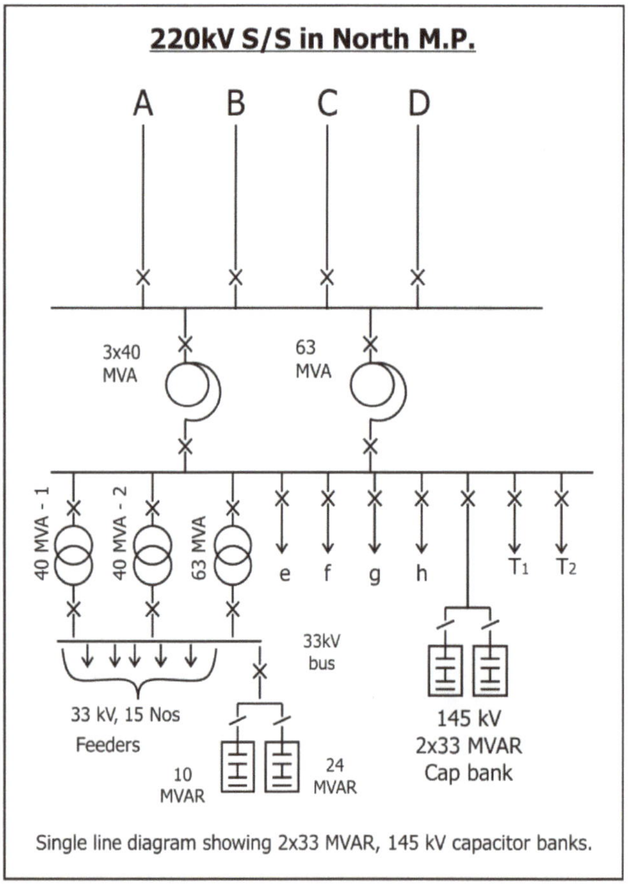

Figure 12: Single line diagram of 220kV & 132kV buses at 2 20kV S/s Gwalior under study.

The scenarios has since been changed over a period of say 6 to 8 years due to the addition of 400kV and 765kV long Transmission lines (long and lightly loaded EHV Transmission lines generate Reactive Power) in the WREB Grid, now, only 1No-33 MVAR Shunt Capacitor bank is taken into service as and when required in this 220kV S/s Gwalior, under- study.

To highlight the benefits of taking the 33 MVAR Shunt capacitor bank into service, the readings taken on dated 27.11.2013 (reference: the S/S control room log-sheets) have been considered. The curves showing improvement in the voltage profile of 132kV bus, 220 kV bus and the MVAR injected by 1 No-33 MVAR, 145kV, Shunt capacitor Bank and its effect in terms of MVAR support on 132kV bus have been shown in the fig.13.

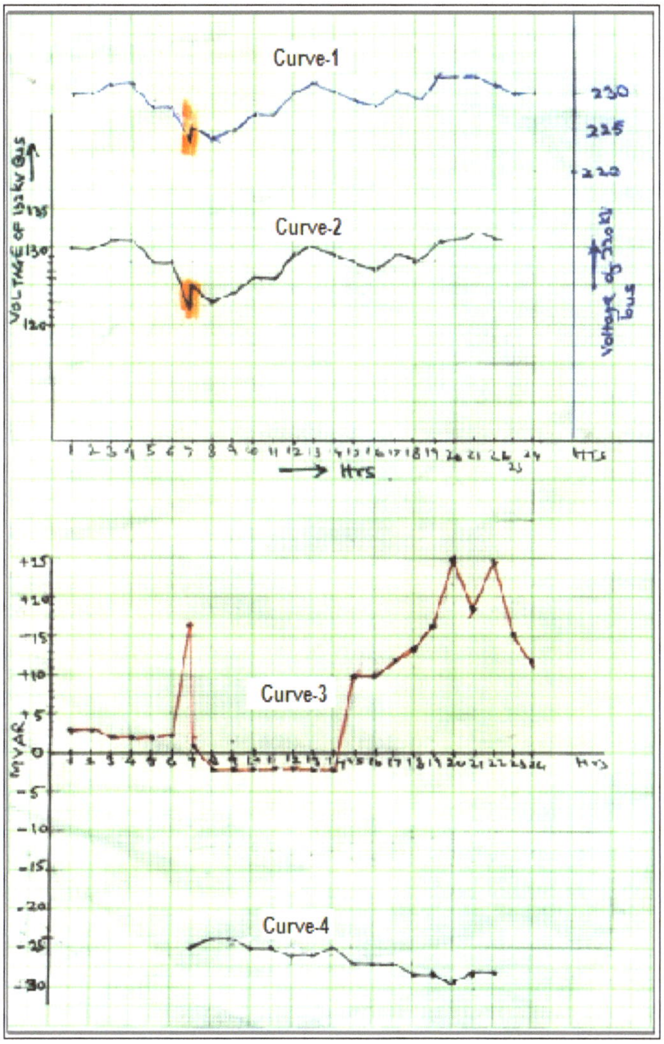

Figure :13: Curves pertain to the data on date 27.11.2013 at the 220kV S/s Gwalior under-study.

Note: (refer fig.13).

Curves from top to bottom are designated serially in numerical order and are as follows;

- Curve No.1: 220kV bus voltage.
- Curve No.2: 132kV Bus voltage.

- Curve No.3 (red): Reactive loading in MVAR, on the 220/132kV Transformers on 132kV Bus.
- Curve No.4 (black): Injection of Reactive MVAR generated (MVAR Support) by 1x 33MVAR 145kV shunt Capacitor bank.
- As per convention, Reactive Support from Capacitor bank has been taken as (–) negative and Reactive loading (inductive) has been taken as (+) positive

As could be seen from the curve No.3 (red) that, though reactive compensation (capacitive) was being injected into the System, from 08.00 hrs to 14.00 hrs with a resultant of (–) 2 MVAR (Capacitive) at 132kV bus, suddenly Reactive MVAR loading increased to 12 MVAR (inductive) between 14.00Hrs to15.00 Hrs and thereafter the inductive/reactive loading further rose to 25 MVAR (inductive).

The curves shown in fig.13 have been judiciously drawn (on the basis of actual meter readings taken from the control room log-sheet) and looking to the same, another Switched 33 MVAR Shunt Capacitor bank should have been connected on to the 132kV Bus at 15.00hrs on 27.11.2013. However, the same could not be exercised, as out of the 2 Nos of 33 MVAR capacitor banks, only 1 No. Capacitor Bank was in service (The other 1No.Capacitor bank was kept out of service by isolating the same through a manually operated 145kV Isolating Switch).

To over-come such situations, an additional 145kV SF6 Circuit breaker (ND type) should be installed exclusively for controlling the 2nd-33 MVAR Shunt Capacitor Bank. The proposed Circuit Breaker's control circuit should be equipped with **Controlled Switching Device** for eliminating the hazards of the transient inrush currents and the transient over voltages, which otherwise would have generated, as such a switching is categorized as Back to back Switching of Capacitor banks.

9.0 Conclusion:

As could be seen from the above case-study, the M. P. Power Transmission Co. Ltd. should consider to install separate 145kV, SF6 Circuit breakers {ND-Type (Normal duty type)} with **Controlled Switching Device,** on the 2nd, 33 MVAR 145 kV Shunt Capacitor banks at the strategic 220kV S/Ss for switching onto the132 kV bus as and when the system demands on Back to bank switching mode.

Since the 145 kV, SF6 Circuit breakers are suitable for Capacitor Bank opening/tripping applications by way of having Re-strike free characteristics, the facility for tripping on faults /opening of Circuit breakers through Controlled switching Device is not essential. The other State Transmission Utilities and also the Govt. & Private entrepreneurs should follow the suit, wherever required.

10.0 Acknowledgement: The author expresses his gratitude to the MD, M.P. Power Transmission Co Ltd. who had kindly permitted him to use some of the details of the 220kV S/S, Gwalior and also expresses his gratitude to officers and the staff of MPPTCL, who helped him in preparing this article.

11.0 References:

- A paper presented in the 8th International Conference "CAPACIT-2014"organized by IEEMA, Mumbai, in Nov.2014 at Delhi, related to Reactive compensation through Shunt Capacitor banks by the Author himself.
- A tutorial and case study on Shunt Capacitor Bank Transients by Govinda Gopa Kumar & fellow authors of Michigan Technological University, Houghton, MI, 49931 and Kalyan K. Mustaphi of Northern States Power Co, Minneapolis, MN 55401.
- D.N.Konidaris, Sr. Member, IEEE, Public Power Corp., S.A., Greece.-"Shunt Capacitor Bank Switching: Analysis and Experience."
- Leaflets of ABB & Areva T&D.
- In consultation with Dr. K. Rajamani.

12.0 Author:K.K.Murty

End

16. Transmission Line Constructed for Higher Level Voltage but Used for Lower Level Voltage-Importance of Insulation Co-Ordination

1.0 Synopsis: Delivering un-interrupted Power to Mega/large industries, Small & medium industries and to public at large for domestic & commercial purposes is the prime responsibility of the Power Utilities.

BIL is a measure of the insulation's ability to resist voltage stress and prevent insulation failure under various operating conditions viz; most frequent type of such over voltages caused by switching operations, transients due to lightning stroke etc. BIL is typically expressed in kilo-Volts $_{peak}$ (kV $_{peak}$) and is a critical parameter for the design, development, manufacturing, and operation of electrical equipment and Transmission lines.

2.0 Case Study: Prevention of frequent failure of 33kV Metering equipment (combined CT & PT unit), fed through 132kV line to Ferro-manganese HT consumer, CD 4.5 MVA.

2.1 Supply arrangement: A 8 km, 132kV Transmission Line was being used for feeding 33kV supply to a Ferro- manganese HT consumer from 132/33kV Substation, at Balaghat (M.P.)

2.2 Occurrence: Failure of 33kV metering equipment took place 3-4 times at the HT consumer's HT metering point.(metering equipment is installed at the point of supply to the HT consumer)

2.3 BIL Levels; (Please see table no.1)

- BIL of 132kV is 650kVp and recommended no. of insulators in a stringing are 9 nos. discs and in the tension location 10 no. of Discs of 70kN.
- BIL level of 33kV is 170kVp. No.of discs in a string is 3Nos. and 4 nos.in the tension location.

Note;

Ques. A general question is raised as to why 9 Discs are used for 132kV Transmission line to form a string of insulators?

Ans. Insulators are supporting the live conductor from earth and thus shall be rated for Line-to-Earth voltage. For a 132kV line, L-E voltage is ≈76kV. Considering highest system voltage of 145kV, it is ≈84kV.

Since, each disc is rated for 11kV and with 9 discs, the string can withstand 99kV, ie. 99kV > 84kV.

SCHEDULE -4
[See clause nos. 5.2.2 (iv)]

SALIENT FEATURES OF BUS POST INSULATORS[(1)]

	Details	Unit	765kV system	400kV system	220kV system	132kV system	33kV system
1)	Nominal/ Rated voltage	kVrms	765	400	220	132	33
2)	Highest system voltage	kVrms	800	420	245	145	36
3)	Minimum total creepage distance	mm	20,000	10,500	6125	3625	900
4)	Lightning (Full wave) impulse withstand voltage (1.2/50 μs)	kVp	2100	1425	1050	650	170
5)	Switching impulse withstand voltage (250/2500μs) – wet	kVp	1550	1050	NA	NA	NA
6)	One minute power frequency withstand voltage- dry and wet	kVrms	830	680	460	275	70
7)	Minimum corona extinction voltage	kVrms	508	320	156	105	NA
8)	Maximum radio interference voltage between phase and ground	μ-volt	2500 (at 508 kVrms)	500 (at 305 kVrms)	500 (at 156 kVrms)	500 (at 92 kVrms)	NA
9)	Type		Solid core	Solid core	Solid core	Solid core	Solid core
10)	Total minimum cantilever strength	kg	800[(2)]	800	800	800	800
11)	Minimum torsional moment		As per IEC-273	As per IEC-273	As per IEC-273	As per IEC-273	As per IEC-273

[(1)] Parameters given here are for installations at altitudes not exceeding 1000m above mean sea level (MSL). For higher altitudes, necessary correction factor shall be considered as per relevant standards

[(2)] Value indicated for 765 kV is minimum, however same is to be increased as per layout requirement.

Table.1

1.0 Cause of failure of the 33kV metering equipment & 30kV LA; Since the 33kV supply to the consumer was being fed through 132kV Transmission line ie the BIL level of the 132kV Transmission line conductor hung with a string containing 9nos.disc at each suspension location was 650kVp. It was observed on occurrence of lightening strokes at any location and during switching operations in the concerned 132/33kV substation, the 33kV metering equipment and 33kv LA failed simultaneously.

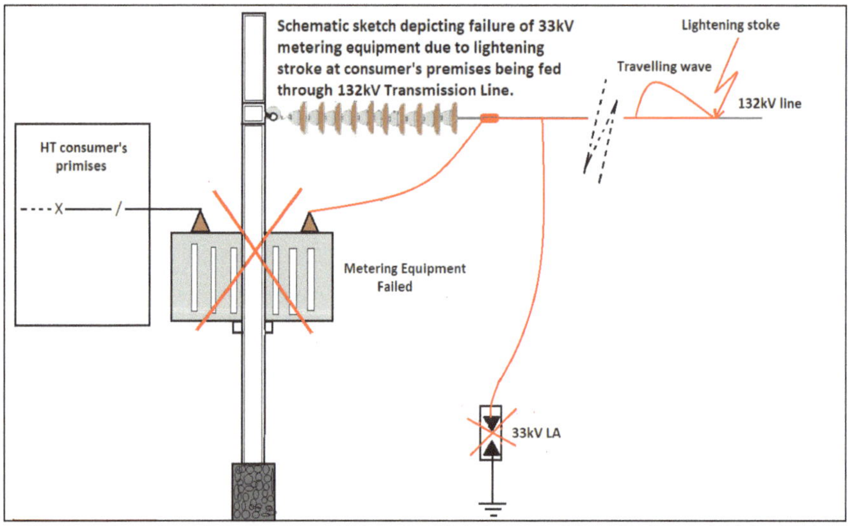

Figure 1: Failure of 33kVmetering equipment & 33kV LA due to heavy lightening stroke.

Transient over voltages occur on all power systems caused due to switching operations and due to lightning stroke, they shall discharge to the earth

through the softest target ie the 33kV Metering equipment the BIL of 33kV metering equipment is 170kVp and 33kV LA, whereas the BIL of the 132kV Transmission line is 650 kVp. Resulting into failure of 33kV metering equipment and 33kV LA simultaneously.(Refer Figure 1).

4.0 Remedial measures: Since, the power to the 33kV HT consumer on was released through 132kV Transmission line therefore it was essential to reduce the BIL Level of The 132kV Transmission line.

i. **In case of Suspension Tower locations**: To short 6 nos. discs out of 9 nos. discs from the suspension String Insulator such that only 3nos of disc are remaining which is adequate for the lowering down the BIL of 132kV transmission line. This remedial action was taken to shorting 6 nos. discs at least on 8 to 10 no, locations at the 132/33kV Substation end and similarly to short 6 no. discs on the 8-10 no. suspension towers 8 to 10 no.at Receiving end.(Refer Figure 2)

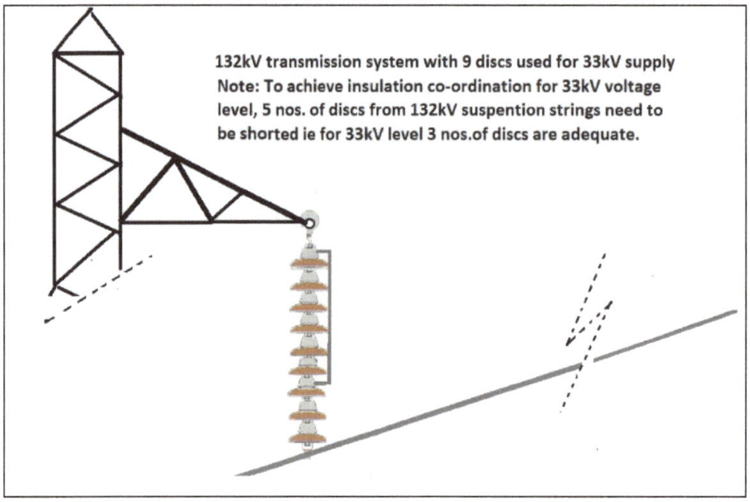

Figure 2: shorting of 6discs from 132kV suspension string.

ii. **In case of tension locations of 132kV towers**: Out of 10 no discs to short 6nos Discs such that remaining only 4 no. Discs shall take care of insulation coordination;(Refer Figure 3)

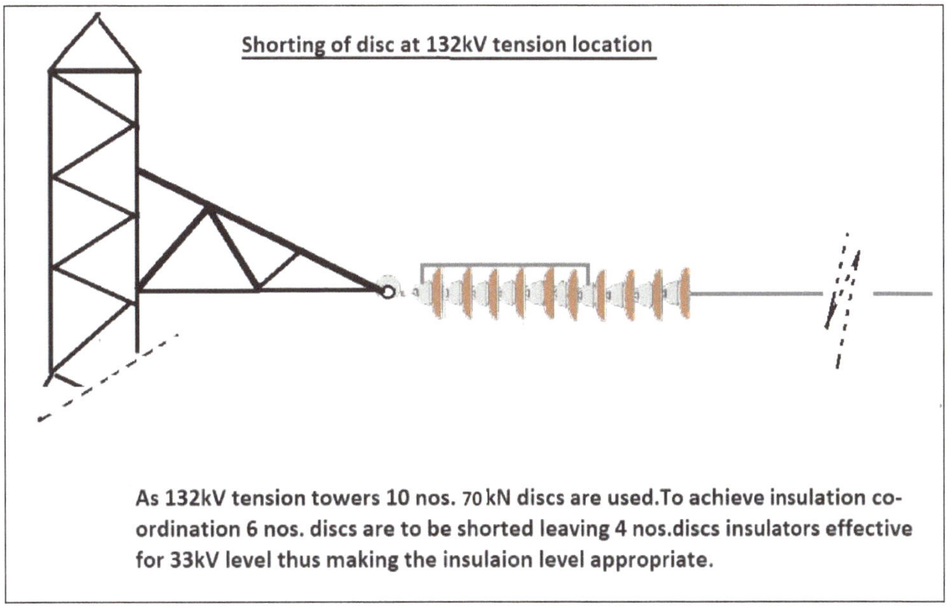

Figure 3: Shorting of 6 nos. discs out of 10 nos. discs at the tension locations of 132kV towers

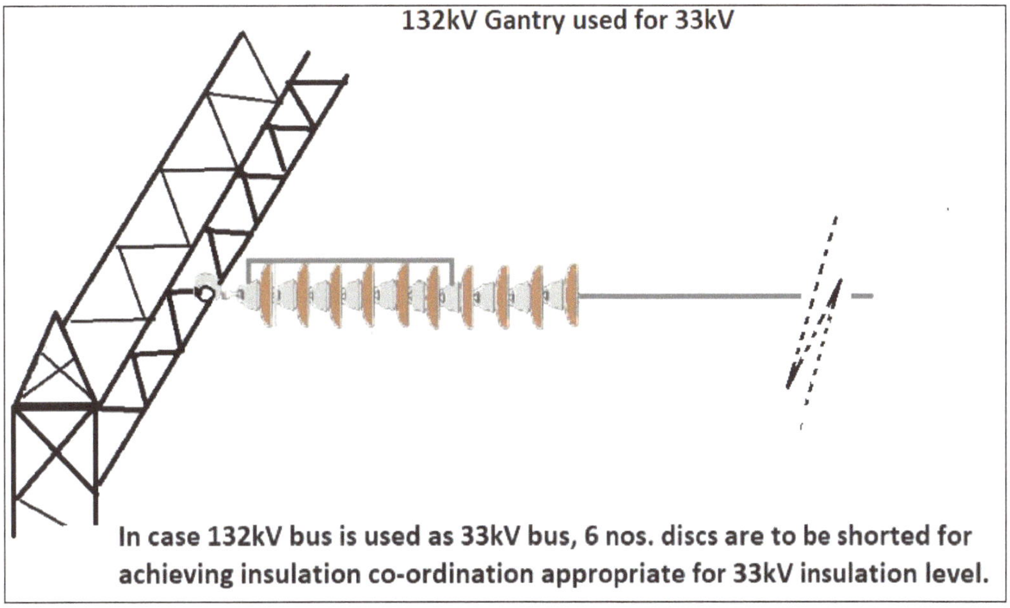

Figure 4: Shorting of 6 nos. discs out of 10 nos. discs at the takeoff gantry of 132kV Transmission line at the sending end 132/33kV substation.(Refer Figure 4)

5.0 Observation:

- After shorting of discs of the said 132kV Transmission line so as to make its required insulation level appropriate for 33kV level, failure of 33kV metering equipment and LA at the HT consumer's end has been arrested there after.

6.0: Author: K.K.Murty

17. Understanding Importance of Basic Insulation Level for Electrical Equipment

Introduction

Electrical equipment plays a vital role in delivering Power to Mega/large industries, Small & medium industries and to public at large for domestic & commercial purposes. To ensure the integrity and proper functioning of electrical equipment, the Power Utilities follow Basic Insulation Level (BIL) as a yard stick. This concept is fundamental to understanding and implementing effective insulation, a key factor in achieving safety and reliability in electrical systems.

Basic Insulation Level, is often abbreviated as BIL, it defines the ability of insulation to withstand voltage stresses.

In this article, the concept of Basic Insulation Level (BIL) is dealt with. Its significance, the factors that influence BIL, test voltage levels, wave shapes, international standards, and its impact on the design and operation of electrical equipment, with a specific focus on transformers.

1.0 DEFINITION:

Bil is a term used in electrical engineering to describe the highest transient voltage level an insulation system can withstand without breaking down or experiencing a dielectric failure. in essence, bil is a measure of the insulation's ability to resist lightening impulse voltage stress and prevent electrical arcing or insulation failure under various operating conditions. bil is typically expressed in kilo-volts peak (kvp) and is a critical parameter for the design, development, manufacturing, and operation of electrical equipment.

2.0 BIL Requirements and Voltage Surges in Power Systems

Power system components, such as reactors, transformers, circuit breakers, and power cables, are designed not only to operate at their rated voltage levels but also to withstand voltage surges or transients that can occur in the power system, such as those generated by lightning strikes or switching operations. For instance, a power transformer with a primary voltage of 132 kV and a BIL of 650 kV can endure voltage surges of up to 650 kV without experiencing insulation failure.

Figure 1: Lightening stroke on132kV Transmission line.

Furthermore, it is critical to note that BIL requirements vary depending on the type of electrical equipment, and they must be met to ensure reliable and safe operation. Failure to meet the equipment's BIL requirements can cause serious damage to the equipment and the electrical system. This is why BIL must be taken into account when designing, selecting, and maintaining electrical equipment.

3.0 The Importance of BIL in Power Systems

i. **Safety:** One of the primary reasons BIL is so important in electrical systems is safety. Inadequate insulation can result in failures, fires, and electrical shocks. Adequate insulation helps prevent electrical faults and ensures the safety of those who work with or are exposed to electrical systems.
ii. **Reliability and Continuity of Service**: BIL is directly linked to the reliability and continuity of service of power systems. Adequate BIL ensures that equipment can consistently operate under normal conditions and transient voltage surges.
iii. **Equipment Lifetime**: Electrical equipment is a significant investment for any organization. Proper insulation not only ensures the reliability of this equipment but also extends its lifespan. Equipment that can withstand the voltage stresses it encounters over time is less likely to require frequent maintenance, repairs, or replacement, resulting in cost savings.
iv. **Compliance with Standards and Codes:** Various international and local standards and codes like IS, IEC and IEEE or any other standard in vogue in other countries. These standards specify BIL requirements for different types of electrical equipment. Complying with these standards is not only a legal requirement but also a way to ensure the safety and reliability of the electrical system.

4.0 Factors Influencing BIL

The determination of BIL for electrical systems takes into account several key factors:

i. **Voltage Levels:** The system voltage or operating voltage of the electrical system is a fundamental factor in establishing the appropriate BIL. Systems with higher voltages require higher BIL ratings.

ii. **Environmental Conditions:** Environmental factors such as humidity, temperature, and pollution can influence insulation properties. Equipment in harsh or contaminated environments may require higher BIL levels to maintain performance.

iii. **Voltage Transients**: The presence of voltage transients, such as lightning strikes or switching surges, can stress the insulation system. Equipment exposed to such conditions may require higher BIL to prevent dielectric failure.

iv. **Specific Equipment Needs:** Different types of electrical equipment may have unique BIL requirements to ensure proper operation under specific conditions. For example, power transformers need to withstand high impulse voltages, while power cables should have a sufficient power frequency withstand voltage.

5.0 Test effect of lightening impulse on out door Power equipment.

The standard test voltage levels, which are determined by the highest equipment voltage (Um) of a winding, are outlined in Table 2 of IEC 60076-3.

5.1 Wave shape of Full Lightning Impulse Voltage

The Lightening impulse voltage is the transient voltage that typically exhibits a rapid rise to its peak value in 1.2μs followed by a slower decay to 50% of Lightening Impulse voltage in 50μs. As shown in Figure 2.

As per IEC 60067-3 the test tolerances of time in % for rapid rise to peak lightening impulse voltage and slower decay to 50% lightening impulse voltage is mentioned as 1.2 μs ± / 50 μs±,

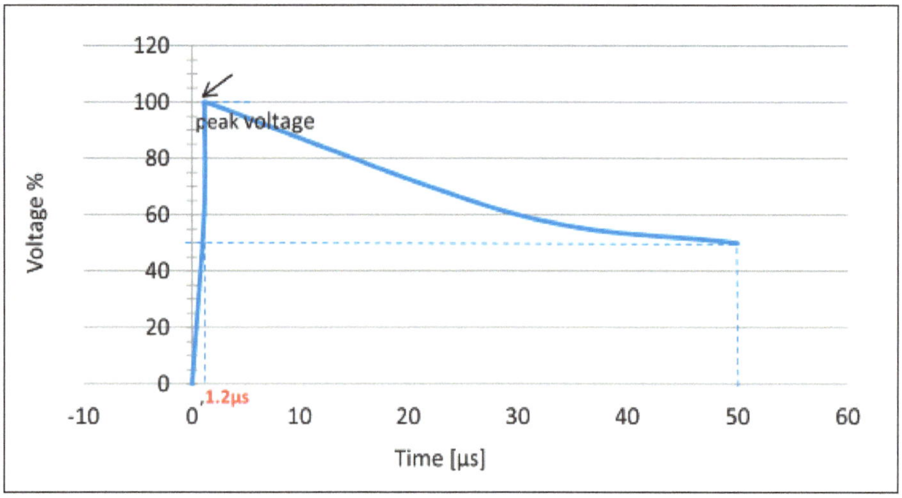

Figure 2

5.2 Insulation Levels

Transformers should be designed in accordance with IEC 60076-3. The key identifier for a set of coordinated levels is determined by its maximum system voltage and basic lightning impulse insulation level (BIL).(please refer to CEA's Recommendation as stated the table no.1.

General Guidelines for
765/400/220/132kV substations and
switchyard of Thermal/ Hydro Power Projects

SCHEDULE -4
[See clause nos. 5.2.2 (iv)]

SALIENT FEATURES OF BUS POST INSULATORS[1]

	Details	Unit	765kV system	400kV system	220kV system	132kV system	33kV system
1)	Nominal/ Rated voltage	kVrms	765	400	220	132	33
2)	Highest system voltage	kVrms	800	420	245	145	36
3)	Minimum total creepage distance	mm	20,000	10,500	6125	3625	900
4)	Lightning (Full wave) impulse withstand voltage (1.2/50 μs)	kVp	2100	1425	1050	650	170
5)	Switching impulse withstand voltage (250/2500μs) – wet	kVp	1550	1050	NA	NA	NA
6)	One minute power frequency withstand voltage- dry and wet	kVrms	830	680	460	275	70
7)	Minimum corona extinction voltage	kVrms	508	320	156	105	NA
8)	Maximum radio interference voltage between phase and ground	μ-volt	2500 (at 508 kVrms)	500 (at 305 kVrms)	500 (at 156 kVrms)	500 (at 92 kVrms)	NA
9)	Type		Solid core	Solid core	Solid core	Solid core	Solid core
10)	Total minimum cantilever strength	kg	800[2]	800	800	800	800
11)	Minimum torsional moment		As per IEC-273	As per IEC-273	As per IEC-273	As per IEC-273	As per IEC-273

[1] Parameters given here are for installations at altitudes not exceeding 1000m above mean sea level (MSL). For higher altitudes, necessary correction factor shall be considered as per relevant standards

[2] Value indicated for 765 kV is minimum, however same is to be increased as per layout requirement.

Table1 [5]

REFERENCES

[1] IEC 60076-1: 2011, Power transformers – Part 1: General.

[2] IEC 60076-3: 2013, Power transformers – Part 3: Insulation levels, dielectric tests and external clearances in air.

[3] IEC 60076-4: 2002, Guide to Lightning Impulse and Switching Impulse Testing for Power Transformers and Reactors.

[4] Standard specifications and technical parameters for transformers and reactors (66 kV & above voltage class) issued by the Central Electricity Authority, Govt. of India in April 2021.

18. Various Clearances in Laying Power Lines & Minimum Safety Working Clearance as Per CEA's Safety Regulation-2023

Preamble; Transmission line AC & DC are the life-line of any country's socio – economic development. It is of prime importance that Power should be transmitted properly in healthy condition. In order to maintain safety of Power system as well as safety of human being, line clearances play vital role. In this Article some of the clauses of CEA's Safety regulation-2023 have been reproduced.

Key words: Ground clearance, Phase to phase clearance, CEA's Safety Regulation 2023,

1.0 CEA's safety regulation -2023 Cl.63. Clearances from buildings

Voltage exceeding 650 V.–

1. An overhead line shall not cross over an existing building as far as possible and no building shall be constructed under an existing overhead line.
2. Where an overhead line of voltage exceeding 650 V passes above or adjacent to any building or part of a building it shall have on the basis of maximum sag a vertical clearance above the highest part of the building

Immediately under such line, of not less than;

   i. for lines of voltages exceeding 650 V and up to and including 33 kV - 3.7 metre;
   ii. for lines of voltages exceeding 33 kV - 3.7 metre plus 0.30 metre for every additional 33 kV or part thereof.

3. The horizontal clearance between the nearest conductor and any part of such building shall, on the basis of maximum deflection due to wind pressure, be not less than, –

   i. for lines of voltages exceeding 650 V and up to and including 11 kV - 1.2 metre;
   ii. for lines of voltages exceeding 11kV and up to and including 33 kV- 2.0 metre;
   iii. for lines of voltages exceeding 33 kV - 2.0 metre plus 0.3 metre for every additional 33 kV or part thereof.

4. For high voltage direct current systems, the vertical and horizontal clearances, on the basis of maximum deflection due to wind pressure, from buildings shall be maintained as below:

Sl.no.	High Voltage Direct Current	Vertical Clearance(m)	Horizontal Clearance (m)
1	100kV	4.6	2.9
2	200kV	5.8	4.1
3	300kV	7.0	5.3
4	400kV	7.9	6.2
5	500kV	9.1	7.4
6	600kV	10.3	8.6
7	800kV	12.4	10.7

2.0 CEA's safety regulation-2023 C.71. Lines crossing or approaching each other and also lines crossing streets and roads. –

Where an overhead line crosses or is in proximity to any telecommunication line, the owner of the overhead line or the

telecommunication line, whosoever lays his line later, shall arrange to provide for protective devices or guarding arrangement as per code of practice or guidelines issued by Power and Telecommunication Coordination Committee and shall observe the following provisions, namely;

i. When it is intended to lay a telecommunication line or an overhead line which will cross or be in proximity to an overhead line or a telecommunication line, as the case may be, the person proposing to lay such line shall give one month notice of his intention so to do along with the relevant details of protection and drawings to the owner of the existing line;
ii. Guarding arrangement shall be provided where a line of voltage not exceeding 33 kV crosses a road or street or elecommunication line;
iii. Where an overhead line crosses or is in proximity to another overhead line, guarding arrangements shall be provided so as to safeguard against the possibility of their coming into contact with each other; and
iv. Where an overhead line crosses another overhead line, minimum clearances as provided in the table below shall be ensured:

Minimum clearances in metre between AC Lines crossing each other

Sl.no	Nominal System Voltage (kV)	11-66 kV	110-132 kV	220 kV	400 kV	765 kV
1	Below11kV	2.44	3.05	4.58	5.49	7.94
2	11-66	2.44	3.05	4.58	5.49	7.96
3	110-132	3.05	3.05	4.58	5.49	7.96
4	220	4.58	4.58	4.58	5.49	7.94
5	400	5.49	5.49	5.49	5.49	7.94
6	765	7.94	7.94	7.94	7.94	7.94

3.0 Minimum clearance in metre between AC & DC lines crossing each other

SL.No.	Syatem voltage AC/DC	100kV DC	200kV DC	300kV DC	400kV DC	500kV DC	600kV DC	800kV DC
1	Below 11kV AC	3.05	4.71	5.32	6.04	6.79	7.54	9.04
2	11-66 kV AC	3.05	4.71	5.32	6.04	6.79	7.54	9.04
3	110-132kV AC	3.05	4.71	5.32	6.04	6.79	7.54	9.04
4	220kV AC	4.58	4.71	5.32	6.04	6.79	7.54	9.04
5	500kV DC	4.71	4.71	5.32	6.04	6.79	7.54	9.04
6	300kV	5.32	5.32	5.32	6.04	6.79	7.54	9.04
7	400kV AC	5.49	5.49	5.49	6.04	6.79	7.54	9.04
8	400kV DC	6.04	6.04	6.04	6.04	6.79	7.54	9.04
9	500kV DC	6.79	6.79	6.79	6.79	6.79	7.54	9.04
10	600kV DC	7.54	7.54	7.54	7.54	7.54	7.54	9.04
11	765kV AC	7.94	7.94	7.94	7.94	7.54	7.54	9.04
12	800kV DC	7.94	7.94	7.94	7.94	7.94	9.04	9.04
13	1200kV AC	10.44	10.44	10.44	10.44	10.44	10.44	10.44

Note: Where two lines cross each other, the crossing shall be made as nearly at right angles as the nature of the case admits and as near the support of the line as practicable, and the support of the lower line shall not be placed in the right of way of the upper line.

4.0 Minimum clearance in air above ground and across road surface of Highways or roads or railway of an alternating current overhead lines, including service lines of nominal voltage system.

[See sub-regulation (1) of regulation (60)]

Nominal voltage system	Clearances above ground		Clearances between conductor & road surface across High way Normal OHE(Where No double Stack containers are to be run on railway track.	Clearance between Conductor & Rail level cross Railway Corridor **Note:** (OHE= Over Head Equipment-Cantilever)	
	Across Street(m)	Else where		#High rise OHE for running of double stack containers on railway tracks.	
33kV	6.50.	5.20	11.60 or U/G cable	U/G cable	U/G cable
66kV	6.50.	5.20	11.60 or U/G cable	U/G cable	U/G cable
110kV	6.50	6.10	11.60	15.56	17.56
132kV	6.50	6.10	11.60	15.56	17.56

220kV	7.02	7.02	12.52	16.46	18.46
400kV	8.84	8.84	14.00	18.26	20.26
765kV	18.00*	18.00*	18.80	21.86	23.86

* Higher clearance due to predominantly induction effects and time varying electric field (ICNIRP limit: 10kV/m for occupational exposure) at voltage exceeding 400 kV.

5.0 The minimum clearance in air above ground and across road surface of Highways, or Minimum clearance between conductor and Rail Level of nominal voltage system.

[See subregulation (2) of regulation (60)]

Sl.no.	DC Voltage	Ground clearance(m)	Clearance road surface across Highway(m)	Minimum Clarence between conductor and Rail level
1	100	6.50	11.25	#
2	200	7.30	12.05	#
3	300	8.50	13.25	#
4	400	9.40	14.15	#
5	500	12.50	17.25	21.23
6	800	18.00	22.75	25.74

1. Highway clearances required 4.75 m higher then ground clearances (considering the vehicle hight is 4.75, as mentioned in the Indian Road Congress documents, 1983).

2. # Railway clearances required 10% higher value than HVAC values (HVAC values are mentioned in Indian Railway document: IRSOD, 2004)

SAFETY CLEARANCES							
	Minimum clearances	Unit	765kV System	400kV System	220kC System	132kV System	33kV System
1.	Phase to phase	mm	7600 9400	4000 4200	2100	1300	320
2	Phase to earth	mm	4900 6400	3500	2100	1300	320
3	Safety working clearance (sectional clearance)	mm	10300	6500	5000	3800	3000
4	Ground clearance	mm	14000	8000	5500	4800	3700
5	Hight of insulator's bottam from ground	mm	2440	2440	2440	2440	2440

Minimum safety working clearance where electricity at voltage 12kV & above.

Sl.no.	Highest system voltage	Safety working clearance(Meters)
1	12	2.6
2	36	2.8
3	72.5	3.1
4	145	3.7
5	245	4.3
6	420	6.4
7	800	10.3

1. The above are valid for altitude not exceeding 1000m.A correction factor of 1.25 per cent per 100 m is to be applied for increasing the clearances for altitude more than 1000m and up to 3000m.
2. The above safety working clearances are based on an insulation height of 2.44 m which is the height of lowest point on the insulator, where it meets the earthed metal, from the ground;
3. "Safety Working Clearance" is the minimum clearance to be maintained in air between the live part of the equipment on one hand and earth or another piece of equipment or conductor on which it is necessary to carry out the work, on the other;
4. The "Highest System Voltage" is defined as the highest rms phase to phase voltage which occurs under normal operating conditions at any time and at any point of the system. It excludes voltage transients (such as those due to system switching) and temporary voltage variations due to abnormal system conditions (such as those due to fault conditions or the sudden disconnection of large loads).

Conclusion; While erecting power lines either AC or DC, the above clearances must be followed scrupulously.

19. Fatal & Non-Fatal Accidents in EHV S/Ss.: Causes & Prevention/Remedial Measures Thereof

1.0 Preamble: While EHV sub stations and Transmission lines constitute Grid network and are very important to Power Transmission Utilities, safety to the equipment & human life is of prime importance. Unlike equipment damage, if the workmen sustain non-fatal accidents with major disability or lose their lives in fatal accidents, the affected families suffer for the life-time. Therefore, it is the primary responsibility of an EHV Substation Manager /Engineer to observe safety rules and measures scrupulously to avoid such incidences. The authors have highlighted in this paper some of the safety measures to be essentially observed by the EHV Sub station Manager/Asstt. Engineer prior to and during maintenance job on EHV equipment in an EHV S/S, which would curb the fatal/ non-fatal accidents of the workmen in the EHV Sub stations. Most critical hazard is because of the ill-effect of Electricity-Induction due to charged EHV lines/ Substation in the vicinity of work-spot/area.

"ELECTRICITY":
Despite being an essential commodity,
is lethal at HV & EHV levels, if not dealt with
properly and carefully.

1.1 Applicable Rules/Regulations/Act: Generation, Transmission, Distribution and Utilization of electrical energy in India are governed by;

- IE Rules-1956(amended up to 25th Nov.2000),
- Electricity Act 2003.
- Central Electricity Authority Regulations 2010 /
- National Building Code etc.

However, in case of EHV S/Ss, excepting National building code other rules/regulations are applicable.

1.2 Statistics-the Eye-opener: In order to highlight the need for implementing safety measures scrupulously, National Safety Council has published the following Statistics which is Eye-opener.

Year	Total Accidents	Human Fatal	Human Non-Fatal	Animal Fatal	Animal Non-Fatal
2012-13	8869	3925	2336	2597	11
2013-14	14290	5517	3534	5182	57
2014-15	14729	5828	3437	5382	92
2015-16	14513	5808	3244	5244	217
2016-17	12713	5488	2718	4499	8
2017-18	14158	5799	2638	5624	97

Table 1– Statistics of Electrical Accidents

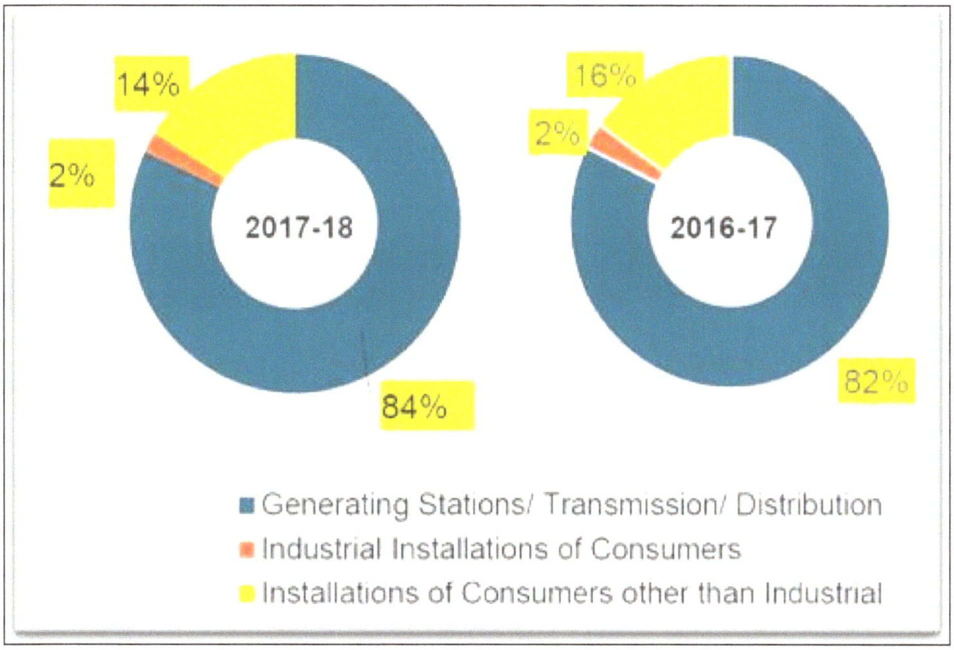

Figure 1: Accident Statistic Utility Wise.

2.0 Key words; Step potential, touch Potential, Earth Switch, Discharge rods, Induction. OSHA(Occupational Safety Health Administration), BHM (Bare Hand Method) &HSM (Hot Stick Method) & Faraday's cage.

3.0 Reasons/Causes for fatal/non-fatal accidents and precautions/remedial measures thereof:

Sr. no	**Causes for fatal/non-fatal accidents in EHV S/s**	**Precautions /Remedial measures to be taken**
1	Erroneous design of Earthing system.	i. Sub-station earthing should be designed such that Step potential, Touch potential and Earth potential rise (EPR) are within permissible limits. ii. Spreading of 40 mm sized metal150mm above the ground level in the switch yard excepting over the cable trenches and earth-pits.
2	Improper earthing after issue of permit to work.	The Supervisor (operations), in-Charge of the shift should • Identify the bay. • Verify whether the area /bay is properly earthed/ discharged through the Discharge rods.

3	Due to back feed of supply from the remote end.	It's the prime and important duty of the Operating staff to; • First, get the EHT feeder switched-off from the remote end (other end). • He should obtain permit from the remote end Substation on Telephone/ Mob./Wireless/PLCC as per procedure laid down by the Utility. • The permit from the remote end must indicate "No Back Feed". • Then he should issues permit to work.
4	Ignorantly, hanging the discharge rods on the EHV Line or EHV equipment first and then connecting the lower end of the conductor of the discharge rod to the earth and vice' versa.	S/s Manager/ Asstt.Engineer/ Supervisory Staff in- charge maintenance must; • Ensure that the integrity of earthing system is already verified as per periodicity laid down by the Utility. • **While discharging** The Discharge Rod with "G" clamp should be preferably used. ➢ Firstly: The lower end of the discharge rod conductor should be earthed by scratching the paint layer of the Structure properly and then tighten it firmly. ➢ Then "G" clamp or inverted "U" hook must be tightened properly/hanged onto the equipment's metallic portion or on the Transmission line conductor. • **While normalizing** ➢ Firstly, remove "G" clamp or inverted "U" hook end from the equipment's metallic portion or from the Transmission line conductor of all the 3-phases. ➢ Then, remove the earthed end.

<table>
<tr><td>5</td><td>Confusingly climbing up on similarly looking live equipment in the neighboring bays by the workman.</td><td>The Sub Station Manager/Asstt. Engineer/ J.E./ Supervisor maintenance must ensure the following;
• Apprise the maintenance staff engaged for the maintenance job in the bay on which the workis to be carried out indicating,
➢ Voltage class.
➢ Name of the bay.
➢ bay number.
• The bay/location for work should be identified and enclosed with Red coloured band of cloth (about 1 foot width).
• Placards/ boards should be hanged on the equipment indicating as dead; “SAFE TO WORK”</td></tr>
<tr><td>6</td><td>Falling of workmen from outdoor equipment, gantry structure and from EHT lines etc.</td><td>The Maintenance Manager/ Asstt. Engr./ Maintenance Supervisor should ensure that
• The workmen entrusted with such jobs should use safety belts/ wrap the rope around waist, use helmets, Insulated hand gloves and safety foot –ware etc.,
• The workmen entrusted with such jobs should not be suffering from ailments like Diabetes, High Blood-pressure, and Vertigo etc. Such workmen must be medically examined for their physical fitness every year</td></tr>
<tr><td>7</td><td>Use of discarded moist and tracking-marked Hot line tools/operating sticks.</td><td>Maintenance officer in-charge, Hot line Maintenance should,
• Ensure the personnel in Hot Stick Method operations are well trained.
• Must ensure healthiness of the hot-line tools used in HSM
• Storage of hot line tools is done in a dry place.</td></tr>
</table>

<table>
<tr>
<td>8</td>
<td>Infringement by workmen in to unsafe working zone while executing work HSM(Hot Stick Method)</td>
<td>
<ul><li>The Hot Stick Method workmen must be apprised about the safe zone/distance for</li></ul>
Working as per voltage class.
<ul><li>As defined below;</li></ul>
Safe working distances as per OSHA Std.(Occupational Safety Health Administration) are as follows;(Table:2)

Voltage Meters Feet
<table>
<tr><td>Voltage range Ph- Ph in kV</td><td>Meters</td><td>Feet</td></tr>
<tr><td>72.6 to121</td><td>1.02</td><td>3’4”</td></tr>
<tr><td>138 to145</td><td>1.07</td><td>3’6”</td></tr>
<tr><td>230 to242</td><td>1.52</td><td>5’</td></tr>
<tr><td>345to362</td><td>2.13</td><td>7’</td></tr>
<tr><td>500 to552</td><td>3.35</td><td>11’</td></tr>
<tr><td>700 to 765</td><td>4.47</td><td>15’</td></tr>
</table>
Table:2
<ul><li>The workmen entrusted with such jobs should not be suffering from ailments like Diabetes, High Blood-pressure and Vertigo etc. Such workmen must be medically examined for their Physical fitness every year.</li></ul>
</td>
</tr>
<tr>
<td>9</td>
<td>Not maintaining safe clearance from the ground/earth of the Faraday cage/ conductive suit clad workmen performing job on live EHT line or EHV bus through BHM (Bare hand maintenance method).</td>
<td>
While getting the EHV bus attended to through BHM(Bare Hand Method), the Officer in-charge of Hot line maintenance must ensure that;
<ul><li>Conductive suits and Faraday cage are tested for their conductivity.</li>
<li>Since linemen work at the same potential as the EHV bus, every care should be taken to keep the conductive suit clad work men/ Faraday cage at a safe distance away from any earthed structure/ equipment in the EHV substation(Table:2)[5.(1)];</li></ul>
</td>
</tr>
</table>

<table>
<tr><td></td><td></td><td><table><tr><td>Voltage class in kV</td><td>Clearance form ground in Metres</td></tr><tr><td>145</td><td>3.7</td></tr><tr><td>245</td><td>4.3</td></tr><tr><td>420</td><td>6.4</td></tr><tr><td>800</td><td>10.3</td></tr></table>*Table:3 [5(1)]*</td></tr>
<tr><td>10</td><td>Discharging EHV Shunt Capacitor banks prior to inherent discharge time.</td><td>The Maintenance Manager/ Asstt. Engr./ Maintenance Supervisor should ensure that the workmen entrusted with such jobs should ensure that EHV Capacitor banks should be discharged only after a lapse of 5 mins. After switching it off.</td></tr>
<tr><td>11</td><td>Randomly removing and reconnecting Bushing jumpers of EHV Transformers especially of 220 Class and above.</td><td>For working on the Transformers, sequence for removal of all the phase and neutral jumpers from Transformer bushings should be strictly followed as mentioned here under for the safety of workmen against induction effect.
1. **Removal of jumpers(opening of Jumpers)**
a. All the HV & LV phases should be individually earthed by the discharge rod properly.
b. All the jumpers of the HV& LV bushings should be removed first.
c. Thereafter remove Neutral bushing jumpers. **(Note:** If we remove Neutral bushing jumper of transformer first, all HV & LV bushing have different potentials due induction effect from nearby / adjacent Lines or Transformers which may causes severe induction to maintenance staff, if he touches any of the phase bushings.)
2. **Reconnection of Jumpers to the bushings**: Connect jumper of Neutral bushings first and thereafter connect jumpers of all HV & LV phase bushings.</td></tr>
</table>

<table>
<tr>
<td>12</td>
<td>Effect of induction while working on midpoint CT in Main-bus and Auxiliary bus configuration of EHT feeder strung on DCDS line, while the otherEHT line being live.</td>
<td>
The S/s maintenance manager/Asstt. Engineer/JE/ Maint. Supervisor must ensure the followings;

- Proper Discharging/ earthing of theTransmission lines and EHV equipment is done in addition to Closing the earth-switch provided with the Line side/load side isolators.
- In case of working on the T clamp of mid-point Line CTs of an EHV feeder emanating from Main bus and Auxiliary bus Systemconfiguration, Cares should be taken toearth/discharge the Midpoint CT on MainBus side, Aux bus side and on the Line side Schematic diagram depicted in Fig;2.

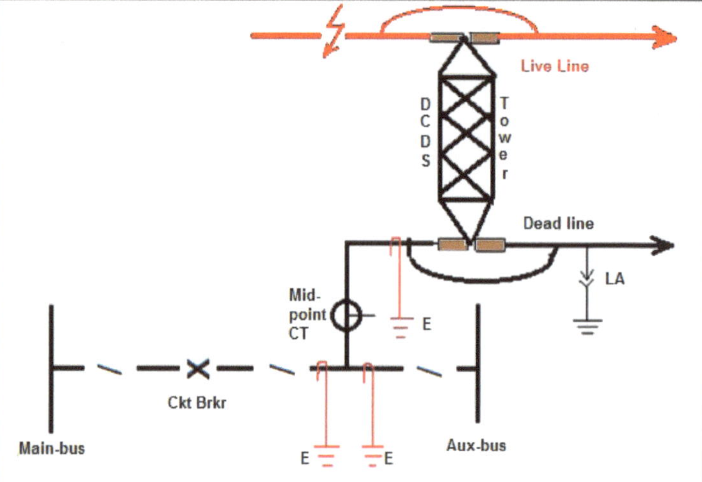

Locations where dischaging/earthing is essential while attending to mid-point CT clamps of a dead Transmission line while the other parallel Transmission line on the DCDS tower is in live conddition

Fig-2: Location of hanging discharge rods while working on midpoint CT in Main bus & aux bus configuration feeders strung on the DCDS towers.

(Note: For clarity, enlarged figure may be seen on page after the narrations in 2-colmns are over.)
</td>
</tr>
</table>

13	Effect of severe induction while working in live 400kV Switch-yard. [Note: In one of the 400 kV S/Ss, the workmen climbing up through the support insulators for attending to the 400kV Pentagraph-isolator underneath the 400kV main bus, experienced severe shock when he touched the 4"IPS tube or any part of pantograph isolator first and also when he touches the IPS tube 2nd time after removing his hand from the IPS tube, immediately after the 1st shock, despite taking precaution of providing earthing on both the IPS tubes, in addition to closing in-built earth-switch of the Pantaraph isolator (Pl see fig.3)	The S/s maintenance manager/Asstt. Engineer/JE/ Maint. Supervisor must ensure the followings; • Proper Discharging/ earthing of the Transmission lines and EHV equipments is done in addition to closing of the earth-switch provided with the Line side/load side isolators. **Cause for experiencing shocks:** • As he climbs up, his body gets charged with static-electricity and he experiences shock as soon as he touches the earthed IPS tube. • He experiences the shock again when he touched the IPS tube 2nd time. This phenomenon is due to his body again getting charged with static-electricity. **Remedial measures:** 1. If Bus configuration is such that, possibility of availing shut down on the concerned bus without hampering Powers System Stability. Then work could be taken up under the dead 400kV bus duly earthed on either side of the work-area, by the workmen using all the safety appliances. 2. A conductive Elevated work Platform(EWP), should be used as depicted in figure no;4; i. IPS tube on either side of the Pentagraph isolator should be properly discharged through discharge rods. ii. Platform should be earthed properly. iii. A bonding lead between the Discharge rod and the Pentagraph isolator should also be connected to the boom of the platform. Above earthing / discharging done to create equipotential work area.

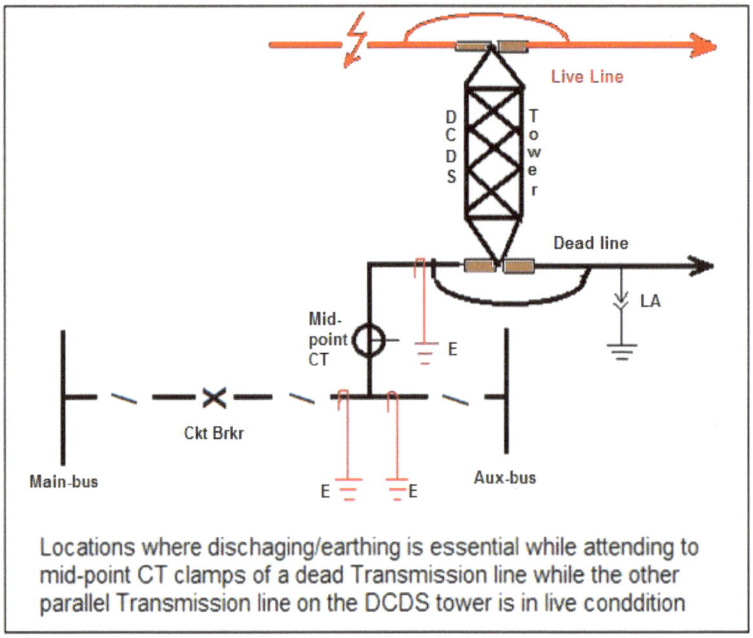

Fig-2: Location of hanging discharge rods while working on midpoint CT in Main bus & aux bus configuration feeders strung on the DCDS towers.

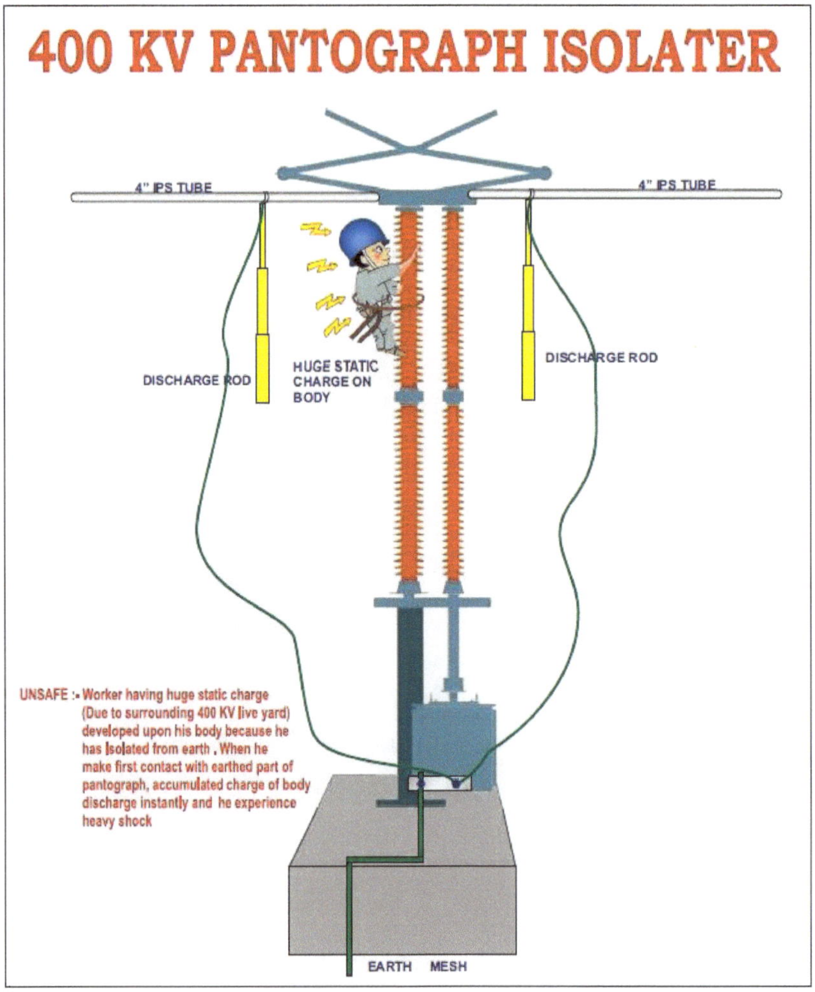

Fig-3: Workman climbing up the 400kV Pentagraph isolator duly discharged on either side.

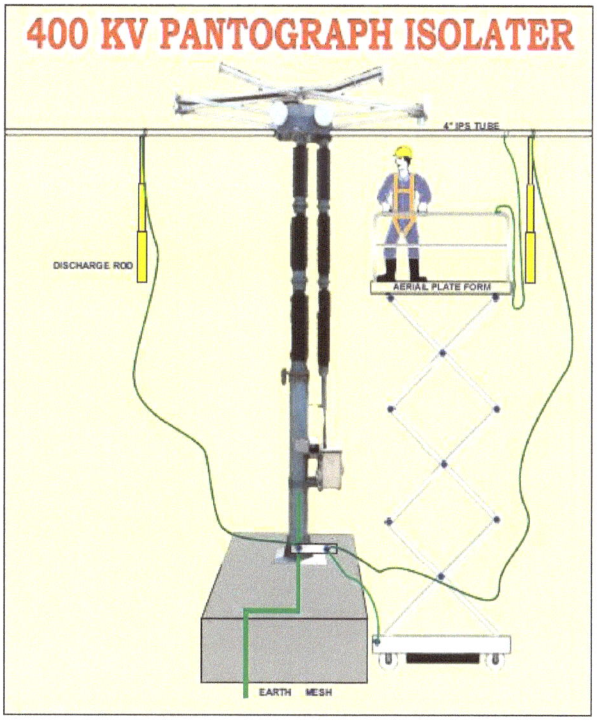

Fig- 4: Use of Elevated work Platform (EWP). [5(2)]

Fig-5: 400kV Pentagraph isolator in open condition.

4.0 Recommendations;

a. "Induction Detector" with insulated stick, should be made available in each 132kV & above EHV S/Ss and should be in variable used to verify if the equipment, Bus, EHT feeders are properly earthed and do not carry any induction.

b. Elevated Work Platform (EWP) for other works in EHV S/Ss in 220kV and 400kV Switch yards should be invariably used establishing equipotential status of the EWP with the work-area as depicted in the figure No.6 [5(ii)].

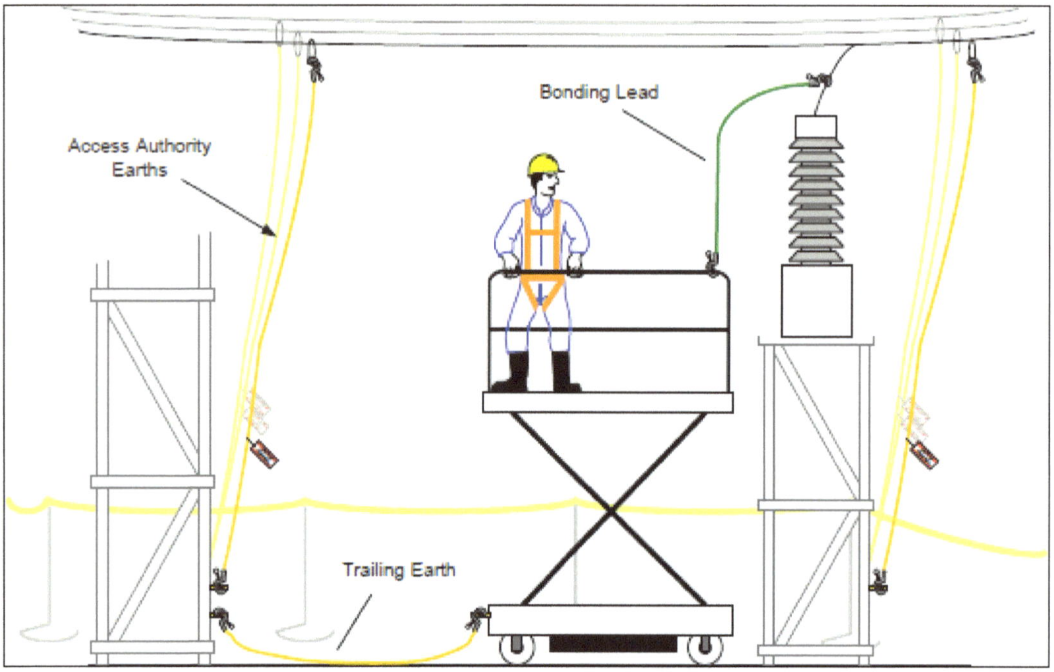

Figure 6 [5(ii)]

c. Due to catching of fire in the Transformers, Catching of fire & bursting of other outdoor equipment like EHV CTs /PTs/CVTs which may result into fatal/no-fatal accidents of any human being walking in the vicinity in the switchyard. To avert such mishaps regular periodical maintenance of the Transformers and other outdoor equipment should be scrupulously done regularly.

d. Integrity of Earthing System should be checked regularly as defined by the utility.

e. Training for workmen on Electrical Safety should be arranged including use of first aid kit/ medicines and Artificial respiration (mouth-to- mouth resuscitation).

5.0 Conclusion:

Our paper focuses on fatal & non-fatal accidents in HV & EHV level sub-stations and measures for prevention thereof up to and including 400kV only. However, the Power Professionals in different wings/discipline in different organisations, Captive Power Plants/ IPPs /Private Entrepreneurs must follow the practice on following counts and visualize the irreparable loss due to fatal and non-fatal accidents to the organisation and to their families at large.

a. It should be our endeavour to have zero tolerance for any accident. A placard should be displayed in each EHV S/s. as depicted in fig. 7

Figure 7

b. Even one death should be the cause of severe pain and alarm to all of us.
c. The plight of families suffering on this account is un-imaginable.
d. Strict monitoring should be exercised and analysis on the scale of accidents on. Regular basis should be done and very effective preventive measures should be taken to bring it to zero level.

6.0 References;

i. IE Rules-1956.
ii. Safe work practices on High voltage Substation Apparatus by TRANSGRID.

6.0 Authors;K.K.Murty, Santosh Dubey & S.K. Chaturvedi.

SECTION-2

ARTICLES RELATED TO TRANSFORMERS.

1. Power Transformers & Reactors-An Over-View

1.0 Introduction (With respect to MPPTCL);

The Power Transformers are very important and costliest equipments in any Power Grid network including the Grid Network of M.P. Power Transmission Co Ltd. They are basically installed in Grid Sub stations. The Grid network of MPPTCL consists of 400kV and 220kV and 132kV transmission lines and the Power Transformers. The Power Transformers in Grid S/S are used for transforming voltages from one level to another without changing the supply frequency (*Definition of the Transformer*). In Our system the Voltages are transformed from 400kV to 220kV, 220kV to 132kV and from 132kV to 33kV.TheTransformers are generally are 3 Phase units(Physically smaller in size than the older units due to optimization of their design etc.), However, the older units of 220kV and 400kV voltage level are single phase units Connected to form a bank of 3 phase unit;

Example: //33kV, 3x40MVA,(1-Ph units) which constitute,220/132kV, 120MVA Transformer bank Mitsubishi, Japan Make and BHEL make Transformers in 220kV S/s Jabalpur, Likewise in 400kV Indore S/s the 400kV/220kV,315 MVA Transformers (2-No.banks) too are 3x105 MVA(1-Ph.units) each. Their physical sizes are huge.

1.1 Capacity of Transformers in MPPTCL:

i. 400/220kV- 315 MVA (Auto Transformer
ii. 220/132kV-100 MVA, 120 MVA and 160 MVA 3 Ph units
iii. 132/33 – 20, 40,50 & 63 MVA.

1.2 Ref fig: 1, It is of 3 Ph Transformer and fig: 2, It is of Single Ph Transformer.

Fig:1 - 3 Ph Transformer.

Fig: 2- 1Ph Transformer.

2.0 Theory of Transformers:

Note: The Trainees are deemed to have learnt about the principle of the transformer including its EMF equation (E_2= 4.44 x f x Ømax x N_2) and the vector diagram etc in their curriculum at their Diploma level courses, therefore the same have not been included in this lecture. However a sketch defecting the same is as below.{fig:3(a)}.

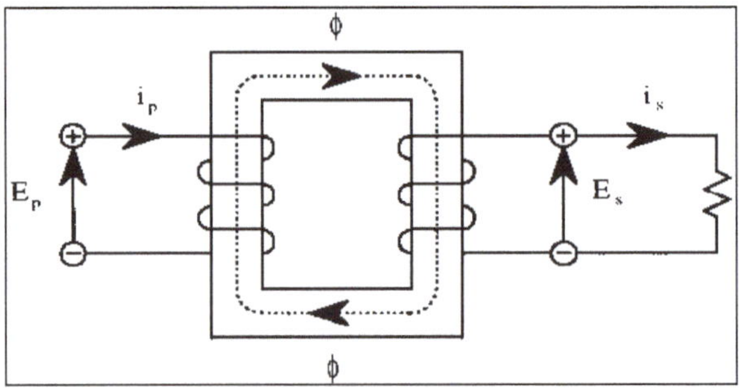

Fig; 3.(a)

1.0 Ratio of Transformation:

The Transformer's Voltage ratio, Current transformation ratios are asfollows;

$$\frac{kV1}{kV2} = \frac{I2}{I1} = \frac{N1}{N2} = k$$

Where kV_1 - Primary Voltage, kV_2-secondary Voltage,

I_1- Primary current, I_2- Secondary Current and N1- Primary No. of turns, N2-Secondary No. of turns.

k- Transformation ratio.

Example: kV_1=132kV, kV_2= 33 kV, the transformation ratio k would be132/33 = 4 and if, I_1=87.5 A then the Current of LV side would be 87.5 x 4=350 A.

4.0. Calculation of Full Load current of a 3 ph Transformer:

As VA= V x I, Therefore MVA= kV x I x10^{-3}

Therefore $I = \frac{MVA}{\sqrt{3}kV \times 10^{-3}} = \frac{MVA \times 10^3}{\sqrt{3}kV} A.$

4.1 Example: Calculate Full Load Current of 132kV side of 160 MVA, 220/132kV Transformer?

$$I_{FL} = \frac{MVA \times 10^3}{\sqrt{3}kV} = \frac{160MVA \times 10^3}{\sqrt{3}132kV} = \frac{160 \times 1000}{\sqrt{3}132} = \frac{16000}{1.732 \times 132}$$

$$= \frac{16000}{228.634} = 699.839 \text{ say } 700A.$$

5.0 Components and accessories of the Power Transformer:

5.1 Main Tank: It is made of suitable grade steel. It should be strong enough to wiyh standthe whole assembled Transformer with Oil with a crane without any damage or distortion.The main Tank body should be capable of withstanding a vacuum of 760mmHg.

There are 2 types of Transformer tanks. (i) Conventional one with top cover and (ii) Bell type.

Bell type is used for large transformers having very heavy core- winding assembly. The shape resembles the shape of a bell therefore the name. The core – coil assembly is installed and fixed on a trolley like bottom portion of the Transformer main tank and the Bell shaped cover is placed and fixed on to the lower trolley like portion of the tank with suitable Gasket in between the Trolley portion and the Bell shaped top cover. The top cover must be provided with Turrets for the bushings, PRD/ PRV (min 2 no.s), Appropriate kV rating bushings slot /box for 3 No.s bushings for core earthing arrangement and OTI & WTI pockets and man holes(2-3 Nos)

The Bell type cover should have slanting top (Hut- roof type slants).

5.2 Core: It is made of high grade CRGO (Cold rolled Grain oriented) silicon steel, having very thin laminations with Carlite insulation coating over it. The thickness of the lamination should be 0.27 mm. The Laminations are formed into core by arranging them one over the other in 3 - limb formation or in 5 – limb formation as per the design of the Transformer. (Ref: fig. 3 showing core under assembly).

Fig: 3 Core of 315 MVA,400/220kV Transformer under assembly.

Hi-B Grade Steel of 0.27mm thickness is mostly used for the core application.The 27MOH has max loss of 1.03watts /Kg@1.7T/50Hz.

Note: The Core is to be earthed only at one place. To ensure its integrity MPPTCL (Erstwhile MPSEB) has insisted the manufacturers to bring 3 leads, one each from the core, from the end frame and from the main tank to the 17.5 kV bushings separately installed on the top of the Transformer and then the connection to earth shall be made externally.

5.2. Windings: The Windings are wound with of electrical grade copper strips (with Kraft paper insulation wrapped over the copper strips). The coils are formed in the cylindrical form and are inserted over the core limbs. (ref.fig:4C).The number of turns and the cross section area is decided as per the current carrying capacity /Capacity of the Transformer

(Design parameter). (ref, fig: 4A&4B).

Figure 4(A): Winding being wound.

Fig: 4B: Core-Coil Assembly.

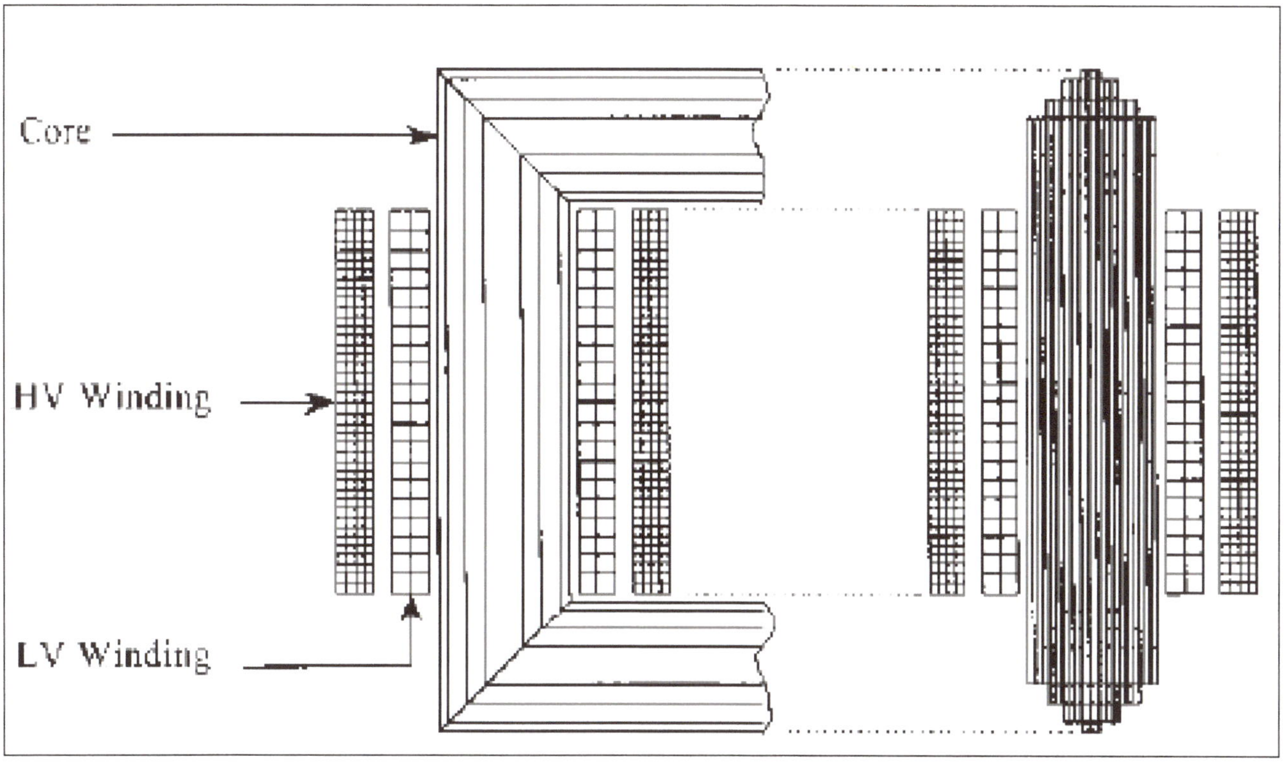

Fig:4C: Placement of winding with respect to the core limb.

4.3 Insulation Materials: (i) Following insulation materials are used;

- Press Board.
- Kraft Paper.
- High density particle board.
- Glass fiber boards.
- Paper and adhesive tape rolls.

4.3 (ii) Liquid insulation: Mineral oil confirming to IS: 335-1993. The Transformer oil may have following major parameters;

Sl No.	Particulars of tests	Limiting Values
1	Density(gm/cm$^{3)}$	0.89(Max)
2	Tan Delta at 90°	0.002 (Max)
3	Resistivity(Ohm-cm) at 90°	35x10^{12}(Min)
4	Water Content(ppm)	50(Max)
5	BDV(kV)rms	
	a.(New oil unfiltered)	30kV(Min)
	b. After filteration	60kV(Min)

The Transformer Oil plays dual role. (i) Provides insulation inside the Transformer & (ii) acts as a cooling medium too.

For ready ref. the Oil Parameters as per IS 335 are tabulated here under;

IS 335 : 1993

Table 1 Schedule of Characteristics

(*Clauses* 5.1 *and* 9.1)

Sl No.	Characteristics	Requirement	Method of Tests	Remarks
(1)	(2)	(3)	(4)	(5)
i)	Appearance	The oil shall be clear and transparent and free from suspended matter or sediments	A representative sample of the oil shall be examined in a 100 mm thick layer at 27°C	—
ii)	Density at 29.5°C, *Max*	0.89 g/cm³	IS 1448 [P : 16] : 1977	*See* NOTE 1
iii)	Kinematic viscosity, *Max*, at		IS 1448 [P : 25] : 1976	—
	a) 27°C	27 cSt		
	b) 40°C	Under consideration		
iv)	Interfacial tension at 27°C, *Min*	0.04 N/m	IS 6104 : 1971	—
v)	Flash point Pensky-Marten (closed), *Min*	140°C	IS 1448 [P : 21] : 1970	—
vi)	Pour point, *Max*	−6°C	IS 1448 [P : 10] : 1970	—
vii)	Neutralization value			
	a) Total acidity, *Max*	0.03 mg KOH/G	IS 1448 [P : 2] : 1967	Alcoholic potassium hydroxide solution of 0.02 N should be used in place of 0.1 N indicated in test method
	b) Inorganic acidity/alkalinity	Nil	do	
viii)	Corrosive sulphur	Non-corrosive	Annex B	—
ix)	Electric strength (breakdown voltage)		IS 6792 : 1972	—
	a) New unfiltered oil, *Min*	30 kV (rms)		
	b) After filtration, *Min*	If the above value is not attained, the oil shall be filtered 60 kV (rms)		*See* Note 2
x)	Dielectric dissipation factor (tan δ) at 90°C, *Ma*	0.002	IS 6262 : 1971	*See* Note
xi)	Specific resistance (resistivity)		IS 6103 : 1971	*See* Note 2
	a) At 90°C, *Min*	35 × 10^{12} ohm-cm		
	b) At 27°C, *Min*	1 500 × 10^{12} ohm-cm		
xii)	Oxidation stability		Annex C	—
	a) Neutralization value after oxidation, *Max*	0.4 mg KOH/g		
	b) Total sludge, after oxidation, *Max*	0.1 percent by weight		

Table 1 (*Concluded*)

(1)	(2)	(3)	(4)	(5)
xiii)	Ageing characteristics after accelerated ageing (open beaker method with copper catalyst)	—	IS 12177 : 1987 Method A	—
	a) Specific resistance (resistivity)		IS 6103 : 1971	
	1) at 27°C, *Min*	2.5×10^{12} ohm-cm		
	2) at 90°C, *Min*	0.2×10^{12} ohm-cm		
	b) Dielectric dissipation factor (Tan δ) at 90°C, *Max*	0.20	IS 6262 : 1971	
	c) Total acidity, *Max*	0.05 mg KOH/g	IS 1448 [P : 2] : 1967	
	d) Total sludge, *Max*	0.05 percent by weight	Annex A of IS 12177	
xiv)	Presence of oxidation inhibitor	The oil shall contain anti-oxidant additives	IS 13631 : 1992	*See* Note 3
xv)	Water content, *Max*	50 ppm	IS 13567 : 1992	—
xvi)	SK value	Under consideration	Annex D	—

NOTES

1 Density of the oil may be measured at ambient temperature and converted to 29.5°C using the following equation:

$$_{29.5}\rho = \rho_t \left[1 + X (t - 29.5) \right]$$

where

t = ambient temperature (in °C),

ρ^t = density measured at temperature t, and

X = correction factor (equal to 65×10^{-5} approximately).

5.4. Bushings: The Bushings provide insulation to the HV, IV and LV Leads coming out from the transformer tank, to the transformer tank. The Bushings used in the transformers are designated as for Oil to Air application. The Bushings are manufactured with hollow Porcelain or ceramic body containing insulation material inside it and having rain shades outside it. These rain shades increase the creapage distance of the Insulation. The Condenser bushings provide uniform potential gradient from live conductor to the fixing flange. Presently most of the HV bushings used are OIP (Oil Impregnated Paper) Condenser bushings. (ref: figure below: Regarding concept thereof)

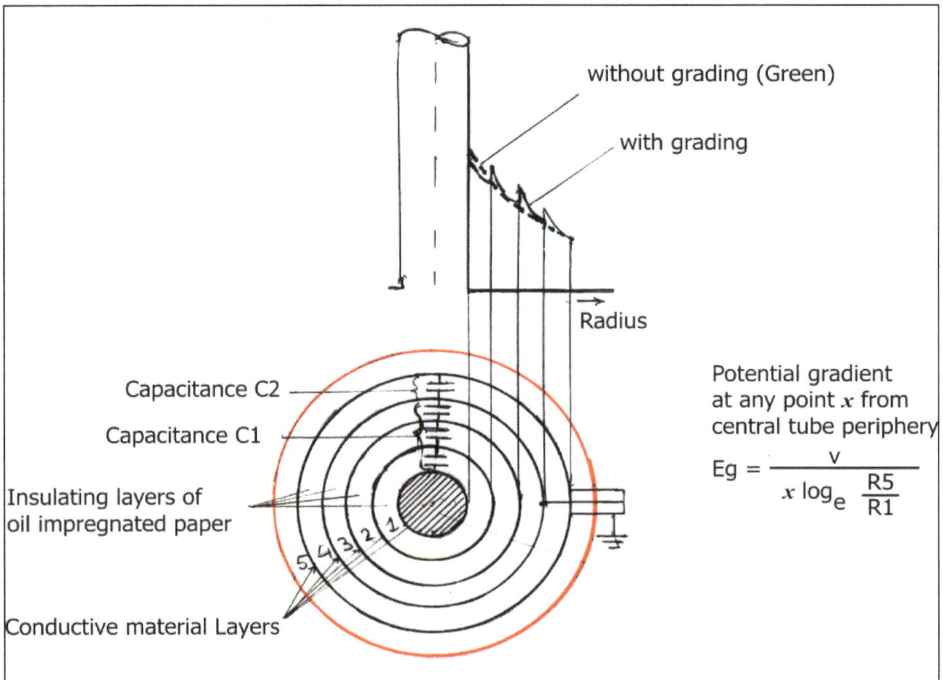

Fig. 1 (a)

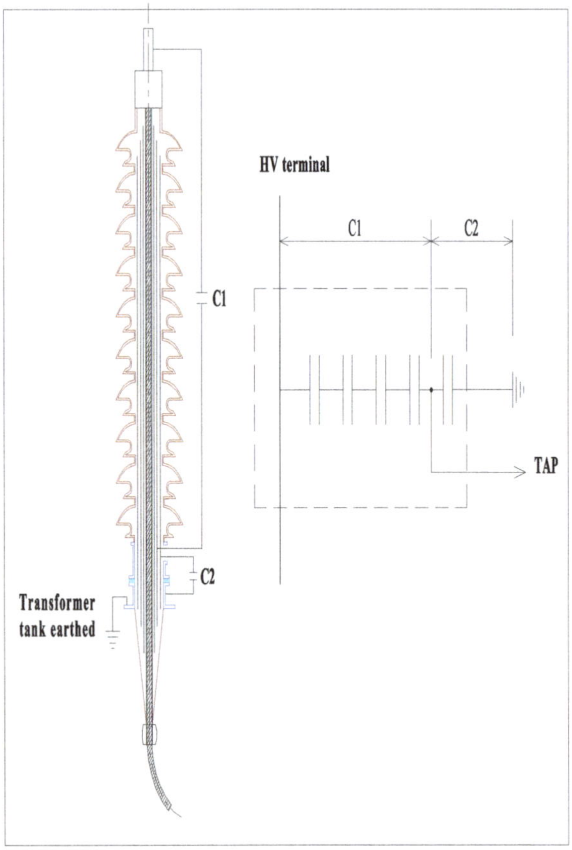

The Bushings are generally made to the highest system voltage ratings. For 400kV - 420kV Bushings, 220kV - 245k V Bushings, for 132kV-145kV Bushings and for 33kV -36 kV bushings.

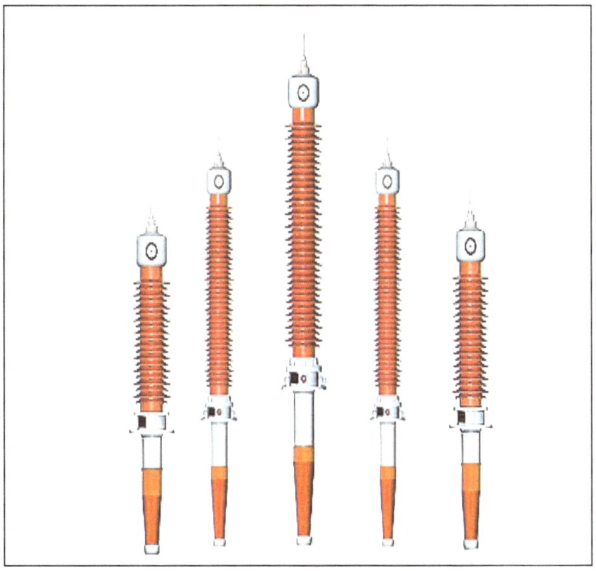

Fig: 5: OIP bushings.

5.5. Conservator and Air cell: It is connected to the main tank of the Transformer through a pipe with Buchholz relay fitted in between, contains transformer oil to a certain level. It is a cylindrical vessel fitted at the top of the Transformer with equalizer pipes. Its function is to give room to the oil filled in the Transformer main tank for expansion and contraction. It has OIL Level Gauge (Indicator). In Larger Transformers MOG (Magnetic Oil level Indicator) is used and in smaller units oil level is indicated through transparent prismatic glass indicator.

The Aircel is made of synthetic flexible material. It is placed inside the Conservator and is inflated with air inside. It provides separation of the Transformer oil from the atmosphere. The breather breaths through the air cell only. (Ref: 6,7).

Fig: 6: Conservator.

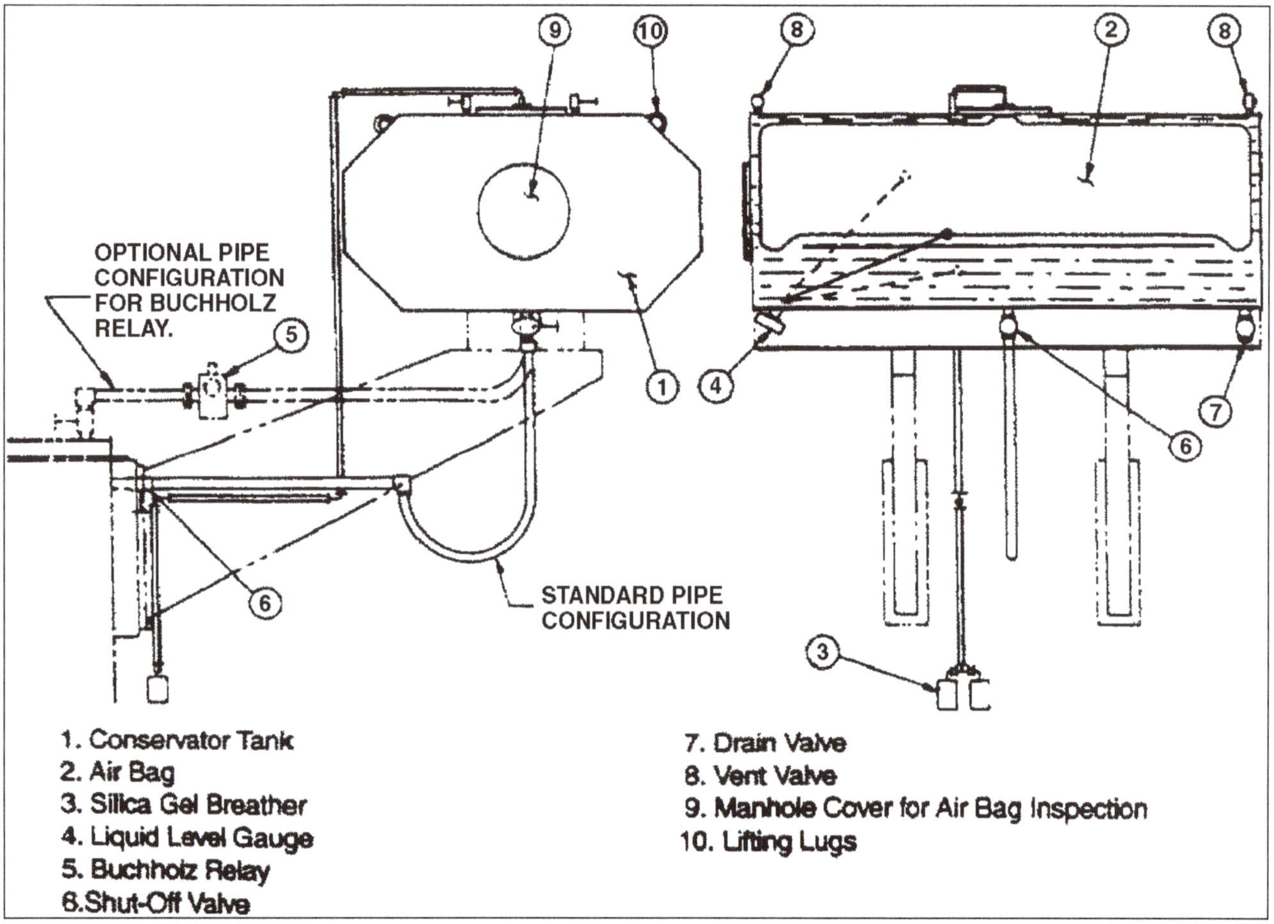

Fig: 7 Air cell

5.7. Buchholz Relay: This is a float operated relay. The floats are hinged at one end. In the normal condition the transformer oil is occupies he space inside the Buchholz relay and the floats float therein due to buoyancy. As soon as some gas is developed due to the internal fault in the transformer, the same travels up to the conservator. It gets trapped in the Buchholz relay making it operate. There are two floats in the relay, one for alarm and one for Trip. Buchholz relay also operates when the Transformer Oil level goes down lower than the Buchholz relay.

Note: (i) ensure that fitting of the Buchholz relay is such that the arrow embossed on the Buchholz relay is pointing towards the conservator. (ii) The angle of inclination of the pipe connecting Buchholz relay and the Coservator should be inclined at about 5° with the horizontal plane.

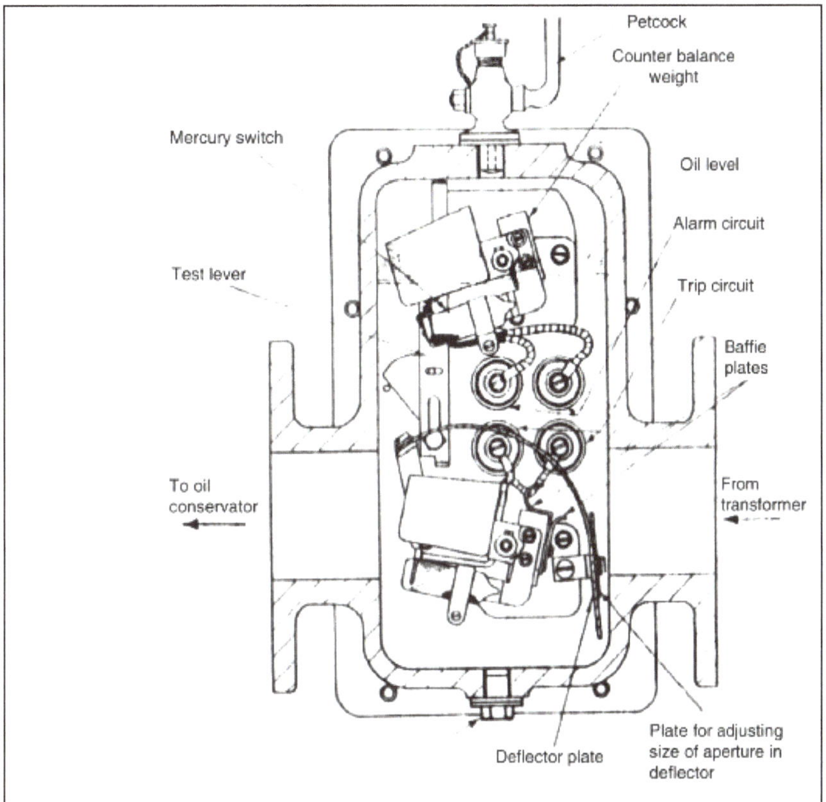

Figure 12
Details of the Bucholtz relay construction

Fig: 8 Buchholz Relay

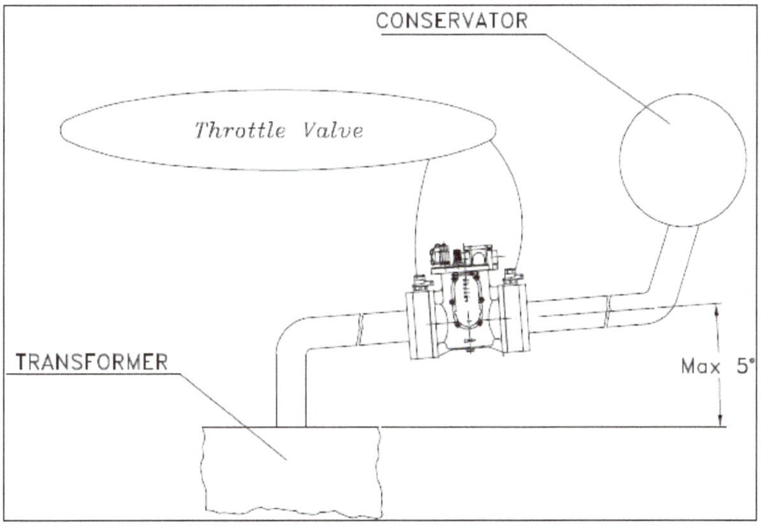

Figure 9: Location of Buchholz Relay.

5.8. On Load Tap Changer (OLTC): It is provided for raising or lowering the taps to match the supply voltage at HV side so as to transform the same to LV to maintain uniformity of the LV side voltage profile. In EHV system generally OLTC are used. The range of variation in the steps of 1.25% from (+) 5% to (–)10 % are generally provided. The Tap variation is through motor driven mechanism.

The basic concept of on load tap changer operation is shown in fig No 10, The Tap operation takes place such that there is no break of continuity of supply while operation of Tap changing is in progress a transit resistances are introduced in between, during the operation.(Ref Figure 10).

The upper half is called **DIVERTER SWITCH** and the lower part is called the **SELECTOR SWITCH** (Ref: Figure 11). The whole assembly is inserted inside the main tank of the transformer. The selector portion is immersed in the transformer Oil inside the main tank where as the upper portion wherein the taps are actually changed through spring loaded mechanism and also contacts make and break, is separate and the oil therein is also separate from the main tank oil.

A RTCC (Remote Tap changer Control Cubicle) is also installed duly wired up for remote control of OLTC from the control room.

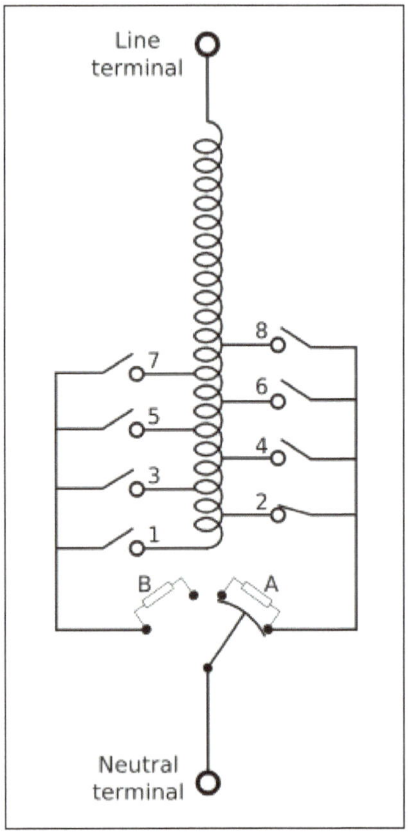

Fig;10. Concept of OLTC

Fig: 11.MR type OLTC unit.

Externally mounted Type on-load Tapchangers: These type of OLTCs are used for smaller Power Transformers The tap leads are connected to the OLTC terminals through barrier board made of Bakelite which contains small bushings which insulate tap terminals.(ref fig:12)

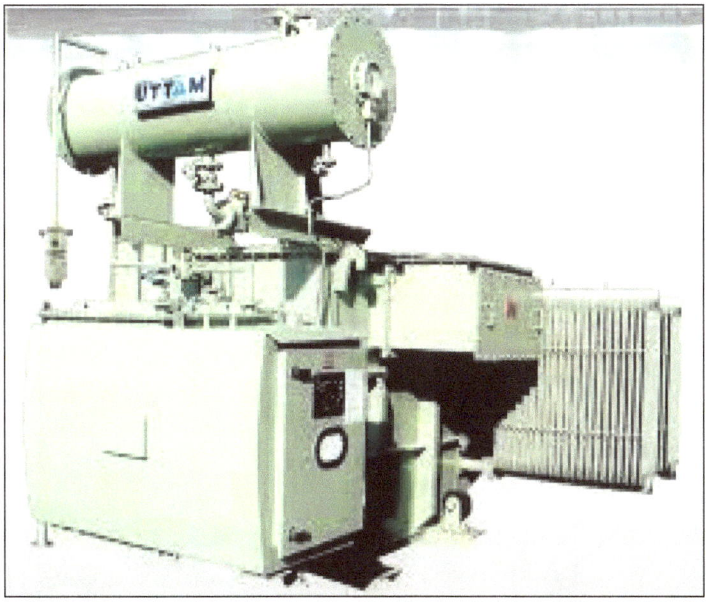

Fig: 12. Externally mounted Type Tap changer

5.9. Oil Surge Relay (OSR): The protective Oil Surge Relay (OSR) is used to protect the malfunction developed inside the On Load Tap Changer (OLTC). It is installed in the pipe between OLTC head and oil conservator.(Fig: 13)

Fig; 13 OSR

5.10. Radiators: Their function is to provide cooling to the hot oil circulating through them while transformer is in service and feeding the load. They are generally connected to the body of the Transformer. In very big transformers say 315 MVA transformer, the Radiators are separately mounted on the structure with piping arrangement to the Transformers.(Fig:14 A&B).

The radiators of very large Transformers are installed separately on a pedestal/structure as designated by group etc.

Fig14 (A) Radiators

Fig: 14 B(separately mounted Radiators on pedestals)

5.11. Cooling Fans: They are used for accelerating the rate of cooling. They are required to blow / blast of air over the radiators. The cooling fans start is set at 50°C, the provision for setting is available in Oil temperature Indicator. (fig:15).

Fig;15- Cooling fans mounted on radiator

5.12. Transformer Oil Pumps: In OFAF type Cooling of large transformers with separately mounted radiators, Oil pumps of adequate size and rating are installed in the pipe between the Lower portion of the Radiators and the bottom portion of the main tank of the Transformer ie Cold oil is circulated in the Transformer through the pumping by the Oil pumps.

5.13. Silica-gel Breather: It contains Silica-gel granules (dehydrating agent) and it is connected to the Air cell/ Conservator through a pipe. The conservator breathes-in air from the atmosphere through silica-gel breather (dehydrating granules are normally blue in colour in dry condition) when the transformer oil contracts as the load reduces and at also contracts at night hours when the ambient temp goes down and vice–versa, when the Transformer oil expands due to raise in temperature inside the Transformer due to loading of the transformer or due to hot weather. (Ref Fig: 16..)

Figure 16.:Silica-gel breathers.

5.13. PRD/PRV: Pressure Release Device is a spring loaded lid on the Transformer (such Device are generally provided on the Transformer main Tank at least at two locations. The operating pressure thereof is 0.49kg/cm^2 ±.07kg/cm^2.This will operate if the pressure inside the transformer increases consequent to some fault/ arching inside the transformer.(Ref fig; 17)

Fig: 17. PRV

5.14. Oil Temperature Indicator (OTI): This indicates the oil temp of the transformer oil inside the main transformer. The pocket for the thermometer is oil filled and provided at the top of the transformer tank. The settings are generally provided w.r.t. temperature rise mentioned on the R&D plate. The OTI meter is housed inside a marshalling box which in turn is mounted on the transformer body.

5.15. Winding temperature Indicator (WTI): Its function is to indicate the temperature of the winding similar to OTI. The WTI is supplemented with a WTI CTs which are connected in the LV, HV side of the Transformer. (Ref; R&D plate Drg, The CT for supplementing the temp readings are provided). The settings are generally provided w.r.t. temperature rise mentioned on the R&D plate. The WTI meter is housed inside a marshalling box which in turn is mounted on the transformer body. (Ref Fig:18)

The settings generally provided are as follows;

SN	Particulars	Fans Start	Alarm	Trip
1	OTI	50°C	80°C	90°C
2	WTI	-	90°C	100°C

Figure 18.: OTI & WTI.

5.16. Thermo Siphon: It's a sort of a filter provided in the transformer to absorb the moisture from the transformer oil while it is in service. It contains dry alumina.

6.0 Difference in auto-transformer and the conventional two winding ransformer:

The auto-Transformer does not have separate winding but its lower voltage lead is tapped from the main HV winding. Auto transformers are used where the transformation ratio is less than 2, such transformers are economical than two winding transformer with the same transformation ratio and capacity.

7.0 Forming 3 phase bank with 3 –nos. 1ph transformers. The 3 phase transformers are generally Star/star connected windings with their neutral grounded. The single phase transformers are connected between Phase and the Ground.3 nos of single phase transformers thus form star point that is neutral in the ground. Since the phase to phase voltage is required for Power transmission purposes, Thus to obtain the same the Voltage ratio of the single phase transformer has to be Voltage divided by root 3 for primary as well as secondary Voltages. Thus as an example let's take the example of 220kV. The transformers have to be /kV. (Ref Fig:19), 33 kV winding Tertiary winding is connected in delta formation.

Note: In practice the Delta tertiary is formed of the third winding of the 3 single phase Transformers over a 33 kV bus as per the vector group required by the user.

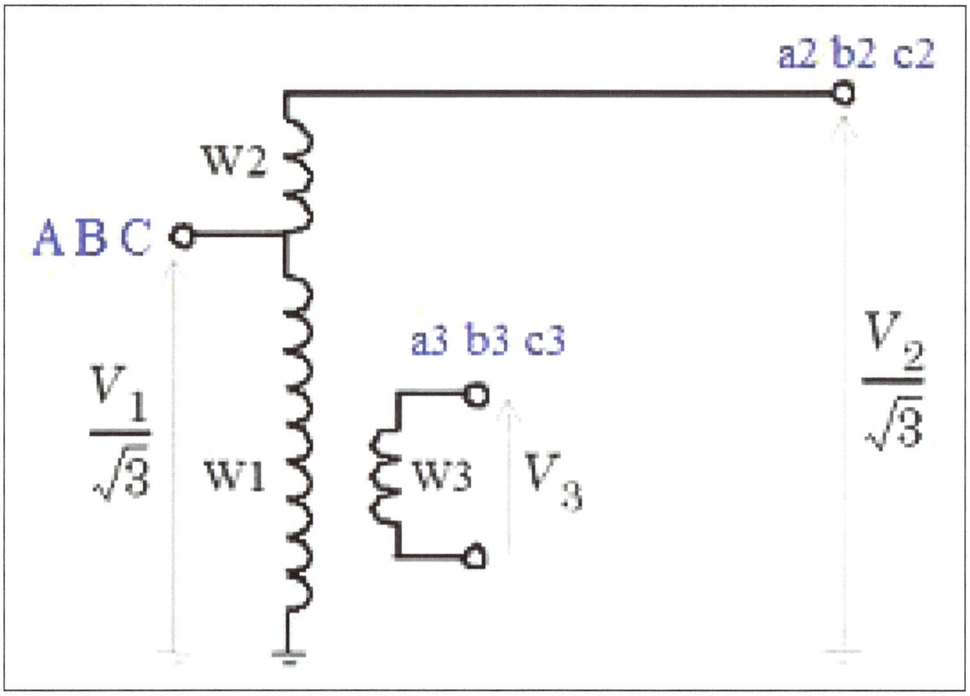

Fig: 19(A). Single Phase Auto-Trans former single line diagram.

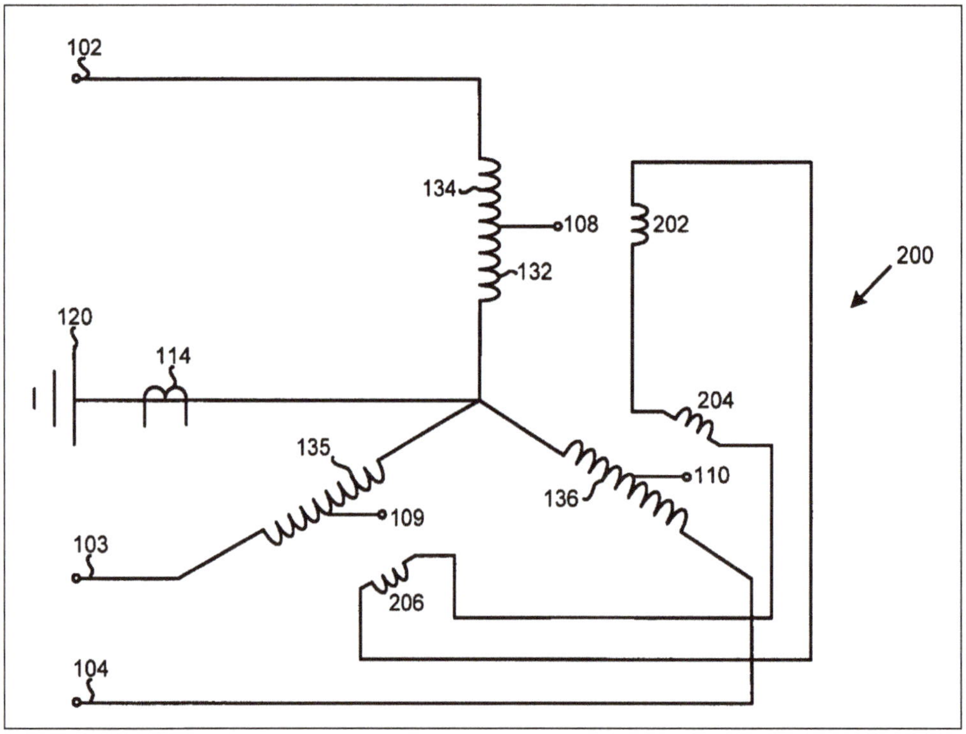

Fig: 19(B). Single line diagram of 3- Single Ph Auto Transformers connected to form 3-Ph bank.

8.0. Tertiary winding: This is the third winding in a Star -Star transformers above 100MVA capacity. This winding is delta connected and the capacity of the winding is 1/3rd that of the Transformer. This winding provides;

Suppresses Harmonic voltages.

- To limit voltage unbalances when main load is asymmetrical.
- Allows sufficient earth fault current (Zero sequence component current) to flow in the HV or LV windings for operation of protective relays.
- It can feed loads at its voltage, compensation devices like shunt capacitor banks/ reactors also be fed from this winding provided the same is designed, capacity wise for these purposes.(ref fig: 20)

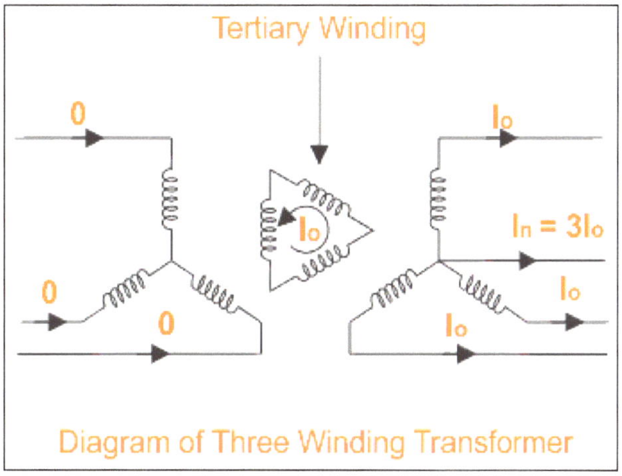

Fig: 20.

9.0 UVW Convention: All the transformers manufactured in India do not have terminal marking in RYB convention though the 3 Ph supply has RYB convention but instead they have UVW convention.

10.0 KYT (Know Your Transformer): The R&D plate (Rating and Drawing plate) of the transformer gives its self-introduction.(Ref attached Photo Copy of R&D Plate of 132/33kV,63 MVA Transformer).

Most of the parameters are self-explanatory, however, following important parameters mentioned therein are explained hereunder;

- **HV nominal voltage**-132kV.
- **LV nominal voltage**- 33kV.
- **Cooling:** ON AN/ON AF.

Oil Natural Air Natural/Oil Natural Air Forced.

Oil Natural Air Natural – Means circulation of oil inside the transformer shall be normal and cooling shall take place in prevailing atmospheric conditions.

Oil Natural Forced Air.-Circulation of oil inside the transformer shall be normal and the cooling fans would blow air on the radiators to accelerate the cooling process.

	Nominal Power(MVA)		Nominal Current(A)	
	ONAN	ONAF	ONAN	ONAF
HV	50	63	218.69	272.55
LV	50	63	874.79	1102.24

It can be seen from the table above, that the capacity of 63 MVA is attained only with fans on. Otherwise the transformer can be loaded up to 50 MVA only.

Note: In Case of **315 MVA Transformer**: Since cooling is OFAF, that means Oil forced and Air forced type. That is Oil is also forced into the Transformers with Motor – pump sets fitted in the pipes between the Radiators, separately mounted on the pedestals and bottom of the main tank.

The ratings vis a vis cooling arrangement is tabulated hereunder;

Voltage(kV)	Nominal Power(MVA)			Nominal current(A)		
	ONAN	ONAF	OFAF	-	-	OFAF
HV- 400kV	189	252	315	-	-	454.66
IV- 220kV	189	252	315	-	-	826.66
LV(Delta Tertiary) - 33kV	63	84	105	-	-	1837.02

Mean Oil Temperaturerise-45°C, means the oil temp of Transformer can rise to 45°C max above the ambient Temp.

- **Mean Winding Temp rise-** 50°C, means the winding temp of Transformer can rise to 50°C max above the ambient Temp.
- **Guaranteed No Load Loss (kW)-** 25(max), against this the actual measured one shall be engraved.
- **Guaranteed Load Loss (kW)-**140(max), against this the actual measured one shall be engraved.
- **Vector Group- YNyn0.**

This means that there is no phase shift between HV and LV voltages. Both the windings are Y N means-Star connected HV winding with HV Neutral and yn means- LV winding too is star connected with neutral.

Note; In our Grid system all the transformers are star connected with neutral solidly earthed.

Meaning of Other vector groups; (Pl.Ref fig below)

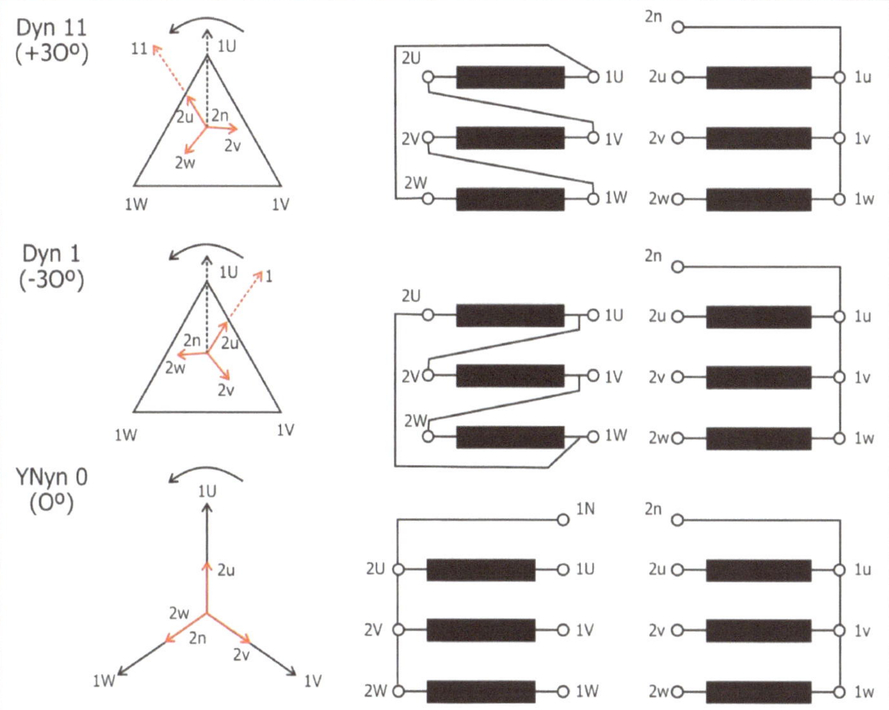

Note : (i) In MPPKVVL Only Dyn11 vector group Power Transformers are installed/are being procured.
(ii) Figure showing windings connected in Dyn1 and Ynym0 for academic interest only
(iii) Other vector group connected windings like YNd11, YNd1 etc. have not been shown.

Dyn11-HV winding is connected in delta formation whereas LV winding is in star formation with neutral, with 30° lead from the HV winding Voltage.11 indicates 11 o' clock position on the clock dial with respect to12 o'clock. Each 5 mins graduation on the clock with respect to each other is of 30°.

Dyn1- HV winding is connected in delta formation whereas LV winding is in star formation with neutral, with 30° lag from the HV winding Voltage.1 indicates 1 o' clock position on the clock dial with respect to12 o'clock. Each 5 mins graduation on the clock with respect to each other is 30° apart.

Note: 220/132kV, 160 MVA Auto transformers has vector group depicted as-YNa0d11. Here "a" stands for Auto transformer. HV and IV have 0° phase difference between them having common neutral - HV neutral N. There is a phase difference between HV and LV winding of 30°.The LV voltage leads HV by 30°.

- **Diagram of Transformer**: The diagram shows the position of the Bushings (HV, LV and Neutral) with respective windings and tap positions.1U,1V,1W & N means HV RYB terminals and HV neutral.

  2U, 2V, 2W & 2N mean LV r y b terminals and LV neutral.

Note: In case of Auto Transformers the terminals are HV, IV, LV. Here IV indicates the tapped connection from the HV windings. The neutral terminal of the HV and IV shall be N only (common terminal).

WTI CTs: WTI CTs are also show indicating their ratios.

Taps- The tap positions are also shown therein, with the location of the OLTC on the transformer top.

Note: It should be noted that generally the Tap changers are located in the HV side of the Transformer as the load current on HV side is less as compared to the LV side.

- **Variation of Voltages with OLTC operations**: Tap position Vs LV Voltages in a tabular form is shown on the R&D plate. As could be seen there from that the variation of taps are from (+) 5% to (-)15% in 1.25% variation steps on HV side which shall give raise/lower on 33kV (LV) side.

That is when the HV voltage goes down, by raising the taps, there shall be increase in the voltage on LV side.

- **Impedance Volts in%: 9.63. %** (As per actually found out for the Transformer installed at 220kV S/S Jabalpur).

Signification of % imp volts: It is the % of Terminal voltage (HV) which shall make full load current to flow in the windings if applied to the HV side terminals,

keeping the LV winding shorted together at their terminals.

Once the % Z is known, the value of fault current for dead 3 phase short at LV side can be calculated as under;

(100/9.63) x $I_{\text{full load}}$ =10.384x $I_{\text{full load}}$ = 10.384x874.79= 9084A.on HV side.

11.0. Nitrogen Injection Fire Prevention and Extinguishing System:

There are Two conditions

i. For Prevention;
   (a) Differential Relay.
   (b) Buchholz Relay.
   (c) Pressure Relief Valve.
   (d)Rapid pressure rise relay.

ii. For Extinction:;
   (a)Fire detector.(Set at 140°)
   (b) Buchholz Relay.
   (c) Pressure Relief Valve.
   (d)Rapid pressure rise relay

The N/O contacts of above protection are arranged in series. That means all contacts should make then only this system shall come in to action. As soon as this system comes in action,

- Non return valve shuts off oil supply from the conservator.
- 10% of Oil is drained by opening the butterfly Valve at the bottom of the Transformer Tank.

- N_2 (inert gas) is injected through the bottom of the transformer Tank.
- N_2 stirs the oil as it goes up and then forms an insulting layer above the remaining oil.
- N_2 keeps the oil temp below 140 °C (Flash point of Transformer oil.)

With the operation of this system the Power Transformers are saved from disastrous fire.(ref.fig: 21)

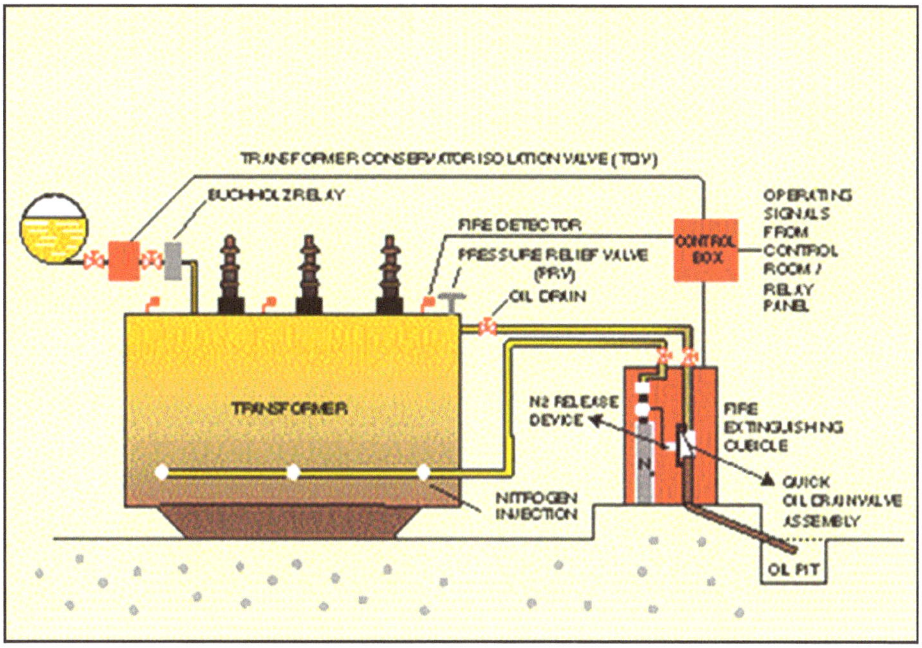

Fig: 21:Schematic sketch of N_2 Fire protection system.

EHV SHUNT REACTORS:

Variable shunt reactor, 120-200 Mvar 420 kV.

Fig: 1-3Ph EHV Shunt Reactor

INTRODUCTION: The EHV Shunt reactors and the EHV Power Transformers apparently look alike.(ref fig;1). They do not have bushings of any other voltage other than the HV (Highest System) Voltage. They are installed in the shunt position of the Long EHV transmission lines (400kV and above) at either ends of the Transmission line in a Grid network. On our 400kV Transmission Lines, 50MVar, 420kV Shunt reactors are installed.

1.0 Reference Standards: The EHV shunt reactor should confirm to IS: 5553(Reactors) and IEC: 60289(1988), Reactors.

2.0 Performance of duty: It lowers down the receiving end voltage of long unloaded/lightly loaded transmission line which ha s developed due to the phenomena of Ferranti effect ie due to the Capacitances formed between the Transmission line conductors and the ground. In other words, it performs the duty of absorption of the Reactive power so generated by the Line to ground capacitances depending upon the line geometry. Reactive power is designated and denoted as "Q" in MVAR.

3.0

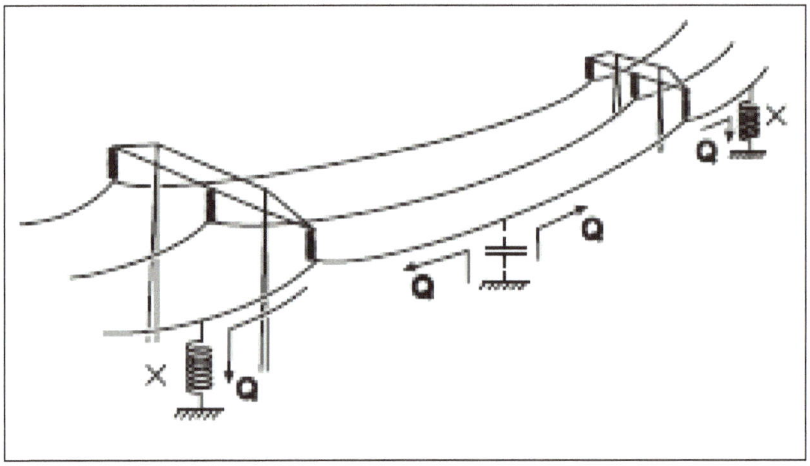

4.0

Fig:2-Shunt reactors connected on Transmission Line.

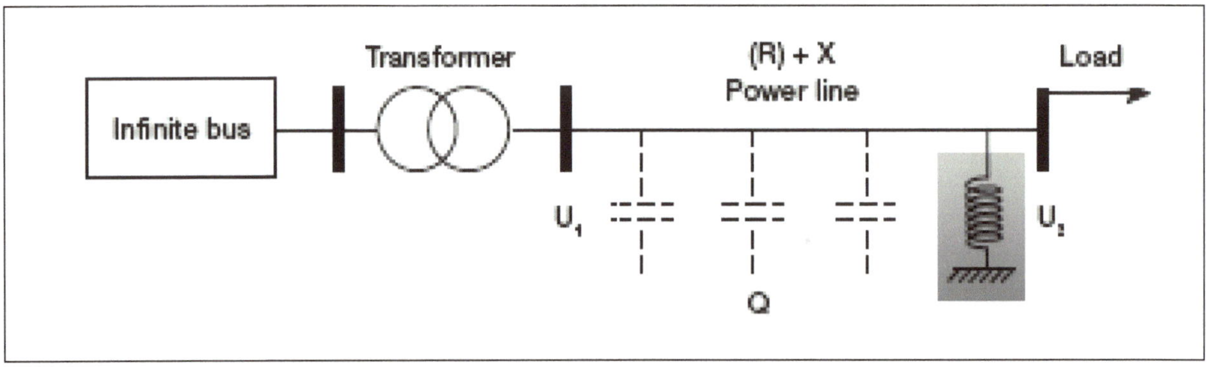

Fig:3 Shunt reactor-U2 on Transmission Line.

5.0 Theory: Unlike Transformer it has a gap in the magnetic circuit and It is known as gapped Core shunt reactor.

Constructional feature of Fixed shunt reactors

A simple schematic diagram of gapped core reactor is as under

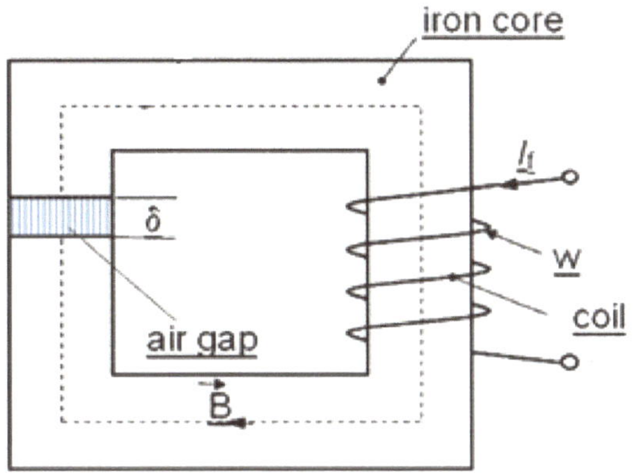

Fig4:Gapped core shunt reactor schematic drawing.

Empirical formulae for basic design of shunt reactors are as under

Parameter	Example 1	Example 2
Reactive Power (Q)	50,000 kVAr	100,000 kVAr
Nominal voltage (Un)	420 kV	420 kV
Frequency (f)	50 Hz	50 Hz
Number of phases	3	3
Nominal current (In)	=Q / (√3xUn) = 68.732 A	=Q / (√3xUn) = 137.464 A
Impedance (Z)	=Un / (√3xIn) = 3528.0 Ω	=Un / (√3xIn) = 1764.0 Ω
Reactance (X)	≈ Z = 3528.0 Ω	≈ Z = 1764.0 Ω
Inductance (L)	= X / (2·Л·f) = 11.230 H	= X / (2·Л·f) = 5.615 H
(for large reactors as Z >> R)		

From above, it can be concluded that impedance / inductance is to be decreased for increased in power rating of reactors.

Table:1

As could be seen from the table: 1, above that with decrease of Impedance/Inductance the Q rating increases.

Please refer fig; 4 above.

B=flux density,

w=No. of turns

δ= Air gap.

Inductance

In order to increase the Q rating L has to be reduced, and for reducing the L air gap δ needs to be increased.(ref;Table:1).

4.0 CORE: Following fig depicts about cores of the Shunt reactors;

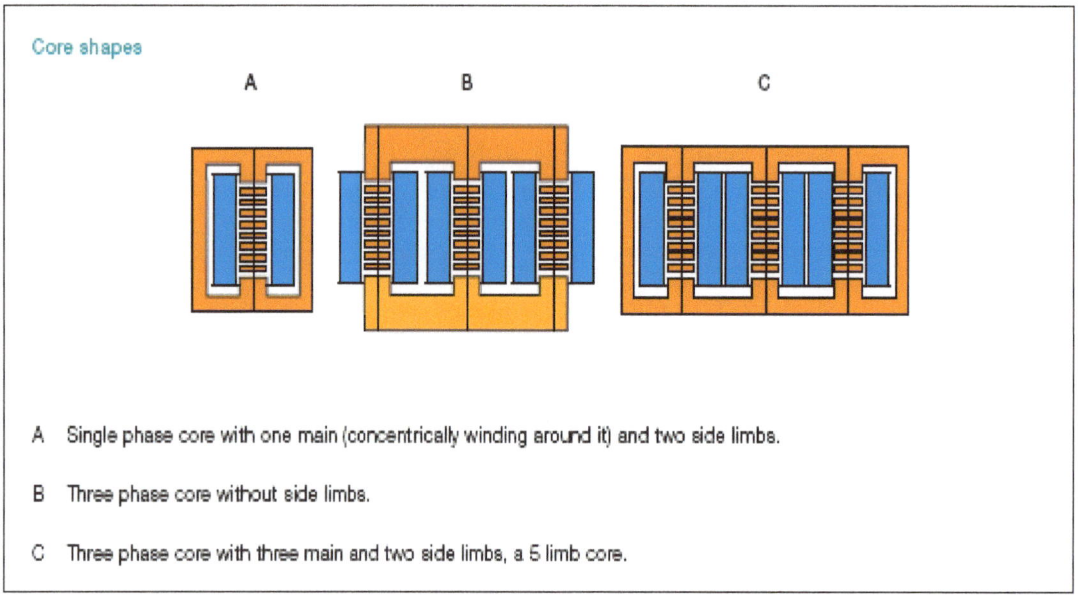

Fig:5- Gapped cores of shunt reactors.

For making Air gaps, stiff and inert spacers are provided one over other.(Ref.fig:6)below.

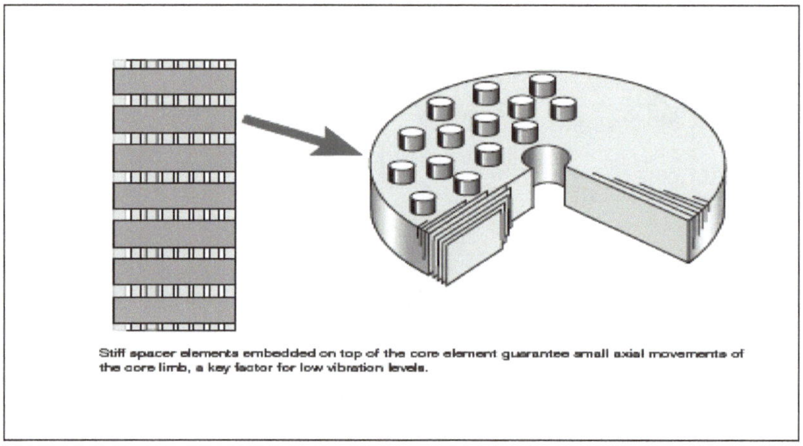

Fig: 6 Spacer Pack.

5.0 Auxiliaries and Components: The Auxiliaries and Components are exactly similar to those of Power Transformers.

6.0 Main differences between Power Transformer and EHV Reactor: The main differences of the Power Transformers and the EHV Reactors are tabulated below;

SN.	Power Transformer	EHV Shunt Reactor
1	**Has HV, IV & LV windings (IV incase of Auto Transformers).**	**Has only HV winding.**
2	**It is delta/star, Star/star, Star connected auto transformer with delta tertiary winding. All star connected windings have neutral bushings.**	**It is Star connected with neutral bushing**
3	**Has closed magnetic Core.**	**Has gapped magnetic Core.**
4	**Due to remenance flux in the core the inrush current of the Transformer is pronounced.**	**There is no remanance flux thus the inrush current is less.**
5	**Performs duty for lowering or raising the voltages as per the system voltage at the EHVS/S of the Grid net work. Like 400kV to 220kV,220kV/132kV etc.**	**Performs duty for absorbing Reactive Power in the Grid system and lower the Receiving end voltage, which increases due to Ferranti effect on the unloaded/ lightly loaded long Transmission lines.**
6	**Rating in MVA.**	**Rating in MVAR.**
7	**Load current varies as per the load demand.**	**Fixed current as per the system voltage & rating of the Reactors.**

2. Pre-Delivery final Inspection of EHV Transformers

- Points to be taken care of by the Third Party Inspector/Surveyor.

The EHV Transformers are the costliest and vital power equipments in the Transmission and Distribution system/Public-Private Generating Power houses/Large factories etc. Perhaps, No authority from Transmission/Distribution /Generating/companies/Electricity Utilities of Private entrepreneurs responsible for purchasing this equipment would ever think of waving off of pre-delivery inspections at Vender's/manufacturer's works. In the wake of shortage of staff most of the Transmission/Distribution/

Generation Utilities, like private entrepreneurs are now availing the services of TPIA (Third Party Inspection Agencies) for such inspections.

The whole responsibility lies on the inspector of TPIA, whether to accept or to reject, the offered transformer.

The activities related to the Inspection need to be divided in three parts.

i. Part (i): Preliminary preparation prior to arrival at the vendor's works.
ii. Part (ii): Activities prior to starting the inspection Job at the works of the vendor.
iii. Part (iii): Actual Inspection activities/witnessing the tests etc.

Part (i): Preliminary preparation prior to arrival at the vendor's works.

The TPIA Inspector deputed for inspection to the works of the manufacturer should go through the letter of inspection offer. Verify the date *and* place of inspection to be carried out, followed by confirmation thereof and the readiness of the offered Transformer. This is because most of the manufacturers have the tendency to forward the inspection call in anticipation of the completion of the assembly of the transformer prior to the anticipated visit of the inspector. The Specs(Specifications)in the PO(Purchase Order)/letter of Indent should be carefully studied by him and should note down the salient points thereof like GTP(General technical Particulars). Study the relevant drawings and note down the salient points thereof. This study should also include any amendments incorporated post PO and also any other points raised/pointed out by previous inspectors during stage inspections etc.

Part (ii): Activities prior to starting the Inspection Job at the works of the vendor.

Following points need to be remembered /taken care of, prior to carrying out the inspection at the works of the Vendor/manufacturer of EHV transformers.

i. Physically confirm the readiness of the Transformer and its placement in the Testing Bay.
ii. Re-study the Purchase order/indent, the GTP (General technical Particulars), the Specs, the approved drawings and the approved QAP (Quality Assurance Plan). Note the salient points and other points modified subsequently as amendments etc. List of special check points if any asked by Client to be noted.
iii. Note the national/international standards of specifications, the offered transformer and its accessories are required to confirm to, including the shade of paint specified in the PO.
iv. The Power Transformer should confirm to the Indian standard IS:2026-1977 part I, II, III, IV & V with latest amendments or IEC-60076 part I, II, III & V with latest amendments (as per clients requirement). Other points other than specified in the national/international standards as mutually agreed by the purchaser and the supplier/manufacturer.
v. If the offered transformer is one of the units of transformers ordered in a lot, it should be verified as to whether witnessing of Type and other special tests in addition to the Routine tests/ FAT (Factory Acceptance Tests) are also involved or not.

Let the offered transformer be such a unit that on which Type tests and Special tests are required to be witnessed.

i. Ask for the plan/schedule the activities of inspection date-wise. Review and approve/ modify if not suitable.
ii. Verify and confirm that all the testing instruments/equipments and all the measuring instruments/ gauges are calibrated by a NABL approved laboratories and are within validity period covering the date of inspection.
iii. Ask for compliance reports to the points raised, if any, during stage inspection approved by the purchaser. In case if the manufacturer has not forwarded the compliance report to the purchaser, the same needs to be reviewed by the inspector/surveyor prior to carrying out the inspection.
iv. Ask for approved QAP (Quality Assurance Plan).Review the same. The inspection should be carried out accordingly.
v. Whether all the national/ international Standard Specifications as referred in the PO, GTP and QAP pertaining to the transformer and its accessories with latest amendments are readily available with them.
vi. Verify that all the amendments have been incorporated and complied with in writing by the manufacturer/vendor to the purchaser.
vii. Ask the manufacturer to provide with test certificates of bought out items/accessories.
viii. Ask for Test results of the tests performed (if any) on the unit prior to the inspection.

Part (III):

Actual Inspection activities/ witnessing the Tests etc.

i. Verify the physical dimensions with respect to GAD (General Arrangement Drawing).
ii. Verify and confirm physically that all the amendments have been incorporated and

The points observed/raised by the earlier inspectors are complied with.

iii. Start witnessing the Tests chronologically, as per date wise approved program.
iv. Following Tests are carried out on the Unit being subjected to Type tests/special tests etc.;

A. **Routine Tests: As per IS 2026-1977/ IEC-60076**

i. Vector group verification (polarity and Phase relationship).
ii. Measurement of Turn Ratios.
iii. Measurement of Winding Resistance.
iv. Measurement of Insulation Resistance.
v. Measurement of No load Loss/ No Load current.
vi. Separate source High Voltage with stand Test.
vii. Load loss and Impedance Voltage Measurement.
viii. Test on aux LT windings at 2 kV rms for 60 secs.
ix. Load loss and Impedance voltage measurement.
x. Induced over Voltage withstand Test with PD measurement.
xi. Magnetic Balance Test. (Not covered in IS: 2026-1977). (Not for 1ph Transformers).
xii. Magnetic Circuit Insulation Test. (Not covered in IS: 2026-1977).
xiii. BDV of Transformer oil samples.
xiv. OLTC operational test.
xv. Oil Leakage Test.

B. **TYPE TESTS AND SPECIAL TESTS;**

i. Lightening Impulse Test.
ii. Switching Impulse Test.(Wherever asked for)
iii. Temp. Rise test.
iv. DGA Test: To be conducted before and after Temp Rise Test.
v. Measurement of Tan delta and Capacitance of Transformer Winding.
vi. Tan delta and Capacitance of HT OIP bushings.
vii. SFRA (Sweep Frequency Response Analysis).
viii. Measurement of Acoustic Noise level. (As per NEMA Std.).
ix. Measurement of Power taken by running cooling fans and oil pumps.

Measurement of Zero sequence Impedance (for 3Ph Transformer only).

C. **OTHER CHECKS AND TESTS;**

i. Pressure test on Pressure relief Device units.
ii. Operational checks of Oil surge Relay and Buchholz relay.
iii. Checks on Control Panels/RTCC.
iv. Dimensional Checks on assembled transformer.
v. Verification of Shade as per PO.
vi. Rating and Diagram plate should be checked for correctness.

Note: It has been experienced that some of the manufacturers/vendors at times get the Transformer inspected by installing OIP bushings of some other unit. The Inspector/ Surveyor may ensure that the OIP bushings earmarked for that particular unit must be used during testing/inspection.

To ensure that the unit is not meddled with the Inspector/surveyor must affix a security seal on the top cover of the transformer bolt/bolt hole.

Numbered Sticker seals with hologram duly signed by the Inspector/Surveyor should also be affixed on its accessories etc.

While preparing the final inspection report the TPIA Inspector, without any prejudice, should mention/ highlight the short comings and discrepancies observed by him during the inspection in the final Inspection report and clearly mention his opinion thereon.

3. Power Transformer Commisioning Check List

Commisioning checklist

After Satisfactory manufacturing of the Power Transformer, its assembly, carring-out type tests & special tests and review of the test result of the brought-out items in the works according to the IEC or IS and the specifications of the order & proper transport, the process of site inspection, checks, tests are initiated some before and some other after its placement on plinth and then its proper connectivity with the auxiliary equipments like circuit breakers, control and relay panel, current and potential transformers, lightning arrestors, earthing system, bus-bar etc is done.following minimum checks & tests shall be ensured by the site commisioning engineer before putting the power transformer into service.

1. Thorough study of the manufacturer's instruction manual and relevent drawings as per the IEC, IS & order's specification.
2. Thoroughly study, review and records of the factory test results of the transformer, auxiliary (bought-out) items for future references and comparision with the site results.
3. Before unloading, have an overall review of the equipment, oil level, oil leakage, gas pressure, gas leakage, any sign of breakage or damage.Check main & all auxiliary equipments for any shortage.
4. Check IR value between core, clamping & tank..
5. The transformer along with its accessories shall be erected in acordance to the drawing. In case of roller fitted transformer on rail proper stopper shall be used to prevent its accidental movement.
6. In case of oil filled transport. sample of oil shall be got tested from the standerd & accredited laboratory for review & records. In case of satisfactory gas filled transport, oil filling shall be done as per standerd practice under vacuum and oil sample be sent to te laboratory for testing.In case of gas filled transported followed by leakage of positive gas pressure from the tank the moisture content shall be ascertained by dew point measurement method for drying out process, if necessary.Drying- out could be done with the help of streamline filter machine at a vacuum around 740 to 750 mm of Hg and temperature of 60 degree C in transformer and 70 degree C in filter machine. To prevent the heat loss & improving drying efficiency insulated lagging could be provided on main transformers. After drying out process & filling of oil a sample shall be sent to laboratory.Trapped air or gas shall be removed from all the pockets.Oil samples from all parts of transformer like OLTC, radiators, tank bottom, conservator etc.shall be tested for di-electric or BDV values with help of site test kit.

7. Fixing of radiators, conservator, OLTC conservator & pronal bag separator, Buchholz relay, HV, MV & LV bushings, pressure relief device, breather, on line gas analyzer, cooling fans, fan supports brackets, OLTC & OLTC driving mechanism, control cubicles & marshalling boxes, WTI & OTI, pipes & valves be checked for alignment & correctness. Core- clamp- tank connection should be tight.
8. Test of fan control with pickup & reset temperature setting.
9. Test of pick-up & reset pressure setting of pressure relieg device.
10. A calibrated HV Insulation Tester shall be used.Measurement of insulation between Core-Clamp-Tank shall be repeated and compared.
11. Isulation test and polarizing index shall be recorded between HV, MV & IV and wrt tank and compared with the factory test results.
12. Insulation value of all the condenser bushings shall be measured, recorded and compared.
13. Check & test of control & relay panel, inter connected control cable.
14. CTs testing for ratio, polarity and its circuit for relays and metering secondary and primary injection to ensure correctness of operational& stabilty of differential for through faults.
15. With the standerd supply voltage i.e. 3 Ph.415 V following low voltage electrical tests shall be performed.

i. HV Magnetizing current, 3-Phase, 3-Wire Method

Supply Volts	**Tap No.**	**1U-Phase**	**1V-Phase**	**1W-Phase**
1U-1V.........	....	....	....	....
1V-1W.........	....	....	....	....
1W-1U.........	....	....	....	....

ii. HV Magnetizing current, Single Phase, Method

Supply Volts	**Tap No.**	**1U-Phase**	**1V-Phase**	**1W-Phase**
1U-N.........	....	....	....	....
1V-N.........	....	....	....	....
1W-N.........	....	....	....	....

iii. IV Magnetizing current, 3-Phase, 3-Wire Method

Supply Volts	**Tap No.**	**2U-Phase**	**2V-Phase**	**2W-Phase**
2U-2V.........	....	....	....	....
2V-2W.........	....	....	....	....
2W-2U.........	....	....	....	....

iv. IV Magnetizing current, Single Phase, Method

Supply Volts	**Tap No.**	**2U-Phase**	**2V-Phase**	**2W-Phase**
2U-N.........	....	....	....	....
2V-N.........	....	....	....	....
2W-N.........	....	....	....	....

v. HV & IV Megnetic Balance Test at Normal Tap No……..

Supply Volts	1U-N	1V-N	1W-N	2U-N	2V-N	2W-N
1U-N………	………..	………..	………..	………..	………..	………..
1V-N………	………..	………..	………..	………..	………..	………..
1W-N………	………..	………..	………..	………..	………..	………..

vi. L

LV Megnetic Balance Test at Normal Tap No……..

Supply Volts	3U-3V	3V-3W	3W-3U
3U-3V ………	……………	……………	…………….
3V-3W ………	……………	……………	…………….
3W-3U ………	……………	……………	…………….

vii. Ratio Test By………..Make……….Type………Meter Sl.No………

Tap No.	Calculated Ratio	Measured Ratio		
		1U-N/2U-N	1V-N/2V-N	1W-N/2W-N
1	………………..	………………	……………...	………………..
2	………………..	………………	……………...	………………..
3	………………..	………………	……………...	………………..
.	………………..	………………	……………...	………………..
.	………………..	………………	……………...	………………..
.	………………..	………………	……………...	………………..
15	………………..	………………	……………...	………………..
16	………………..	………………	……………...	………………..
17	………………..	………………	……………...	………………..

viii. Core Insulation Testing

Test Voltage:-………………kV Megger……………..Make, S.No………..
Ambient Temperature:-…………………………°C..

1	**Core toTank……………………………….**	**………………………..…MΩ**
2	**Core to Core Clamp……………………**	**………………………..…MΩ**
3	**Core clamp to Tank …………………..**	**………………………..…MΩ**

ix. Leakage current Measurement at Tap No……….

3 Phase Shorting Done on LV Side and Phase

Supply Given on HV Side

Voltage Given On HV Side	Current Measured On HV Side	Current Measured
1U-1V………………………..	**1U………………………………**	**Core Clamp-Tank**
1V-1W………………………..	**1V………………………………**	**Core-Core Clamp**
1W-1U………………………	**1W……………………………..**	**Core -Tank**

x. Calibration of Temperature Indicators

i. Oil Temperature Indicator – Done on

ii. Winding Temperature Indicator

c. HV…………………Phase Done On………………..

d. IV………………….Phase Done On………………..

e. LV…………………Phase Done On………………..

xi. 3Phase AC Supply Given On HV with 3Phase Shorting on IV Side

SUPPLY VOLT AGE	TAP. NO.	HV Current in (AMP)			IV Current in (AMP)			2N-SC Ternimal		WTI CT Current in (mA)		
		1U	1V	1W	2U	2V	2W	N CONNECTED TO EARTH	N NOT CONNECTED TO EARTH	HV-PH	IV-PH	LV-PH
1U-1V	1											
	2											
	3											
	4											
	5											
1V-1W	6											
	7											
	8											
	9											
	10											
	11											
1W-1U	12											
	13											
	14											
	15											
	16											
	17											

xii. Vector Group and Polarity Test

a. **YNa0d11**

3Phase supply Given on HV side (with 1U-3U Shorted):-

Voltage Given	
1U-1V	**...............Volts**
1V-1W	**...............Volts**
1W-1U	**...............Volts**

Voltage Measured	
1U-1N	**...............Volts**
3W-N	**...............Volts**
1U-3W	**...............Volts**
1U-2U	**...............Volts**
2U-N	**...............Volts**
2W-3W	**...............Volts**
1W-3V	**...............Volts**
3W-2V	**...............Volts**
3W-2W	**...............Volts**

Conditions for YNa0d11

1. 1U-N = (1U-3W)+(3W-N)
2. 1U-N= (1U-2U)+(2U-N)
3. 1W-3V>1W-3W

   3W-2V=3W-2W

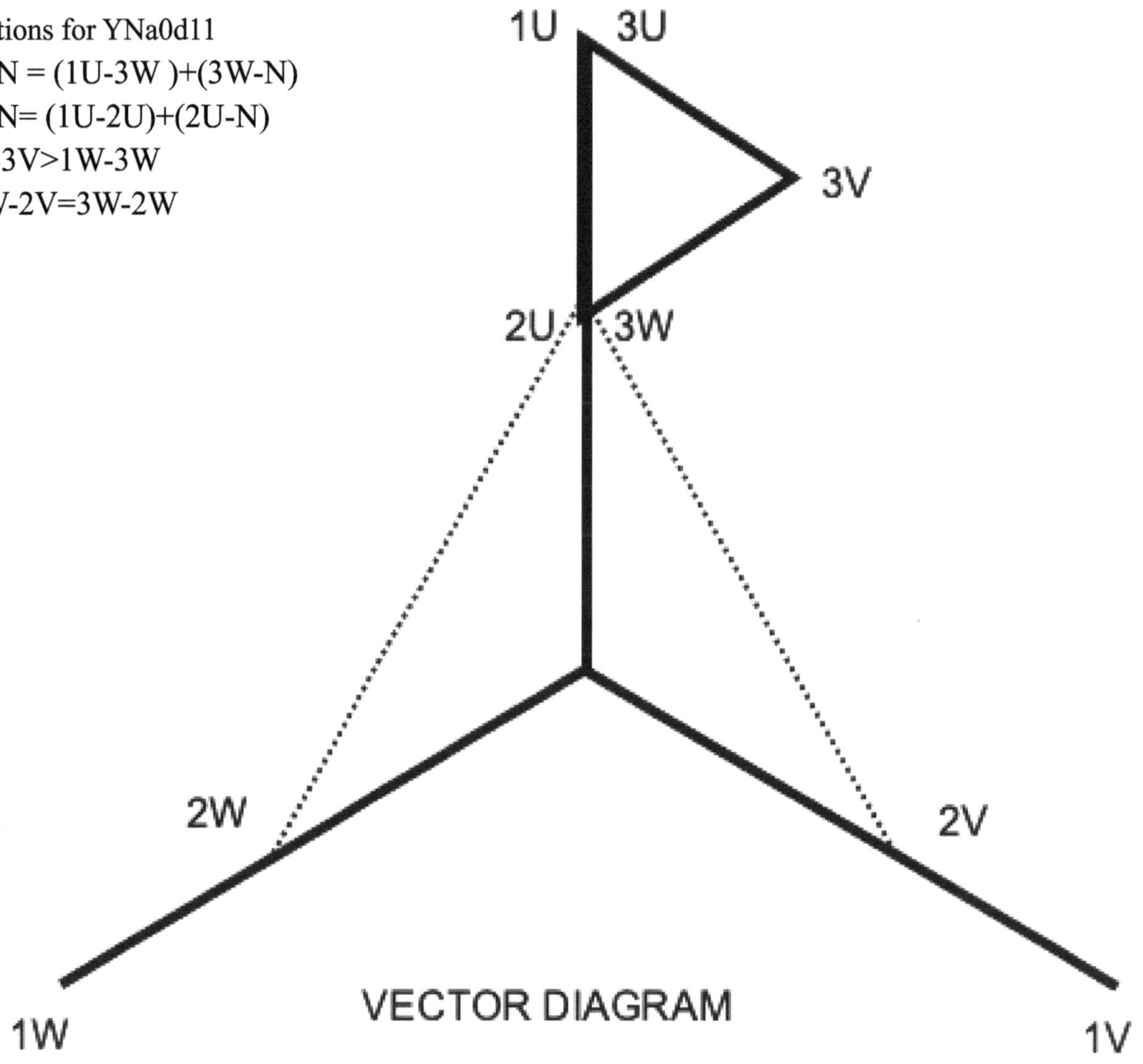

VECTOR DIAGRAM

b. YNyn0

3Phase supply Given on HV side (with 1U-2U Shorted):-

Voltage Given	
1U-1V	……………Volts
1V-1W	……………Volts
1W-1U	……………Volts

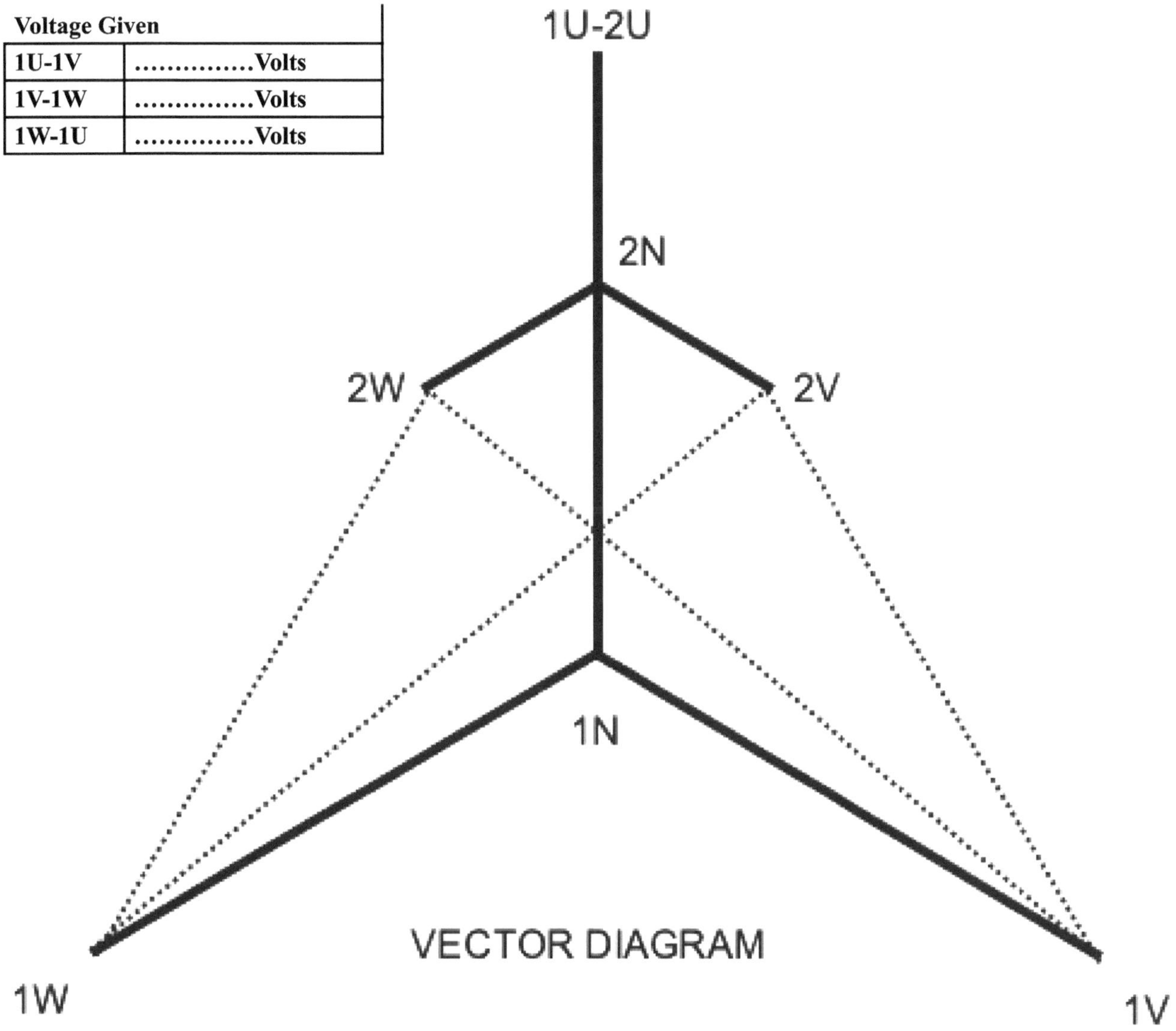

Conditions for YNyn0

1. 1V-2V =1W-2W,
2. 1V-2W=1W-2V
3. 1V-2V <1V-2W
4. 1W-2W< 1W-2V
5. 1U-1N= (1U-2N) + (1N-2N)

1U-2W	……………Volts
1W-2V	……………Volts
1V-2V	……………Volts
1V-2W	……………Volts
2W-1W	……………Volts
2N-1N	……………Volts
1U-1N	……………Volts
1U-2N	……………Volts

XIII. Winding Temperature and oil Temperature Indicators Temperature Settings

INDICATOR	PHASE	ALARM	TRIP	FAN AUTO START	
				GROUP 1	GROUP 2
OTI		...........°C	...........°C	...........°C	...........°C
WTI HV	HV PHASE.........	...........°C	...........°C	...........°C	...........°C
WTI IV	IV PHASE..........	...........°C	...........°C	...........°C	...........°C
WTI LV	LV PHASE.........	...........°C	...........°C	...........°C	...........°C

xiv. Insulation Resistance Measurement

Megger.........kV Make...... Type ....S.No.................................

Oil Temperature ....°C Conversion Factor at 60°C.....................

Terminal	Insulation resistance in M-Ohm			PI	Converted 60 sec. Value at 60°C
	15 Sec.	60 Sec.	600 Sec.	600 Sec/ 60 Sec	
HV-E	....	....	....	....	....
LV-E	....	....	....	....	....
HV-LV	....	....	....	....	....

Note:- When measuring IR Values for any windings all other winding are to be connected to Earth.

Table of Multiplying Factor for conversion of IR Value at 60°C at Page No......12......

xv. OLTC CONTINUTY CHECKS..............DONE ON DT..............

Applying a low voltage on one winding, checks for any discontinuity of induced voltage in the other winding while tap changer to operate throughout the range in the raise and lower direction.

xvi. Tan-Delta And Capacitance Measurement Of Bushing:

Measuring instrument Make................. S.NO..........................

Voltage of Measurement.....................Ambient Temp...........°C

Weather Condition Fair/Rainy..................

Particulars	HV Bushing			IV Bushing			LV Bushing		
	1U	1V	1W	2U	2V	2W	3U	3V	3W
Serial No. of Bushing.	....	....	....	....	....	....	....	....	....
Tan-Delta (%)	....	....	....	....	....	....	....	....	....
Capacitance(pf)	....	....	....	....	....	....	....	....	....
IR Values by ....kV range of 5KV Megger	....	....	....	....	....	....	....	....	....

xvii. Oil Sample Test Results After Last Filtration Cycle Of Transformer

(Before Commissioning of Transformer)

Particular	**Main Tank**	**OLTC Tank**	**Particular**	**Main Tank**	**OLTC Tank**
Break Down Voltage (BDV)			**Kinematic Viscosity at 27°C**		
Moisture Content (PPM)			**Interfacial Tension (N/M)**		
Tan-Delta at 27°C			**Resistivity at 27°C**		
Tan-Delta at 90°C			**Resistivity at 90°C**		
Acidity			**S.K. Value (%)**		
DGA of Main Tank					

xviii. Test On Tap Changer Checklist

Whether following Checked	**Yes**	**No**	**Whether following Checked**	**Yes**	**No**
Whether wiring of DM & RTCC Panel is complete			**DM Delay (Timer MPT) Checked Settings**		
Oil Level in DM Gear Box			**Out of Setp Time (OST) Checked Setting**		
Tap Selector Operates at Vmin			**Overload Relay (MPR) Checked Setting**		
Dirverter Switch Operates at Vmin			**Parallel Operation with other Transformer**		
Mechanical Limit & Clutch Operates at			**Whether locking is available at Tap 1**		
Abnormal sound, if any, during Operation			**Whether locking is available at Tap 17**		
Over Travel for motor operation			**Phase Sequence of Supply of DM**		
Electrical Limits at extreme taps (DRLS/DLLS) Checked			**In case of Oil filled Gear Box oil is filled up to required level**		

xix. **Test On Cooler Unit**

a. **No. of Fans - (i) Bank-I: ....ii) Bank-II: ....iii) Stand by: ....**
b. **Overload tripping of Fans, Setting given ....A**
c. **Operation of Fans Checked for**
(i) Manual (ii) Local Auto (iii) Remote Auto
d. **Auto Operation Of Fans set at**

i. **i) Bank-I: ....°C ON .... °C OFF**
ii. **ii) Bank-I I: ....°C ON .... °C OFF**
iii. **iii) Stand by: ....°C ON .... °C OFF**

c. **Direction of Fans Checked & Found...............on Dt..................**
d. **Cooler Supply Failure alarm Checked & found...............on Dt..................**
e. **Fans Trip Alarm Checked & Found...............on Dt..................**
f. **Cubicle Heater Checked for operation on Dt...............**
(i) Manual (ii) Auto Setting

xx. **Protection Circuit Checking**

a. **Oil temp.indicator alarm trip checked & setting given Alarm ....°C trip........°C**
b. **Winding temp.indicator alarm trip checked & setting given Alarm ....°C trip........°C**
c. **Main Buchholz Relay tested for Alarm .... Tested for Trip .... on Dt. ....**
d. **OLTC Surge relay Tested for Trip .... on Dt. .... any Maloperation of the relay during OLTC operation .... (Yes/No)**
e. **Conservator Magnetic oil level gauge checked for operation on Dt. .... Checked for Alarm at oil level .... on Dt. ....**
f. **Intertrip of Circuit Breaker Checked on Differential Relay Signal ....**
g. **Differential Relay**

Type .... S. No. .... Make ....

Bias Setting ....

h. **Backup 3 O/C Relay on HV Side: -**

Type .... S. No. .... Make .... PS .... TMS .... DMT .... CT Ratio ....

i. **Backup 2 O/C + 1 E/F Relay on LV Side: -**

O/C

Type .... S. No. .... Make ....

PS .... TMS .... DMT ....
CT Ratio ....
E/F
Type .... S. No. .... Make ....
PS .... TMS .... DMT ....
CT Ratio ....

j. **OVER FLUXING RELAY: -**

Type .... S. No. .... Make ....
Alarm .... V/F Setting .... Time Setting ....
Trip .... V/F Setting .... Time Setting ....

k. **All Annunciation circuit checked & Found .... on Dt. ....**

xxi EARTH RESISTANCE OF EARTH PIT: -HV Side .... IV/LV Side ....

HV-LA .... IV-LA .... LV-LA ....

xxii TESTING OF OVER CURRENT & RECORDING OF PICK UP VALUE

HV SIDE 3 O/C RELAY SETTING				LV SIDE 2 O/C+1 E/F RELAY SETTING			
	R-Phase	Y-Phase	B-Phase		R-Phase	E/F	B-Phase
Pickup Current				Pickup Current			

xxiii DIFFERENTIAL RELAY

Pickup/Operation Current .... A

xxiv CIRCUIT BREAKER TIMINGS

		R-Phase	Y-Phase	B-Phase
HV CB TIMINGS CHECKED FOUND	Closing			
	Trip Coil - I			
	Trip Coil - II			
IV CB TIMINGS CHECKED FOUND	Closing			
	Trip Coil - I			
	Trip Coil - II			
LV CB TIMINGS CHECKED FOUND	Closing			
	Trip Coil - I			
	Trip Coil - II			

xxv STABILITY TEST OF TRANSFORMER DONE ON Dt. .... FOUND ....

(Ok/Incorrect)

xxvi TAP OPERATION FROM 1 TO .... AND .... TO 1

Checked from local and remote found interlocks OK at extreme taps on Dt……………..

xxvii MULTIPLE CORE EARTHING CHECKED AND NO MULTIPLE CORE EARTHING FOUND ON DT. ....

xxviii. a) Air cell inflated fully by giving Nytrogen Pressure ……………..Kg/Cm^2

b) Air cell deflated by pushing oil from bottom after deflating of air cell oil level in conservator maintained to 1/3rd level at 30 °C in MOG and full level in plain glass gauge.

c) Breather properly connected to outlet of air cell on Dt…………………

d) Oil filled in cup provided at bottom of breather on Dt…………………..

e) Breathering of transformer checked and found on Dt…………………

xxix. Air released from all parts at……………Hrs. on Dtd………………

xxx. Openong of all valves checked on Dtd…………………

xxxi. Oil filled in WTI and OTI pockets after cleaning of pockets on Dtd…………..

xxxii. a) HV neutral earth Arrangement checked

i. **Conductor Size …………….**
ii. **No.of conductors**
iii. **Both/all the conductors terminated into ……………side earth pit.**

b) LV neutral earth Arrangement checked

i. **Conductor Size …………….**
ii. **No.of conductors**
iii. **Both/all the conductors terminated into ……………side earth pit**

c) Transformer body earthing Arrangement checked

i. **Conductor Size …………….**
ii. **No.of conductors**

d) HV & IV/LV LA earthing Arrangement checked on Dtd……………

xxxiii. Tripping of HV, IV & LV Circuit breakers checked through main Buchholz protection found ………………….(CORRECT/INCORRECT)

Multiplying Factors for Conversion of I.R. Values at 60°C Oil Temperature

Oil Temp ^{0}C	60 ^{0}C Converting M.F.	Oil Temp ^{0}C	60 ^{0}C Converting M.F.
20 °C	0.088	42 °C	0.340
21°C	0.094	43 °C	0.360
22°C	0.100	44 °C	0.380
23°C	0.108	45 °C	0.400
24°C	0.116	46 °C	0.430
25°C	0.125	47 °C	0.450
26°C	0.132	48 °C	0.490
27 °C	0.138	49 °C	0.530
28 °C	0.144	50 °C	0.560
29 °C	0.152	51 °C	0.580
30 °C	0.160	52 °C	0.625
31 °C	0.174	53 °C	0.660
32 °C	0.180	54 °C	0.700
33 °C	0.190	55 °C	0.740
34 °C	0.210	56 °C	0.780
35 °C	0.225	57 °C	0.825
36 °C	0.240	58 °C	0.880
37 °C	0.255	59 °C	0.950
38 °C	0.270	60 °C	1.000
39 °C	0.285	61 °C	1.100
40 °C	0.300	62 °C	1.200
41 °C	0.320		

The Periodicity of Testing / Maintenance of Transformer is as under

1	IR Value of transformer	Two times in a year Pre & Post Mansoonm
2	IR Value between Core, end Frame and tank.	Two times in a year Pre & Post Monsoon.
3	Transformer oil test Results Top and Bottom.	Once in a year
4	Tan delta & capacitance Measurement of bushing	Once in a year
5	Transformer oil dissolve gas analysis	Once in a year
6	HV & LV Magnetizing current measurement.	Once in a year

7	Magnetic balance test	Once in a year
8	. Transformer winding resistance	Once in two year
9	Measurement of vibration & noise	Once in two year
10	OLTC & OLTC drive mechanism Maintenance.	Once in a year

The Periodicity of Maintenance of associated equipments are as under

1	Differential Relay Testing	Two times in a year Pre & Post Monsoon.
2	Back up O/C Relay Testing	Two times in a year Pre & Post Monsoon.
3	Annunciation Testing	Two times in a year Pre & Post Monsoon
4	OTI & WTI operational Checks	Two times in a year Pre & Post Monsoon.
5	Inter trip test	Two times in a year Pre & Post Monsoon

Apart from the above the transformer is required to be tested to ascertain, its healthiness & cause of tripping, after every fault tripping on OLTC Buchholz, Main Buchholz, PRV & Differential indications.

The recommended tests are as under

1. Measurement of insulation resistance and polarization index of transformer.
2. Measurement of voltage ratio on service & normal taps.
3. Measurement of HV winding resistance at all taps.
4. Measurement of IV, LV winding resistance.
5. Continuity Test on On Load Tap Changer.
6. Magnetic balance test with voltage applied on HV side and induced voltages measured on HV as well as on LV side.
7. Magnetic balance test with voltage applied on LV side and induced voltages measured on LV side.
8. Measurement of HV side exciting current at low voltage by 3 phase-3 wire method & single phase method on normal & service taps.
9. Measurement of LV side exciting current at low voltage by 3 phase-3 wire method & single phase method.

Authors: K.K.Murty and J.J.L.Kapil

4. KYT (Know Your Transformer)

1.0. Preamble: It is of prime importance for an EHV S/S and Sub transmission S/s Managers/ Engineers to know about each and every transformer installed in his / her Sub Transmission S/s, EHV S/s or Switch yard. This Article captioned as "KNOW YOUR TRANSFORMER" has been attempted by the principal author on the analogy of KYC (Know Your Customer) by various banks. It is advisable to the S/S Managers and Engineers to maintain history and technical data of Transformers for any referral

2.0 KYT (Know Your Transformer):

2.1 R & D Plate: The name plate of a transformer is known as the R&D plate (Rating and Drawing plate).It introduces the Transformer in many aspects which are important for the custodian/ owner/ manager/Engineer to know.

3.0 Type of Transformers commonly used in the Power sector;

i. 1 phase transformer,
ii. 3 phase transformer.
iii. Auto transformers.

4.0 Manufacture's Details;

i. Logo & Name of Manufacturer.
ii. Maker's type, Serial No. and Year of manufacture.
iii. P.O. no.

5.0 Standard Specification; The transformer must confirm to following latest standards in vogue in India.

i. *IS: 2026 & (ii) **IEC 60076.
*IS: Indian Standard specification issued by Bureau of Indian Standards.

**IEC: International Electro-technical Commission

6.0 Rating; The rating of the transformers depends on the cooling System adopted for cooling the Oil inside the transformer. Generally the highest rating is achieved when the highest designated rating of the cooling system of the transformer is in operation.

Example;

i. **160MVA, 220/132/33kV auto transformer:** 160MVA is achieved when "ON AF" is in operation.
ii. **315 MVA, 400/220/33 kV autotransformer:** 315 MVA is achieved when "OF AF" is in operation.
iii. **135 MVA,240/11kV GSU transformer in Hydel power Stations:** Since these transformers have OFWF cooling. They have only one rating.

6.1 Cooling System; The cooling system is defined with an abbreviation. of 4-alphabets.First 2 – alphabets denote the Oil cooling and the following 2-alphabets denote the status of the air /water for cooling;

Abbreviation.	Status of oil	Status of air for cooling	Status of water
ONAN	Oil Natural	Air natural	-
ON*AF	Oil Natural	Air forced	-
**OFAF	Oil forced	Air forced	-
***OFWF(Coolers)	Oil forced	-	Water forced

Table-1

* AF: Such cooling arrangements are used where the radiators are mounted directly on the body of the transformer tank and also separately mounted structure/pedestal support. Cooling of the hot oil takes place while it is passing through the radiators and cooling fans mounted on these force the air on to the radiators.

**OFAF: In such cooling arrangements the bank of radiators are externally mounted on structure/ pedestal equipped with adequate no. of cooling fans. The radiators are connected to the tank through top and bottom headers. The suitable in-line type oil cooling pumps are connected at the outlet of the bottom headers in the cooling pipe work at the oil in let end of the tank. The cooled oil through the radiator banks is forced into the transformer tank through the pumps.

***OFWF (Coolers) also known as Oil to Water Heat exchanger; This type of cooling arrangement is generally provided where water is available in abundance particularly in Hydel Power Stations It is a general practice to provide 2x100% OFWF coolers (includes one 100% standby).The oil pressure is more than the water pressure. The standby cooler can be taken over in service by operation of / regulating certain valves with either of the pumps. The operation of such cooling arrangement is kept in service for 24x7.

6.2 Typical transformer ratings: Typical ratings mentioned here under are the standard ratings as specified in CBIP Manual no. 317 for Indian Conditions at frequency of 50Hz.

Sr.No	Maximum declared rating	ONAN	ONAF	OFAF
1	25 MVA, 33/11kV Transformer. Full Load current HV LV	20 MVA 349.9A 1049.7A	25MVA 437.4A 1312.1A	-
2	63 MVA, 132/33, Transformer Full Load current HV LV	50MVA 218.69A 874.79A	63 MVA 275.55A 1102.24A	-
3	160MVA,220/132/33kV, Auto Transformer, Full Load current HV IV *LV Tertiary (Rating-MVA) LV Line current.	128MVA 335.92A 559.87A 42.66 MVA 746.44A	160 MVA 419.89A 699.82A 53.33MVA 933.03A	-
4	315 MVA,400/220/33kV Auto Transformer, Full Load current HV IV *LV Tertiary(Rating-MVA) LV line current	189 MVA 273A 496A 63MVA 1102A	252 MVA 364A 661A 84MVA 1470A	315MVA 455A 827A 105MVA 1837A
5	105 MVA//33kV, Full Load current HV IV *LV(Tertiary) Rating in MVA **LV current	42MVA 181.9A 330.6A 14 MVA 424.24A	63MVA 272.8A 496.0A 21 MVA 636.36A	105 MVA 454.7A 828.8A 35 MVA 1060.60A
	GSU transformer in Hydel Stations	-	-	OFWF
6	***80MVA,220/11kV, YNd11, Full load current HV LV	-	-	80MVA 209.94A 4198.91A
7	***130MVA,235/11kV, YNd11, Full load current HV LV	-	-	130MVA 319.38A 6823.43A

Table-2

Note:

* 1) Normally, the MVA rating of the Delta-connected tertiary winding is 1/3rd of the MVA rating of the main transformer or Autotransformer.

2) i. Full load currents are calculated as under, based on the 3 Phase MVA rating.

ii. Calculation for full load current;

$$\text{Ifl}\frac{\text{MVA}}{\sqrt{3}\times\text{kV}}\text{A},\ I_{fl}=\frac{315\times1000}{\sqrt{3}\times400}=454.67\text{A}$$

**3) i. In a single phase unit of 105 MVA, $\frac{400}{\sqrt{3}}/\frac{220}{\sqrt{3}}/33\text{kV}$, Since the 33kV winding is of 35 MVA, the current that shall appear in the 33kV winding can be calculated as follows:

$$\text{Ifl}\frac{\text{MVA}\times1000}{\sqrt{3}\times\text{kV}}\text{A},$$

ii. $I_{fl}=\frac{35\times1000}{33}\text{A}=1060.60\text{A}$

***4) i. OFWF type of cooling system is adopted in GSU Transformers of the Hydel Stations only where water is available in abundance.

ii. The operation of this type of cooling has to be for 24x7.

iii. Since it has to be in operation for 24x7, thus it has 100% spare cooling system which can be taken in service as and required by operation of certain valves and changing over to the spare pump.

7.0 Formation of 3-phase bank out of 3 units of 1-Ph Transformers:

A sample case of 3 x 105 MVA, $\frac{400}{\sqrt{3}}/\frac{220}{\sqrt{3}}/33\text{kV}$,1-phase transformers to be connected to form a 3-phase bank of 315MVA 400/220/33kV transformer with vector group YNa0d11is shown in figure 1.

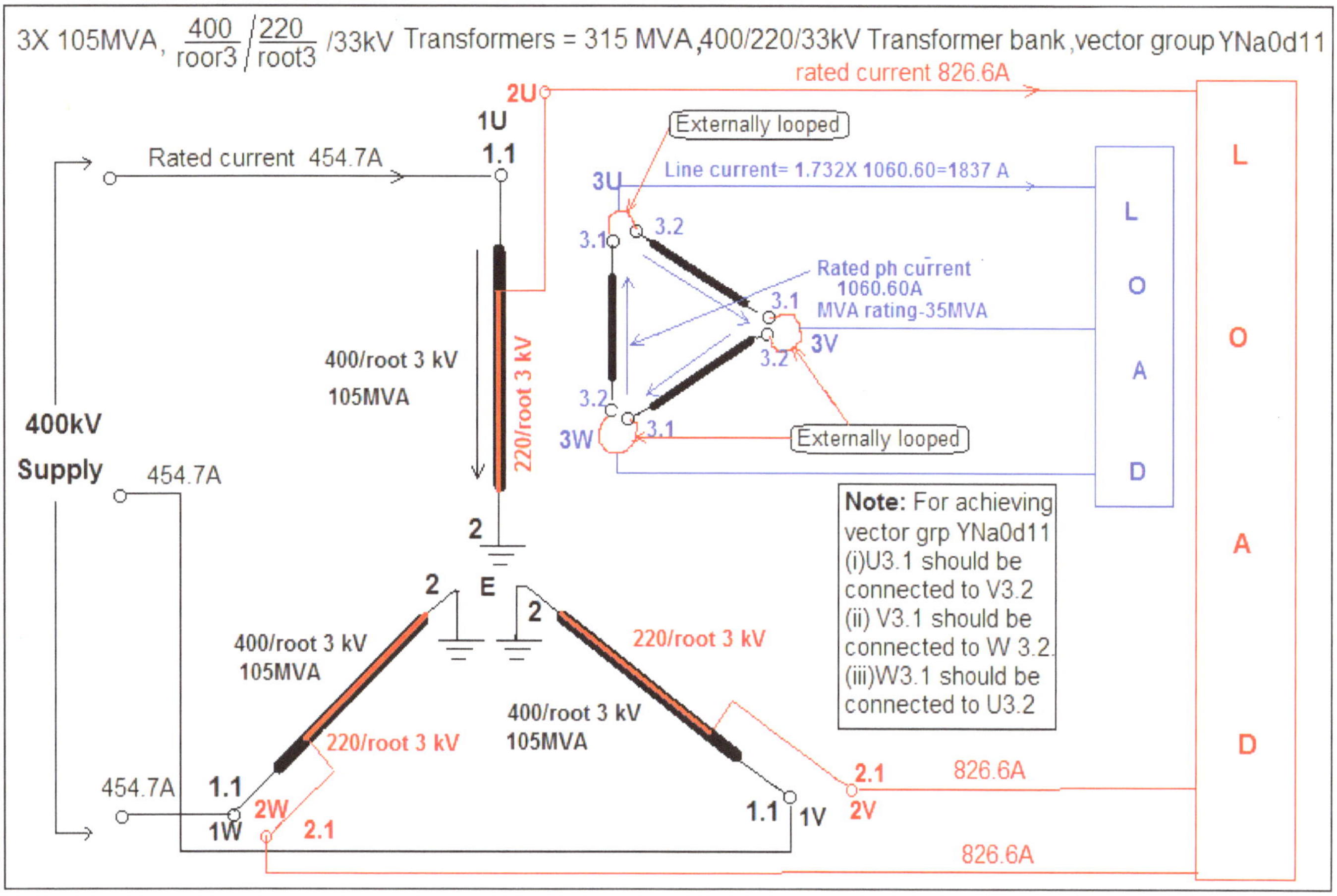

Figure 1: 3x105 MVA,//33 kV, 1-phase units connected to form 315 MVA, 400/220/33kV, 3-phase Transformer bank with vector group YNa0d11.

8.0 Vector group: Vector Groups generally used in Indian power network:

i. **Dyn11**-HV winding is connected in delta formation whereas LV winding is in star formation with neutral with 30° lead from the HV winding Voltage vector;11 indicates 11 O' clock position on the clock dial with respect to12 O'clock. Each 5 min graduation on the clock with respect to each other is of 30° apart.{ref.fig.3(i)}

ii. **Dyn1**- HV winding is connected in delta formation whereas LV winding is in star formation with neutral, with 30° lag from the HV winding Voltage vector 1 indicates 1 O' clock position on the clock dial with respect to 12 O'clock. Each 5 min graduation on the clock face with respect to each other is 30° apart. {ref.fig.3(ii)}

iii. **YNyn0**- This means that there is no phase shift between HV and LV voltages. Both the windings are Y N means-Star connected HV winding with HV Neutral and yn means- LV winding too is star connected with neutral. {ref.fig.3(iii)}

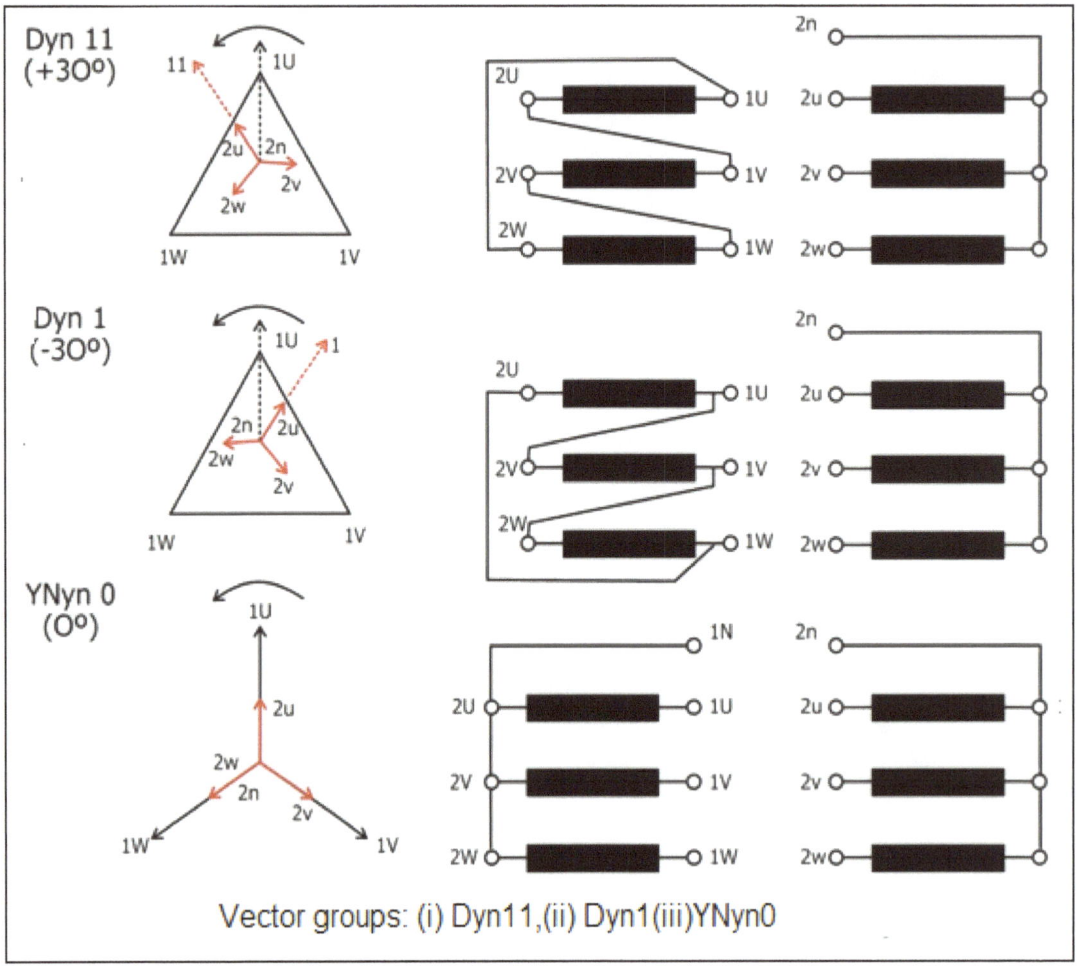

Figure 3: Vector representation: Dyn11, Dyn1 & YNyn0.

iv. **YNa0d11:** Auto transformers have vector group depicted as-YNa0d11. Here "a" stands for Auto connection. HV and IV have 0° phase displacement between them having common neutral - HV neutral N. There is a phase displacement of 30° between HV and LV voltage vectors. The LV voltage vector leads HV by 30°.(ref fig 4)

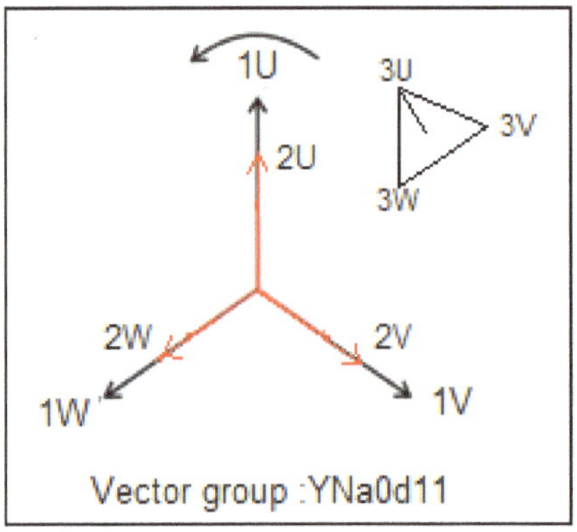

Figure 4: Vector representation of Autotransformer

v. **YNd1 or YNd11 -**The GSU (Generator Step Up) Transformers are generally of YNd1 group however in some Generating Utilities YNd11 are also used. (ref.fig. 5)

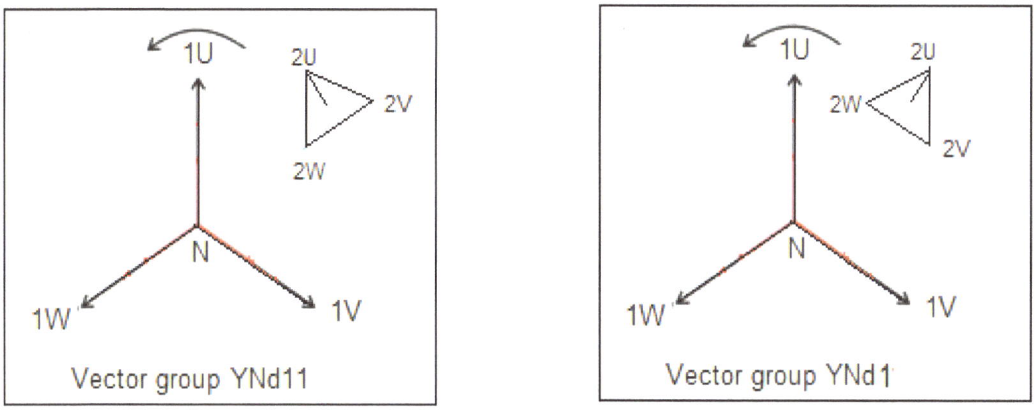

Figure 5: Vector representation of Generator step-Up Transformers.

Note: (i) It may be noticed that while representing the vector group of any Transformer, the vector representation should be HV vector first then the LV vector. This is generally the case with the GSU (Generator Step-Up) Transformers. The terminal voltage of any Generator is in LV Delta only.

Example: Normally, 200MW Thermal Power Generator has the terminal voltage of 15.75 kV (Delta). The GSU of this is 250 MVA, 240/15.75kV which is used to connect the 220kV supply to the Grid network. The vector group thereof is YNd1.

9.0 Insulation levels; The insulation levels of the windings are mentioned in Power frequency and BIL (Basic Insulation Level, which is Lightning Impulse withstand voltage, applicable as per IS: 2026 (part-3)/IEC 60076 -3.[1]

9.1 Typical insulation levels

c. 315MVA, 400/220/33kV, Auto Transformer
d. 160 MVA, 220/132/33kV Auto-Transformer.
e. 63MVA, 132/33kV Transformer.

Transformer	**SI (Switching Impulse)**	**LI (Lightning Impulse) withstand Voltage (BIL)**	**Power frequency withstand voltage**
a) 315MVA,400/220/33kV, Auto Transformer.			
HV	1050 kVp	1300 kVp	570 kV
IV	750 kVp	950 kVp	460 kV
LV(Tertiary)		250 kVp	95 kV
HV neutral		250 kVp	V

b) 160MVA,220/132kV/33kV, Auto transformer			
HV	750kVp	950 kVp	395 kV
IV		550 kVp	230 kV
LV(Tertiary)		250 kVp	95 kV
HV Neutral		250 kVp	V
c) 63MVA,132/33kV Transformer			
HV		550 kVp	230 kV
HV N			95 kV
LV N			95 kV

Table 3

10.0 BIL of Bushings: Though it is not a part of the R&D plate but it is essential to note the BIL of bushings. It has been stipulated in the CBIP manual no.317 on Transformers that the BIL of Bushing should be one level above the BIL of the Transformers.[2]

As per CBIP's recommended specification for Power Transformers as stated in Manual on Transformers-317 (Rev.-2013)		
Voltage rating	Rated Lightning with stand voltage of Transformer	Basic Insulation Level of Bushings
420kV	1300/1425kVp	1425/1550kVp
245kV	950kVp	1050kVp
145kV	550/650kVp	650/750kVp

Table 4

Note: Prior to installation of bushings the S/S Manager/Engineer must verify the above aspect.

11.0 Impedance and significance of % Impedance Volts;

11.1 Definition: The percentage impedance of a transformer is the volt drop in terminal voltage on full load due to the winding resistance and leakage reactance expressed as a percentage of the rated voltage.

11.2 Impedance volts in% at rated MVA base:

It is the % of the rated HV terminal voltage required for circulating the full load current, keeping all the 3 LV terminals shorted.

Example;

Say, full load current flows from the source if 13.86 kV voltage applied to the HV side (132kV) keeping the LV (33kV) side short circuited of a say 40 MVA 132/33kV Transformer.

$$\%Z = \frac{13.86}{132} x100 = 10.5\%$$

On the R&D plate, it is designated as %Impedance volts at base MVA.

$$Isc = \frac{Ifl}{\%Z} x100.A = IFIx(\frac{100}{\%Z})A,$$

Where Ifl = Full load current and Isc = Short Circuit current

Say full load current of a 132kV side of 20MVA, 132/33kV Transformer is 87.5 A, and the %impedance volts is say 10.5%.

Therefore,

$$I_{Sc} = = 87.5x \frac{100}{10.5} = 833.33A \text{ on HV side and Isc on LV side} = 3333.32 \text{ A.}$$

Note: % Impedance volts of individual transformer varies as per its winding configuration (resistance and leakage reactance) therefore its true value obtained after the transformer's complete assembly and testing at OEM's works is inscribed/engraved on the R&D plate indicating the % Impedance volts at base MVA.

12.0 Temperature Raise;

i. **Top Oil Temperaturerise**-45°C, means the oil temp of Transformer can rise up to 45°C max above the ambient temp.

ii. **Mean Winding Temp rise-** 50°C, means the winding temp of Transformer can rise to 50°C max above the ambient temp.

Note:

a. In case of OFWF cooled transformers which are generally used as GSU Transformers in Hydel Power Stations. Such cooling is recommended where water is available in abundance. The temperature rise as compared to equivalent rating Transformer with conventional cooling is almost 5°C above. A sample case of 130 MVA 235/11kV Transformer is Oil temp. rise: 55°C and Winding temp. rise is 60° C.

b. Considering the average ambient temperature as 50° C, Oil temp and Winding temperature Alarm and trip are set.

13.0 Guaranteed losses; Typical guaranteed" No Load loss and Load loss" are defined by the purchaser. The Transformers are designed and manufactured accordingly. Actual measured losses should be within the defined max losses and should be engraved on the R&D plate of the transformer;

14.0

Sr.No.	MVA rating of Transformer	No-load Loss (max) in kW	Load Loss (max) in kW
1	40MVA 132/33kV, IMP Make	20	98
2	160MVA,220/132/33kV Auto Transformer, AREVA make	35	190
3	315MVA 400/220/33kV Auto Transformer, TELK make	66	401.5

14.0 Other Details;

i. **Sample 1;** This pertains to160MVA 220/132/33kV Autotransformer of BBL make.

Weight of core	kg	26,472
Weight of copper(bare)	kg	8,378
Weight of Insulation	kg	990
Weight of core& winding	kg	39,000
Weight of of steel(total) (Tank & steel acessories, Radiators etc)	kg	15,000
Totalweight of oil	kg	87,300
Weight of Untanking (top tank)	kg	11,500
Weight of transportation(with N2 filled)	kg	56,000
Oil for first filling	Ltrs	28,342

Table-5,

ii. **Sample 2**; This pertainds to 25 MVA 33/11kV Transformer EMCO make;

Wt of core and winding	kg	25,050
Tank and fitting mass	kg	12,535
Core & frame mass	kg	13,985
Winding mass	kg	11.065
Mass of oil	kg	12,180
Volume of oil	Ltr	14,850
Transport mass (heaviest) (oil filled)	kg	45,600
Radiator mass	kg	4,500
Radiator volume/mass	Lit/kg	1800/1475
Total mass	kg	49,765
Type of OLTC		EQ16/CTR
Guarantee period of from date of dispatch		**66 months**

Table-6

Note: Some of the manufacturers mention Oil in kgs only on the R&D plate, in such cases oil in kgs needs to be converted in to volume considering Density of oil as $0.89 g/cm^3$ max at 29^{0} C. as per IS:335 (this can be assumed between0.82 to 0.89) and thus the MF would be 1/0.89 = 1.123.

Alternatively,

Take some oil in a beaker (already weighed) or a jar with graduation in ml (volume) then measure its weight and find out its Density at the ambient temperature. Then Use the formula 1/ Sp. gravity of the Oil.

$$\textbf{Sp. gravity of Oil} = \frac{\text{Wt of oil / volume of Oil}}{\text{Wt of water / Eqivalent volume of Oil}}$$

15.0 Tap Changers;

a. **OLTC (On line Tap Changer):** It is generally used in Transmission and Sub transmission transformers.
b. **OCTC (Off Circuit Tap Changer):** It is generally used in GSU (Generator Step Up) Transformers and Low Rated Sub Transmission Transformers.

15.1 OLTC;As the name signifies, it is equipment which is very essential for raising / lowering of transformer-taps so as to maintain the voltage profile on load of the system.

15.2 Type of OLTCs;

a. **Externally mounted type;** OLTC is externally mounted on the Transformer body type with its tap changing action namely selection and then connection of taps is conducted in this externally mounted compartment. The tap leads are brought to the tap changer through epoxy bushings and Bakelite barrier board. This is generally used in small rating Transformers..

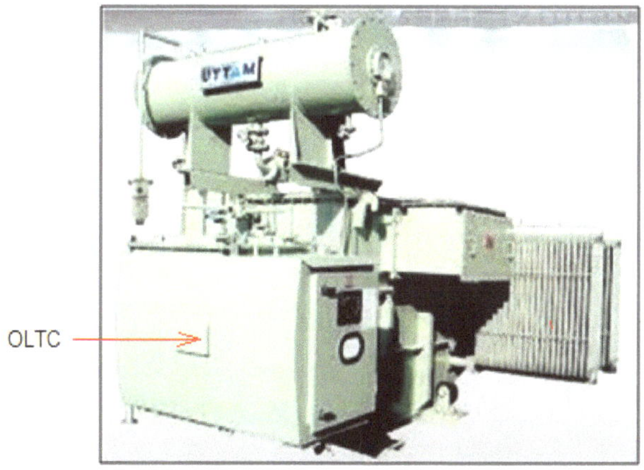

Extenally mounted type OLTC

Figure 6: Externally mounted OLTC

b. **In-tank type;** This type of Tap changer is in two parts the Tap selector is dipped in the transformer tank oil selects the taps therein. The Diverter Switch is encompassed in a separate Insulated cylinder containing transition resistance etc. please ref figure no 7.

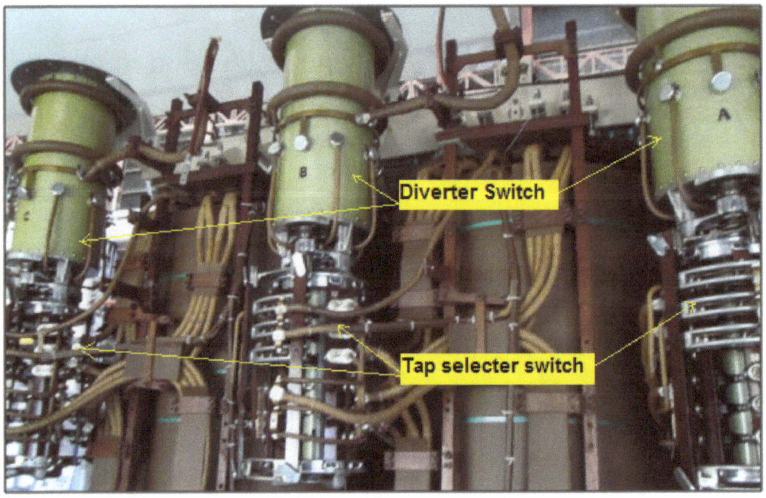

Figure 7: In tank type OLTC.

15.3 Placement/Location of OLTC: It is generally located in the HV winding as the handling of the same as compared to LV winding is easy and due to the fact that it performs its duty in a lower current side of the Transformer ie the HV side.

15.4 Placement of the OLTC in Y connected Transformer: In a 2-winding Transformer, the Tapchanger is generally placed close to the starpoint of the 3 phases, which is subsequently formed into the neutral point and in practice is the said neutral is grounded.The potential near the star point is very low.

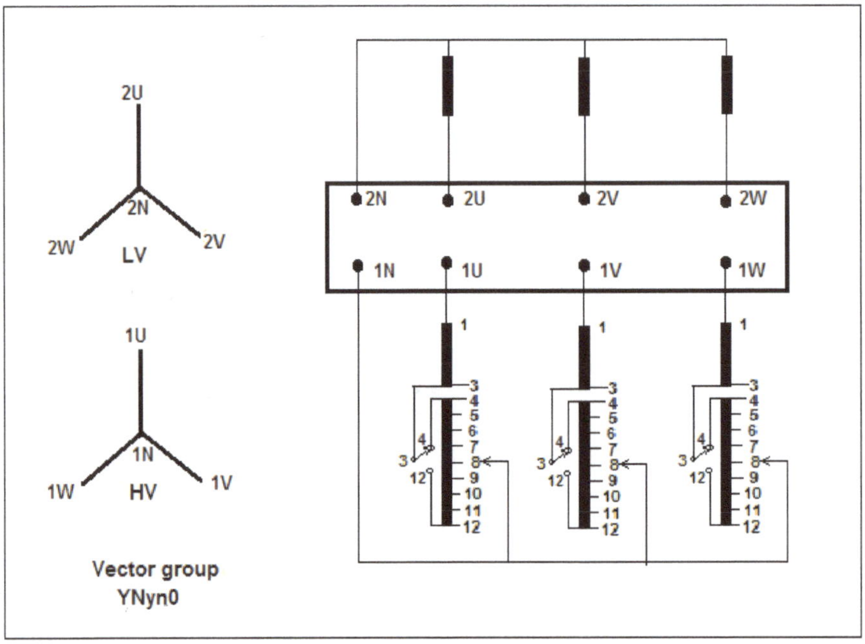

Figure 8: Tap winding located at the HV-Neutralof a Star/Star (YNyn0) Transformer

15.5 Placement of Tap winding in Dyn11 Transformer HV Delta side;

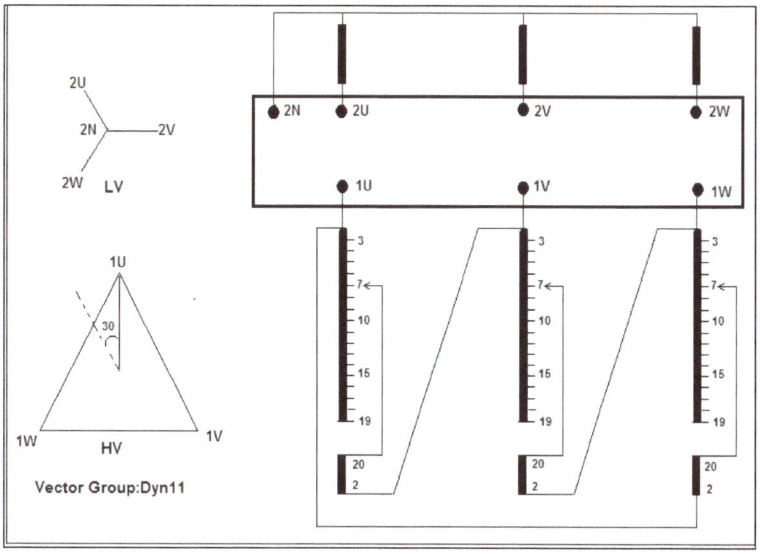

Figure 9: Tap winding in a Delta/Star (Dyn11) transformer.

15.6 Location of OLTC in an Auto-Transformer: In the Auto –Transformer unlike the 2-winding transformer, there is no separate winding for IV (intermediate voltage), as it is called. This Transformer also has LV winding in Delta-called as tertiary winding. The taps are selected from a tap winding which is separately provided between the HV and IV windings.ref. fig.10

15.7.1 As a practical example the tap connection of 220/132/33kV, 160 MVA 3 – Ph Auto Transformer, BHEL make is shown in figure no.10; There are as many as 13-taps with single reversing changeover selector switch.

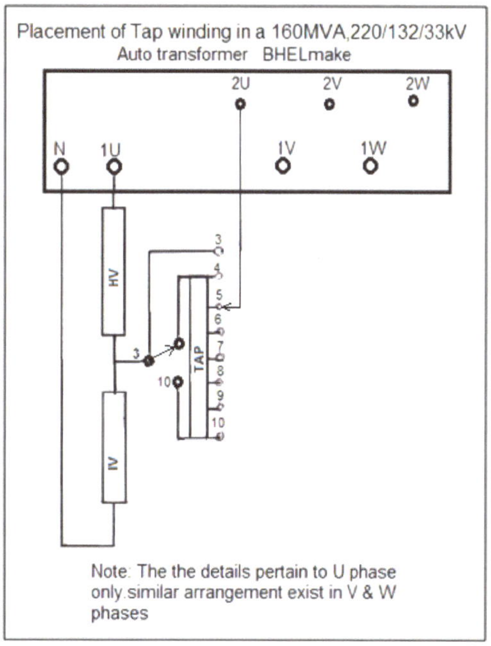

Figure 10: Tap Winding of 160 MVA, 220/132/33kV Auto Transformer (YNa0d11)BHEL make.

15.7.2 Intermediate Voltage (IV) at different tap positions achieved through different tap connections thereof;

HV line terminals1U,1V,1W	IV line terminals:2U,2V,2W		
HV kV	Tap position	Lead connections	IV kV
220kV	1(max)	11-4 3-10	145.2
220kV	2	11-5 3-10	143.6
220kV	3	11-6 3-10	141-.9
220kV	4	11-7 3-10	140.3
220kV	5	11-8 3-10	138.6
220kV	6	11-9 3-10	137
220kV	7	11-3 (3-10)(3-4)	135.3
220kV	8	11-5 3-4	133.7
220kV	9(Normal)	11-6 3-4	132
220kV	10	11-7 3-4	130.4
220kV	11	11-8 3-4	128.7
220kV	12	11-9 3-4	127.1
220kV	13(Min)	11-10 3-4	125.4

There are 13 No. of taps (–5% to +10% with a variation of 1.25%. per step.

Table - 7.

15.8 OCTC (Off Circuit Tap Changer): It can be seen from the fig 12, that, since it is a YNd1 GSU transformer, the tap winding is at the neutral only. There are only 5 no taps. Tap no 3 is the Normal tap with variation of 2.5%. This is of a GSU Transformer 250 MVA, 240/15.75kV TELK make transformer.

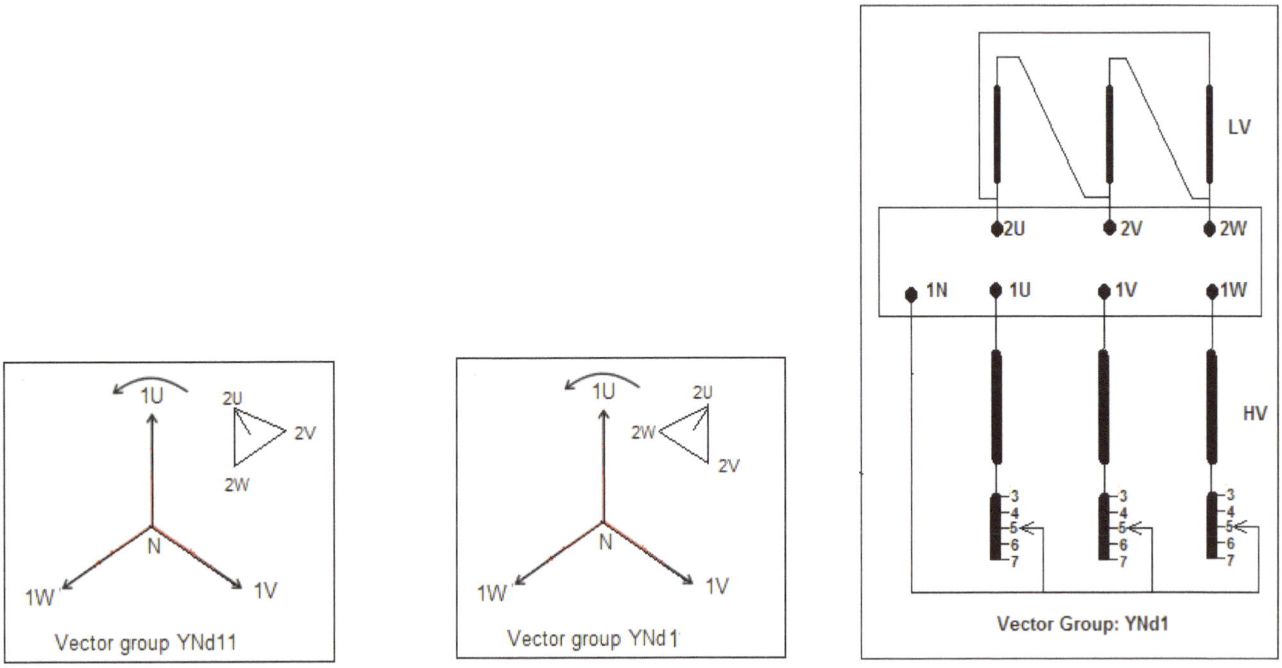

Figure 11: OCTC Taps and Tap winding of 250 MVA 240/15.75 kV,(YNd1) GSU Transformer.

16.0 CTs; Following are the CT ratios and Accuracy class, knee point voltage& VA burden etc used of a160 MVA 220/132/33 kV YNa0d11 Auto transformer as specimen.

(a)Protection CTs:These ring type CTs are placed encircling the Bushing in the turrets. The CTs are bushing CTs and the out door mounted Neutral CT of PS classof used for REF(Restricted Earth Fault) protection.In an Auto transformer HV & IV winding being YNaod11 vector group, have one Neutral only.

Protection CTs (Bushing CTs)						
Phase	core	ratio	Acc. Class	Knee point Voltage;Vk	Secondary resistance (Ω)	Burden VA
HV	Core1	800/1 A	PS	1000V	<1	-
IV	Core1	800/1 A	PS	1000V	<1	
LV	Core1	600/1A	5P10	-	-	15
LV	Core2	600/1A	1	-	-	15
N(outdoor)	Core1	800/1 A	PS	1000V	<1	-

Table- 8.

a. **WTI- CTs**: They are ring type CTs Winding temperatures applications of different windings.

WTI CTs				
Phase	core	ratio	Acc. class	Burden (VA)
1U	Core2	420/2.0-1.8-1.7A	3	15
1W	Core2	420/2.0-1.8-1.7A	3	15
2V	Core2	778/2.4-2.3-2.1A	3	15
3U	Core1	934/2.6-2.5-2.3A	3	15

Table -9

17.0 Oil Preservation Systems.

i. **Conventional Conservator:** This system is common in Smaller Transformers /Sub Transmission Transformers where the Air space above the oil level is vented through a De-humidifier unit known as Silica gel breather. The transformer breaths the atmospheric Air through it when the oil cools down during night hours or when the load is reduced during off peak hours.If proper monitoring of condition of Silica-gel breather is not done there is every posibility of moisture entering the Transformer.

ii. **a. Air-cell conservator:** It is almost two decades since this oil preservation system came into practice.This prevents possibility of transformer oil coming in contact with the external breathing air. In this system, an Air-cell (a synthetic rubber bag) occupies the space above the oil, the interior of the bag is then connected the atmosphere so it can breathe in air when the transformer cools down and oil volume is reduced and breathe out when the transformer is heats up.(fig.12)

b. **Oil filling procedure:** The plate shown in figure. 12 is also affixed on to the transformer tank indicating the procedure to be followed as narrated in chronologic manner/order thereon; (fig.12)

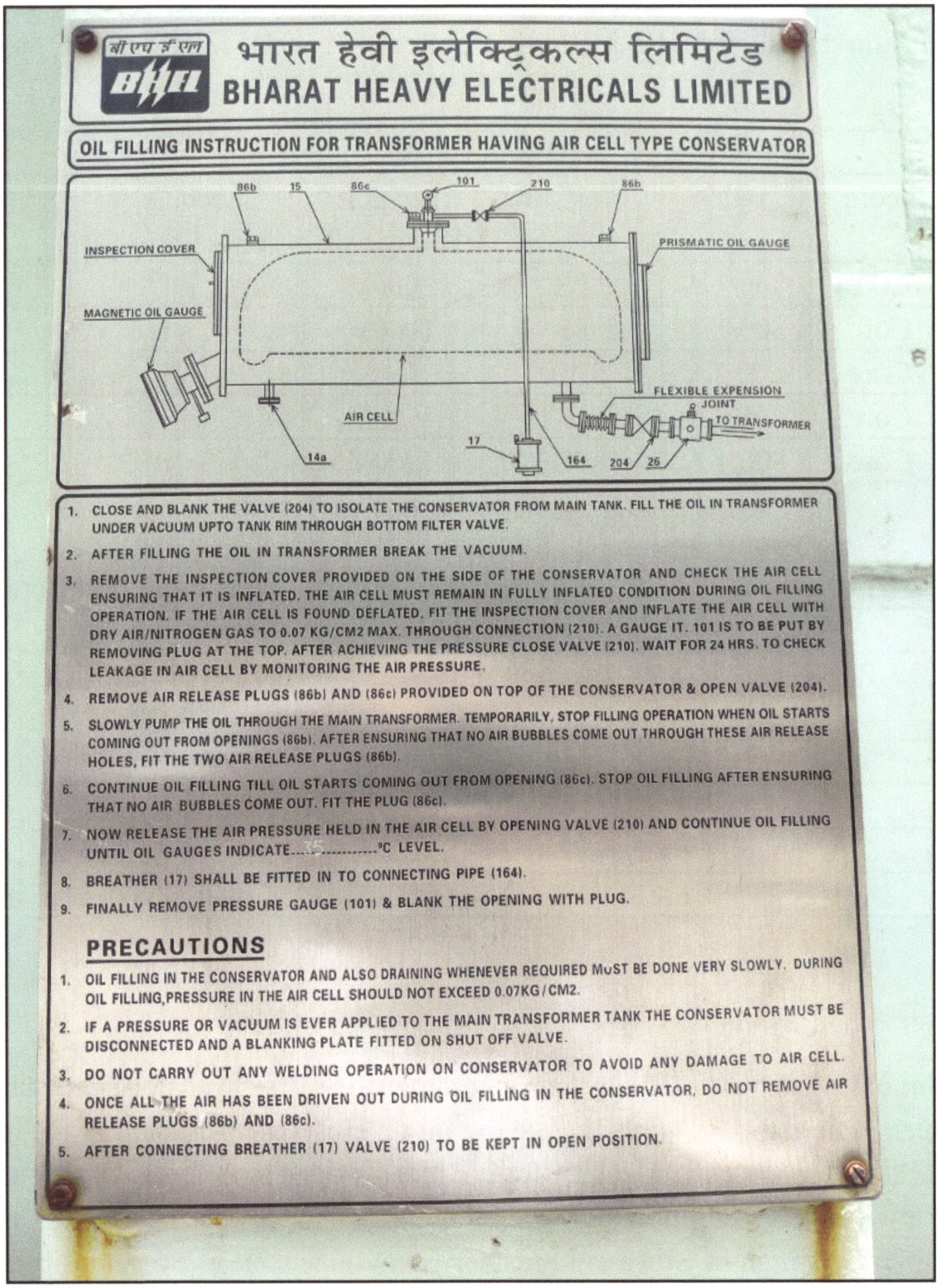

Figure 12: Oil filling Instruction for Transformer having Air cell type Conservator.

18.0. Conclusion; In any switch yard/Substation, AIS or GIS, there are several type of equipment which are important at their places, thus, like transformers they should also be given enough importance and details of each and every equipment should be noted down from their respective name plates bay-wise and record thereof should be maintained.

29.0 References:

1. IS: 2026(part-3)
2. BHEL's 2nd edition, Transformers.
3. CBIP manual no.317 on Transformers.

20.0 Some of the R&D plates are annexed herewith as sample R&D plates.

Annex-I

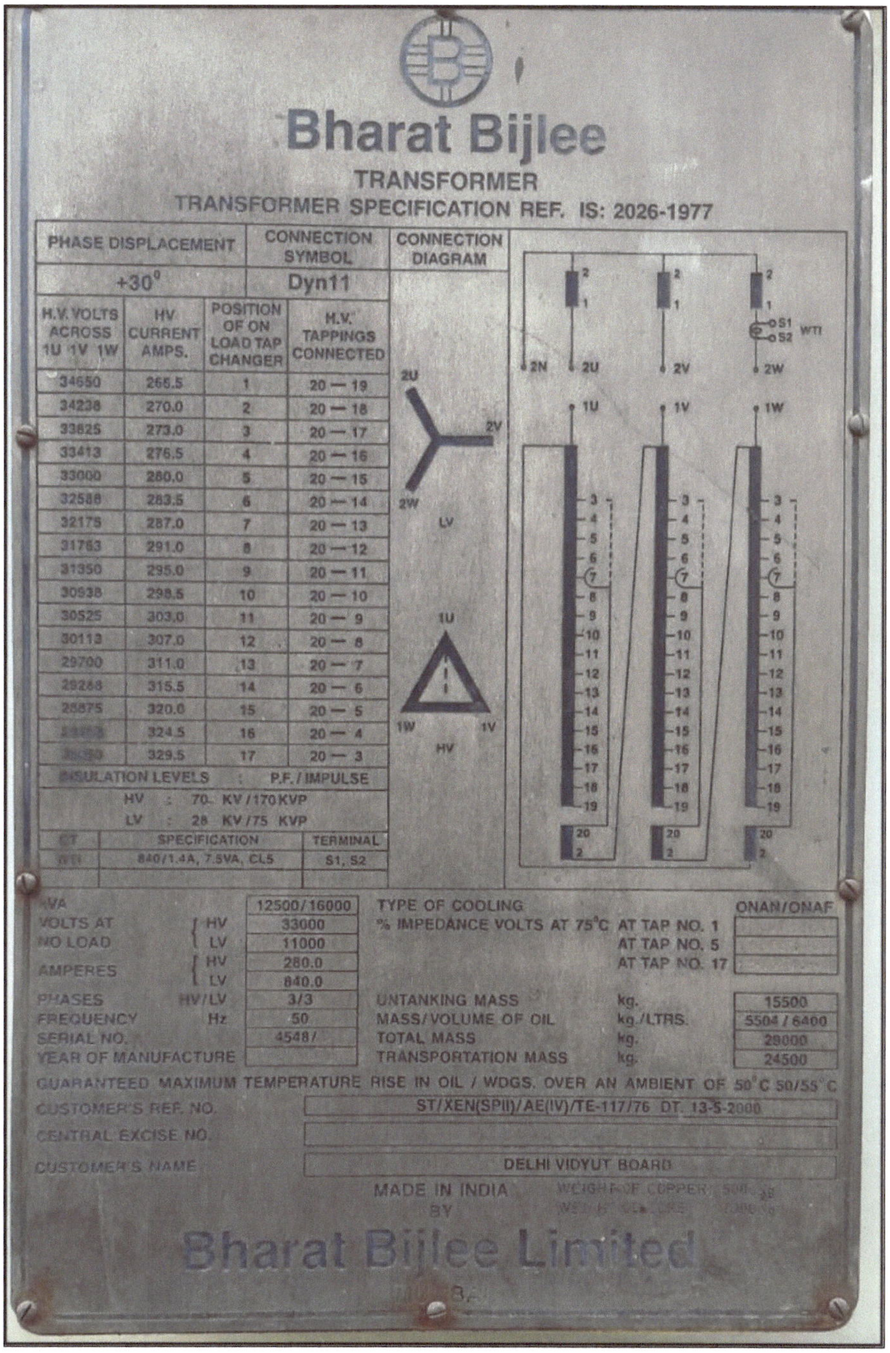

Annex-I: 16 MVA, 33/11kV, Dyn11 Transformer Bharat Bijlee Ltd. make.

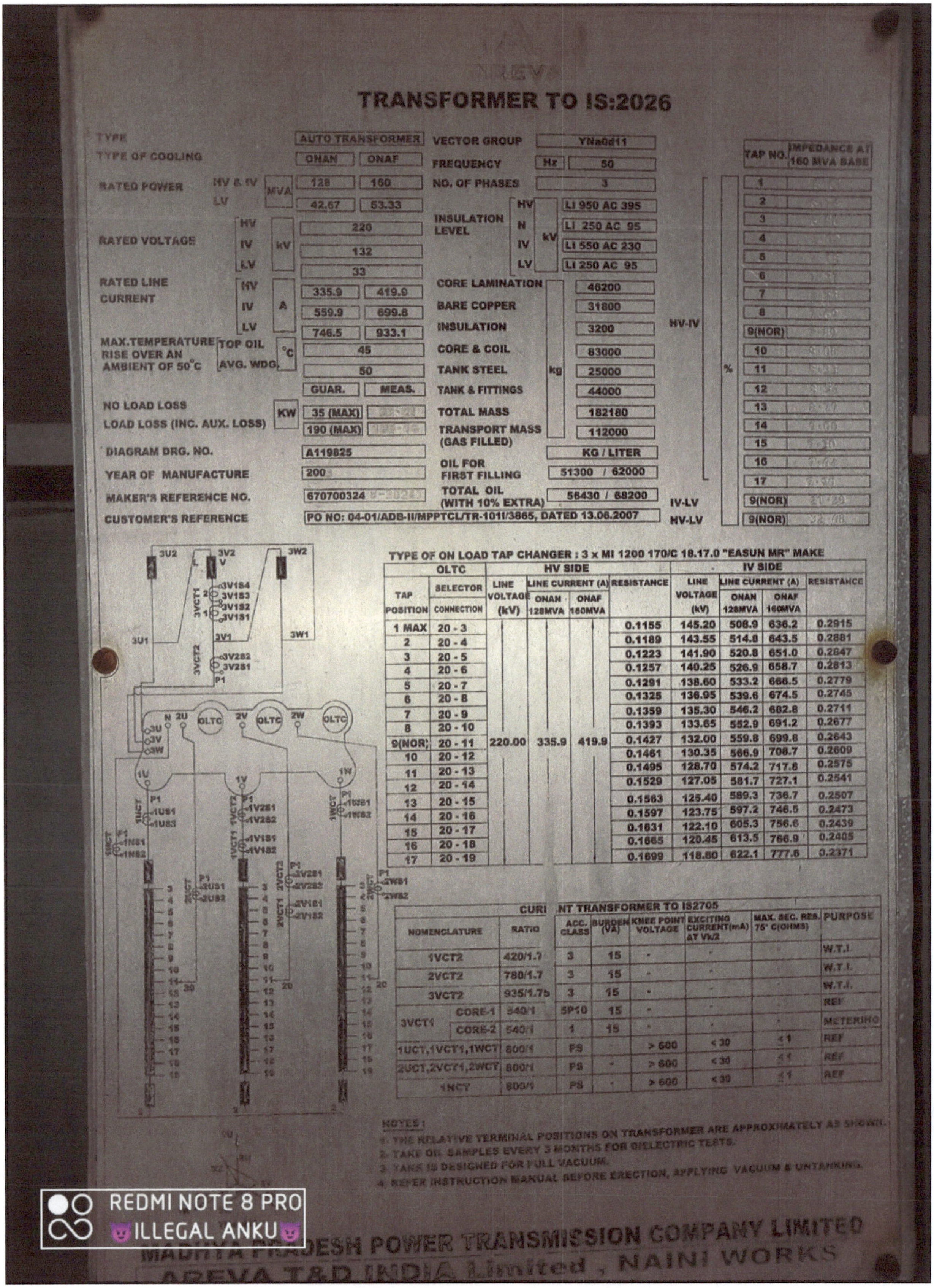

TRANSFORMER TO IS:2026

TYPE		AUTO TRANSFORMER	
TYPE OF COOLING		ONAN	ONAF
RATED POWER (MVA)	HV & IV	128	160
	LV	42.67	53.33
RATED VOLTAGE (kV)	HV	220	
	IV	132	
	LV	33	
RATED LINE CURRENT (A)	HV	335.9	419.9
	IV	559.9	699.8
	LV	746.5	933.1
MAX.TEMPERATURE RISE OVER AN AMBIENT OF 50°C (°C)	TOP OIL	45	
	AVG. WDG.	50	
		GUAR.	MEAS.
NO LOAD LOSS (KW)		35 (MAX)	[illegible]
LOAD LOSS (INC. AUX. LOSS) (KW)		190 (MAX)	[illegible]
DIAGRAM DRG. NO.		A119825	
YEAR OF MANUFACTURE		200[illegible]	
MAKER'S REFERENCE NO.		670700324 [illegible]	
CUSTOMER'S REFERENCE		PO NO: 04-01/ADB-II/MPPTCL/TR-1011/3865, DATED 13.06.2007	

VECTOR GROUP		YNa0d11
FREQUENCY	Hz	50
NO. OF PHASES		3
INSULATION LEVEL (kV)	HV	LI 950 AC 395
	N	LI 250 AC 95
	IV	LI 550 AC 230
	LV	LI 250 AC 95
CORE LAMINATION	kg	46200
BARE COPPER	kg	31800
INSULATION	kg	3200
CORE & COIL	kg	83000
TANK STEEL	kg	25000
TANK & FITTINGS	kg	44000
TOTAL MASS	kg	182180
TRANSPORT MASS (GAS FILLED)	kg	112000
		KG / LITER
OIL FOR FIRST FILLING		51300 / 62000
TOTAL OIL (WITH 10% EXTRA)		56430 / 68200

	TAP NO.	IMPEDANCE AT 160 MVA BASE (%)
HV-IV	1	[illegible]
	2	[illegible]
	3	[illegible]
	4	[illegible]
	5	[illegible]
	6	[illegible]
	7	[illegible]
	8	[illegible]
	9(NOR)	[illegible]
	10	[illegible]
	11	[illegible]
	12	[illegible]
	13	[illegible]
	14	[illegible]
	15	[illegible]
	16	[illegible]
	17	[illegible]
IV-LV	9(NOR)	[illegible]
HV-LV	9(NOR)	[illegible]

TYPE OF ON LOAD TAP CHANGER : 3 x MI 1200 170/C 18.17.0 "EASUN MR" MAKE

OLTC		HV SIDE				IV SIDE			
TAP POSITION	SELECTOR CONNECTION	LINE VOLTAGE (kV)	LINE CURRENT (A) ONAN 128MVA	LINE CURRENT (A) ONAF 160MVA	RESISTANCE	LINE VOLTAGE (kV)	LINE CURRENT (A) ONAN 128MVA	LINE CURRENT (A) ONAF 160MVA	RESISTANCE
1 MAX	20 - 3				0.1155	145.20	508.9	636.2	0.2915
2	20 - 4				0.1189	143.55	514.8	643.5	0.2881
3	20 - 5				0.1223	141.90	520.8	651.0	0.2847
4	20 - 6				0.1257	140.25	526.9	658.7	0.2813
5	20 - 7				0.1291	138.60	533.2	666.5	0.2779
6	20 - 8				0.1325	136.95	539.6	674.5	0.2745
7	20 - 9				0.1359	135.30	546.2	682.8	0.2711
8	20 - 10				0.1393	133.65	552.9	691.2	0.2677
9(NOR)	20 - 11	220.00	335.9	419.9	0.1427	132.00	559.8	699.8	0.2643
10	20 - 12				0.1461	130.35	566.9	708.7	0.2609
11	20 - 13				0.1495	128.70	574.2	717.8	0.2575
12	20 - 14				0.1529	127.05	581.7	727.1	0.2541
13	20 - 15				0.1563	125.40	589.3	736.7	0.2507
14	20 - 16				0.1597	123.75	597.2	746.5	0.2473
15	20 - 17				0.1631	122.10	605.3	756.6	0.2439
16	20 - 18				0.1665	120.45	613.5	766.9	0.2405
17	20 - 19				0.1699	118.80	622.1	777.6	0.2371

CURRENT TRANSFORMER TO IS2705

NOMENCLATURE		RATIO	ACC. CLASS	BURDEN (VA)	KNEE POINT VOLTAGE	EXCITING CURRENT(mA) AT Vk/2	MAX. SEC. RES. 75° C(OHMS)	PURPOSE
1VCT2		420/1.7	3	15	-	-		W.T.I.
2VCT2		780/1.7	3	15	-	-		W.T.I.
3VCT2		935/1.75	3	15	-	-		W.T.I.
3VCT1	CORE-1	540/1	5P10	15	-	-	-	REF
	CORE-2	540/1	1	15	-	-	-	METERING
1UCT,1VCT1,1WCT		800/1	PS	-	> 600	< 30	≤1	REF
2UCT,2VCT1,2WCT		800/1	PS	-	> 600	< 30	≤1	REF
1NCT		800/1	PS	-	> 600	< 30	≤1	REF

NOTES :
1. THE RELATIVE TERMINAL POSITIONS ON TRANSFORMER ARE APPROXIMATELY AS SHOWN.
2. TAKE OIL SAMPLES EVERY 3 MONTHS FOR DIELECTRIC TESTS.
3. TANK IS DESIGNED FOR FULL VACUUM.
4. REFER INSTRUCTION MANUAL BEFORE ERECTION, APPLYING VACUUM & UNTANKING.

MADHYA PRADESH POWER TRANSMISSION COMPANY LIMITED

AREVA T&D INDIA Limited , NAINI WORKS

Annex-II:40 MVA, 132/33kV YNyn0 Transformer IMP make.

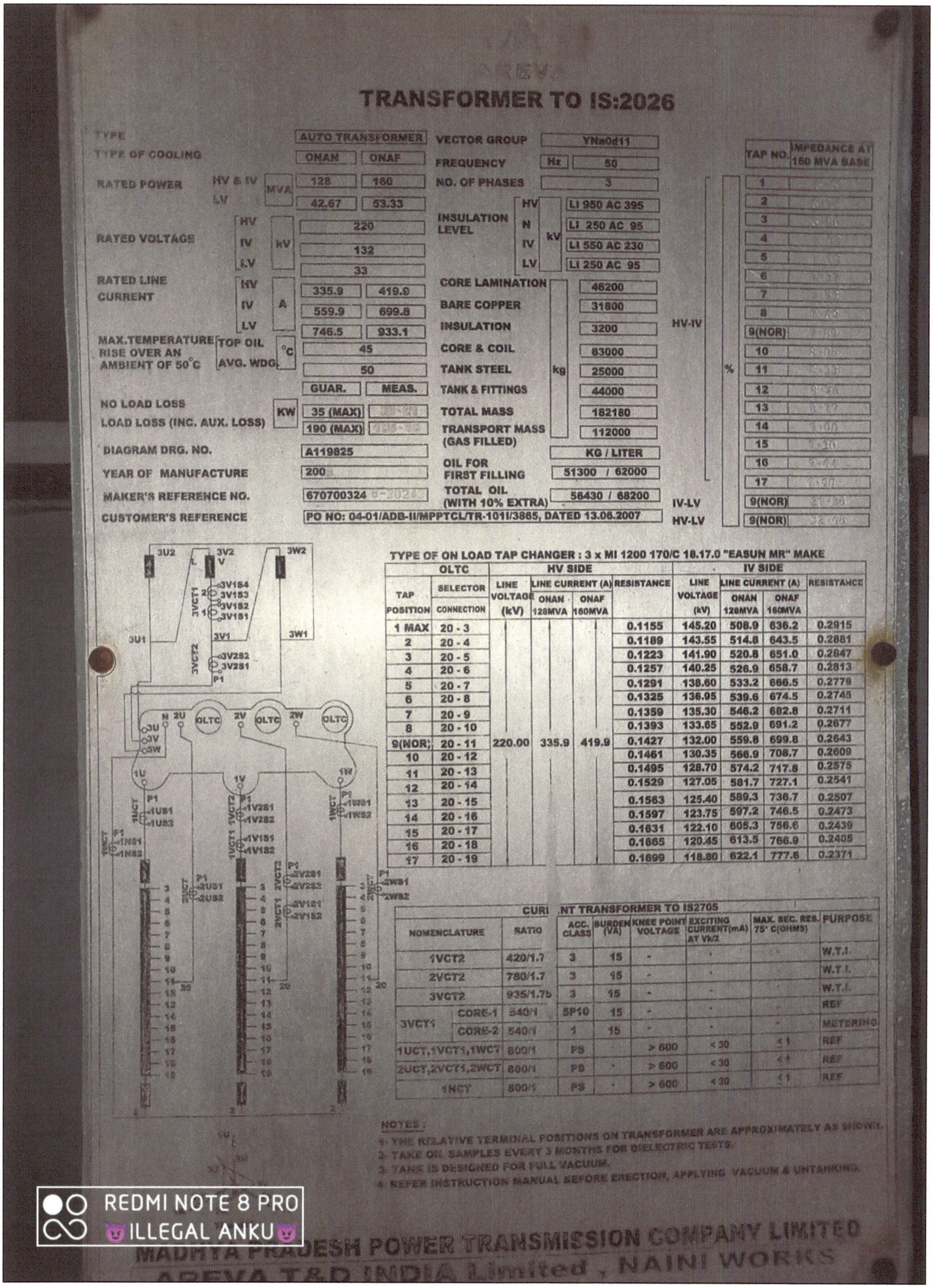

Annex-III:160MVA, 220/132/33kV Auto Transformer, AREVA make.

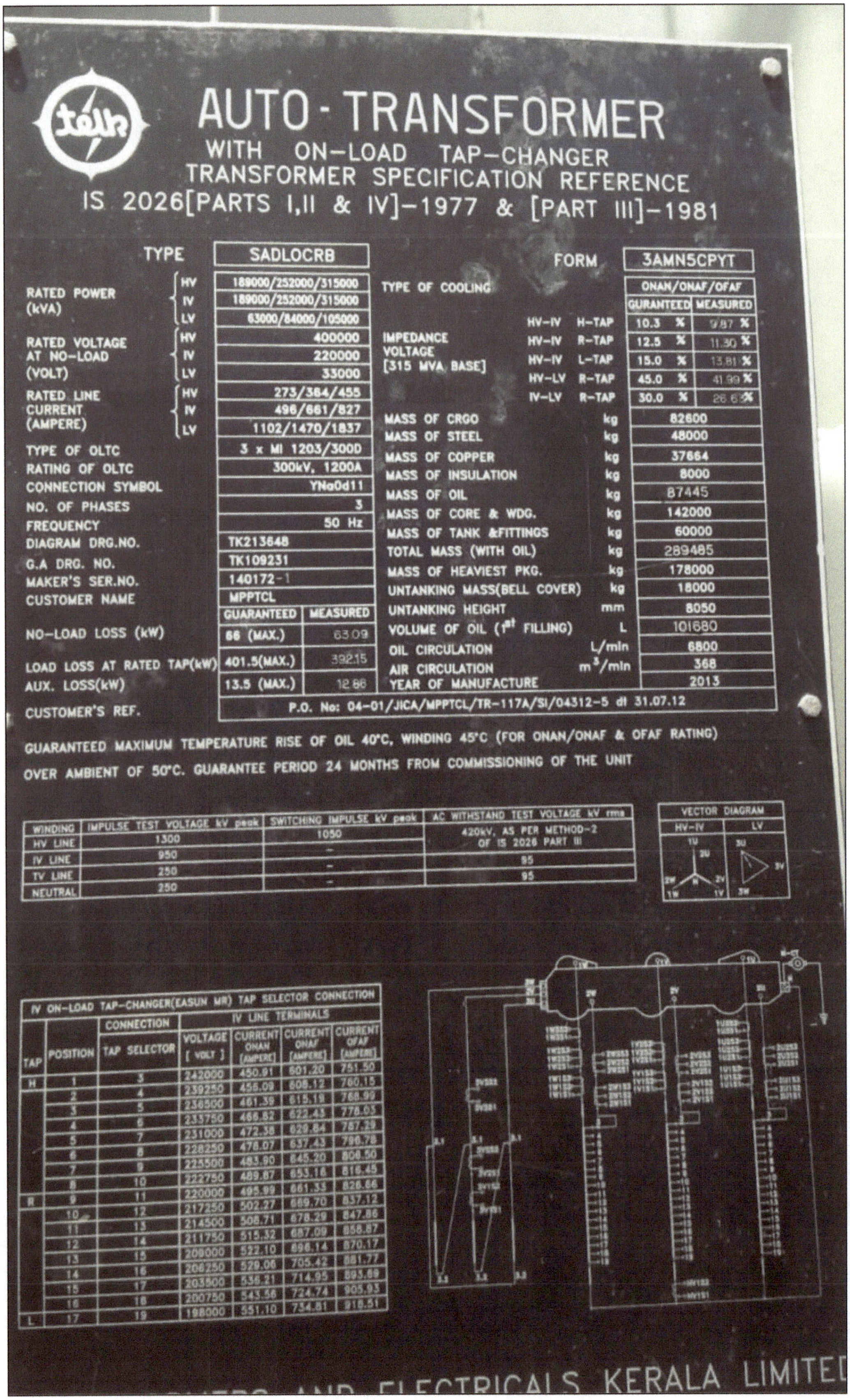

Annex-IV: 315 MVA, 400/220/33kV Auto Transformer TELK make.

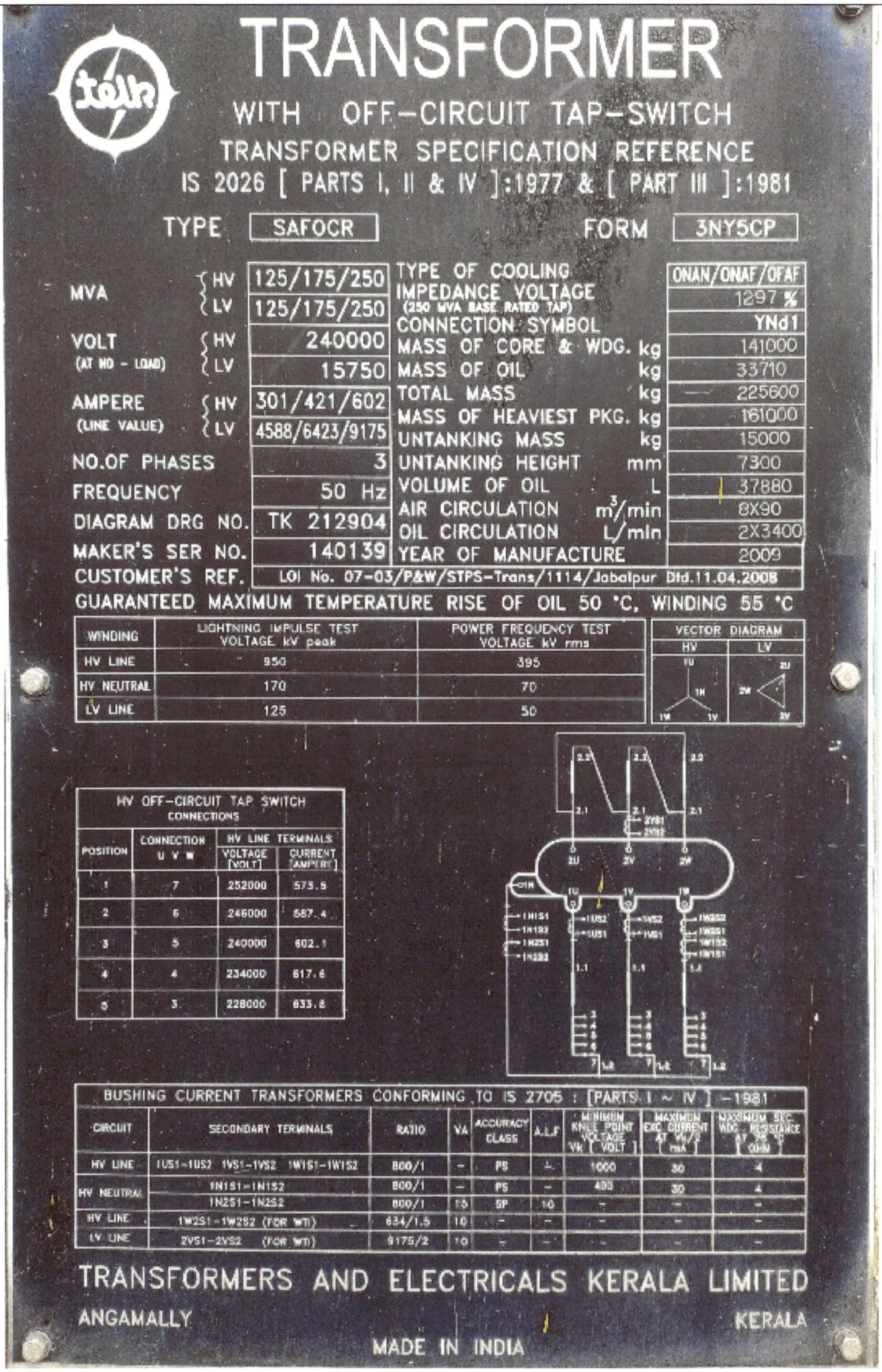

: Annex V:250 MVA, 240/15.75 kV GSU Transformer, TELK make.

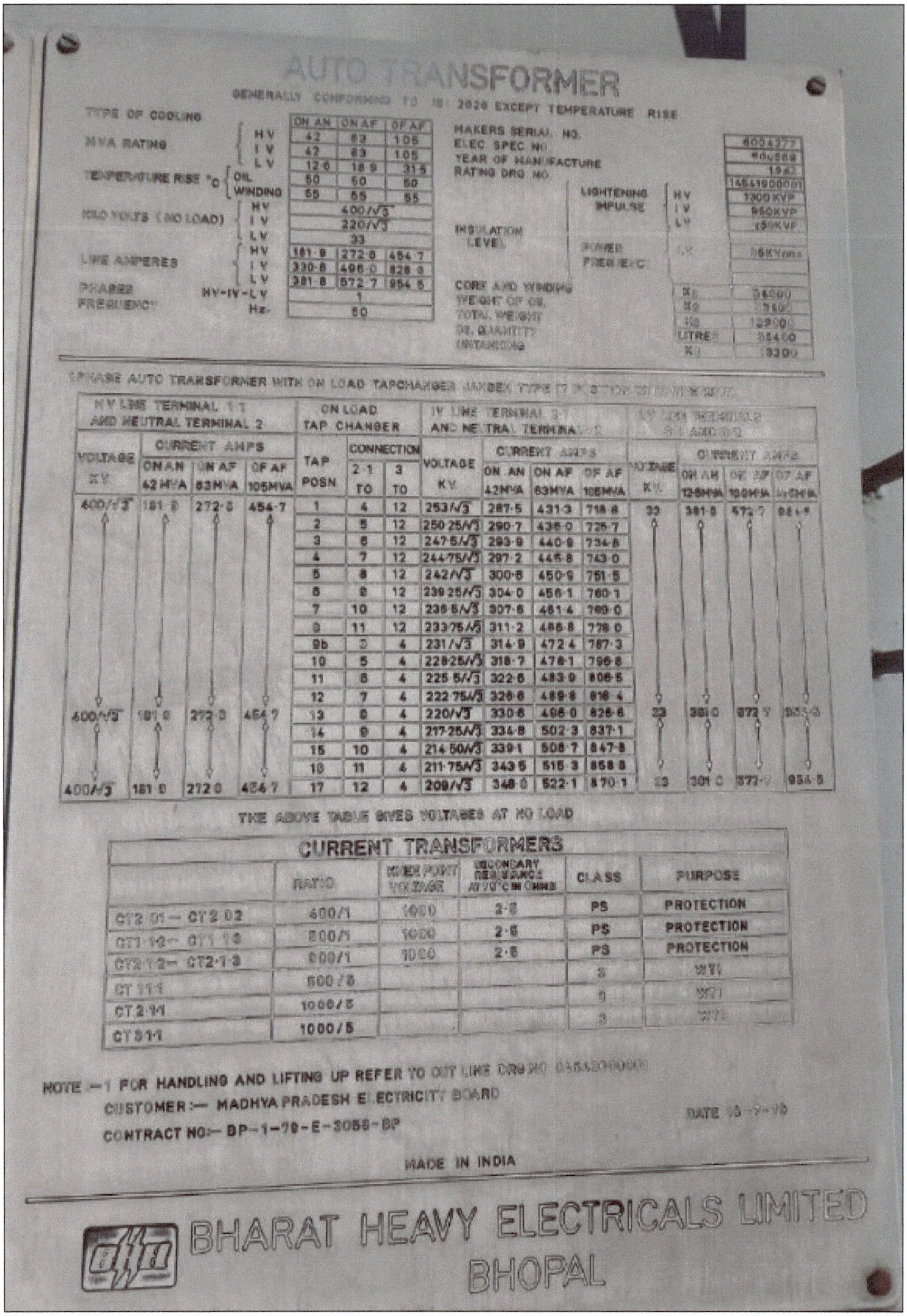

Annex-VI: 105 MVA $\frac{400}{\sqrt{3}} / \frac{220}{\sqrt{3}} / 33\text{kV}$ *Single Ph Transformer, BHEL make*

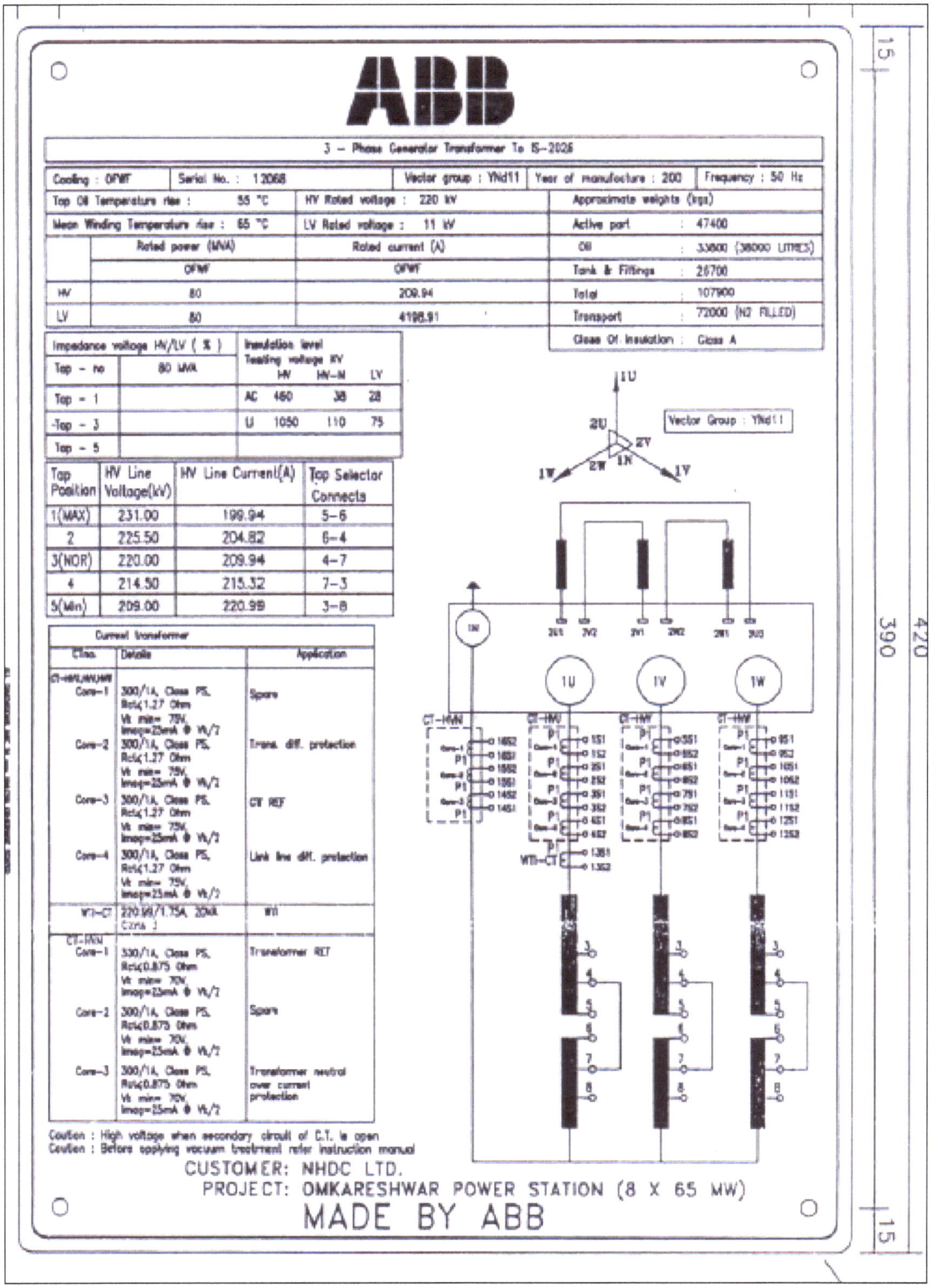

Annex-VII: 80 MVA, 220/11kV, GSU-Transformer with OFWF cooling system of Omkareshwar Hydel Power Station of NHDC.

21.0 Authors; (1) K.K.Murty,(2) Santosh Dubey (3)S.K.Chaturvedi &(4)Iqbal Khan.

5. LT-Magnetic Balance & Magnetizing Current Tests-Prima Facie Diagnostic Tools for Assessing Healthiness of the Transformers

1.0 Preamble: In the olden days there were no equipment/instruments for verification whether the EHV transformer has sustained any physical damage such as dislocation of core/coil assembly during its transportation to the site. Generally, the sites for EHV Substations are situated in the remote places. The prima-facie diagnostic tool for such occurrences were the LT magnetic balance tests and magnetizing current tests only.

Now, in the present scenario, each consignment of a transformer is equipped with an Electronic Impact recorder fitted on to the body of the transformer for verification, if the transformer has sustained any impact/jerk etc during transportation to site can also be verified with the signature graph of SFRA (Sweep Frequency Response Analysis) taken prior to transportation of the Transformer.

Apart from the initial pre-commissioning tests of the transformer, the magnetic balance and the magnetizing current tests are also performed with LT supply at sites after severe faults on the transformers or failure due to severe seismic condition. These tests may be designated as prima-facie diagnostic tests to assess the health-condition of the transformer post-sever faults as stated here-under;

a. Inter-turn shorting,
b. Buckling/bulging of the windings,
c. Open circuit in the windings,
d. Bushing failure,
e. Insulation failure,
f. Any external loops around the core,

The Magnetic balance and magnetizing current tests with LT supply are neither contained in the national standard nor in any other international standards recommended for the Power transformer tests, however most of the State Transmission Utilities insist for these tests during FAT (Factory Acceptance Tests) of the Power Transformers so as to keep the test reports as the bench-mark.

**The Authors have narrated the methodology and importance of these tests highlighting some of the case-studies with test results thereof as annexed in the Annex-I.

2.0 Key Words: FAT (Factory Acceptance Tests), SFRA (Sweep Frequency Response Analysis), PRV (Pressure Release Valve), Magnetic balance test, magnetizing current test.

3.0 Magnetic balance test; The 3-ph Transformer core generally has3 limbs. When a 3-ph balanced voltage is applied to HV or LV bushings, the fluxes are produced in the respective limbs and they add-up to zero at the yoke. This condition is said to be magnetically balanced core. The core loses its magnetic balance due to erratic flux distribution in the core, consequent to following conditions;

ii. Sustain jerks & jolts during transit to site and/or re-location of the transformer at site.
iii. While in service, tripping of the transformer on severe internal faults involving core and windings.
iv. Due to severe seismic disturbances (Earth quakes).

In order to assess the status of health of the transformers, LT magnetic balance Test is conducted at site.

3.1 LT- Magnetic balance test;

If an LT voltage say, V-volts is impressed across the ends of the coil (Phase bushing & the neutral bushing) on the outer limb, then the voltage developed in the middle limb winding would be of the order of 2/3rd V and that in the outer limb would be 1/3rd V on pro-rata basis of the flux distribution based on approx. reluctances of the magnetic paths. Similarly, if the other outer Limb winding is subjected to V- volts then the voltages induced would be 2/3rd V in the middle limb and 1/3rd V in the other outer limb. When the middle limb is applied V-volts, the flux distribution will be 1/2 &1/2 in the outer limbs and shall induce 1/2-V & 1/2-V-volt, theoretically.

LT Magnetic balance test of YNyn0 transformer

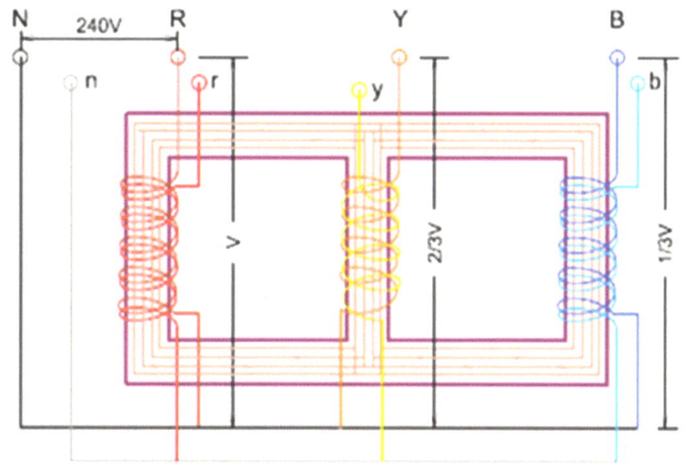

Sample readings (Theoritical)

	Voltage applied	Measured between		
		RN	YN	BN
RN	240	240	160	80
YN	240	120	240	120
BN	240	80	160	240

Figure 1:LT Magnetic balance test.

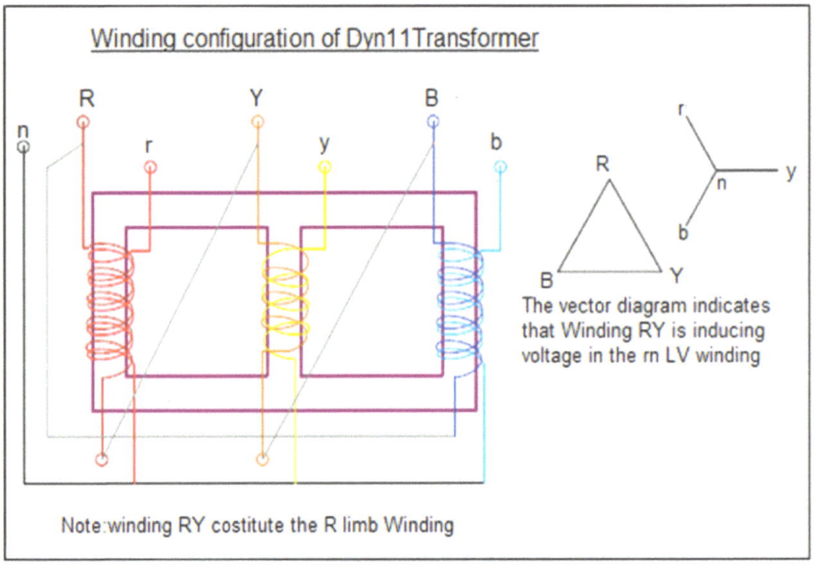

Figure 2: Winding configuration of Dyn11 Transformer.

In the above Dyn11 transformer, the R phase coil shall be between R & Y ph bushings. Likewise, the Dyn1 transformer shall have R phase coil between R & B bushings.

However, in practice some deviations would be observed as follows;

a. when middle phase is excited, the fluxes and voltages induced in the outer phases should be in the range of 40% to 60% since the reluctance values corresponding to two outer phases may differ slightly on account of minor differences in air gaps at the joints.
b. When any of outer phases is excited, the induced voltage developed in the middle phase may be as high as 90% and voltage induced in the other outer phase may be as low as10% for a healthy core condition.

If the deviations are within tolerable limits as mentioned above, the subject transformer is declared healthy. If the voltage values are observed to be abnormally different from the desired values, it can be concluded that the transformer has sustained some jerk/winding fault affecting the flux distribution of the magnetic circuit.

Suppose, there is an inter-turn fault in the R phase and when an LT voltage is applied to Y phase winding (between Y-ph bushing and Neutral bushing), a much higher voltage will be induced in B phase as compared to that in R phase, giving an indication of a fault in the R phase winding. A high circulating current in the faulty section, opposes the magnetizing flux thereby reducing the induced voltage in the faulty phase.

In practice, the values taken as reference are either pre-commissioning values or the latest pre-fault values taken during last maintenance as the case may be.

4.0 Magnetizing current:

4.1 Concept of Magnetizing current; The current required to magnetize the core when the Transformer is charged from the HV side keeping the LV side open is known as Magnetizing Inrush current which predominately contains 2nd harmonic current. The steady- state magnetizing current is the no load current of a transformer. It ranges from 0.2 to 2% of the full-load current.[1]

The 3-phase core type transformer has 3 limbs. The LV windings are woundencircling the core limbs and the HV windings are wound encircling the LV winding on each limb.

As per the design of the core the magnetizing current of the middle limb should be theoretically in the ratio of IR: IY: IB=1: 0.718: 1, as the reluctance of the central phase is nearly half that of outer phases [1].

4.2 **LT Magnetizing current test:**

Keeping in view, the fundamentals of flux distribution due to the reluctance as mentioned in para 4.1, Magnetizing current test with LT supply has been adopted and practiced in all the Utilities in India. It is as a prima-facie diagnostic test when the transformer is first received at site and also post tripping of the transformer due to severe earth-quake or severe internal faults actuating the Differential protection along with Buchholz and PRV protections.

It identifies disturbance in magnetic circuit due to dislocation of core during transit or due to severe fault in any of the windings.

In this test a 3 phase LT-supply is applied to the transformer under test and the results are observed. The magnetizing current test results are in milli Amperes. In practice it is experienced that magnetizing currents drawn by the outer limbs are almost equal, however, the magnetizing current drawn by the middle limb is approximately 70% of the magnetizing currents drawn by the outer limbs.

Due to non-availability of balanced 3-ph LT supply at sites, now a days, this test is carried out with single phase LT supply. It is applied across each phase and neutral bushings of the Transformer. Pronounced current in any of the windings shall decisively indicate the faulty phase. It is very useful in locating problems such as defect in magnetic core structure, shifting of winding, turn to turn fault or problem in tap changing device (OLTC or OCTC). These conditions result in change in the effective reluctance and flux distribution in the magnetic circuit, resulting into requirement of a larger current, to force a given flux through the core.

4.21 Stipulation in IEEE Std: It is not out of place to mention here that the Excitation Current Test is narrated in Cl.7.2.11.2.2 of IEEE std. C 57 152-2013, the Guide for diagnostic field testing of fluid-filled Power Transformers, Regulators and Reactors.[3] This test is performed with 1-ph, HT voltage up to 10 kV at rated frequency, however in our national context, availability of 1-Ph HT supply up to 10kV in the field is a limiting factor.

Note:

i. Winding resistance test being a DC test may further disturb the as-found magnetic circuit due to remnant flux, therefore, it should be carried out after the magnetizing current test and magnetic balance test.
ii. The magnetizing current test and the magnetic balance test are complementary to each other.

5.0 Other tests to substantiate health status of the Transformer: Following are the tests which substantiate the authenticity of magnetic balance test and magnetizing current test carried out with LT supply as the prima-facie diagnostic tests to verify the status of the of the transformer at site;

a. LT Short circuit rest,
b. IR Values and PI,
c. winding resistance test,
d. Ratio test,
e. tan δ of the windings,
f. tan δ of the bushings,
g. DGA (Lab test on the oil sample collected post- tripping).

6.0 **Case Studies:** The case studies are based on the actual test results of different instances are mentioned here under;

Case Study.1

- Sub-station: 132 kV S/S. AA
- Severe Tripping of 40 MVA, 132kV/33kV transformer.
- Date of tripping: 21/09/2020.
- Relay indications:

i. Diff relay indication C (Differential relay, ABC) trip ie B-phase in RYB convention,
ii. High set differential trip,
iii. Buchholz Relay Alarm & Trip.

Observation:

- 145kV, OIP bushing of B-ph failed (bursted) and simultaneously due to excess pressure oil gushed out penetrating through the cementing of Y-ph bushing porcelain with the metallic base damaging it, beyond repairs.
- Splashing of Transformer oil all around the transformer.
- Slight bulging of the transformer tank at one location at the bottom portion of the transformer body.

Testing: Transformer tested at tap No 4.

i. **LT Magnetizing current test;**

a. LT 1- ph, 2 Wire method, HV side;

LT -1 Ph allied-200V	Previous test results			Post fault test		
	R-N	Y-N	B-N	R-n	B-N	C-N
	1.8mA	1.4mA	1.9mA	1.9mA	1.6mA	2.1mA

b. LT-1Ph,2 wire method, LV side

ii. **LT- Magnetic balance test;**

a. LT-1ph 2 wire method, HV side;

Applied 200V	Pre fault condition			Post fault condition		
	R-N	Y-N	B-N	R-N	Y-N	B-N
RN	200	173	27	200	163.8	38.3
YN	105	200	195	109	200.3	92.5
BN	30	170	200	40.5	161.7	201.1

Note: As could be seen from the post-fault magnetizing current test and magnetic balance test, it was decisively confirmed that the transformer was found in healthy condition. The as found other tests to substantiate the findings of magnetic balance test and the magnetizing current test were performed as stated here-under;

i. LT short-circuit test,
ii. IR values and PI test,
iii. Winding resistance test,
iv. Ratio test,
v. tan δ and capacitance measurement of the windings,
vi. tan δ and the capacitance of the remaining HV R-ph bushing.

Action Taken; All the as found tests mentioned in Cl. 1.2.2 were found within acceptable limits and following decisions were taken and executed successfully;

i. for replacing the damaged HV B-ph & Y-Ph 145kV bushings with new ones,
ii. Re-fitting of PRV by removing a dummy circular blanking plate.
iii. Removal of debris of damaged porcelain of L2 of the HV-Bph bushing, pieces of aluminium foils and burnt kraft paper with the help of locally made a spade like tool out of iron scrap material, after decanting the transformer oil.
iv. filtration of the Transformer oil after decanting and refilling the oil into the Transformer,
v. performed pre-energization tests.

After the above activities the transformer was re-energized successfully and it is still in service.

Case study 2;

- Sub-station:132 kV S/S.
- Failure 40 MVA 132/33kV Transformer
- Date of failure: 28/06/2020

Indications: HV side;

a. Differential relay Trip: A phase, (ABC Convention.)equivalent to R phase (RYB convention). Buchholz Trip.

LV side: O/C R phase, E/ F high set.

Testing; Transformer tested at tap No 6.

i. **LT Magnetizing current test; Post tripping;**

Phase	HV side				LV side			
	3Ph, 3W		1-Ph		3-Ph, 3W		1-Ph	
	Voltage	Current	Voltage	Current	Voltage	Current	Voltage	Current
R-Y	397	47.6mA	R-N 226	73mA	R-Y 397	0.71A	R-N 226	1.06A
Y-B	398	23.8 mA	Y-N 226	2.2 mA	Y-B 398	0.34 A	Y-N 226	17.8 mA
B-R	395	23.4 mA	B-N 226	2.3 mA	B-R 395	0.33A	B-N 226	17.9 mA

ii. **LT Magnetic Balance test; Post tripping**

HV Side						LV side		
RN	YN	BN	rn	yn	bn	rn	yn	bn
228	200	28	56	50	6	226	183	45
4	228	224	1	57	56	2.6	228	199
2	225	228	0.5	56.5	57	1.14	226	228

Analysis

a. Relay indication of ABB RET 650 Relay shows that "A" Differential relay has operated in RYB convention it is "R" Phase winding getting involved along with Buchholz trip indication. This means that the fault is of severe nature.

b. It is observed from 3ph 3Wire and 1ph 2 W Testing methods the magnetizing current in R phase winding (HV&LV) were observed higher. It is anticipated that there must have been inter-turns shorting higher magnetizing current is to balance AT as the turns must have got shorted.

c. the magnetic balance too is disturbed which is evident from the lower value of voltages appearing against the R ph winding limb in HV and LV magnetic balance tests.

Joint inspection at repairer's works; The faulty transformer was jointly inspected by the Transmission Utility's Engineer and repairer's Engineer at the works of the repairer. It was revealed that "R phase LV winding had sustained severe inter-turn shorting which confirms our findings that fault laid in "R-ph" winding.

Following photographs of joint inspection are depicted hereunder as fig1 to Figure 3;

Figure 1: Inspecting the core-coil assembly of the transformer.

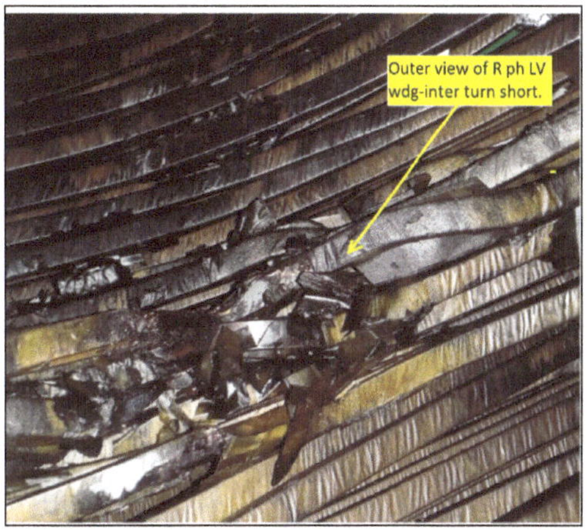

Figure 2: Outer view of R ph LV winding showing inter turn shorting.

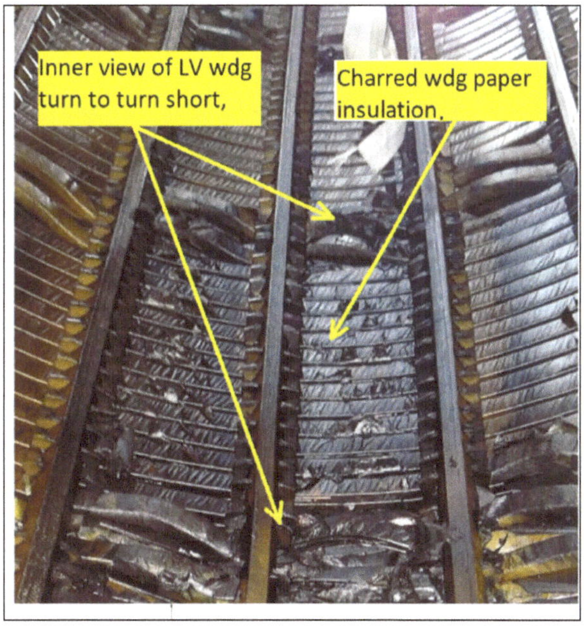

Figure 3: Inner view of R ph LV winding showing turn to turn short and charred insulating paper wrapped around winding.

Observation; As could be observed from the above LT test results;

a. In the magnetic balance test, there will be much reduced induced voltage in the faulty phase as compared to remaining healthy phases.
b. In the magnetizing current test, pronounced magnetizing current in LT 3-ph,3 wire and 1-ph, 2 wire system is observed in faulty winding which indicates turn to turn fault in that particular winding.

Note: Pronounced magnetizing current would appear in that particular phase in which minimum voltage is observed to have induced in the faulty limb in magnetic balance test.

Conclusion;

a. The test results obtained from the above case studies with the Magnetizing Current Tests and LT Magnetic balance tests with LT 3-ph and 1-ph supply, may be designated as the prima-facie reliable diagnostic tests to assess the health status of a Transformer which had tripped due to severe faults at site. The joint inspection photographs in sample case study no. 2 confirms the same.
b. Looking to the observations, the LT Magnetic balance tests and the LT Magnetizing Current tests may be called as complementary to each other.
c. Since, none of the failed transformers were repairable at site, they were sent for repairs to the OEMs or repairers works. The transformers which had lived more than 20 years were survey-reported and were auctioned.

7.0 References;

1. Prof.Dr.S.V.Kulkarni and Prof.S.V.Khaparde-Transformer Design and practice 2nd edition.
2. V.K.Lakhiani and Dharamvir (CGL Ltd.M'deep, Bhopal):Experience in using magnetic balance test as diagnostic test for Transformers, paper presented in the International Conference on 6th-7th April 2000 at CBIP, New Delhi.
3. IEEE std C57 152-2013; Guide for diagnostic field testing of fluid-filled Power Transformers, Regulators and Reactors.

8.0 Authors; K.K.Murty, N.S.Lodhi & Ramesh Singh

6. Dos & Don'ts for EHV Transformers

1.0 Preamble: The EHV Transformers are the costliest and vital equipment in any Power network /large Industries. It is the prime duty of an EHV Sub-Station Manager / Sub-Station O&M Engineer to look after any EHV Transformer right from its receipt at site to its life-cycle scrupulously with mother-like feelings, as mother nurtures her child as soon as the baby is born.

In the absence of any standards in India, guide-lines of the OEMs are followed, which are limited to the transformers only. This Article is significant and important contribution of experienced Engineers in which they have specifically narrated/ deliberated all the aspects from receiving the consignment of an EHV Transformer at site to its life-cycle including safety aspects pertaining to men & material.

This may be used as guide-lines by the Sub-Station Managers/ Sub-station Engineers in the absence of any such guide-lines issued /in vogue in their Utilities/ Industries.

2.0 Key Words; N2 (Nitrogen Gas), SFRA (Sweep Frequency Response Analysis), FAT (Factory Acceptance Test) and TCIV (Transformer Conservator Isolation Valve) and OEM (Original Equipment Manufacturer) and Torr: unit of pressure equivalent mm of Hg. DGA (Dissolved Gas Analysis) and Furan analysis.

3.0 Dos

1. On arrival of the consignment of the EHV transformer at site and before unloading it from the multi-wheeled low bed tractor trailer.

Figure 1: Transformer loaded on multi wheeled low bed Tractor Trolley.

a. Before unloading of transformer from tractor trolley, it should be ensured that trailer is placed in such a way that when it is unloaded on plinth its HV Bushings are on HV Side Bus Bar.
b. Check the Transformer for any signs of physical damages during Transportation thereof.

c. Get Impact recorder and its curves provided by factory officials.
d. Check, N2 Pressure is found to be in +ve pressure.
e. Check the dew point of N2, it should not be more than -30°C.(in case of 400kV Transformers).
f. Check and verify that the Transformer-core of the Transformer is single-point earthed.

Figure 2: Earthing of Tank, Core and Core frame on the top cover of the transformer.

g. Measure the IR value between core-frame-tank with 500 V DC range megger.
h. Check and verify the isolation of test leads from tank by taking continuity with multi-meter.

Figure 3: Depicts epoxy test bushings on the blanking plates of the Transformer turrets and test leads are connected to them for LT tests of the transformer on the low bed Trailer.

i. Measure the magnetic balance, magnetizing current and verify from FAT.
j. Verify the vector group of Transformer as per R&D plate.
k. Prepare Transformer's Life sketch-book as soon as the transformer is received at site. Enter all FAT results of the transformer in that life sketch book.
l. Connect N2 gas cylinder with regulator, if the Transformer is to be stored for long, so as to maintain **(+*ve*)** pressure.

3.1 Erection, installation and assembly;

1. Shift the transformer to the plinth and lock all rollers/wheels with rails on the plinth.
2. Care should be taken for ventilating the Transforms main tank by opening blanking plates on the top of the transformer tank and at any other location in the lower level of the tank.
3. It should be ensured that any person either from supplier's side or departmental person does not carry pen/pencil or any other material in his upper pocket before entering the tank of the Transformer for internal inspection.
4. Internal inspection should be carried out carefully. All core bolts, core-frame-tank leads, selector switch leads tightness should be checked.

[**Note:** In case of internal inspections of large Transformers, Dry compressed Air /Oxygen Cylinders (4-5-Nos) are kept ready at the site. There is always a threat of insufficient oxygen inside the tank which may cause breathlessness / giddiness to the personnel who would be entering the tank for internal inspection. In order to obviate such condition, dry compressed air/Oxygen is injected through the bottom valve, keeping one blanking plate open on the top of the transformer tank diagonally opposite to the bottom valve through which dry air /Oxygen will be injected for proper ventilation. Ensure continuous flow of dry air/ oxygen during entire internal inspection period]

5. Fit equalizing pipes between main tank and OLTC compartment/compartments.
6. Before oil filling, all residual oil in the Transformer tank should be drained out and the tank bottom should be cleaned with cotton cloth.
7. After the activities as mentioned in item nos.8 & 9 are over, Transformer Tank (excluding cooler banks and conservator)shall be evacuated to a pressure of vacuum 1.00Torr and should be held for 48hrs.

[**Note:** (a) Now a days some OEMS design the cooler banks vacuum proof in that case cooler banks can be included.

(b) torr to mmHg conversion: 1 torr = 1 mmHg.]

8. Fill up Oil to the core level. Before filling oil ensure all oil valves are in correct position either closed or opened as required.

[**Note** (a) Mineral oil confirming to IS: 335 in case of Mineral Oil filled Transformer or

(b) Synthetic Ester fluid confirming to IS-16081-June 2013(IEC equivalent IEC-61099)]

12. In case of 400kV Transformers before filling oil, absolute vacuum of 1.00 torr should be taken and held for 72 hrs. Then N2 gas of less than –50°C dew point should be pushed in till such time the vacuum pressure comes to the normal value and then to 0.2kg/sq.cm pressure and to be kept in this condition for next 24hrs, then release N2 and measure dew point of N2 coming out from the transformer. It should not be more = 40°C. If it is observed that it is > 40°C, then, N2 purging cycle to be repeated ones again.

[Note: Care should be taken to blank all the holes of PRVs and shut the valves of external Radiators etc.]

13. Before erection of bushings tally and confirm the Bushing nos. from FAT results that they are the same ones used during FAT.
14. Erect: Radiators, Thermo siphon, PRV, Pipeline from main to conservator including Buchholz relay,

[**Note:** (a) In case the transformer is equipped with NIFPS system, while assembling the pipe from conservator to the main tank containing Buchholz relay, Transformer conservator isolation valve (between Buchholz relay and conservator side valve.ref sample fig below), and the steel bellow should be fitted in between them carefully.

Figure 4: installation of TCIV (Transformer Conservator Isolation valve)

b. Before erection of conservator, check leakage of the Air-cell and then erect the conservator.]

15. Before filling oil turn TCIV to filtration mode.
16. Fill Oil to Conservator level.
17. Charge the air-cell as per the procedure of OEM.
18. Get SFRA check done at site and verify the curves with the signature curves taken during FAT.

[Note: There are two schools of thoughts, in this regard. Some Utilities get this SFRA test during pre-commissioning tests]

19. Fill Thermometer pockets with oil.
20. Carry on Transformer oil filtration job till the BDV is in the range of 75 to 80kV and ppm-5 keeping in view that the temperature of Transformer tank oil does not increase beyond 50^{0} C.

[Note:

a. During filtration work Radiator Valves should be kept in closed condition.
b. After the filtration work is over, push the oil from the bottom and open radiator valves one by one]

21. In case of 400kV transformers, The OEM's Erection Engineer shall take care during Erection and Oil filling/ drying out job of the 400kV transformer as per the procedure laid-down by the respective OEMs (Original Equipment Manufacturer).The Sub Station manager/S/S Engineer should extend all possible help and Support. to the OEM's erection Engineer
22. OLTC operating mechanism and OLTC conservators erection: Connecting of Operating rod and the worm gear arrangement to be fitted and also fitting of OSR (Oil Surge relay) etc.
23. While connecting Tap Driving Mechanism Box with tap changer care should be taken so that both DM Box and TC are on same tap number.
24. Neutral Grounding flat or conductor of suitable size must run through support insulators and connect to two separate earth pits.
25. Verify the resistance of Neutral Grounding pit/pits. It should be below 0.5 Ohms
26. Verify the integrity of the earthing system.
27. Ensure proper earthing of the body of the transformer at earmarked 2 no. places.
28. **Delta Tertiary winding/bushings; (3U, 3V &3W): Regarding the protection against transferred surges;**

a. Follow the instructions of the OEMs.
b. In the absence of any instructions from the OEMs, in case of 220/132/33 kV Auto transformers, then earth one terminal either 3U or 3W of the bushing.
c. In case of 400/220/33kV Auto Transformers and above, strictly follow the instructions of the OEMs.

[Note: To protect the Tertiary winding against transferred surges, they may recommend Surge Protection Capacitors in conjunction with LAs on each phase of Tertiary winding].

ii. In case, 33kV Delta tertiary winding is used for feeding loads for Sub-station supply, immediately after the bushings, 30kV LAs should be installed on each phase. A DTR of 33/0.433 kV, Dyn11 is to be installed of appropriate capacity with suitable isolation & protection arrangements in 33 kV as well as on 0.433kV side.

iii. In case, 33kV Delta tertiary winding is used for feeding loads to DISCOMs then Zig-Zag Neutral Grounding Transformer of appropriate capacity should also be installed with associated Relays, Circuit breakers, CTs, Isolators and LAs.

iv. In case, 33kV Delta tertiary winding is used for feeding reactive compensation devices such as Capacitor banks or Reactor, necessary LAs, Isolators, CTs and Circuit breakers of adequate capacity should be installed. Some Utilities also use NDR Protection (Neutral displacement protection) using 3ph, 33kV RVTs.

29. Ensure that 1U,1V & 1W designated terminals of Transformer are matching with R,Y,B phase of buses respectively of switchyards on HV and 2U,2V & 2W on IV(in case of Auto Transformers)/ LV side(in case of Two winding transformer) and 3U,3V & 3W in delta tertiary winding in case of Auto-Transformer.

[Note: The nomenclature used for Transformer winding/bushings is in U, V,& W. form with prefix 1,2 & 3 respectively as per IS:2026-Part (iv)1977 reaffirmed in 2001.]

In a Transformer bank of three single phase units, ensure tertiary winding connection as per vector group YNa0d11 of the Auto-Transformer and connect them accordingly externally either on 33kV bus drawn near the Auto transformer or through 36 kV, XLPE cables.

[Note: If Delta Tertiary is not used either for Reactive compensation equipments, Station supply or feeding 33kV loads, then one corner of Delta bushing should be earthed.]

30. **Pre Commissioning Tests;** In some Utilities**,** the Job of relay settings and testing thereof are responsibilities of the Manager/ Engineer S/S, However in most of the Utilities in India, the Job responsibility lies with a separate wing/department/division which are regally designated as Testing Divisions, MRT divisions and Equipment & Relay protection Divisions etc

31. A. Prior to pre commissioning Tests, the responsible authority should calculate and arrive at appropriate settings for each and every protective relay and shall arrange to test the protective relays accordingly.

    B. Perform the pre-commissioning tests as recommended by the respective Utilities in particular; however, following Pre-commissioning tests are recommended and may be performed in agreement with the Commissioning Engineers of different OEMs.

    a. Core Insulation Test.
    b. Magnetic balance test.
    c. Magnetizing current Test:
    d. Vector group test,
    e. Short Circuit test.
    f. SFRA (Sweep Frequency Response Analysis) Test.
    g. Capacitance and Tan **δ** of Transformer Windings.
    h. Capacitance and Tan **δ** of Bushings.
    i. Insulation Resistance IR value measurement and PI.
    j. Ratio Test.

k. Winding Resistance Measurement.
l. Through Fault Stability test.
m. REF Stability tests.

[Note:

i. Carry-out all the Pre-commissioning tests mentioned above before charging /commissioning of the Transformer and verify the test results with the FAT results.
ii. SFRA (Sweep Frequency Response Analysis) Tests should be carried out and the curves obtained during pre-commissioning test should be essentially verified with the signature curves of SFRA test taken during FAT.
iii. Ensure testing of the transformer with AC and thereafter DC tests should followed, otherwise the test results taken at the time of FAT would not tally.]

C. Ensure that all the Protective relays, Circuit breakers, CTs, Isolators and LAs, etc are tested and are ready for charging.

[Note: In case of SPR Breaker (Single-Pole Reclosure breaker), Pole Discrepancy protection should be checked and the recommended setting is 300 ms to 500 ms]

D. **OTI &WTI Settings**: Sample procedure for setting OTI & WTI settings:

The Guaranteed temperature rise of Oil and Winding are embossed on the R&D plate of each Transformer.

Say Guaranteed Oil temperature raise is 40 °C and Guaranteed Winding temperature raise as 50 °C as mentioned on the R&D plate of the Transformer.Assuming maximum ambient temperature as 50 °C;

OTI& WTI Alarm/Trip & Fan start	Guaranteed Temp. rise deg. C	Amb. Temp deg. C.	Calculation: Guaranteed temp rise+ Amb. temp. deg C	Temp. Settings in deg C
OTI Trip	40	50	40+50=90	90
OTI Alarm	40	50	May be set at	80
WTI Trip	50	50	50+50=100	100
WTI Alarm	50	50	May be set at	85
Fan start			May be set at	60

[Note: It is a general practice and logical also that the Winding temp alarm should be set prior to OT Trip]

E. All the protections on the Transformer body itself as stated hereunder must be manually operated or simulated such that they in-turn operate the master trip relay.
   a. Buchholz trip.
   b. PRV trip,
   c. OTI & WTI Trip.
   d. OSR trip (Oil surge relay is installed for OLTC).

[Note1: Check the Alarm fascia for respective operation of the above trippings.]

F. All the annunciation indications should also be checked through simulating the conditions of each relay;
   a. Oil level low,
   b. Buchholz alarm,
   c. Oil temperature alarm.
   d. WTI alarm.

G. All other annunciations pertaining to the Transformer's external Protections are to be checked on the operator's desk for their genuinity by manually operating or simulating the respective relays such as Differential relay, restricted earth fault relay and Over-fluxing relay as required.

H. Trip test to be taken through the master-trip relay or module and ensure tripping of HV and LV breakers simultaneously through simulation of Master trip relay or Module for master trip

3.2 Preparation before energizing the Transformer;

32. Ensure readiness the NIFPS system.
33. Release air from all the earmarked points including bushings before energize.
34. Release air from each radiator or bank of radiators as the case may be.
35. Ensure that connectivity of all tan-**δ** points and core-frame-tank terminals are earthed properly.
36. Before energizing of the Transformer release the air through the vent plug of the Diverter switch.
37. Check the Oil level in the Diverter switch.
38. Check the healthiness of Protection circuits.
39. Check the Aux DC system.
40. Check the pointers of all gauges for their free moment.
41. Check the Buchholz relay and readjust the float/ switches etc if necessary.
42. Check the healthiness of the air-cell by oil levels in Prismatic oil level indicator on the conservator tanks.
43. Ensure removal of the discharge rods from HV, IV/LV bays wherever they were hanged.

3.4 Energizing the transformer:

44. To be on safer side, charge the transformer at tap no. 2 in case of the CFVV (Constant Flux Voltage Variation) system type of OLTCs.

[Note: In the CFVV, the value of HV voltage shall remain constant, thus the flux shall also remain constant in the Core irrespective of the tap position of the LV voltage]

45. At the time of charging of transformer, first switch-on the stuck breaker protection in/out switch to "in" position.
46. Put the NIFP system in service (on) position.
47. Energize the transformer and keep it without load for at least 24 hrs.

48. Now a days, since numerical relays are used for protection hence curve of Inrush/charging current Vs time may also be captured.
49. After 24hrs of switching off of the Transformer release air from all the earmarked points including bushings before taking load.
50. Switch on the Transformer and load the Transformer.

3.4 Routine maintenance;

51. Earmark the working area of the transformer's HV & LV bays, including all the outdoor equipments with red-colored cloth strips about 1foot wide.
52. Every precaution to safe-guard men and material at site should be ensured before and after taking shut-down on the Transformer
53. Carry out routine maintenance jobs as per the schedule defined by the Utility
54. **Transformers equipped with NIFPS;**

After the transformer is switched off, following action /activities are to be performed to facilitate periodic maintenance and testing of the transformer protections etc.

a. Auto/manual Switch of NIFPS to be turned to "Off" position on the control panel in the S/S Control room.
b. Lock Nitrogen Injection solenoid valve and Oil drain solenoid valve by properly inserting the respective lock pins in the Control Cubicle in the Switch yard.

[Caution: If above activities are not done, there is every possibility that NIFPS may mal-operate during testing of protections of the Transformer resulting in to Nitrogen injection into the Transformer and simultaneously draining of 10% of oil thereof.]

55. a. **DGA (Dissolve Gas Analysis)**: Recommended frequency for DGA is Quarterly. The results thereof should confirm to IEC 60599, however if there is an appreciable abnormality observed, in the DGA test results, the Matter should be taken up with the OEMs. In such cases, the frequency of DGA should be carried out as per the OEM's recommendations and report thereof should be appraised to the OEM for further action/decision.

b. **Evaluation of Oil Quality**: Oil sample should confirm to IS: 1186.

[Note:

i. For DGA oil sampling should be done from the top,
ii. For Oil parameters, Oil sampling should be done from the top & bottom of the Transformer]

c. **Furan Analysis**: It is recommended that Furan analysis of Transformer oil may be to be carried out once in every 2 years, of the old Transformers which are continuously rendering service since last 25-30 years. The samples from (bottom and top of the transformers are to be collected and sent to the reputed laboratories like CPRI, Bengaluru, ERDA, Vadodara or any other NABL accredited labs for analysis.

While discharging the switched-off Transformer, care should be taken to connect the loose end of the earthing lead to be firmly tied / tightened on to a metallic structure after scraping the paint or rust thoroughly / earthing point FIRST, and then the earthing Clamp/ G-clamp of the discharge-rods should be connected to the HV- LV jumpers to discharge the Induced residual High voltage.

56. After the EHV Transformer is switched off and discharged as mentioned in previous item, the phase jumpers should be removed first and there-after, the neutral jumper should be removed from the transformer bushings
57. While normalizing the connections of the Transformer, first of all the neutral jumper/clamps should be connected and thereafter the phase jumpers to respective bushings.
58. Attend the leakage in the Bushings immediately if found during operation by taking a shut-down on the transformer.
59. The bushings should be cleaned with dry cotton cloth during each shut downs.
60. Special emphasis to be given to tan δ & capacitance measurement and monitoring thereof of OIP/RIP bushings.
61. Check the Oil in the Transformer and OLTC for Dielectric strength and Moisture content. Take suitable action for restoring the quality.
62. Check the oil level in the Oil-cup and ensure air passage is free in the breather. If oil is less, make-up the oil.
63. Check the oil for Acidity and sludge as per IS 1866 in case of mineral Oil.
64. If any inspection cover is opened for some inspection or work it should be re-fixed with a new Gasket/O -ring
65. Check the WTI and OTI pockets and replenish the oil if found less in their respective pockets.
66. Check all the bearings and driving mechanism and lubricate them as per schedule.
67. Check the healthiness and readiness of NIFPS (Nitrogen Injection Fire-Prevention system).
68. a. Check the integrity of Grounding of HV & LV neutrals in their respective earth pits.
69. b. Check integrity of earthing of all the equipments in the Transformer bays with the help of Primary injection kit.
70. Check the earthing of Tertiary winding bushing.
71. Earth resistance to be checked periodically
72. Thermo-vision survey/Infra Red Scanning of all the Bushings, Bushing clamps, Jumpers of the HV and LV bays should be carried out monthly or periodicity as defined by the Utility.
73. Check cooling fan's functions vis' a vis setting of temperatures thereof.
74. Check air cell in conservator, it should be in inflated position.

Trip/ appearance of Alarms;

75. Check the Transformer thoroughly after any alarm or protection operates.
76. Check the HV/IV/LV bays for damages of equipment, if any.
77. After each tripping of the Transformers, DGA should be carried out and the results should be compared with the results of earlier DGA report.

4.0 Don’ts;

1. **Do not accept the consignment of the Transformer at site,**
    a. If multiple earthing of core found while the transformer is in the loaded condition on a the low bed trailer itself.
    b. If the nitrogen pressure not found having +ve pressure.
    c. If the Impact (vibration) recorder shows some abnormalityIf the LT tests carried out while the transformer is on the low bed Trailer found unsatisfactory.

[Note: Do not forget to inform the OEM about the as found status of the Transformer immediately]

2. While tightening/Loosening any bolt etc inside the Transformer tank through the inspection windows, ensure that the person carrying it out should not keep any article in his upper pocket.
3. Do not use low capacity lifting hydraulic Jacks.
4. Do not forget to tie a cotton tape at one end of the spanner or any other tool and the other end of the cotton tape should be tied to any clamp or hook on with the Transformer top/body while tightening/ loosening operation of any bolt inside the Transformer through the Inspection window.
5. Do not keep Transformer/Reactor N2 gas filled at site for a long period (>6months).
6. Do not commission the Transformer before conducting all the Pre-commissioning Tests.
7. Do not leave Delta tertiary winding terminals unprotected outside the tank. They should be connected to LAs / Surge Protection capacitors in conjunction with LAs of appropriate rating.
8. Do not move the Transformer with Bushings mounted.
9. Do not parallel the Transformer without confirming that the transformer in question fulfills all the conditions/ criteria such as Vector Group, Terminal Voltages, Voltage ratios, tap variations of OLTC and % Z.
10. Do not over load the Transformer beyond the limits as specified in IS: 2026

    (Part-7):2009(IEC equivalent IEC 60076-7:2005).

[Note:

    a. Relative aging rate-If hot spot temperature exceeds from 80 deg. C to 140 deg, C, for every 6 deg. C temperature rise the relative aging becomes double.
    b. Loading Beyond Name Plate detailed Current (pu);
    c. Cyclic Over loading:1.3 pu keeping in view the WTI settings.
    d. Emergency loading (30 min): 1.5 pu keeping in view the WTI settings.]

11. Do not change settings of WTI and OTI, alarm and trip frequently.
12. Do not leave Red pointer behind black pointer in WTI & OTI.
13. Do not meddle with the protection circuits.
14. Do not allow conservator Oil to fall below-1/4 level.
15. Do not allow bushing Oil level to fall, they must be immediately topped-up.
16. Do not leave the marshalling box doors open. They should be locked.

17. Do not switch off heater in the Marshalling Box.
18. Do not leave ladder unlocked.
19. Do not allow unauthorized persons near the Transformers.
20. Do not mix oil to the existing Oil, unless it confirms to IS: 335 in case of mineral oil and IS: 16081- June 20013 in case of Synthetic Ester fluid.
21. Do not continue with deteriorated silica-gel it should be immediately replaced immediately.
22. Do not energize the transformer without thoroughly investigating/observing the Transformer after appearance of Buchholz relay Alarm or Trip indication and tripping on Differential, OSR and PRV.
23. Do not energize the Transformer without thoroughly testing the Transformer for its healthiness and without analyzing the trapped Gas in the Buchholz relay and DGA of the Transformer oil.
24. Do not parallel the transformers without bringing them on appropriate tap position as per operating instructions.
25. Do not charge the incoming of transformer without ensuring that capacitor bank is 'OFF'.
26. Do not leave cap of tan delta measurement point on the bushings in open condition.
27. Do not push the oil through drain valve.

5.0 Conclusion: The activities narrated in this Article may be treated as general guide -lines, if followed the life expectancy of the transformer would be enhanced however, if there are other specific instructions of the utilities, they may be followed scrupulously.

6.0 References;

1. Terminal marking UVW as per IS: 2026-Part (iv) 1977 reaffirmed in 2001.
2. Mineral oil fresh IS: 335.
3. Synthetic Ester fluid confirming to IS: 16081-June 2013(IEC equivalent IEC-61099).
4. Dissolve Gas Analysis IEC 60599.
5. Loading guide for Oil immersed Power transformers IS:2026-part-7(IEC equivalent; 60076-7)

7.0 Authors:K.K.Murty, J J L Kapil, Santosh Dubey & S K Chaturved.

7. In - house Fabrication of Turret-adopter for Retrofitting of 245 kV OIP Bushing Having Longer OIL- End

ABSTRACT: This article focuses on a true case history of replacement of a leaky 245kV OIP Bushing of a 160 MVA Transformer at 220kV Sub-station at Jabalpur, M.P. India with an available bushing having longer L-2 (Oil end length from the flange to bottom) by in- house fabrication of the Turret –adopter (an extension piece to turret) on an innovative thought / idea at site without disturbing its location in minimal down- time.

Key words: Turret, OIP Bushing (Oil Impregnated Paper Bushing).DL Type (Draw lead type), PCD (Pitch-Circle Chordal Distance)

1.0 Introduction: Maintaining large inventory of major components of transformers for the whole population of transformers in any Transmission utility is a very costly and uneconomical affair. With this in view, Bureau of Indian Standards has issued an IS namely IS: 12676: 1989"Oil Impregnated Paper insulated Condenser Bushings-Dimensions and Requirements" for the most sensitive and essential component of the Transformer, "the OIP-Bushings", however some of the manufacturers do not follow the aforesaid IS and design their Transformers at their own convenience and constraints putting the end users/ customers in trouble when ever EHV bushings fail/ need replacement.

One such incidence took place at the 220kV Sub-station Jabalpur and replacement of 1- no.245kV bushing was essential of a 160 MVA 220/132kV Transformer. The available bushing in the Stores was of having larger dimension of L-2 (Oil end length from flange to bottom) to accommodate the same was impossible. The OEM of the Transformer was contacted and was requested to provide us with extension of Turret or a longer turret but they showed their inability to do so. The only option left was to accept the challenge. The proverb "Necessity is the mother of invention" made us ponder and innovate an idea of locally fabrication of the extension cylindrical piece to extend the height of the original turret with available in-house resources.

2.0 Incidence/Occurrence: At 220kV Jabalpur S/s, there are 4 Nos. 220/132/33/KV Transformers

a. 2x160 MVA GEC & BHEL, Make.
b. 2x120 MVA (3x40 MVA, 1-Ph units) Mitsubishi & BHEL Make.

Catering to the demand of Jabalpur and neighboring places.

Out of the 4 No.s, 220/132/33 kV Transformers as indicated above, 160 MVA GEC make Transformer (Year of manufacture-1993) developed profuse leakage of oil from "U – Phase 245kV OIP bushing from the jointing location of the porcelain bushing and the bushing flange sometime in June 2016.

The seepage of oil could not be arrested by pasting /application of homogeneous paste of porcelain powder mixed in Araldite. Every time topping up of transformer oil into the Bushing was needed. However this kind of remedy used to last about 4-6 months only. The risk of failure of the said bushing prevailed though the tan δ and capacitance values were within limits.

Looking to the ensuing Rabi-season (Irrigation season) and the risk of catastrophic failure of the Transformer, replacement of the Transformer or replacement of the said leaky bushing was essentially felt.

Figure1: Leaky 245kV OIP bushing with application of paste of powdered porcelain and araldite.

3.1 Details of the leaky Bushing; Relevant Details of original leaky 245 kV OIP bushing are as follows;

i. Make: BHEL,
ii. Voltage: 245kV
iii. Current rating: 800Amps.
iv. Length from the flange to the Stress ring (L-2):830 mm.
v. Type: DL(Draw lead type)
vi. Year of Manufacture of the Transformer: 1993.

4.0 Activities and constraints: A thorough search for a substitute bushing with similar physical dimensions and current rating was made in the whole state of M.P. but in vain.

Following were the Options;

a. Replace the Transformer.
b. Procure a tailor made OIP Bushing to the required physical dimensions.
c. Manage for using the available OIP bushing of following rating and L-2 dimension;

i. Make: CGL.
ii. Voltage:245kV

iii. Current 1250Amp.
iv. Type: DL type.
v. L-2:1130 mm.(Confirms to IS: 12676:1989 Reaffirmed in 2001)

Out of the above options, the option (c) was found to be appropriate by way of economy, practicability and the minimum down-time.

As noted from the name plate that the year of manufacture of the said Transformer was 1993 ie 4 years later than the issuance of the IS: 12676:1989 by the Bureau of Indian Standards.

Contrary to the Standard specification IS: 12676:1989, pertaining to the physical dimensions of the OIP bushings, the manufacturer did not follow the same and designed the Transformer with bushing of oil- end length L-2 of 830 mm, jeopardizing and defeating the option of interchangeability of major/ main components of 160 MVA Transformers putting the M.P. Power Transmission Co. Ltd. in trouble.

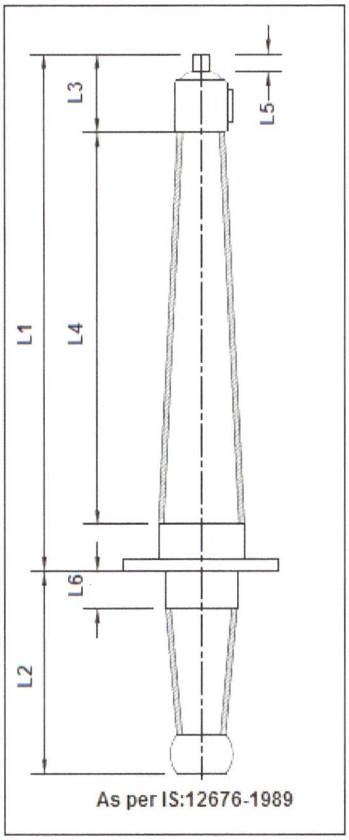

Figure 2: OIP bushing dimensions (in erect position) as per IS: 12676-1989 [1]

1.1 Note: Figure No 2 has been specifically depicted for the information of a common reader as the bushing manufacturers abroad, name the portion – L 2 differently for example;

- ABB has designated it as L-1.
- TRENCH, China (THVS) has designated it as L-3.
- CGL as - H2.
- Alstom as – L5.
- MGC(Moser Glaser), Switzerland. as-L

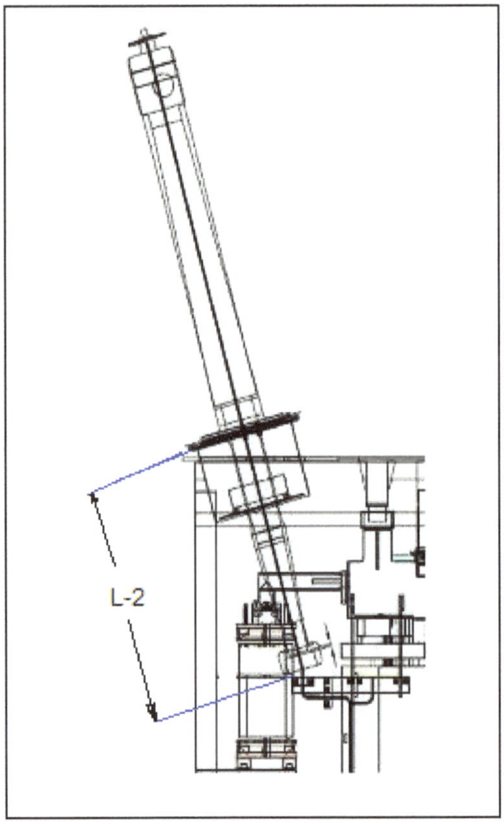

Figure-3: Longer L-2 fouling with Core-clamping frame.

5.0 Challenge: The challenge was “How to use the available bushing?”

The physical dimension of the length of L-2 was 300 mm more than the original bushing (1130mm - 830mm), therefore it was obvious that the Oil end portion of the new bushing would foul with the core-coil assembly clamping frame(end frame).(Refer figure-3)

As the saying goes ”NECESSITY IS THE MOTHER OF INVENTION” such a challenge has led us to explore/innovate some ideas keeping in view the feasibility/practicability without ousting the Transformer from the system in minimum possible time and shut downs/down times.

5.1 Innovative Idea:

- To fabricate an adopter on the turret such that enough clearance could be achieved from the clamping frame of core-coil assembly and to seat the bushing flange thereon.
- To extend the HV lead length of Draw lead type bushing.

6.0 Action to be taken;

i. Designing
   a. A turret adopter (Turret extension piece)
   b. An extension piece for the lead for DL type Bushing.
   c. Extension of equalizing pipes.

ii. To plan Number of shut downs.

iii. To measure the physical dimensions of the turret such as internal & external diameters, PCD and number of bolting holes etc.
iv. Fabrication of Turret-adopter, DL extension piece and equalizer pipe extension.
v. Final installation and testing.

6.1. a. Designing of Turret adopter(Turret Extension piece):

1. Turret - adopter should be designed such that;

- it should withstand vacuum filling of oil at 10 milli tor in the transformer with all fittings.
- It should withstand continuous internal pressure of 35KN/m^2 over normal hydrostatic pressure of oil.

To fulfill the conditions mentioned above, the wall thickness and the flange collar of wrought iron should be such that it should be equivalent to the original turret wall thickness and the flange collar.

The turret - adopter was fabricated;

a. with wrought iron plates 300x 8 mm. purchased from the local market and got them rolled/ curved through the hydraulic rolling machine. They were rolled/curved as half of 618 mm diameter into 2-halves of chord and welded together with precision to form a cylinder of 618 mm Dia. and the flange on both the ends were fabricated with 800 x 800x 8 mm wrought iron plate with internal diameter of 618 mm and external diameter of 720 mm. having 24- nos bolting holes of 19 mm diameter at diameter of 690 mm(PCD) and hole to hole distance of 90 mm. The turret-adopter finally fabricated is shown in figure 4.(refer figure.4).

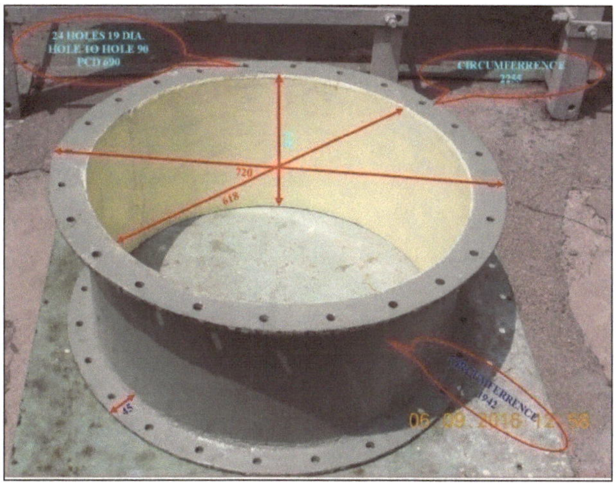

Figure 4: In-house fabricated Turret- adopter.

6.2 Extension of Draw Lead:

The length of the draw lead was extended by a 300 mm using a 360 mm copper rod available in the S/s of a draw rod type-bushing by suitably turning the treads on either ends for fixing the Lead-end and the terminal-end.(Refer figure:5)

Figure5: Copper extension piece showing suitably turned threads at both ends.

The fabricated piece for extension having connected to the original draw lead of the bushing and the extended turret-adopter are depicted in the figure No. 6.

Figure 6: Fitting of Extended turret- adopter and extension of draw lead.

6.3 Extension of equalizer pipe:

The equalizer pipe was extended and connected to the main pipe connecting to the conservator.

6.4 Final installation of Bushing on the Turret -Adopter:

The Bushing was finally installed on the Turret-adopter with Neoprene rubberized gasket of ½" thickness. (Refer figure: 6).

Figure 7: Completion of Installation of new bushing with L2-1130 mm on the bushing turret –adopter.

After completion of installation of the new bushing with L2=1130mm, over the Turret-adopter, oil topping up with filtered transformer oil was done.

7.0 Pre- energization tests: Following tests were performed satisfactorily prior to pre-energization of the Transformer.

i. IR values taken with 5kV megger.
ii. Tan δ and capacitance of the Bushing.
iii. Transformer magnetizing current Test.
iv. Magnetic balance Test.
v. Winding resistance test on all taps.

8.0 Total down time:

i. Measurement of dimensions- 1 ½ hrs shut down taken.
ii. For disconnection of jumpers and preparation etc- 1 day.
iii. For finally removal of old bushing and installation of Turret -adopter, lead extension, new bushing fitting and piping-1 day.

The said Transformer was re-energized successfully on 06.09.2016.

9.0 Acknowledgement: The author expresses his gratitude to the MD, M.P. Power Transmission Co Ltd for permitting him to write this article. The author also expresses his gratitude to the Chief Engineer (T&C), MPPTCL and his team for extending all possible help and support.

10.0 Reference:

1. IS: 12676:1989.(Reaffirmed 2001), Oil Impregnated Paper Insulated Condenser Bushings-Dimensions and Requirements.

11.0 The Authors: 1) K.K.Murty 2) JJL Kapil & 3) Santosh Dubey.

8. Auto-Transformer and On-Load Tap Changer (OLTC)

1.0. Principle of Auto-transformer(Figure 1):

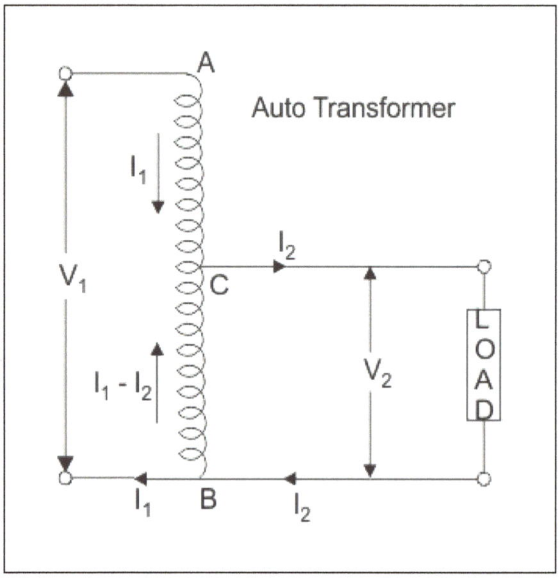

Figure 1

An Auto-Transformer has a common winding and is manufactured when the transformation ratio is less than 2. Unlike Conventional transformer, the HV-IV windings are electrically connected in a fashion shown in the fig. the arrangement is similar to that of a Dimerstat (voltage variac).

N1 and N2 are HV and IV windings (Intermediate winding) respectively. The Transformation equation is as follows;

$$\frac{V1}{V2} = \frac{N1}{N2} = K.$$

As could be seen from the fig: that the portion of winding AC(N2-N1) carries I_1 current and the winding N2 carries I2-I1 current.

2.0 Saving in copper in Auto-Transformer:

The wt of copper in AC section (series winding) is proportional to; $(N_1\text{-}N_2)\ I_1$.

The wt of copper in the section BC is; $N_2\,(I_2\text{-}I_1)$ The wt

So total Wt in auto Transformer s Proportional to; $(N_1\text{-}N_2)I_1 + N_2\,(I_2\text{-}I_1)$.

In a 2- winding Transformer the total Wt of copper is Proportional to; $(N_1I_1 + N_2I_2)$,

$$\frac{\text{Wt of Cu in AutoTransformer}}{\text{Wt of Cu in 2Wdng Transformer}} = \frac{(N1 - N2)I1 + N2(I2 - I1)}{(N1I1 + N2I2)};$$

$$1 - (\frac{2K}{2}) = 1 - K; \text{ where } K = \frac{N1}{N2} = \frac{I2}{I1}.$$

Therefore, Wt of Cu in Auto Transformer (Wa) = (1-K) X Wt of Cu in 2-Wndg Transformer (W_0).

Thus saving = W_0 – Wa = K. W_0.

Therefore Saving will increase as K approaches Unity.

3.0. 3-Ph Auto Transformer.

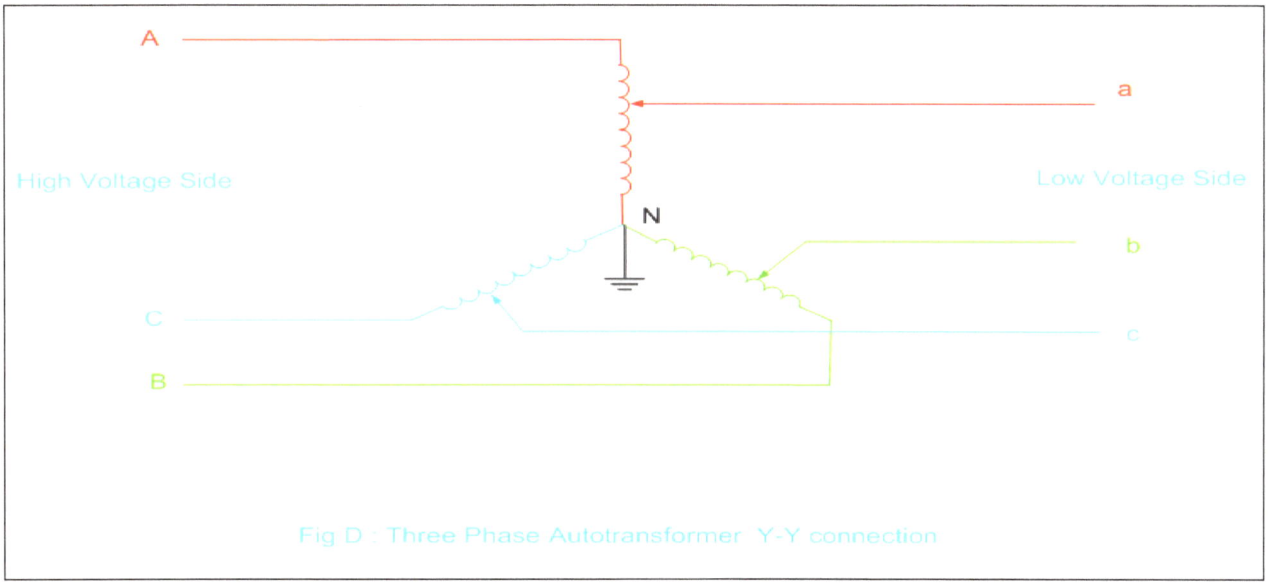

Figure 2

3.1 Formation of 3- Ph bank with 3 Nos., 1-Ph units:

3-single phase units are formed in to 3-phase bank on the external bus and the ground (The Star point and Neutral are formed in the ground) (Figure 2). The capacity of 3-Ph bank shall be 3 x capacity of 1- ph Unit. Such banks exist in some of our 400kV S/s and in many of 220kV S/Ss.

Example: At 220kV S/S Jabalpur, 3x40MVA (1-Ph units) = A bank of 120MVA 3phase Transformers exist. The Voltage rating of 1-Ph units of 40 MVA Transformers is $\frac{220/\sqrt{3}}{132/\sqrt{3}}$ / 33kV.

4.0 ON LOAD TAP CHANGER.

As the name specifies, it is an equipment which is very essential for raising / lowering of transformer-taps so as to maintain system voltage profile on load without disturbing the system/ continuous process of an industry.

There are 2-types of Systems;

1. Externally mounted on the Transformer body type with its tap changing action namely selection and then connection of taps is conducted in one compartment. This is generally used in small rating Transformers.
2. This is known as M-Type Tap changer. This is in 2- parts (i) The Tap selector & (ii) The Diverter switch. Combined together they make one M-Type On load Tap changer. Ref. fig. 3A

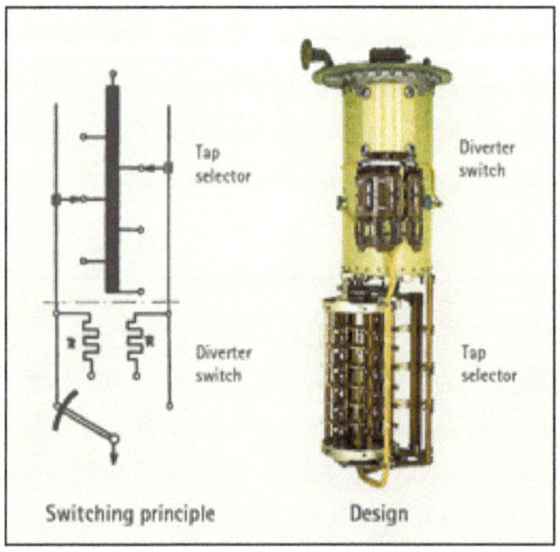

Fig.3A: Switching Principle & Complete unit with Tap seletor & Diverter switch

Figure 3 B: OLTC-Intank type

We shall confine ourselves to the M-Type OLTC only.

i. **Tap Selector** (Figure 3A)**:** It comprises gearing arrangement and with spring loaded fixed and moving contacts. It portion of the M-Type OLTC is suspended in the Transformer main tank and deep into the transformer oil present in the Transformer. The tap leads from the windings are connected to this unit and depending upon the tap positions the contacts are made under spring loaded driving mechanism. During this whole process of tap selection, since it is only a selection process, there is no sparking during its process.

ii. **Diverter switch unit**(Figure 3A**):** This unit comprises of spring operated energy accumulator, the diverter switch itself and the Transition registers. This is housed in to a separate cylindrical compartment. The sparking during tap change operation may take place.

5.0 Understanding Switching sequence: First the selector switch selects the tap and then the switching action takes place in the Diverter switch through transit resistances(to avoid open circuiting) (.ref fig below;)

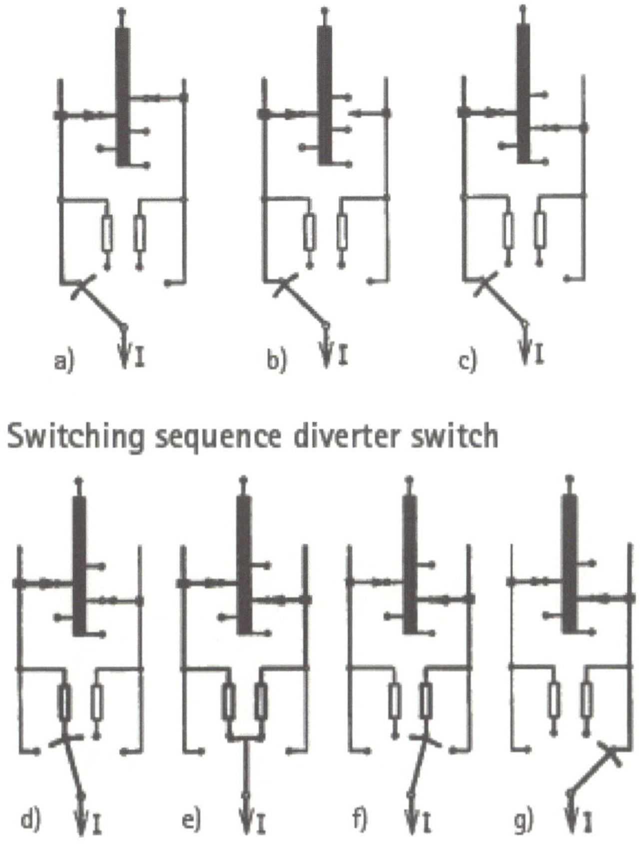

Fig. 12: Switching sequence of tap selector – diverter switch (arcing switch)

Figure 4: Switching Sequence

6.0 Placement/Location of OLTC: It is generally located in the HV winding as the handling of the same as compared to LV winding is easy and due to the fact that it performs its duty in a lower current side of the Transformer ie the HV side.

6.1 Placement of the OLTC in Y connected Transformer(Figure 5:A) In a 2-winding Transformer, the Tapchanger is generally placed close to the starpoint of the 3 phases, which is subsequently formes into the neutral point and in practice is the said neutral is grounded. The potential near the star point is very low.

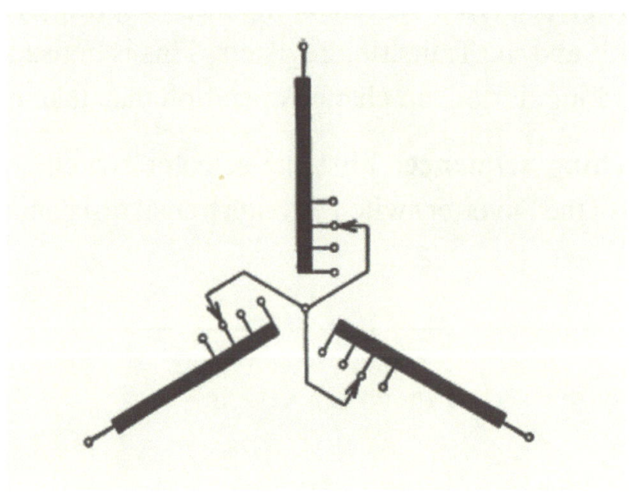

Fig. 6: OLTC with neutral end of tap winding

Figure 5A

6.2. Placement of OLTC in HV Δ winding: Since most of the 2 -winding Transformers, specially the step down Power Transformers in the Industries and in the Sub-Transmission network Sub-stations are of vector group Dy11. The OLTC has to be incorporated in the delta winding only, with full insulation level as that of the HV winding. Ref. fig. 5B

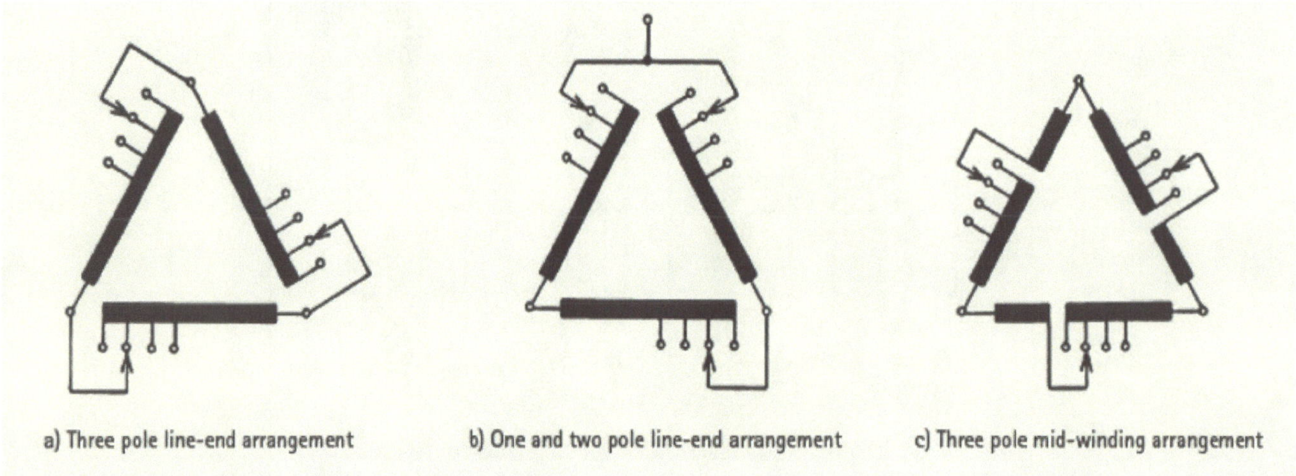

Fig. 7: OLTC with delta-connection of tap winding

Figure 5 B

6.3 Location of OLTC in an Auto-Transformer: Since in the Auto –Transformer unlike the 2-winding transformer, there is no separate winding for IV(intermediate voltage, as is called in an auto Transformer having Delta-tertiary winding also), the taps are selected from a tap winding which s separately provided between the HV and LV windings.ref fig.below;

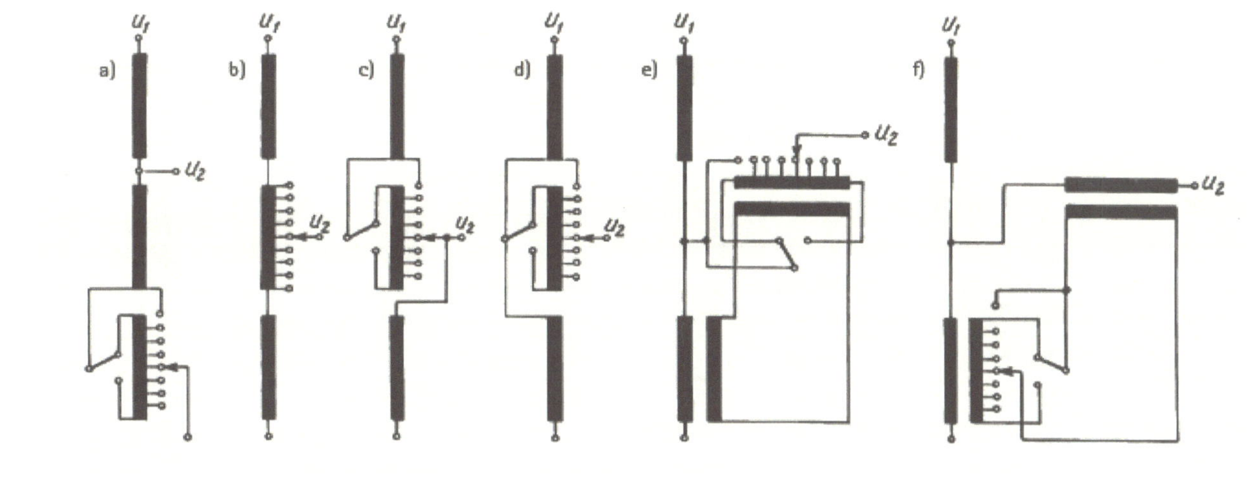

Fig. 8: OLTCs in autotransformers

Figure 5C

In the above fig the winding configurations are shown.

6.3.1 As a practical example the tap connection of 220/132/33kV, 160MVA 3 – Ph Auto Transformer, BHEL make is shown here-under(Figure 6)**;** There are as many as 13-taps with single reversing changeover selector similar to as shown in fig; c) above.

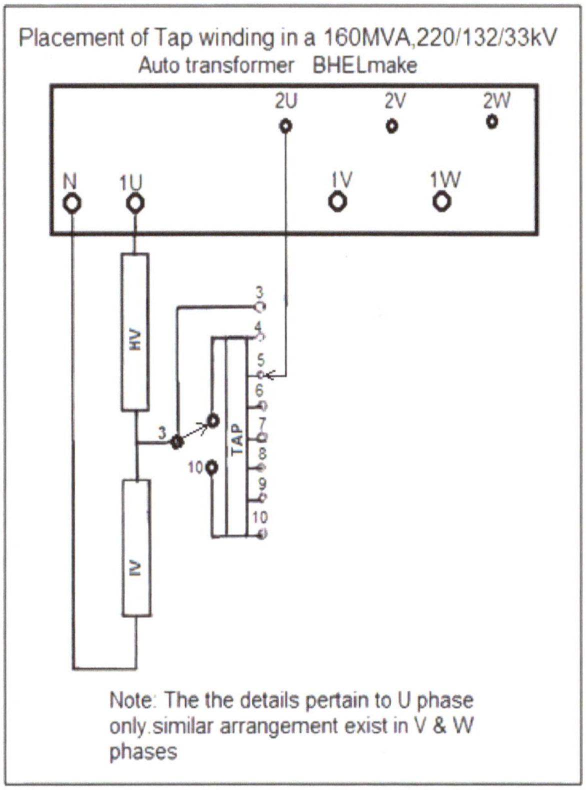

Figure : 6.

6.3.2. IV at different tap positions achieved through different tap connections thereof (Table1);

HV line term;1U,1V,1W	IV line terminals:2u,2v,2w		
kV	Tap position	Lead connections	IV kV
220kV	1(max)	11-4 3-10	145.2
220kV	2	11-5 3-10	143.6
220kV	3	11-6 3-10	141-.9
220kV	4	11-7 3-10	140.3
220kV	5	11-8 3-10	138.6
220kV	6	11-9 3-10	137
220kV	7	11-3 (3-10)(3-4)	135.3
220kV	8	11-5 3-4	133.7
220kV	9(Normal)	11-6 3-4	132
220kV	10	11-7 3-4	130.4
220kV	11	11-8 3-4	128.7
220kV	12	11-9 3-4	127.1
220kV	13(Min)	11-10 3-4	125.4
There are 13 No. of taps -5% to +10% with a variation of 1.25%.			

Table1

7.0 RTCC (Remote Tap changer Control Cubicle)(Figure 7): A RTCC (Remote Tap changer Control Cubicle) is also installed duly wired up for remote control of OLTC from the control room.

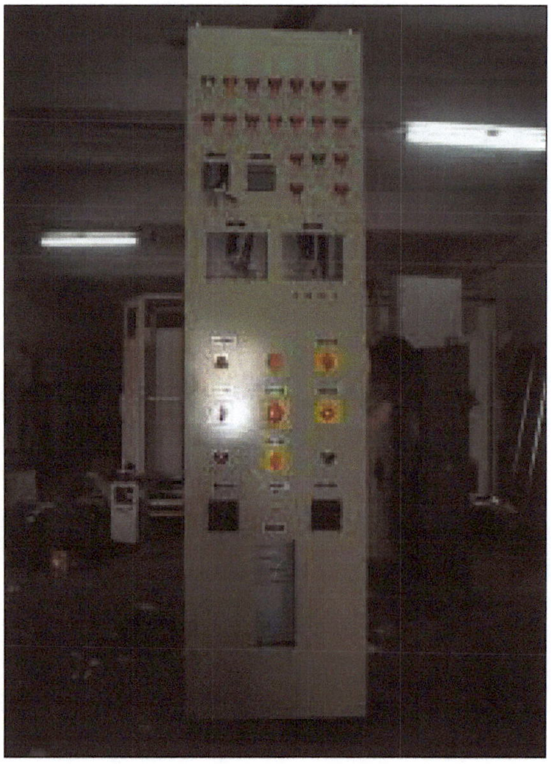

Figure 7:RTCC

8.0 OIL SURGE RELAY:

Figure 8 Oil Surge Relay

The Oil Surge relay (Figure 8) is installed in the pipe connecting the tap changer Divertor switch chamber and the Oil conservator separately provided for the OLTC.

It shall act for any pressure developed due to excessive sparking etc.

9. Identification of Defect in OLTC of 160MVA 220/132kv Autotransformer and In-House Repairing & Retrofitting Thereof

1.0 Abstract:

The 220kV and 400kV EHV Transformers and the Transmission lines of MPPTCL constitute the important part of the WREB (Western Region Electricity Board)along with other constituent members. Therefore, it is of utmost importance to keep all such EHV Transformers and Transmission lines of 220kV and 400kVin healthy condition.

Any outage due to failure of main components etc of the EHV Transformers of 220kV and 400kV Voltage class call for immediate attention by way of detecting the faults and carrying out In-house repairing work etc. In-house repairing work in-turn saves MPPTCL from many related and tangible losses.

There had been a case of failure of a Diverter-switch of OLTC of an old 160 MVA 220/132 kV, Auto-Transformer N0-1,(manufactured in1996) in the 400kV Sub-station at Katni in M.P. Power Transmission Co. Ltd., on dated 24/06/2019.

The fault was diagnosed/ detected by the Testing team of Katni on the 24/06/2019 and was put back in service after its in-house repairing /retrofitting job carried out by the Sub-Station Maintenance team, 400 kV S/s, Katni, in minimal possible time on its own plinth on date 04/07/2019, without availing the services of the OEM's service team.

This paper has been written by the authors highlighting the activities of detecting the fault without DRM and carrying out in-house repairing/retrofitting etc with limited resources and putting it back in service.

2.0 Key words: OLTC (On Line Tap Changer), Diverter Switch, Selector Switch. TWRM (Transformer Winding Resistance Meter), DRM (Dynamic Resistance meter).

3.0 In-tank type OLTC (also known as MR type in Power Utilities):

Figure 1: In-tank OLTC in a Transformer

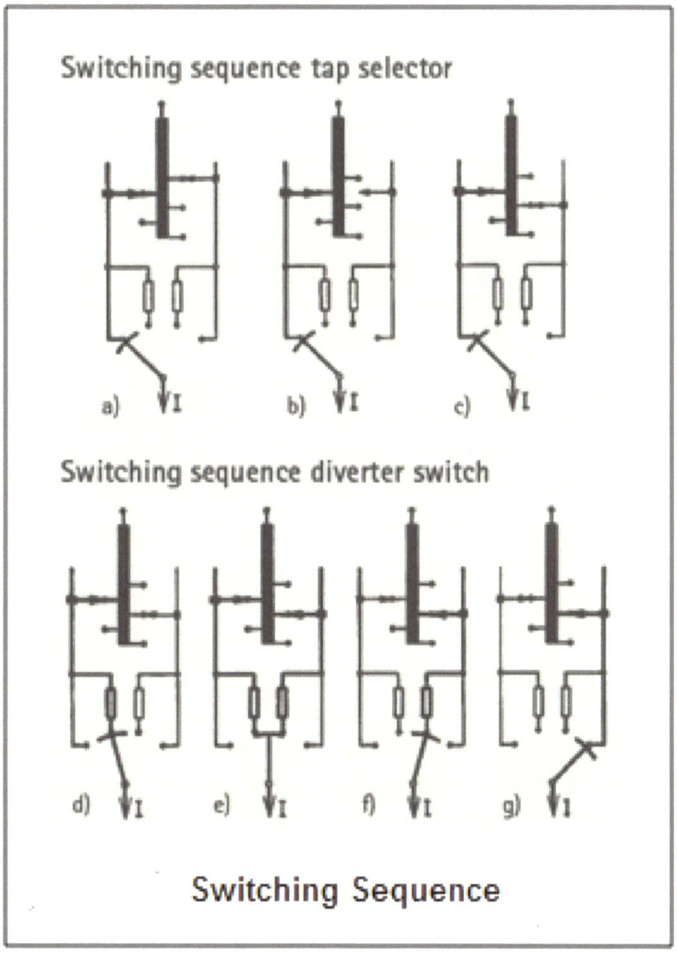

Figure 2: Switching sequence in Diverter Switch

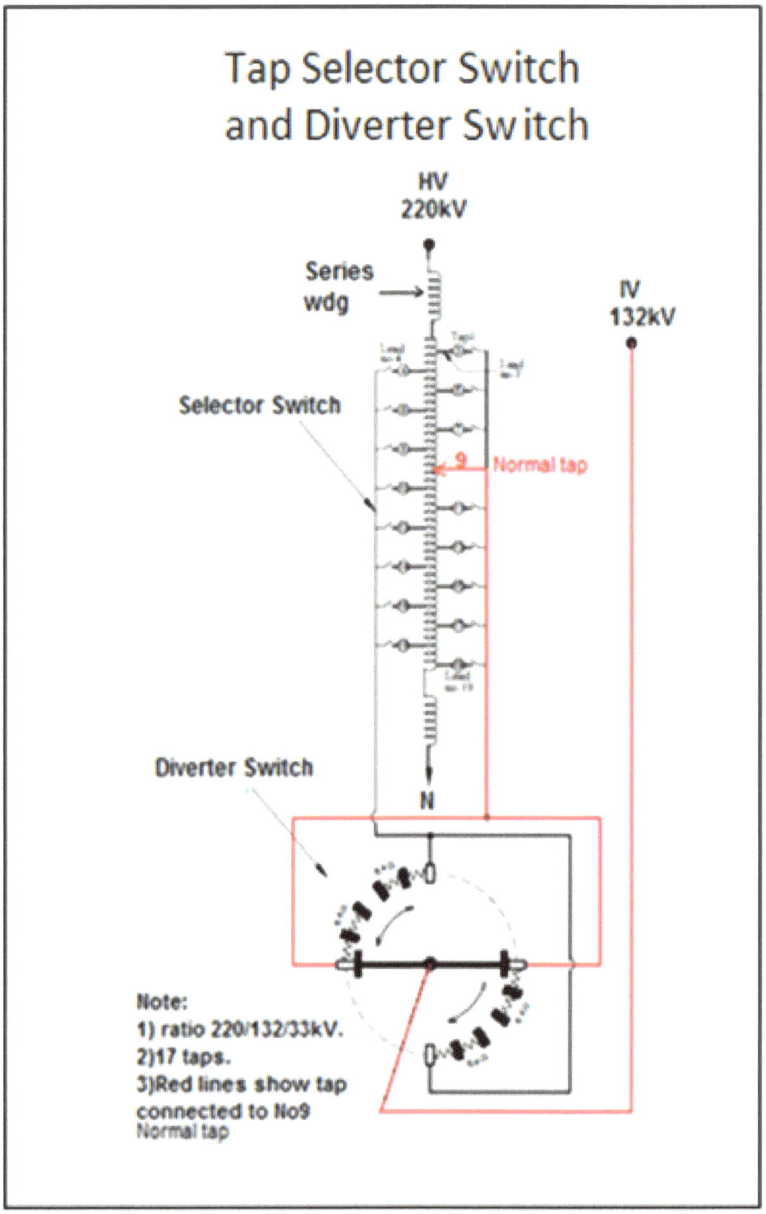

Figure 3: Sketch showing winding configuration of 160MVA, 220/132kV Auto Transformer No.1, made at tap position at Normal tap 9. (Red coloured fonts and lines)

4.0 Brief History:

Details of the Transformer;

- Location of installation: 400kV Sub-station, Katni.
- Rating: 160 MVA, 220/132/33 kV, Auto-Transformer No.1.
- Date of commissioning: 11/02/1998.
- Date of OLTC/diverter switch failure: 24/06/2019.
- How connected: Running in parallel with 160MVA, 220/132/33 kV, Transformer No.2
- Maximum Load recorded on this transformer:136 MVA
- Sharing of total load: 248 MVA along with 160 MVA, Transformer No.2.

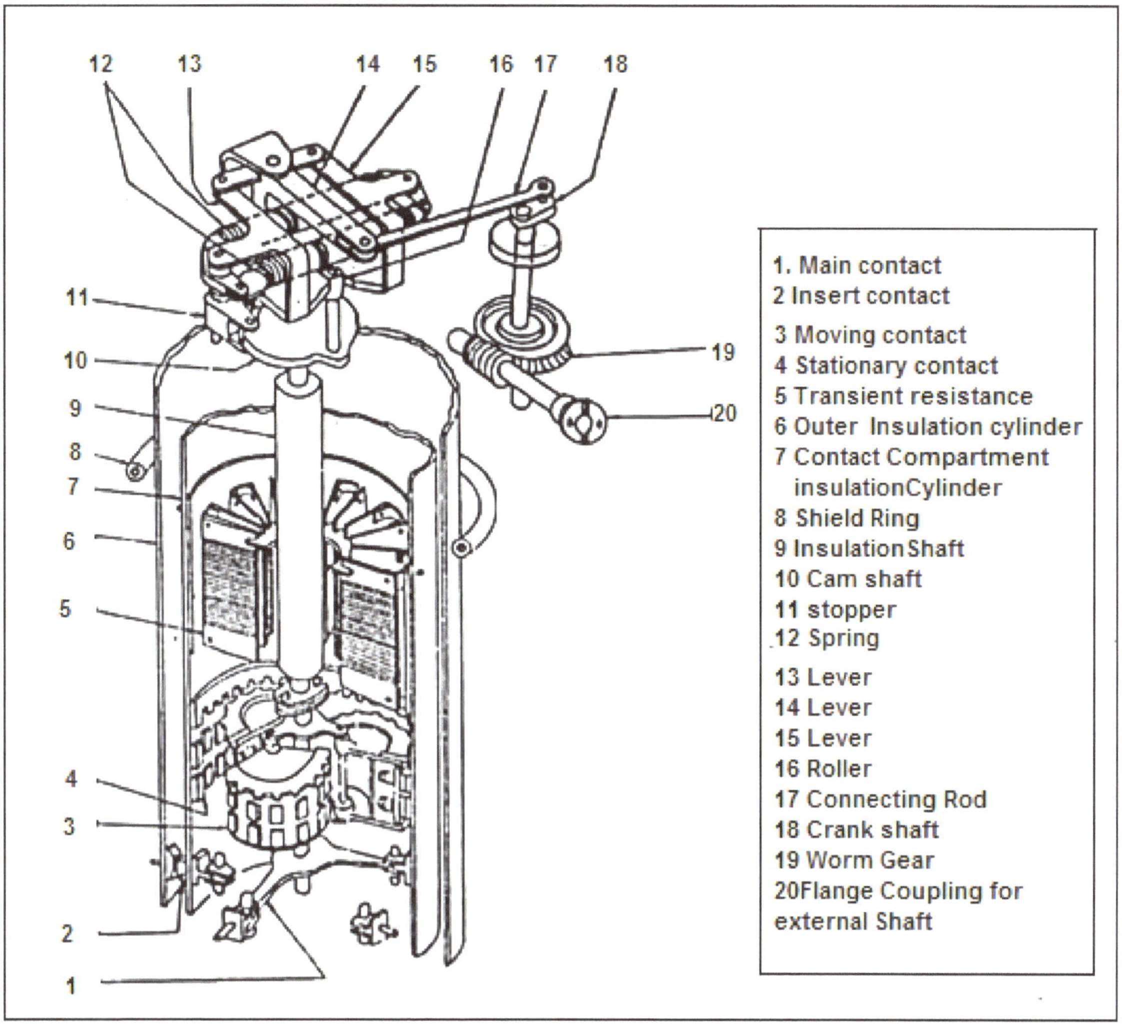

Figure 4: Components of Diverter switch

4.1 Maintenance schedule of the Transformers;

As per the prevailing practice in M.P. Power Transmission Co. Ltd., the Transformer maintenance is carried out twice in a year before and after the monsoon. In one of such maintenances, problem in Diverter switch of this 160MVA 220/132/33kV TELK made-Auto Transformer was detected on dated 24/06/2019.

4.2 Detailed report of fault detection:

A). On 12/03/2018 shutdown was taken to carry out Pre-Monsoon maintenance as per schedule. Winding resistance measurement started after checking diverter switch, make before break operation by moving tap from any odd to even and to odd tap using digital meter as analogue multi-meter was not available. Continuity check was also done by using digital multi-meter.

As a last test, winding resistance measurement was started from 2W phase (B-phase in RYB convention) and resistance of tap no. 1 and 2 measured successfully and while shifting the Tap position from tap no 2 to 3 manually the winding resistance kit developed defect as shown in the table no1.

	Winding resistance of 160MVA,220/132kV Tfr no.1				
DATE	12.03.18			OT	46
				MF	1.103
TAP NO	2U	2V	2W	STEPS	
				Converted Value at 75°C	
				Average Value	Converted value
1	-	-	478.5	478.5	527.79
2	-	-	473.1	473.1	521.83
3	-	-	Kit becomes defective.	- -	-
4-17	-	-		- -	-

Table.1 Failure of Winding resistance meter while measuring Winding resistance of the Transformer on dt.12.03.2018.

Presuming that the failure of the Winding resistance kit might have been due to the internal circuitry of winding resistance kit itself and since no other winding resistance meter was available in the S/S. and as the evening peak-load period was approaching fast, thus the Transformer was charged and put back in service without measurement of the winding resistance.

B). On dated 24/06/2019 shutdown on the same Transformer was again arranged to carry out Pre-Monsoon maintenance as per schedule.

a. All tests were performed successfully except the winding resistance test.
b. Finally, as the last test, the winding resistance measurement was being carried out using the same kit which was used on 12/03/2018(after getting it repaired through the OEM).
c. The winding resistances of the 2U, 2V phases were taken successfully,
d. While measuring the winding resistance of 2W phase, the kit developed defect, while shifting the tap no.2 to tap no.3, though the measurement of the winding resistance was successfully carried out while moving the tap from tap no.1 to tap no.2.

Observation: It was observed that kit again got defective while measuring the winding resistance of same phase ie of 2 W and under the same conditions as prevailed on 12/03/2018.

The winding resistances measured with the same repaired Winding resistance kit by the OEM are tabulated in the Table no 2.

Winding resistance measurement in m ohms of160MVA Tfr No.1					
DATE	24.06.19			OT	48
				MF	1.095
TAP NO	2U	2V	2W	Conversion at 75°C	
				Average Value	Converted Value
1	453.5	455.2	483.3	464.00	508.08
2	449.4	451.4	478.1	459.63	503.30
3	445.1	446.2	Winding resistance not taken due to kit failure. While moving forward from tap no 02 to 03,Kit stops.	445.65	487.99
4	441.4	442.7		442.05	484.04
5	436.8	438.3		437.55	479.12
6	431.8	433.2		432.5	473.59
7	427.4	428.3		427.85	468.50
8	421.6	423.8		422.7	462.86
9	417.9	419.6		418.75	458.53
10	412.6	415.9		414.25	453.60
11	407.0	408.7		407.85	446.60
12	400.5	404.6		402.55	440.79
13	396.2	399.0		397.6	435.37
14	392.7	394.1		393.4	430.77
15	388.8	389.5		389.15	426.12
16	383.6	384.9		384.25	420.75
17	378.5	377.2		377.85	413.75

Table.2: Failure of Winding resistance kit on 24/06/2019 while shifting the tap from 2nd to 3rd Tap of 2W phase.

4.3 Probable reason for failure of Winding resistance kit:

As per practice, the Winding measurement kit was kept in circuit during tap-changing operations; however, in this instant case the kit appeared had failed because of its design constraints.

The TWRM meters of other manufacturers are capable of identifying open circuit and other faults in the Diverter Switch and Selector Switch of OLTC if any, by clearly indicating open circuit if there is any such condition in any of the transition resistances, formation of carbon on the fixed or moving contacts of Diverter switch and/or due to snapping of any connecting lead at the Selector switch end of Tap-winding.

Note: DRM (Dynamic Resistance Measurement) Kits of other manufacturers are equipped with a display screen on which the curve of switching operation including the Transition resistance can be viewed (Figure 5). Any deviation in the curve during subsequent measurements as compared to the initial signature curve shall be treated as faulty Diverter switch.

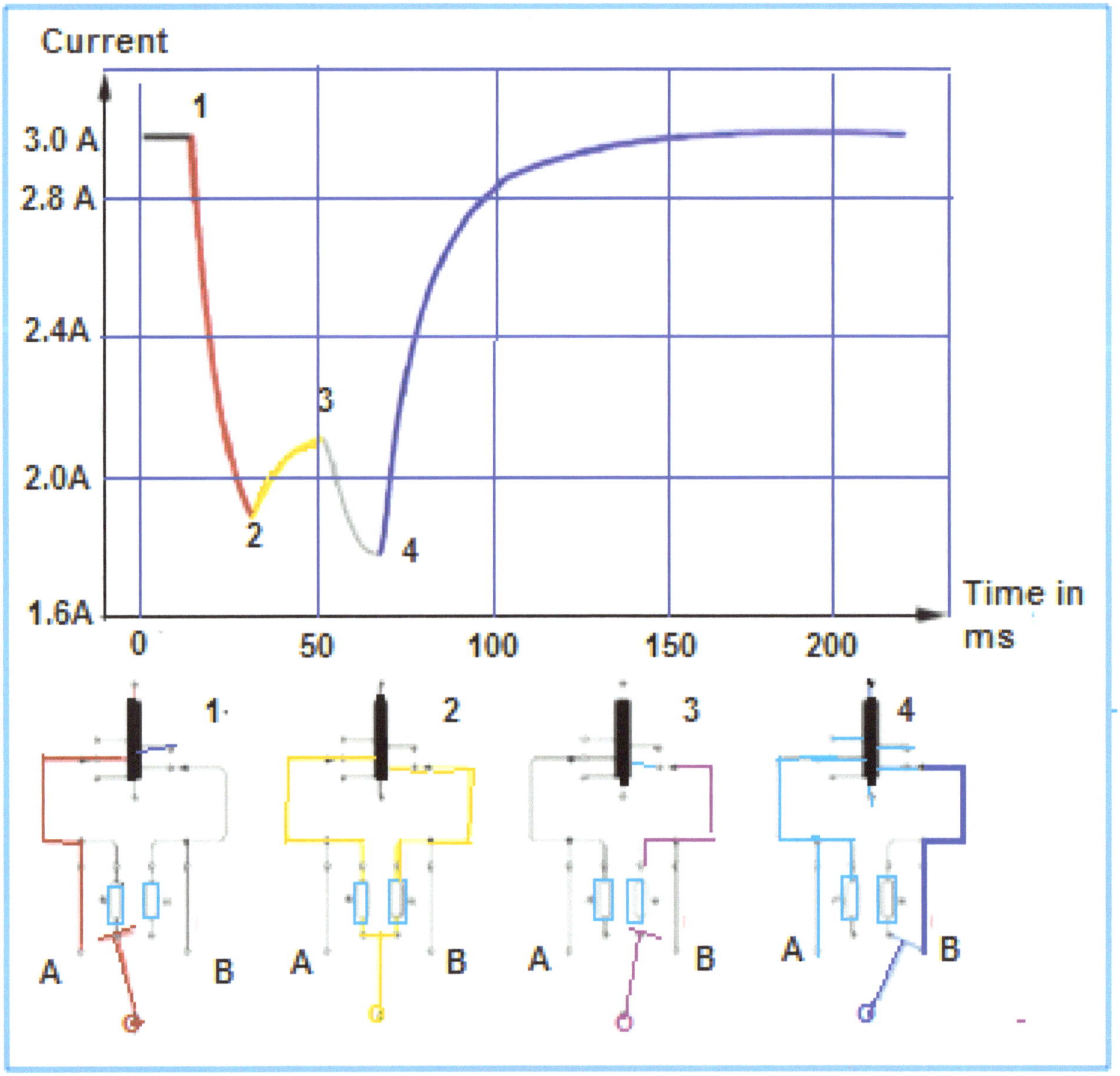

Figure 5: Typical DRM curve of OLTC in operation.

4.4 Checking and verification through other methods.

To confirm whether the fault was due to design constraints in the Transformer Winding Resistance Measurement kit or it was in the winding of phase 2W itself.

i. **DC battery method**; As analogue multi-meter was not available at that time therefore resistance measurement was carried out using 9-volt battery and 2-volt centre zero volt-meter (used in battery maintenance work) and found a flick while moving tap from 2 to 3 and break was also observed in other taps movement.
ii. **Analogue Multi-Meter Method**: Following day, an analogue multi-meter was arranged from the Testing Division Jabalpur and continuity was again checked with it. It was observed that the result was found to be similar to the previous day results as checked using 9-volt battery and battery voltmeter set. The test report is as follows;

S. No	Tap Position	Tap Position while shifting to next.	Continuity observed in phases					
			2U	2V	2W			
						Forward		Reverse
1	1		OK	OK		OK		OK
2		1-2	OK	OK		Flick observed		Flick observed
3	2		OK	OK		OPEN		OPEN
4		2-3	OK	OK		OK		OK
5	3		OK	OK		OK		OK
6		3-4	OK	OK		Flick observed		Flick observed
7	4		OK	OK		OPEN		OPEN
8		4-5	OK	OK		OK		OK
9	5		OK	OK		OK		OK
10		5-6	OK	OK		Flick observed		Flick observed
11	6		OK	OK		OPEN		OPEN
12		6-7	OK	OK		OK		OK
13	7		OK	OK		OK		OK
14		7-8	OK	OK		Flick observed		Flick observed
15	8		OK	OK		OPEN		OPEN
16		8-9	OK	OK		OK		OK
17	9		OK	OK		OK		OK
18		9-10	OK	OK		Flick observed		Flick observed
19	10		OK	OK		OPEN		OPEN
20		10-11	OK	OK		OK		OK
21	11		OK	OK		OK		OK
22		11-12	OK	OK		Flick observed		Flick observed
23	12		OK	OK		OPEN		OPEN
24		12-13	OK	OK		OK		OK
25	13		OK	OK		OK		OK
26		13-14	OK	OK		Flick observed		Flick observed
27	14		OK	OK		OPEN		OPEN
28		14-15	OK	OK		OK		OK
29	15		OK	OK		OK		OK
30		15-16	OK	OK		Flick observed		Flick observed
31	16		OK	OK		OPEN		OPEN
32		16-17	OK	OK		OK		OK
33	17		OK	OK		OK		OK

Table.3 Continuity checked while operating taps in ascending and descending order with keeping analogue multi-meter it in circuit during tap operation

Observations: (Pease refer to Table No3).

Forward Direction;

i. Continuity is observed is written as" OK" while shifting from even to odd position in the ascending order (forward direction) in case of 2W phase.
ii. Flick is observed written as" flick observed" while shifting from odd to even position in the ascending order (forward direction) in case of 2W phase.
iii. While in case of 2U&2V Continuity is observed written as" OK" against each shifting from odd to even and from even to odd.

Reverse Direction;

i. Continuity is observed written as" OK" while shifting from odd to even to odd position in the descending order (reverse direction) in case of 2Wphase.
ii. Flick is observed written as" flick observed" while shifting from even to odd position in the descending order ((reverse direction) in case of 2Wphase.
iii. While in case of 2U&2V Continuity is observed written as" OK" against each shifting from odd to even and from even to odd.

Observation; As observed from the Continuity test as stated in table No.3 At all odd taps Continuity is OK, whereas at all even tap it is found OPEN.

Thus, failure of the TWRM due to its design constraints was over ruled.

From the above observation it seemed that either one of the transit resistances might have been opened (since the Diverter Switch has only 2 positions) or one of its contacts must have got jammed and therefore not changing its position and showing OPEN position in EVEN Tap nos. while moving the position of the Diverters switch "to and fro" manually.

Thus, it was confirmed that the problem was in the Diverter Switch only. Therefore, it was decided not to charge the Transformer.

The DGA of OLTC oil was carried out and high concentration of hydrocarbon gases found in sample. On the basis of LT test and DGA gas analysis it was decided to open 2W phase OLTC Diverter Switch.

5.0 Condition of the Diverter Switch: After removing the Diverter Switch assembly, it was found that Transit resistance found OK and Main contact, insert contact, Moving Contact, stationary contact and 132kV take off contacts damaged very badly and jammed due to sparking and welding/pitting as predicted. Please see the figures.6 to 10.

The condition as-found was beyond repairs.

Figure 6 View showing carbonised worm Gear and other operating mechanism after top cover removed.

Figure 7: Inspecting faulty Diverter switch unit removed from the Transformer

Figure 8A: Condition of the Bottom plate of the Diverter switch. Spring loaded contacts A&B seen badly burnt and melted.

[Note: Bottom plate of Diverter switch insert which rests on the two Silver coated studs at the bottom of the Insulation cylindrical of the Diverter Switch (fig. 8B) on which 2 nos. spring loaded Contacts A & B rest. This cylindrical shell is suspended inside the tank and the132kV lead is connected to the132kV Bushing.

Figure 8B: Hollow insulation Cylinder (item No 6 in fig.4) inner side with 2 Nos. Silver coated Copper studs

Figure 9:No.1&2 are moving contacts (male) 3-Arcing Moving contacts. (Used during transient period).

Figure 10: 1 & 2-Insert fixed contacts(female).3-Arcing Stationary contacts

[Note: a. contact no.1 is seen very heavily welded.
b. Arcing first fixed Contact of no.3 little welded.]

6.0 Details of repairing and retrofitting works:

The damages revealed that the Diverter Switch was beyond repairs (Ref.fig.6 to10).

Similar Diverter Switches 2 nos, mounted on an old 125 MVA, 220/145kV TELK made, abandoned Auto- Transformer in 220kV S/S Damoh were removed and safely transported to the 400kV S/S Katni.

Figure 11: Thoroughly cleaned diverter switch tank.

The Diverter Switch brought from 220kV S/S Damoh was installed after thoroughly cleaning of the Diverter Switch and the tank (fig:11), but it was found that there was no movement in the Diverter switch while trying to change the tap positions.

The said Diverter switch was again removed and was subjected to force to and fro motion manually with the help of a pipe as shown in the fig: 12 but it did not move.

Figure 12: Due to jamming the Diverter switch it did not move when tried manually.

Then the bottom plate was removed for checking the reason for its no- movement. Following were found;

- It was found that 2nos. insert contact and 1 no. stationary arcing contact damaged.
- Main contact shaft bearing (bottom) found clogged.

Figure 13: bottom Bearing

The above defective parts and the bottom bearing as shown in fig.13 were replaced carefully using these parts from another Diverter switch brought from 220kV S/S Damoh, contacts brazing and finishing was done at local engineering work shop and re-assembled it after thoroughly cleaning all the internal parts with fresh oil.

The re-assembled unit was checked for it's to & fro motion manually.

Completely re-assembled Diverter Switch

Fig: 14: Completely repaired and re-assembled Diverter Switch

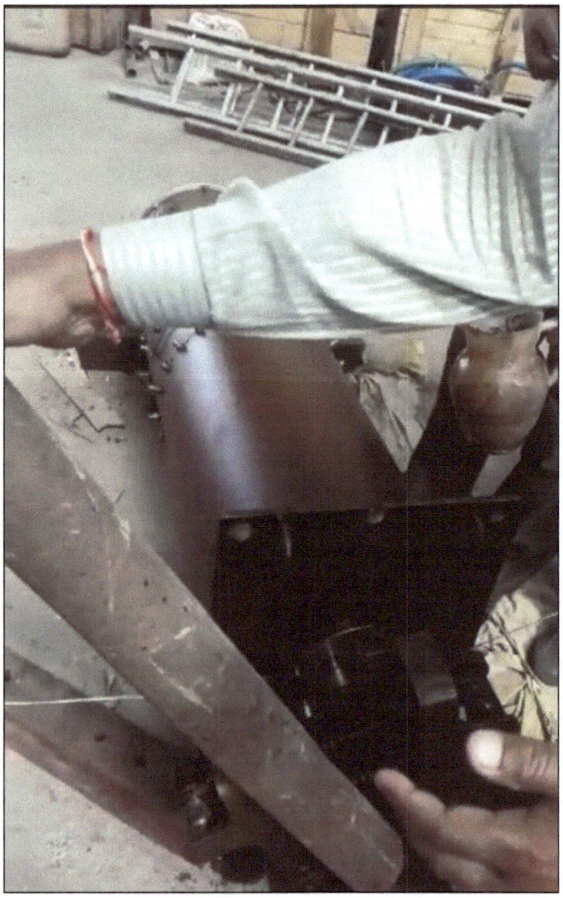

Figure 15: To & fro motion accomplished after repairing and re-assembling of the Diverter Switch.

Finally manual operations were done and found that;

- Resistances between the fixed contacts were checked and found 6.4 Ω each.
- Continuity between the odd taps and even taps to 132 kV take off point checked and found in order.

Above results revealed that the Diverter Switch has becomes healthy in all respects. Therefore, the same has been installed on the transformer. Thereafter filled with fresh oil and tap-operations checked. The Diverter Switch started moving smoothly and due synchronised sound with selector switch positions. Then filtration of all the three Diverter Switches was done with a 2000 LPH filter machine for 4 hrs (8 cycles) each.

The BDV of the oil was checked and found to be 78.5kV.

Then continuity checked with tap operation from 1 to 17 and 17 to 1 carried out and no flick observed. It gave a feeling that the repairing and retrofitting was successful. Then all L.V. Testing of transformer done and found in order. Then winding resistance of each phase taken by TWRM make- DV power, results found satisfactory as could be seen from the test result tabulated below (Table:3) confirming that Diverter Switch has been successfully repaired and retrofitted.

Winding resistantance of 16 MVA 220/132kV tfr no.1, after repairs/retrofitting in m ohms.					
DATE	01.07.2019			OT 28	
				MF 1.178	
TAP NO	2U	2V	2W	Conversion at 75°C	
				Average Value	Converted Value
1	417.9	416	422.1	418.67	493.19
2	414.7	411.9	419.9	415.50	489.46
3	408.1	406.4	412.1	408.87	481.64
4	404.8	400.6	407.3	404.23	476.19
5	398.2	397.9	401.4	399.17	470.22
6	395.2	390.4	396.4	394.00	464.13
7	388.4	386	391	388.47	457.61
8	385	380.6	389	384.87	453.37
9	378.1	375.2	381	378.10	445.40
10	375.1	371.9	377.2	374.73	441.44
11	368.3	365.8	371.5	368.53	434.13
12	365.3	361	367.3	364.53	429.42
13	358.3	356.9	361.7	358.97	422.86
14	354.9	350.9	357.6	354.47	417.56
15	349.2	347.5	358.2	351.63	414.22
16	348.3	340.9	348.2	345.80	407.35
17	350.0	341.9	345.7	345.87	407.43
HV	592.1	589.3	589.2	590.2	695.26
LV	48.89	48.81	48.75	48.82	57.51

Table.3: Winding resistance measurement after finally

Installing the retrofitted repaired Diverter Switch.

All test results were sent to higher authorities for their perusal and approval. After having received their concurrence, the transformer was charged on date 04/07/2019 successfully and put on load on date 05/07/2019.

7.0 Probable causes leading to such failure:

The Diverter Switch comes into action when changing taps on load. It changes position from odd to even and vice-versa by inserting transition resistance during the on-load tap changing operations without break in the circuit. When the Diverter Switch changes its position from odd to even or even to odd on load, arcing take place in arcing contacts, resulting in generation of Hydrocarbon gases and some decomposition of Diverter Switch oil and which is categorized a normal phenomenon.

i. Since we do not operate taps on load therefore there should not be any chance of arcing and oil decomposition. As per previous history taps are operated during periodic maintenance only while taking winding resistance in every 2 years in off- condition. Observation of oil decomposition, clogging of contacts pitting/welding like spots, sludge deposition give rise to a feeling that at some point of time some contact must have become loose and the insert moving contact must have stuck inside the fixed contact and got welded at tap no.9 (N) position. Such incidences lead to failure of Diverter switch in this instant case.
ii. It is also observed that the copper braided flexible leads were found badly burnt and melted as could be seen in fig. nos.9. Perhaps, their current carrying capacity appears to be inadequate.

8.0 Suggestion for consideration: While going through the repairing process it was observed that insert stationary contact of the Diverter Switch is having very small contact area and also it is in two separate pieces assembled together with spring loaded arrangement. It appears to be of complicated design (see fig16), increasing contact resistance to main contacts, while all 132kV current flows through it during normal working condition. Because of very small contact area of insert contact, continuous heating take place during entire working period, results in decomposition of oil near main & insert contact area, continuous deposition of sludge at bottom of the Diverter switch tank which clogged 132kV take off contact (braided copper lead), and clogging main & insert contacts. This appears to be the reason of the Diverter switch getting heated up without its operations.

Figure 16: Insert contact of TELK make (MR Type) DS, having very small contact area and complicated design.

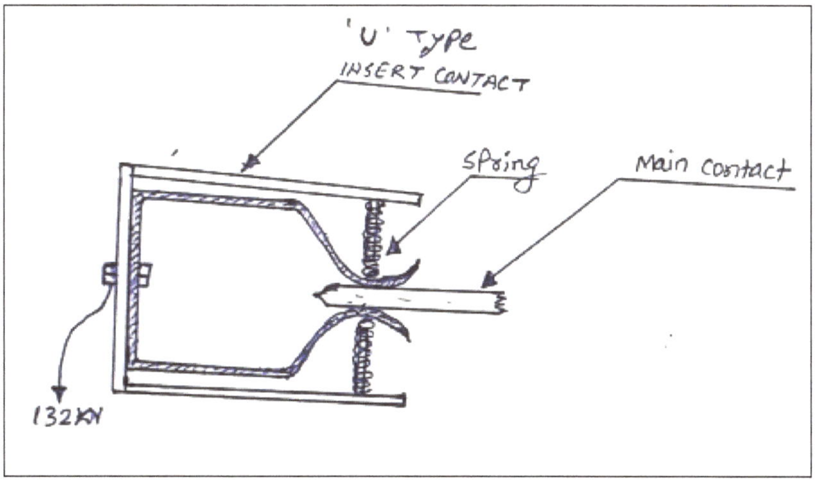

Figure 17: Proposed design of insert fixed contact.

We had inspected 5 No. Diverter Switches of the same make (MR Type) and found similar problem, so it is clear that design of insert contact has to be improved to avoid such deterioration, which takes place without any on-load tap operations. It is felt that the insert fixed contact design may be improved as proposed in fig.17.The OEM may take a note of this.

9.0 Gains/advantages to MPPTCL;

Following are the gains/ advantages to the MPPTCL due to in-house repairing and retrofitting using available means.;

i. Repaired, retrofitted and restored supply in minimal time.
ii. Restoring its importance in the Grid-network
iii. Avoiding Load shedding for long periods, thus avoiding dissatisfaction to consumers at large.
iv. Avoiding over loading of transformers of nearby sub stations for long and thus enhancing their life expectancy.
v. Savings on;

a. Cost of new Diverter Switch and transportation charges to the site.
b. Service charges for repairs etc to the OEM's repair team or other agency.

vi. and many more intangible advantages.

11.0 Authors;(i) K.K.Murty(ii) M.D.Palande(iii) C.B. Kushwaha & (iv) S.K. Chaturvedi

10. RLA of Power Transformers with Case Studies

ABSTRACT:

The useful life of Power Transformers, is assumed to be 35-40 years. However, there is no literature expressively quantifying the same. Un-foreseen failure of old transformers in the Grid-network result into system disturbance or cascade tripping. Such contingencies call for un-scheduled load-trimmings / shedding which in-turn give rise to colossal loss of revenue to the Power Utilities and also dissatisfaction amongst the Industrialists and consumers at large. Therefore, it is of utmost importance to obviate such situations and safe-guard the Grid-system from the aforesaid disturbances and losses.

M.P. Power Transmission Utility has guidelines for selection of all aging transformers for assessing the health-status thereof. The genesis of these guidelines is however, not based on any rigorous data or analyses.

The authors have analyzed many old transformers ranging from 16MVA to 40 MVA, in the M.P. Power Transmission Utility, however, as sample cases only 07 nos. have been enlisted in this paper.

The methodology narrated in this paper has a limited validation of the M.P. Power Transmission Utility guidelines, however, it will serve as an effective guideline for the field-Engineers in India and abroad to assess the health-status of their Transformers. Thus, catastrophic failures and colossal loss to respective Utility could be avoided, if unreliable transformers are identified and are replaced expeditiously.

KEY WORDS AND EXPANSION OF ABBREVIATIONS.

DGA (Dissolved Gas Analysis), Furan-2-Furaldehyde (2FAL)), DP (Degree of Polymerization),

1.0 INTRODUCTION.

Un-foreseen failure of Transformers in an EHV Grid-network result into system disturbance or cascade Trippings. Such contingencies call for un-scheduled load-trimmings / shedding which in turn give rise to colossal loss of revenue to the Utilities and also dissatisfaction amongst the Industrialists and consumers at large. Therefore, it is essential to obviate such situations and safe-guard the Grid-system from the aforesaid disturbances and losses. Such unforeseen incidences generally take place due to failure of the very old transformers in the system.

In-view of above, it is essential to assess the remaining serviceable life of the transformers for rendering faithful service which have rendered continuous service over 20 years in a Grid system and to timely replace them with new ones, if found unserviceable.

Following criteria was adopted by the M.P. State Transmission Utility for selection of all such aging transformers for assessing the health-status thereof;

i. Which had rendered continuous service for more than 20 years
ii. Deterioration trend in IR and PI values observed.
iii. DGA showed increasing trend in the key gases and also increasing trend in CO and CO2 gases with reference IEC 60599.
iv. Percentage of relative water saturation in the transformer oil: > 15 % [1],
v. DP value <250.

The Authors have narrated the methodology to assess the health-status of the aging transformers. While assessing the same, if health-status of any of them was found deteriorated and approaching towards end of its life, was recommended for removal from service and to be replaced with a healthy transformer.

2.0 CAUSES FOR DETERIORATION OF SOLID AND LIQUID INSULATIONS;

Deterioration in overall insulation is attributed to the following:

i. Improper monitoring and maintenance of the breathing system.
ii. Increase in moisture contents in the oil above 20 ppm.
iii. Continuous over loading of transformers beyond 80 °C to 140 °C leads to loss of life to half for every 6°C beyond 80°C as stipulated in the IS: 2026 (Part-7):2009 (IEC equivalent IEC 60076-7:2005) [2], resulting in a fall in DP value nearing 200.
iv. Sustaining many through faults conditions.
v. Sustaining switching and Lightening surges.
vi. Improper/irregular maintenance practices.

3.0 PERCENTAGE OF WATER SATURATION IN OIL;

The quantity of dissolved and dispersed water in mineral oil is significant for two reasons

i. presence of polar water molecules in the mineral oil adversely affects the dielectric properties of the mineral oil.
ii. the amount of moisture in the oil can be reflective of amount of moisture in the paper insulation. The solubility in mineral oil is temperature-dependent. Therefore, a statement as mg/kg of water in the oil without temperature information would not be adequate.

The calculation of percentage of water saturation is [(mg/kg water in insulating liquid)/(mg/kg of water in insulating liquid at saturation)] × 100, in other words (ppm of water/So) has greater significance as it indicates the possibility of free water formation in the oil. [3]

3.1 FREE WATER IN OIL

" Free water" that exists in the form of droplets if the water content in oil exceeds the saturation level. In cellulose materials, free water may exist in macropores. In addition, following are the basic reasons for increasing the water contents in oil [4]:

a. Residual moisture in the thick structural components not removed during factory drying-out or moistening of insulation surface during assembly.

b. Inhaling atmospheric air through fused silica-gel while oil volume shrinks during cold and moist nights and during off-peak hours of the day, resulting into poor dielectric strength of oil.

c. Aging (decomposition) of cellulose and oil.

d. Table 8 of IEEE Std 62-1995 gives general guidelines for interpretation of data expressed in % of water saturation in oil vis-à-vis condition of dryness of paper insulation. [1]

% Of water saturation in oil	Condition
0-5	Dry insulation
6-20	Moderate to wet. Lower numbers indicate fairly dry to moderate levels water in insulation, whereas values towards the upper limit indicate moderately wet insulation.
21-30	Wet
>30	Extremely wet

Table1. Interpretation of data expressed in % of saturation [1].

3.2 FORMULA FOR SOLUBILITY OF WATER IN OIL AND SPECIMEN CALCULATION;[5]

$$\text{Log}_{10} S_O = \left(-\frac{1567}{K}\right) + 7.0895$$

Where:

- S_O is the solubility of water in mineral oil.
- K is the absolute temperature in Kelvins (^{O}C+ 273).
- ^{O}C is the oil temperature in Celsius at the time of sampling

$$\%\ \text{saturation} = \left(\frac{\text{ppm of water}}{\text{So}}\right) X\ 100$$

3.2.1 SPECIMEN CALCULATION FOR % SATURATION;

Moisture content =16 ppm'

Temp. of top oil = 30° C,

Temp. in kelvin (^{O}K) = 30 + 273 = $303^{O}K$

Log_{10}So = (- 1567/K) + 7.0895 = - 5.1716 + 7.0895 = 1.9197

So = $10^{1.9197}$ = 82.775

Therefore % Saturation $= \left(\frac{16}{82.775}\right) \times 100 = 19.32\%$

4.0 SOLID INSULATION IN THE TRANSFORMER;

The Kraft paper is a major solid dielectric material used for conductor wrappings, barrier boards, spacers and clamps (in compressed or resin-bonded forms). The major constituents of the soft wood craft paper are cellulose (80 %), hemi cellulose (12%) lignin (about 8 %) and some mineral substances.

The structural formula of cellulose is as shown in the Figure 1.

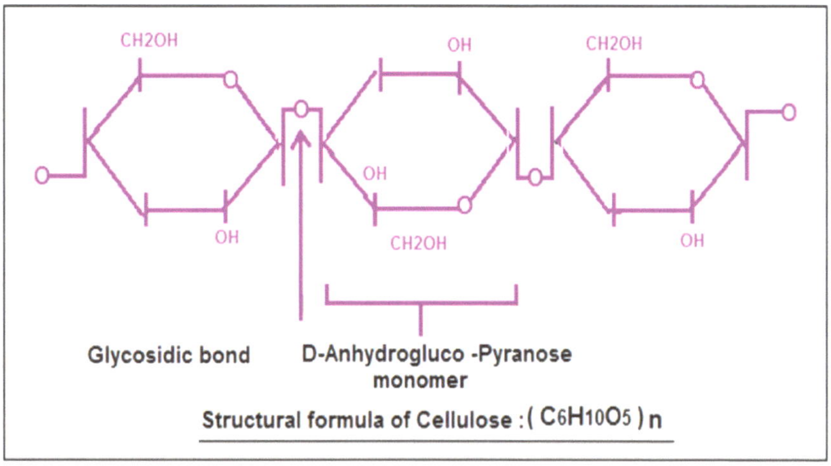

Figure 1. Structural formula of cellulose ($C_6H_{10}O_5$)n. [6]

Three most common degradation factors for cellulose have been identified and they are thermal, oxidative, and hydrolytic. When cellulose is subjected to a temperature of 200°C, the beta linkages (glycosidic bonds) tend to break and open the glucose molecule rings and thereby lose mechanical strength. The by-products of this reaction are [6]:

a. free glucose molecules,
b. moisture,
c. CO & CO2,
d. organic acids.

The presence of oxygen promotes oxidation and cellulose molecules have a tendency to oxidize. The reaction of oxidation on the cellulose causes the glycosidic bond to weaken and it can cause scission of the cellulose molecule chain. The oxidation of hydroxyl produces carbonyl (aldehydic) and carboxyl (acidic) compounds. Moisture is also a by-product of this oxidative reaction.

The moisture produced as explained above is also transferred to the oil.

5.0 FURANS:

As stated in paragraph 4.0, the byproducts of paper insulation (cellulosic insulation) degradations are;

CO, CO2, moisture, organic acids, and free *glucose molecules.

The free *glucose molecules further degrade into aromatic components

known as FURANS.

Following are the Furans;

Following are the Furans,

1. 5H2F (5-Hydroxymethyl-furaldehyde) is an unstable compound and can decompose further into other furans as follows:
2. 2-Furaldehyde (2FAL)
3. 5 Methyl-2-Furaldehyde (5M2F)
4. 2-Acetyl furan (2ACF)
5. Furfuryl alcohol (2-FOL)

However, all the above components except 2 FAL are not very stable under operating conditions in the transformers. Their life-span is a few months only, thereafter they degrade into 2 FAL which remains stable for several years. The molecular structure of 2-Furaldehyde (2FAL) is shown in Figure 2.

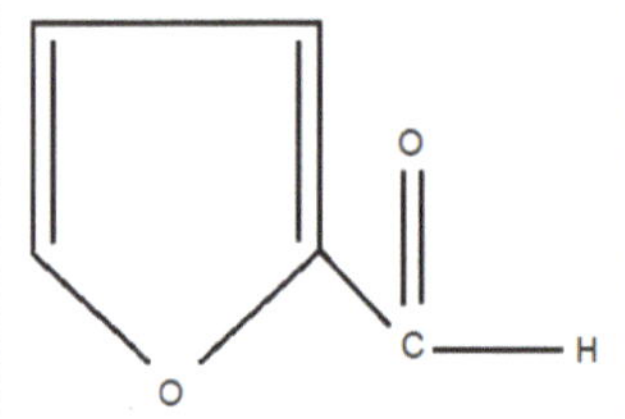

Molecular structure of 2FAL

Figure 2: Molecular structure of 2FAL. [6]

6.0 EVALUATION OF MECHANICAL PROPERTIES OF INSULATING PAPER:

The mechanical properties of insulating paper can be established by direct measurement of its tensile strength or Degree of Polymerization (DP). These properties are used to evaluate the Life expectancy of a Transformer. Direct measurement thereof is not practically possible for the in-service transformers as analysis of paper insulation for its DP value requires removal of a few strips of paper from the aging Transformers. The learned scientists of Transformer- Chemistry have innovated that "when cellulose molecules de-polymerize (break into smaller lengths or ring structures) chemical compounds known as furans are formed" Finally, all the other components degrade to Furan:2-Furaldehyde (2FAL) which remains stable for years together.[6]

7.0 FORMULAE-FURAN CONTENTS VS DP-: Some Scientists of Transformer Chemistry had innovated equations to calculate DP value with the help of quantity of 2 FAL in ppm or ppb found in the mineral oil in the serving Transformer.

The DP is evaluated through following different models (mathematical equations) innovated by the Scientists of Transformer Chemistry as depicted in figure 3., however, there had been lot of debates on selection of a particular model for establishing the relationship between the Furan and the DP value? [7],

1) Chendong's equation;

$$DP = \frac{\log(2FAL) - 1.51}{-0.0035}, \text{ where 2FAL in ppm}$$

2) De Pablo' equation;

$$DP = \frac{7100}{8.8 + 2FAL}, \text{ where 2FAL in ppm}$$

3) Pahavan pour's equation;

$$DP = \frac{800}{[0.186 \times 2FAL] + 1}, \text{ where 2FAL in ppm}$$

4) Stebbin's equation

$$DP = \frac{\log(2FAL \times 0.88) - 4.51}{-0.0035}, \text{ where 2FAL in ppb}$$

Figure 3: Models innovated by the scientists of Transformer Chemistry.

[**Note:**(i) ppm = Parts per million = $1/10^6$, (ii)ppb = Parts per billion =$1/10^9$, (iii)1ppm =1000 ppb.]

7.1 GRAPHICAL REPRESENTATION FURAN VS DP;

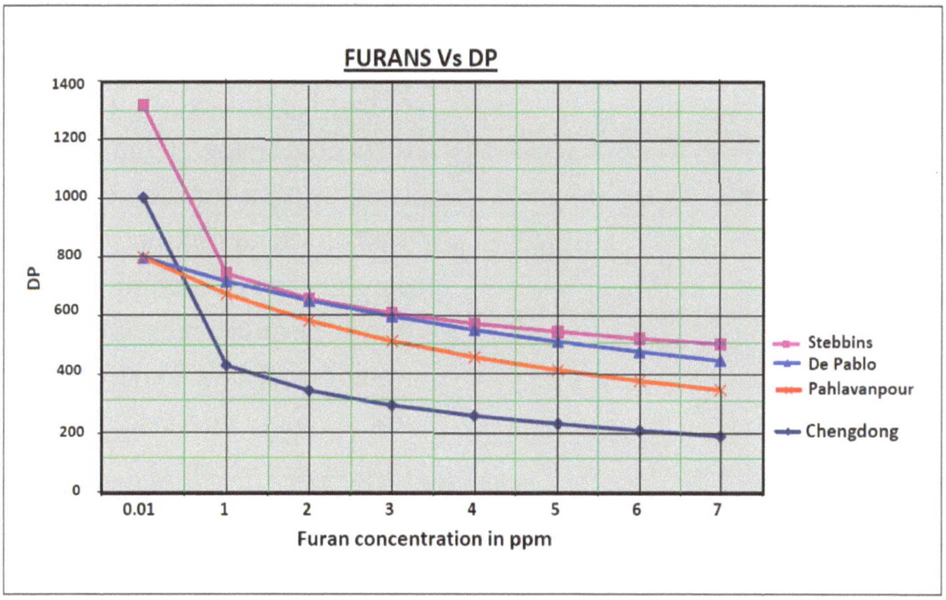

Figure 4: Graphical representation of Furan concentration in ppm Vs DP value.[7]

8.0 REMNANT-LIFE OF TRANSFORMERS IN % WITH REFERENCE TO FURAN (2 FAL) AND DP VALUE;[8]

Since, it is difficult to obtain paper samples for evaluation of the DP from in-service EHV Transformers, the method of estimation of the DP and the remaining/residual life of the Transformers is assessed, by measuring the Furan content 2 FAL (2-Furaldehyde) in ppm/ppb in the Transformer oil of the Transformers. The Chengdong's model has been adopted world-wide for assessing the % of remnant life of the Transformers.

2 FAL (ppb)	Estimated Degree of Polymerization (DP)	% Of Remnant life.	Interpretation
58	800	100	
130	700	90	Mild to minimal paper aging
292	600	79	Significant paper deterioration but still away from critical point.
654	500	66	
1464	400	50	Paper is starting to approach critical condition, suggest re- sampling in 1-2 year's-time.
1720	380	46	
2021	360	42	
2374	340	38	Paper is approaching the critical condition. Suggest re-sample within 1-year to reassess condition.
2789	320	33	
3277	300	29	
3851	280	24	Paper is approaching critical condition. Recommend transformer be taken out of service immediately and thoroughly inspected. Paper samples can be taken from portion for direct DP testing.
4524	260	19	
5315	240	13	Tests Indicates extensive paper degradation exceeding the critical point. Strongly recommend that the transformer be taken out of service immediately and visually inspected.
6245	220	7	
7337	200	0	End of life of paper.

Table 2: % remanent life assessment.[8] [9]

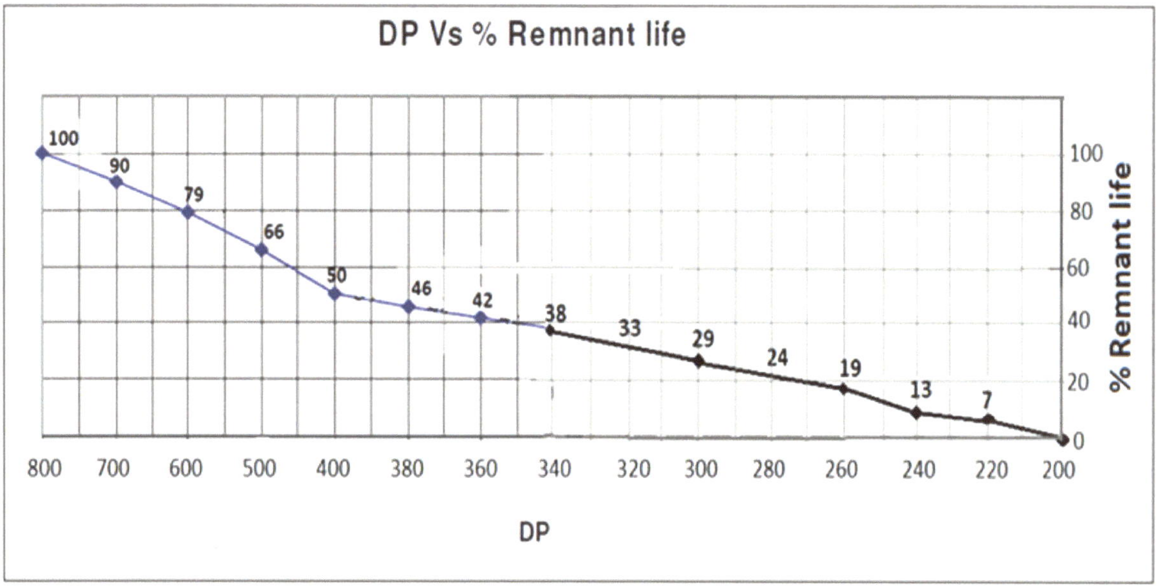

Figure 5: Graphical Representation of the values shown in the table 3.

9.0 FLOW CHART OF PAPER INSULATION DEGRADATION Vis 'a vis %RESIDUAL LIFE;

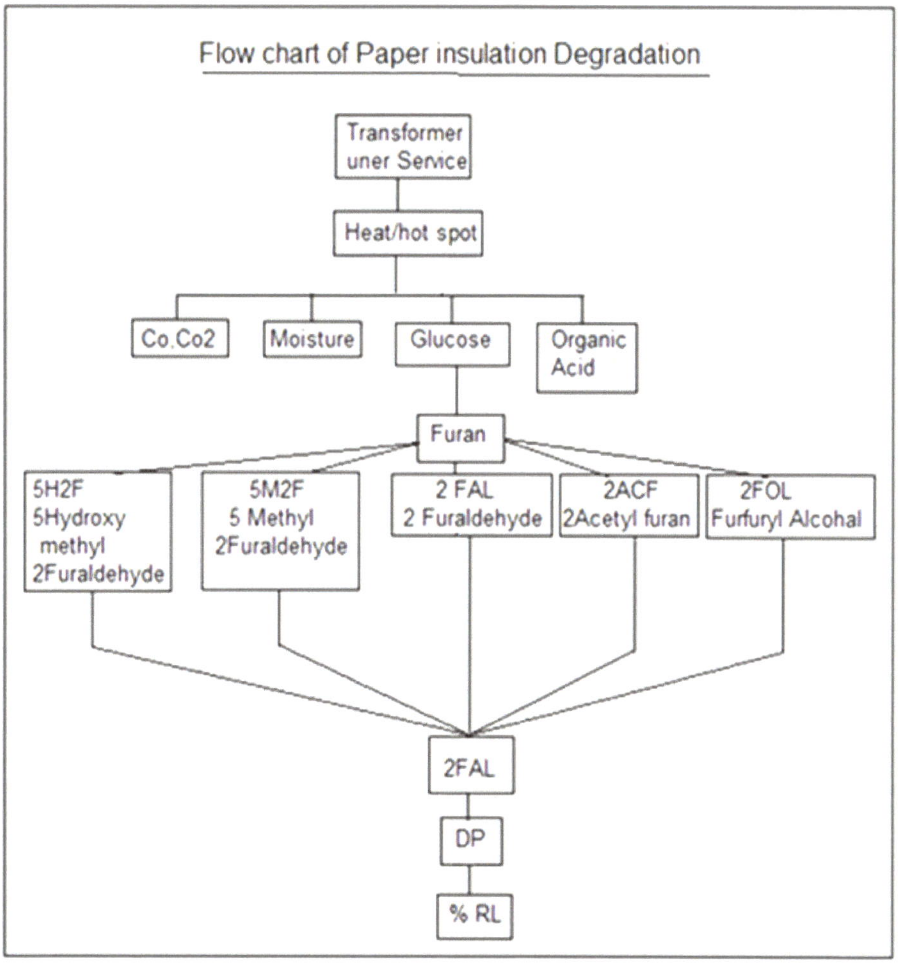

Figure 6: Paper insulation degradation vis 'a vis % residual life flow chart.

10.0 CRITERIA FOR SELECTION OF OLD TRANSFORMERS FOR EVALUATING % REMNANT LIFE;

i. Transformers which have served for more than 20 years.
ii. Deterioration in IR and PI values.
iii. DGA showed increasing trend in the key gases and also increasing trend in CO and CO2 gases with reference IEC 60599.
iv. Water saturation in oil if found >*15 % [1].
v. DP value <250.

*Note: This figure is selected considering the value shown in the Table 1, indicating guidelines for interpretation of % saturation of water in oil.

11.0 SELECTION OF TRANSFORMERS FOR EVALUATING % THE REMNANT LIFE;

old transformers that Sample list of a few have rendered continuous service for over 20 years with the calculated RS % at sampling temperature is shown in the Table 5.

No.	Substation name	Name of transformer	Transformer serial No.	**Water content** (ppm)	Temp. in °C	Calculated R.S. %	Remarks
1	132 KV S/s Patan	40 MVA 132/33 KV BBL	473372	16	46	9.24	CAL at 50 °C
2	132 KV S/s Patan	20 MVA 132/33 KV GEC	1328571	18	38	14.87	CAL at 40 °C
3	132 KV S/s VFJ	40 MVA 132/33 KV BBL	4733/21	15	52	8.67	CAL at 50 °C
4	132 KV S/s Kymore	16 MVA 132/33 KV NGEF	2800034714	21	50	12.13	CAL at 50 °C
5	132 KV S/s Kymore	20 MVA 132/33 KV BHEL	6004548	27	51	15.60	CAL at 50 °C
6	220 KV S/s Jabalpur	40 MVA 220/132 KV BHEL	6004131	20	50	11.56	CAL at 50 °C
7	220 KV S/s Jabalpur	40 MVA 220/132 KV BHEL	6004129	24	52	12.94	CAL at 52 °C

Table 5: A sample list showing transformers which have served for more than 20 years with moisture contents of the oil at the sampling temperature.

11.1 CASE STUDY 1;

(Please refer Cl 3.2 for formula for % Saturation and Sample calculation in Cl 3.2.1)

The transformer at Sr, no.5 of Table 5; 20 MVA 132/33 kV BHEL transformer bearing Sr. no.6004548 at 132 kV S/S, Kymore.

11.1.1 Calculation of % of solubility of water in oil;

Log10So =(-1567/K) +7.0895,

Where;

- So is the solubility of water in mineral oil,
- K is the absolute temperature in Kelvins (^{o}C + 273),
- °C is the oil temperature in Celsius at the time of sampling,
- % saturation = (ppm/So) x 100,
- Oil temperature = 50°C and moisture content in the oil 27 ppm,
- K = 50 + 273 = 323° K

Therefore, Log10 So = (-1567/323) +7.0895 = 2.2382

SO = $10^{2.2382}$ = 173.06135,

Therefore, % saturation = ppm/So x 100 = (27/173.06135) x 100 = 15.601 %

Since *15.601% > 15 %, this transformer is identified for removal from the system.

*Note: 15 % saturation of water in oil was taken as the benchmark for selection of old transformers for evaluating the remnant life.

11.1.2 FURAN ANALYSIS REPORT; It's 2 FAL was reported to be 3570 ppb, the DP is around 260 as per Chengdong's curve ie the % remnant life is 19%.

11.1.3 RECOMMENDATION;

i. Since % saturation of water in oil was 15.601%,
ii. % Remnant life found to be 19% in the range of high risk of failure.

Looking to above, this Transformer was strongly recommended for removal from the System.

11.2 CASE STUDY 2;

Transformer at Sr.no.7 of table 5; 40 MVA, 220/132 KV, BHEL, Transformer bearing sr.no.6004129 at 220kV S/S Jabalpur.

11.2.1 Calculation of % of solubility of water in oil;

Log10So =(-1567/K) +7.0895,

Where:

- So is the solubility of water in mineral oil,
- K is the absolute temperature in Kelvins (^{o}C + 273),
- °C is the oil temperature in Celsius at the time of sampling,
- % saturation = (ppm/So) x 100,
- Oil temperature = 52° C and moisture content in the oil 24 ppm,
- K = 52 + 273 = 325 ^{o}K

Therefore, Log10 So = (-1567/325) +7.0895 = 2.268

SO = $10^{2.268}$ = 185.353,

Therefore, % saturation = ppm/So x 100 = (24/185.353) x 100 = 12.94%

Therefore % saturation is within the bench mark of 15%.

11.2.2 Furan analysis report; Its 2 FAL was reported to be 1520 ppb, the DP is around 380 as per Chengdong's curve ie the % remnant life is 46% indicating accelerated ageing.

11.2.3 Recommendation;

i. Since % saturation of water in oil was 12.94%,
ii. % Remnant life found to be 46 %, paper is approaching critical condition,

Looking to above, this Transformer was recommended for continuance in the circuit, however, monitoring of condition was suggested to be done in 1-2 year's-time.

12: NOTE;

i. The concentration of Furanic component 2 FAL gives an indication of the condition of paper in terms of degree of polymerization, while rate of change of furan concentration can indicate the rate of aging of paper.
ii. High concentration of 2 FAL (2 Furaldehyde) is an indication of aged cellulosic insulation.
iii. Average DP value of new kraft paper is 1,000 to 1,200. Breaking down of cellulose during manufacturing and transformer drying process brings down the DP value to 800.
iv. DP value < 200 indicates extensive loss of paper degradation approaching the critical value ie threat of failure of the aged transformer.

13.0 SOME OF THE SPECIMEN PICTURES ARE DEPICTED IN THE FIGURES 7(A) & (B), 8 & 9.

i. Figure 7: (a) Insulation structure of healthy paper. (b) depicts structure of paper when the DP value fell down to <200.
ii. Figure 8: Showing effect of DP value<200 in a Power transformer.
iii. Figure 9: Showing effect of DP value<200 in a Power transformer.

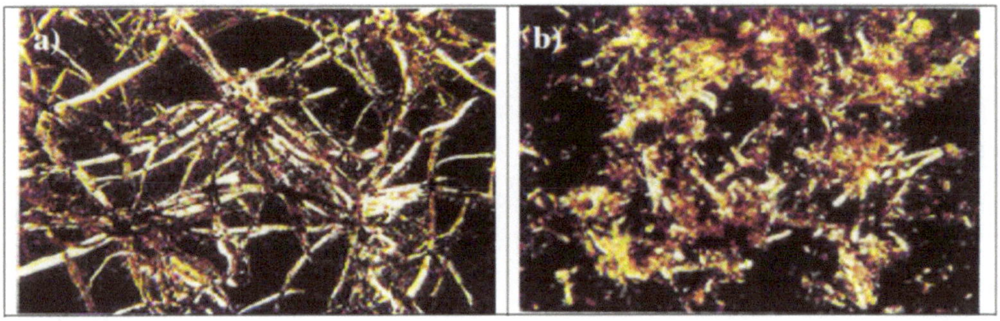

Figure 7(a). Figure 7(b)

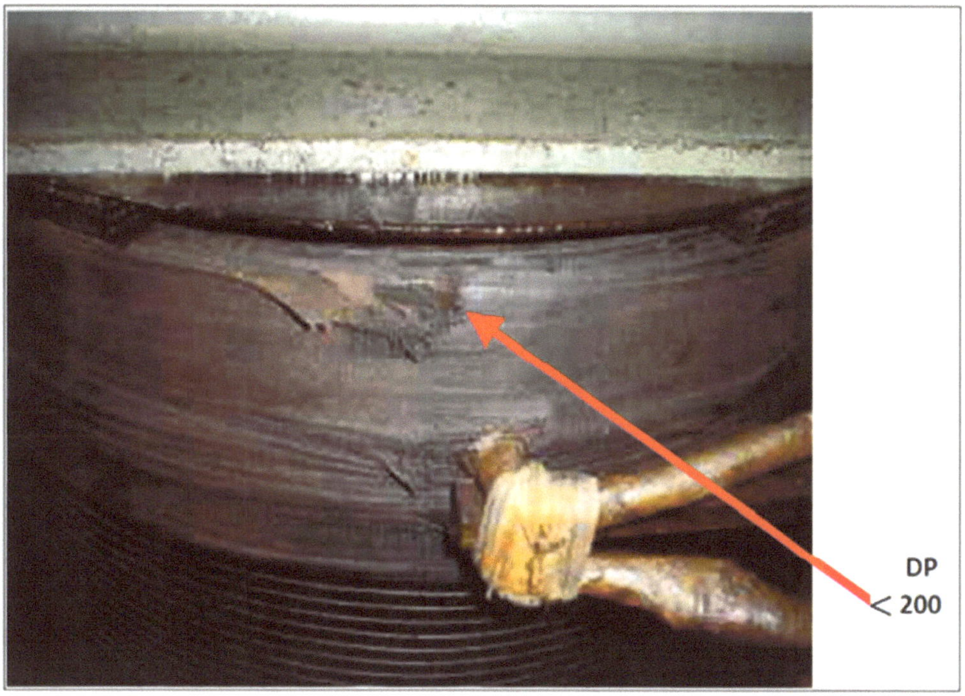

Figure 8

Figure 9

13.0 CONCLUSION AND SUGGESTIONS;

i. This methodology is found to be effective in MPPTCL. It is suggested that the power utilities may follow suit.

ii. Degree of Polymerization (DP) is used to evaluate the Life expectancy of a Transformer through Furan analysis. Direct measurement of DP is practically not possible for the in-service transformers as analysis of paper insulation for evaluation of DP value requires removal of a few strips of paper from the aging Transformers. Therefore, the analysis of oil for furan contents is a "Non-invasive test".

iii. This test should be included in the Transformer's maintenance schedule at least once in 2-years of the Transformers which have served >20 years and data should be reviewed in conjunction with DGA, fluid insulation tests and maintenance history.

iv. The main advantage of using this technique as a diagnostic tool is that these FURAN compounds are degradation by-products specific to paper only which are soluble in the Transformer Oil but cannot be produced by the oil itself.

14.0 **References;**

1. IEEE standard 62-1995, chapter 6.3.12
2. IS: 2026 (Part-7):2009 (IEC equivalent IEC 60076-7:2005).
3. IEEE Std C57.152-2013 guide for Diagnostic field testing of fluid filled Power Transformers, Regulators and Reactors.
4. CIGRE report 349, Working Group A2, 30 June 2008.
5. IEEE-C57-106-2002 Guide-for-Acceptance-and-Maintenance-of-Insulating-Oil-in-Equipment.
6. IEEE Std. C57.152-2013, IEEE guide for diagnostic field testing of fluid filled power transformers, regulators and reactors (Annex-E)
7. Nkosenye Sidwell Metetwa-Research paper "Accuracy of Furan analysis in estimating DP in power transformers.
8. An introduction to the Half Century Transformers by the Transformer Maintenance Institute, Myers Co, 2002.
9. I.A.R. Gray, Transformer Chemistry Services: Evaluation of Transformer Solid Insulation.

15.0 The Authors; 1) K.K.Murty, 2) D.C.Jain.

11. Life Cycle of Transformer

Abstract; *The EHV Transformers are the costliest and vital equipment in any Power network /large Industries. It is the prime duty of EHV Sub-Station Managers, Asset managers / Sub-Station O&M Engineers to look after any EHV Transformer right from its receipt at site to its life-cycle scrupulously with motherly feelings.*

Generally, guide-lines of the OEMs are followed, which are limited to their transformers only. This Article is significant and important contribution of experienced Engineers in the field of Operation & Maintenance and Testing & Commissioning up to 400kV in the EHV sub-stations. They have specifically narrated/ deliberated all the aspects from receiving the consignment of an EHV Transformer at site to its life-cycle including O&M and enhancing life expectancy thereof.This is an Article may be used as guide-lines by the Sub-Station Managers, Asset managers/ Sub-station Engineers in the absence of any such guide- lines issued by their Utilities or in their Industries.

Keywords and Acronyms; N2 (Nitrogen Gas), SFRA (Sweep Frequency Response Analysis), FAT (Factory Acceptance Test) and TCIV (Transformer Conservator Isolation Valve), OEM (Original Equipment Manufacturer), Torr: unit of pressure equivalent to mm of Hg., DGA (Dissolved Gas Analysis), FDS (Frequency Domain Spectroscopy.) and Furan analysis.

1.0 Introduction; The EHV leveled Transformers are the costliest equipment in any substation/ switch-yard of a Transmission/Generating Utilities or industries. Utmost care is required to be taken right from the receipt of the consignment of the EHV Transformer tank carrying Core-coil assembly at the site loaded on a multi-wheeled low-bed trailer. The Transformer is accepted only if all the prescribed tests are found satisfactory without unloading the transformer from the low-bed multi-wheeled trailer. Thereafter, all the activities are carried out right from its unloading from the trailer shifting to the plinth and installation on it. Thereafter, its erection, testing-commissioning and O&M is carried out which fall under the category of life-cycle. All the points are discussed under the sub-clauses, which should be carried out scrupulously to obviate conditions of trouble-shooting or premature failure of the transformers against their predefined life-span of 25 years.

Premature failure of EHV Transformer not only causes a colossal loss to the Utility, disturbance to the Grid-network but also dissatisfaction amongst the consumers at large. In case of continuous-process Industries, the owners/entrepreneurs would suffer enormous loss due to unforeseen failure of their EHV Transformers.

2.0 Activities on receipt an EHV Transformer;

1. On the arrival of the consignment of the EHV transformer at site and before unloading it from the multi-wheeled low-bed trailer, following activities of checking and testing are carried out;

Figure.1: Transformer received loaded on multi wheeled low-bed trailer.

a. Before unloading of transformer from the multi wheeled low bed trailer, it should be ensured that trailer is placed in such a way that when it is unloaded, shifted and installed on the plinth its HV Bushings should face towards the HV Bus-bars.
b. Check the Transformer for any signs of physical damages during Transportation thereof to the destination.
c. Get Impact recorder and its curves provided by factory officials.
d. Take the SFRA test in presence of the OEM's, personnel and tally the same with the one taken at the time of dispatch of the Transformer from their works. [1]
e. Check, N2 Pressure. It should be found in +ve pressure in case of dispatch without oil. In case of dispatch of main body in oil filled condition, check oil level and leakages if any.
f. For oil filled transformers a sample of oil should be taken from the bottom of the tank and tested for BDV and moisture content. If the values do not meet the relevant standards the matter should be taken up with the manufacturer.
g. Check the dew point of N2, it should not be more than -30°C. (in case of 400 kV Transformers).
h. Check and verify that the Transformer-core of the Transformer is earthed at single-point only.

*Figure.2: showing Tank, Core and Core-frame earthed together on the top cover of the transformer, brought-out through individual leads to 2.5kV *bushings.*

[***Note**: As far as kV rating of such bushings is concerned 2.5 kV is adequate, however, some customers ask for 10kV AC one minute as test voltage. None of the international Standards specify this test. These test points are basically for verification, whether core is earthed at single point or is getting earthed through the clamping frame or through the tank due to some reason or the other.]

i. Measure the IR value between core-frame-tank with 500 V-2.5kV DC megger.
j. Check and verify the isolation of test leads from the Transformer tank by taking continuity with a multi-meter.

Figure.3: Depicts epoxy test bushings on the blanking plates of the Transformer turrets and test leads are connected to them for

LT tests of the transformer on the low-bed Trailer.

k. Perform the LT magnetic balance test, magnetizing current test and verify the test results from FAT.
l. Verify the vector-group of Transformer as per R&D plate.
m. Check and confirm that packing case numbers match with the packing list. Check their contents and tally with the packing list, if any of the packing case is damaged.

1.1. Prepare the Transformer's Life sketch-book(register) or computerized Excel- File as soon as the transformer is received at site.

a. enter all FAT results of the transformer.
b. enter all the activities of checking and testing mentioned in clause 2.0-1).
c. enter all pre-commissioning tests and activities.
d. enter all the test results during its 1/2 yearly/annual routine maintenance activities including test reports pertaining to SFRA, DGA, Furan analysis etc.
e. any other emergency repair or replacement of its components etc.

3.0 Storage [1];

3.1 Storage- less than 90 days (With Dry Air or Dry Nitrogen):

The transformer should be kept under positive pressure between 0.1 bar and 0.3 bar (1.5 lb/in^2 to 4.4 lb/in^2) at all times in storage. A reserve gas supply > 10 bar (150 lb/in^2) is required to be coupled with a pressure vacuum regulator.

3.2 Storage—up to six months (Liquid filled).

Some OEMs permit storage in dry nitrogen or dry air for up to six months, but storage with liquid filled is preferred.

An unassembled transformer can be vacuum filled with liquid up to the core level and the space above the liquid is to be pressurized with dry air or nitrogen. An elevated pressure should be applied for several hours and the tank checked for leaks. Both the cylinder pressure and the tank pressure should be monitored and recorded.

3.3 Storage—more than six months. (For substituting faulty unit expeditiously);

The spare Transformer should be installed on its plinth. It should be fully assembled, vacuum filled with liquid, tested and be stored with proper liquid level in the conservators and silica-gel breather should be monitored from time to time such that the spare transformer is kept ready for replacement of the faulty one in case of any emergency.

ii. If the spare transformer has OFAF cooling system equipped with pumps, periodically (3–6 months) one-half of the pumps should be run for 30 min followed by operation of the other half. The fans should be run for approximately 10 min on monthly basis and control cabinet heaters should be checked.
iii. If the spare Transformer has ONAF cooling system. The operation of the fans should be checked at regular intervals.

Note; Where 3+1 single phase units make the bank, then to save on time, in case of replacement of the faulty unit, all the control cables from top of the transformer pertaining to Buchholz, PRV, OSR etc. should be brought into a marshalling box which would facilitate necessary change over expeditiously in control cable connections.

4.0 Erection, installation and assembly;

4.1 Shifting the transformer to the plinth and placing thereon: Unload the transformer with the help of wooden sleepers and Hydraulic-jacks of adequate load bearing capacity, lower down on to the MS 'C' channels placed upside down on a pair of rails poles properly greased (refer figure 4) and drag the transformer with the help of manual tirfor-winch or motorized geared winch-machine of appropriate capacity on to the plinth on the rail center, fix the wheels and lock all wheels with rails on the plinth.(refer figure 5)

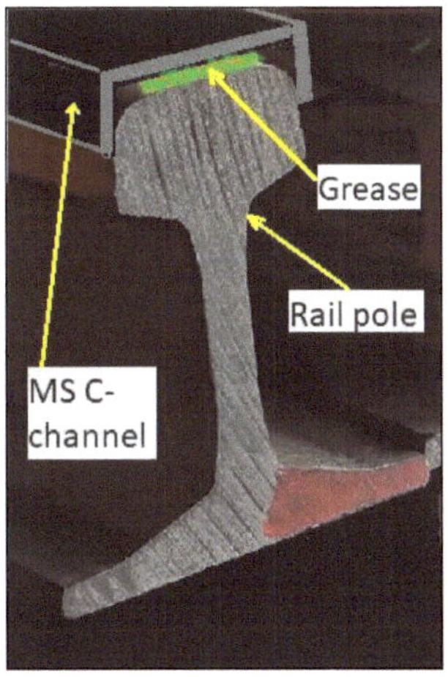

Figure 4: MS- C:Channel placed on the greased rail poles.

Figure 5: picture showing wheels of the transformer locked on the rail pieces grouted on the plinth and the eyes used for hooking wire-rope for dragging.

5.0 Internal inspection;

5.1,1 Preparation and precautions;

5.0.0.1 In case of internal inspections of large Transformers, Dry compressed Air/Oxygen Cylinders (4-5-Nos) are kept ready at the site. There is always a threat of insufficient oxygen inside the tank which may cause breathlessness / Giddiness to the personnel who would be entering the tank for internal inspection. In order to obviate such condition, dry compressed air/Oxygen is injected through the bottom valve, keeping one blanking plate open on the top of the transformer tank diagonally opposite to the bottom valve through which dry air/Oxygen will be injected for proper ventilation. Ensure continuous flow of dry air/oxygen during entire internal inspection period.

5.0.0.2 To facilitate proper internal inspection, arrangement for illuminating inside the transformer tank with non-emitting heat type lamp say LED lamp etc. should be made.

5.0.0.3 Arrangement for videography/Photography with high pixel camera should be made.

5.0.0.4 Before a person is made to enter the transformer tank for internal inspection, following should be essentially ensured by the Engineer

/supervisor executing the job at site that;

1. he is wearing bare minimal clothing.
2. he should not wear any ornamental/religious chain or locket, ring and watch etc.
3. He should tie spanner and wrench of appropriate size with a cotton tape round his waist.

5.0.1 Internal inspection activity;

1. Tightness of all core bolts, core-frame-tank leads, selector-switch leads should be checked carefully, ensuring that none of the bolts are left unattended.
2. Verification of Job no. engraved on the Core-frame assembly matches with the job no. written on the R&D plate.
3. Abnormality /ambiguity found if any during internal inspection, photography/ videography thereof should be taken.

5.1 Oil filling activity:

i. Before oil filling, all residual oil in the Transformer tank should be drained out and the tank bottom should be thoroughly cleaned with cotton cloth.
ii. Fit equalizing pipes between main tank and OLTC compartment/compartments, so that equal vacuum is applied in both ie. in the tank and the tap changer, otherwise diverter-switch oil chamber / barrier board of Tap changer will get damaged.
iii. After the activities as mentioned in item nos.4.2b) are over, Transformer tank (excluding cooler banks and conservator) shall be evacuated to a pressure of vacuum 1.00 Torr and should be held for 48hrs up to 220kV class and for 72hrs for 400kV transformers and above.

[**Note**: a. Now a days some OEMs design the cooler banks vacuum proof in that case cooler banks can be included.
b torr to mmHg conversion: 1 torr = 1 mmHg.]

iv. Fill up the*Oil/ester fluid to the core level. Before filling oil/ester fluid ensure all oil-valves are in correct position either in closed or in opened condition as required.

[*Note;

a. Mineral oil confirming to IS: 335-2018(IEC 60296-5th ed,2020 & IEEE C57.106, 2015) in case of Mineral Oil filled Transformer.
b. Synthetic Ester fluid confirming to IS-16081-June 2013(IEC 61099:2010 describes specifications for unused synthetic organic esters for electrical purposes).
c. Natural Ester fluid confirming to IS: 16659 - 2017 (IEC 62770:2013 describes specifications and test methods for unused natural esters in transformers and similar oil-impregnated electrical equipment).]

5.3.1 (i) In case of 400kV Transformers before filling oil, absolute vacuum of 1.00 torr should be taken and held for 72hrs. then N2 gas of less than −50°C dew-point should be pushed in till such time that the vacuum pressure comes to the normal value and then to 0.2kg/sq.cm pressure and to be kept in this condition for next 24hrs, then release N2 and measure dew point of N2 coming out from the transformer. It should not be more > − 40°C. If it is observed that it is> − 40°C, then, N2 purging cycle to be repeated ones again.

(ii) Dew-point acceptable limits against temperature of insulation: For transformer/reactor with a gas pressure of 2.5-3 PSI, the acceptable limits of dew point shall be as per table 1. [4]:

Temperature of Insulation in °F	Permissible dew point in °F	Temperature of Insulation in °C	Permissible dew point in °F
0	-78	-17.77	-61.11
5	-74	-15.0	-58.88
10	-70	-12.22	-56.66
15	-66	-9.44	-54,44
20	-62	-6.66	-52.22
25	-58	-3.33	-49.99
30	-53	-1.11	-47.22
35	-48	+1.66	-44.44
40	-44	+4.44	-42.22
45	-40	+7.44	-39.39
50	-35	+9.99	-37.22
55	-31	+12.77	-34.99
60	-27	+15.55	-32.77
65	-22	+18.33	-29.99
70	-18	+23.11	-27.77
75	-14	+23.88	-25.55
80-	-10	+26.66	-23.33
85	-6	+29.44	-21.11
90	-1	+32.22	-18.33
95	+3	+34.99	-16.11
100	+7	+37.75	-13.88
110	+16	+43.33	-8.88
120	+25	+48.88	-3.88
130	+33	+54.44	+0.55
140	+44	+59.99	+5.55

Table 1

Table 1- Variation of dew-point of dry air/N2 Gas filled in the transformer/reactor tank w.r.t temperature;

6.0 Erection of bushings

6.1.1 Before erection of bushings;

6.1.1.1 tally and confirm the Bushing nos. from FAT results that they are the same ones used during FAT.

6.1.1.2 Bushing should be thoroughly checked for any crack, oil leakage. Even a hair-line crack over the condenser portion of Bushing renders the Bushing useless, therefore proper handling of Bushing should be ensured and if it is found with small crack it should not be used.

These days all the Bushings are hermetically sealed therefore leaky Bushing should not be used; as the leaky joint is a source of moisture ingression.

6.1.1.3 Before installation of the Bushings, the Bushing CTs (Turret CTs) should be tested for ratio, polarity, winding resistance and knee point voltage using test-kit CT Analyzer confirming to IS:16227-part 2 /IEC61869-2.

[Note: Some OEMs dispatch the Turrets with BCTs fitted inside therein with blanking plates on the either side of the Turrets. In such cases, remove the blanking plates and test the BCTs as narrated above with an additional test; IR value test of Secondary winding terminals to Earth (Turret body), before fitting the turrets on the ear- marked locations on the transformer body/on top of the Transformer.]

Figure 5A: Bushing CTs (Turret CTs).

6.1.1.4 IR value of the Bushings should be taken before installation.

6.1.1.5 Wherever arrangements for tan-δ measurements are available then they should also be recorded.

[Note: tan-δ and C should be measured after having held the bushing in the up-right erect position on a specially fabricated structure for 24 Hrs.]

6.1.2 Lowering down the bushings:

6.1.2.1 Care should be taken during lowering down the bushings into the tank through the turrets. Such jobs are executed through single long-boom crane and using some nylon ropes for handling the Bushings and also for aligning the bushing's lower portion as per the angular deviation of the turrets. Care should be taken such that the bushing's lower portion does not foul with the Bushing CTs.

Figure 5B: Hydra Long-boom crane.

6.1.2.2 Since the bushings of 400kV Transformers, 400kV Reactors and above are physically very long, therefore utmost care should be taken during lowering down the bushings into the tank through the turrets. It is recommended by the OEMs to use 2 nos. long-boom Hydra- cranes to execute such precision jobs.

Figure 6: Lowering down 420kV bushing with the help of 2-nos. Hydra long-boom cranes.

[**Note:** i. As could be seen from the figure 6 that 2-nos. long- boom cranes have been used. One crane is for lifting the Bushing and the other one is exclusively for carrying a belt-man with assigned job of lowering down and pulling up the draw lead /draw-rod and fixing the bushing terminal.

ii. While lowering down the Bushing, care should be taken such that the bushing should not foul with the Bushing CTs (Turret CTs).

d) Procedure:

Pull the lead up from the winding.

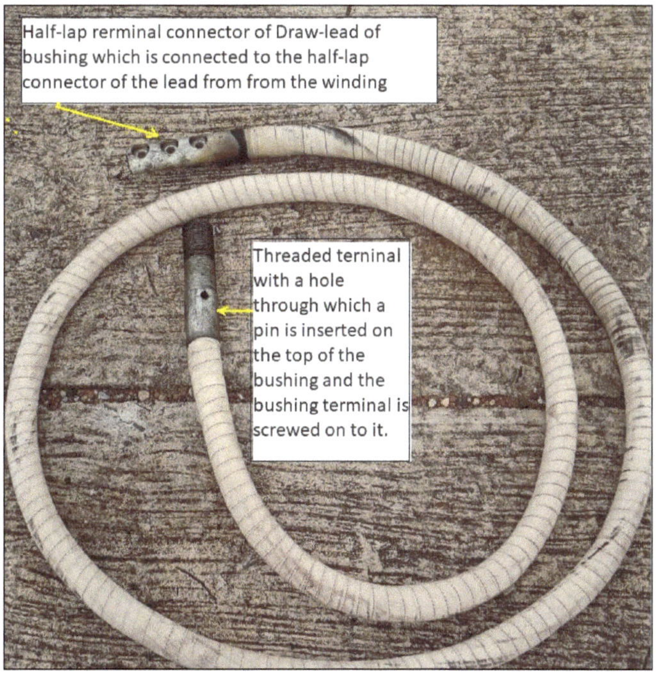

Figure 7A: Draw-lead of the Draw-lead type bushing.

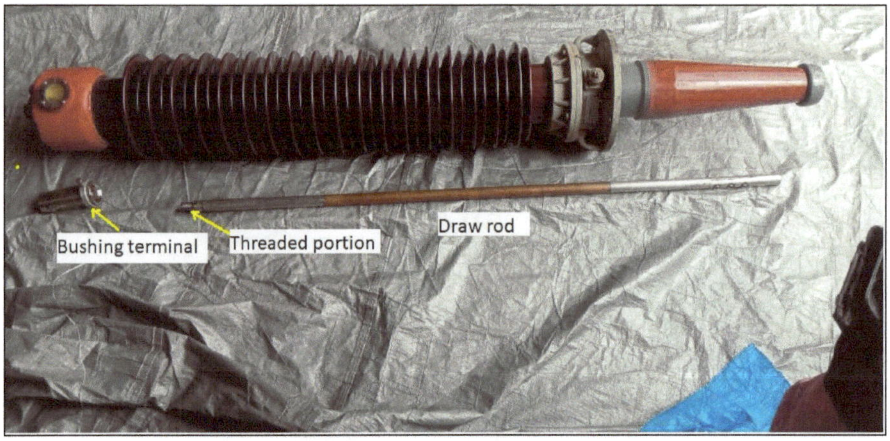

Figure 7B: Draw-rod type bushing

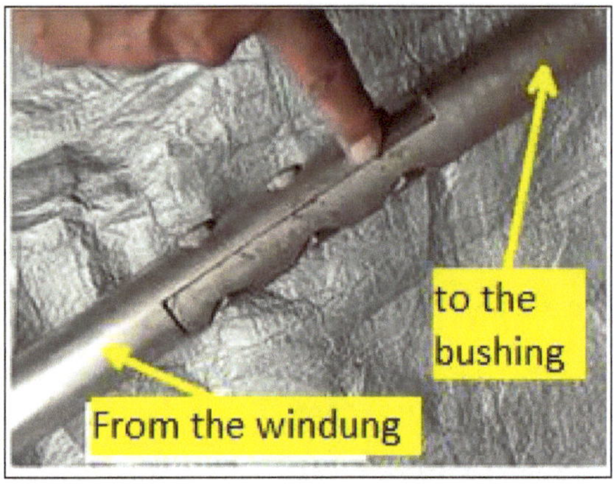

Figure 7C Joint between draw-rod of bushing and the winding lead.

Figure 7D

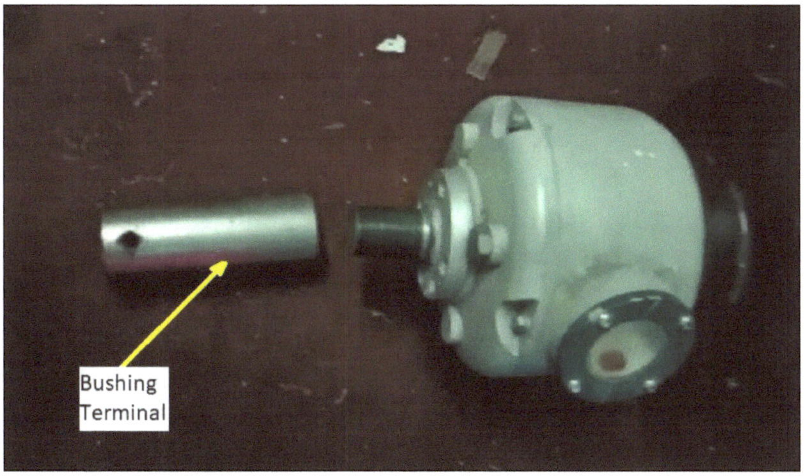

Figure 7E

i. Lower down the draw-lead /draw-rod of the bushing (Refer figures 7A &7B). Couple the half-lap joint clamp of the lead from the winding with the half-lap joint clamp of the draw-lead / draw-rod of the bushing together & tighten them (refer figure 7C). Insulate them with crape-paper tape and wrap cotton-tape over the same. In case of draw-lead care should be taken so that lead is not twisted otherwise lead will fall short.

ii. Pull the draw-lead/draw-rod up and tighten the top terminal of the bushing.

   a. Pull the bushing up and hold in position by inserting a pin through the draw-lead. (Refer figure 7D)
   b. Pull the draw-rod type bushing up and screw the bushing terminal over it.

iii. After the bushings are installed on the turrets, it should be ensured that tan δ measuring test-tap cover provided on the bushings are properly fitted and condenser is properly grounded on the tank-cover of the transformer.

7.0 Erection of Radiators, Thermo-siphon, PRV, Pipelines from main tank to conservator including Buchholz relay,

7.1 Erection of NIFPS fire protection system and Buchholz Relay:

7.1.1 Note: In case the transformer is equipped with NIFPS (Nitrogen Injection Fire Protection System), preparation of Oil sump, supply of equipment, Installation and commissioning is in the purview of the OEM of NIFPS.

7.1.2 In case the transformer is equipped with NIFPS, while assembling the pipe from conservator to the main tank following components are to be fitted serially

7.1.2.1 Buchholz relay,

[Note: (a) Arrow embossed on Buchholz relay must point towards the conservator.

(b) The "Test & Service switch" should be locked in service position properly otherwise it will shift to the test position due to vibration and transformer will trip.]

7.1.2.2 TCIV (Transformer Conservator Isolation Valve) (refer figure 8A),

7.1.2.3 steel bellow pipe,

7.1.2.4 conservator.

7.1.2.5 Care should be taken to fit gate valves as shown in the figures 8A.

Figure.8A: Showing Buchholz relay (with acrylic cover to protect against rains) and TCIV (Transformer Conservator Isolation valve).

Figure 8B: Buchholz Relay.

1.1.1 Fire Protection System-Option; Automatic MulsiFyre System [2];

Figure 9: Automatic MulsiFyre System.

This system is widely used for fire-fighting of out-door Transformers. Fire detectors located at various strategic points are used to sense high temperature near the transformer. If the temperature exceeds the set value, the automatic, MulsiFyre system sprays water at high pressure on the surface of the transformer to control the fire on any burning oil spilled over.

It consists of following main components;

i. Main Hydrant: Main Hydrant carries water which is the main fire extinguisher.
ii. Fire detectors: Fire detectors are generally thermocouples.
iii. Ring mains and nozzles: Ring mains surround the transformer are provided to feed water at high pressure to the nozzles at different levels.
iv. Pumps: Main duty of the pumps is to feed water to the Main Hydrant.

[Note: The supply of equipment, erection and maintenance thereof is generally lies with the OEMs.]

7.3.1 Conservator Erection:

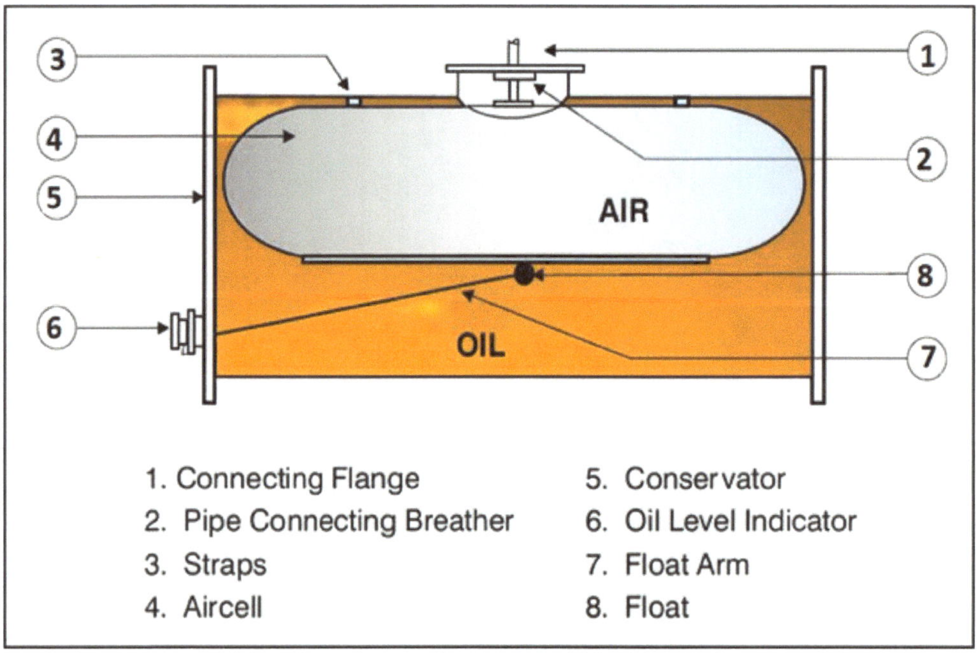

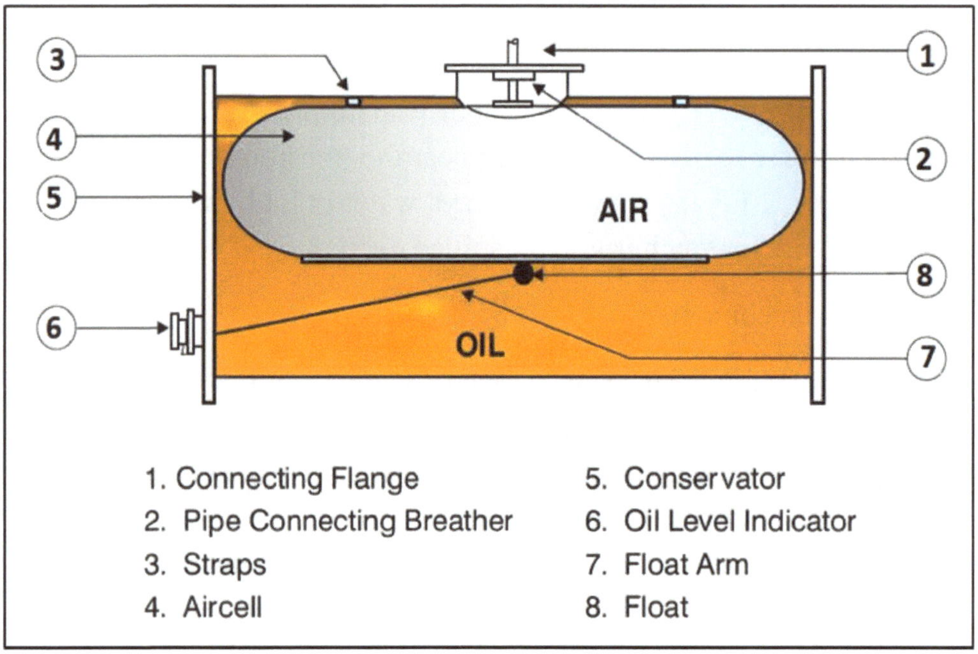

Figure10A: Conceptual figure of Conservator with Air cell showing main components.Before erection of conservator, check leakage of the Air-cell (Refer figure 10B of Air- cell.) and then erect the Conservator.

[Note: N2 filled conservator: This system has an N2 Gas replenishment attachment which continuously maintains N2 pressure above the oil

Figure 10B: Air-cell

Install the air-cell as per the instructions /procedure defined by the OEM, however, the procedure has been narrated in the Sub-clause no.5.4.1 & 5.4.2.

7.3.2 Installation of Air cell;

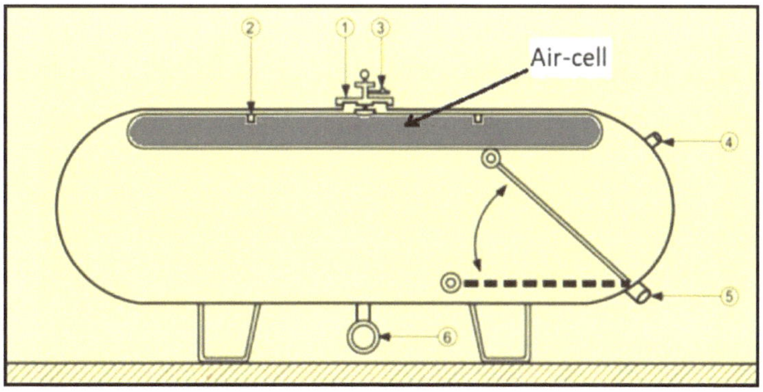

Figure10C: Conceptual figure of installation of Air-cell

Steps to be taken chronologically;

c. Insert the Air cell into the conservator through open end.
d. Suspend the Air cell from the hooks (2) in the ceiling of the conservator using the loops.
e. Close the open end of the conservator.
f. Fill the Air Cell to a pressure of 10kPa (0.1 bar). Close the filling valve (3) while leaving the system pressurized. A relief hole (4) must be on the oil filled side of the conservator so that the Air Cell can freely expandThe Air Cell will gradually stabilize. After 6 hours, adjust the air pressure again to 10kPa (0.1 bar).
g. A 24hours leak test is then to be made on the Air Cell. After 24 hours check the pressure reading again. If the pressure has not dropped significantly, the test is correct. The temperature should be as stable as possibly during the leakage testing period.

1.1.1 **Filling the Conservator;**

Steps to be taken chronologically;

1. Before filling oil turn TCIV to the filtration mode.
2. The pressure of air in the Air cell should be 10kPa (0.1 bar). If not, increase the pressure to the correct level and close the air-filling valve.
3. Check the air vent valves (4) are open at both ends of the conservator
4. Open the valve (6) between the conservator and transformer tank and pump in more oil so that the oil rises to the conservator. The pumping speed should be suitably slow so that the pressure within the Air Cell does not exceed 13 kPa (0.13 bar). Stop pumping, when oil begins to come out of the vent valves and close the vent screws.
5. Regulate the pressure in the Air Cell to normal by opening the plug in the flange or air filling valve from which the pressure hose is removed
6. Reopen the valve (6) between the conservator and the transformer tank and continue filling with oil until the oil level indicator gives the correct reading according to the temperature of transformer.

(**Note**: Instructions towards inflating the Air-cell and Oil filling are generally affixed on the transformer body by some OEMs).

1.1.2 **Magnetic Oil Level Gauge (MOG):**

Inside the conservator tank, a float is used to sense the levels of oil andmove. This is transmitted to a switch mechanism by means of magnetic coupling. The float and the magnetic mechanism are totally sealed. The pointer connected to the magnetic mechanism moves indicating the correct oil level. It should be ensured that the Oil level be indicated on the dial of the MOG as 35°C and also there is provision for Low oil level alarm through micro- switch.

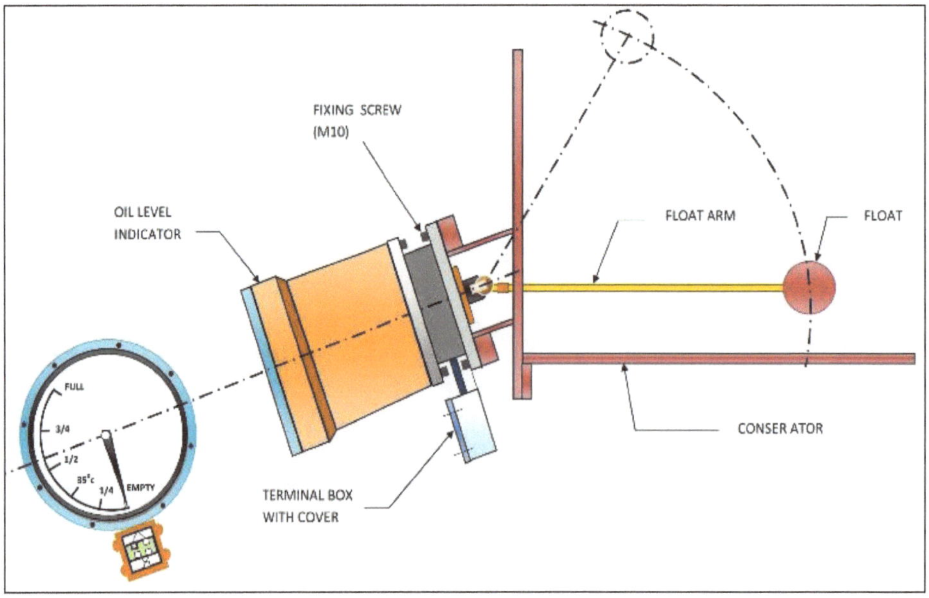

Figure 10D: Conceptual figure of MOG.

1.4 **Erection of breather:** Assemble the pipe-line from conservator to breather mounting (refer figure 11A &11B). The breather is connected to transformer by either threaded or flanged joint (after removing the Seal). Before putting the Breather into service, the silica gel granules are filled in from the top and blanked with a plug. Silica Gel is Blue when Dry; Pink when moist, remove the Oil Cup and fill it up with fresh transformer oil up to the line marked on Cup. Remove the seals on the air holes of the Cup and fit the Cup to the breather.

Figure 11A: Silica gel Breather.

Figure 11B: Twin-Silica gel breather in service.

7.6 Installation of OTI &WTI system;

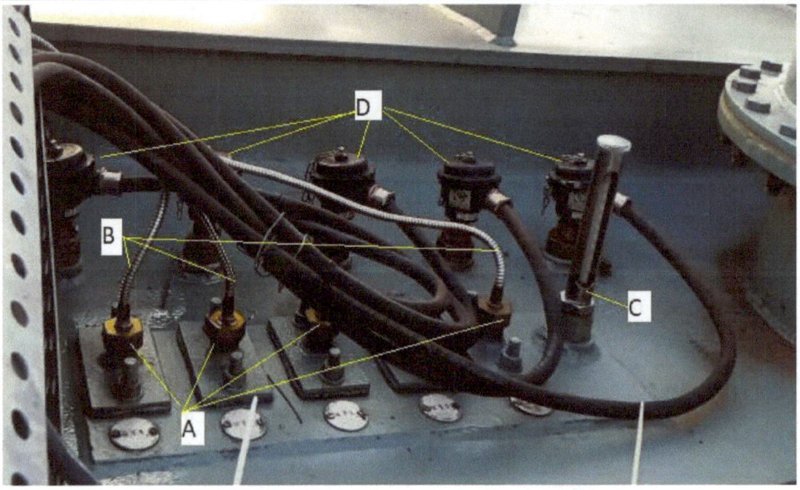

Index

A -OTI &WTI measuring bulbs inserted in the Oil filled Pockets.

B-Capillaries from each bulb to WTI OTI meters

C - Alcohol filled Thermameter

D - RTD Pt-100 for remote display on RTCC panel.

Figure 12A: WTI & OTI Bulbs and RTD Pt-100 units are shown in the respective pockets on the top of the transformer.

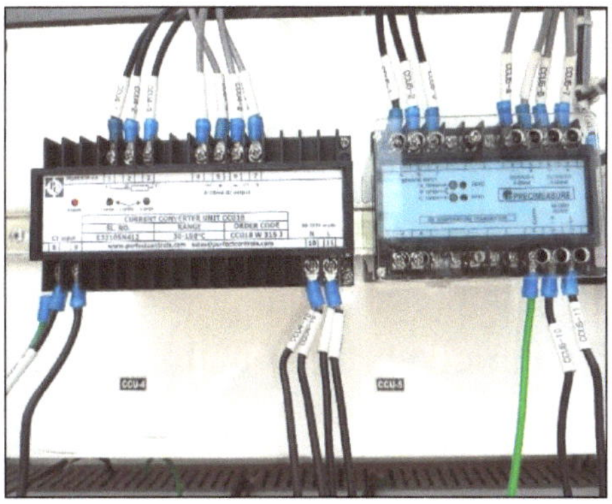

Figure12D: Winding Current Compensation unit (CCU).

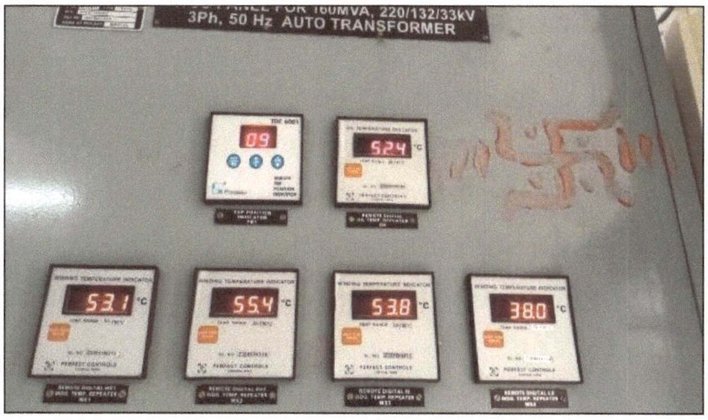

Figure12E: Remote display meters of WTI & OTI on RTCC Panel.

The thermometer bulbs at the end of capillary pipes and RTD Pt-100 sensor units (Refer Figure12 A) are inserted in the earmarked pockets (filled with transformer oil) on the top of the transformer for OTI and WTI. Both the Oil Temperature Indicator and Winding Temperature Indicator meters (Refer figure12B&12C) are provided with maximum pointers and mercury switches. Precaution: The steel capillary tubes are normally quite flexible but they should not be bent sharply and should be supported with clips to prevent unnecessary sagging. The temperature measured through RTD Pt-100 is compensated with WTI CT current in the CCU (Refer figure 12D) and the output thereof is used for remote display (refer figure 12E).

1.5 **Cooling system;**

1.5.1 **Radiators;**

The basic function of radiator of transformer accelerates the cooling rate of transformer. Thus, it plays a vital role in increasing loading capacity of transformer. The working principle of radiator is very simple. It just increases the surface area for dissipating heat of the oil.

Figure 13A: Body mounted radiators.

1.5.2 **Cleaning & Checking;**

a. the radiators must be cleaned externally. These are also required to be clean from inside by flashing with transformer insulating oil.
b. Before cleaning from inside it is necessary to check the leakage of radiator. To check the leakage, one side of radiator is blanked with blanking plate fitted with pressure gauge and another side is fitted with NRV. Through NRV air is injected with pump or compressor up to pressure 0.2 kg/cm2 for 2 to 3 hours, if leakage occurs pressure reduce.
c. Repair the leakage and again check.
d. The valve blanking plate is to be removed only when the radiator is ready for mounting on flanges.

1.5.3 **Procedure for erection of radiators;**

1. Lift the radiator vertically. Bring the radiator nearer to the valve. Remove the blanking plates from valve. Slide the radiator on the bolt from pipe flange the radiator valve and tighten using plain washer, spring washer and nuts. Tighten nuts in sequence applying uniform pressure on the gasket to make a proper leak proof fitting. Assembled all the radiators in the same manner.
2. Clamping arrangement at outer ends of the radiators is to be fitted to minimize the vibrations of radiators.
3. Make arrangements to fill the transformer oil through the main conservator. It is recommended to use hot filtered oil directly from the filter.
4. Fill only one radiator at a time. Open the bottom valve. Slowly unscrew the air release plug on the top of the radiators until the air starts escaping. The oil from the main tank will now flow in the radiator. Start putting fresh oil in the conservator.
5. After filling, open valve at top to equalize with main tank.
6. Fill the remaining radiators in the same manner.
7. After filling the oil in radiators and venting air from top then, radiator valves at top and bottom must be kept in OPEN position and sealed.

1.5.4 **Separately mounted radiators;**

In the large transformers the radiators are sometimes separately mounted on header each at top & the bottom, which are supported on frames. Flanges are provided on these headers for fixing the radiators.

Figure 13B: Externally mounted radiators.

Figure13 C: Exhaust fans and inline axial flow Oil pumps/motor.

1.5.5 Specification of motors & fans;

i. Cooling fans and oil pump motors (Refer figure 13C) shall be suitable for operation with 415 volts, three phase 50 Hz power-supply and shall be of premium efficiency class IE 3 conforming to IS: 12615/IEC 60034-1.

ii. Each cooling fan and oil pump motors shall be provided with starter, thermal overload and short circuit protection.

iii. the motor winding insulation shall be conventional class 'B' or better type. Motors shall have hose proof enclosure equivalent to IP:55 as per IS: IEC:60034-5.

iv. the temperature rise of the motor shall be limited to 70 deg. C above ambient of 40 deg.& shall comply with IS:12615/IEC 60034-1.

8.0 Thermosyphon Filter;

Thermosyphon filter is an ONLINE Transformer Oil Preservation system for Power Transformers. The system basically consists of a container containing absorbent (Normally Activated alumina) which has very high moisture absorption capability. and mounted in the flow line of transformer oil out-side the main tank of transformer While transformer oil circulates through this the absorbent takes away moisture from the oil. The oil circulation can be either under thermal head or through forced pump. The moisture content in transformer oil can be maintained to a level of less than 10 ppm.

Figure 14: Thermosyphons.

9.0 Oil filtration job;

a. Fill Thermometer pockets with oil/ester fluid and dip the calibrated thermometers.

b. Carry on Transformer oil filtration job till the BDV is in the range of >60kV and moisture content to <10 ppm as per IEC 60422-2013(refer table 4) keeping in view that the temperature of Transformer tank oil does not increase beyond 50°C.

[Note:

i. During filtration work Radiator Valves should be kept in closed condition.

ii. After the filtration work is over, push the oil from the bottom and open radiator valves one by one]

c. In case of 400kV and above transformers, The OEM's Erection Engineer shall take care during Erection and Oil filling/ drying out job of such transformer as per the procedure laid-down by the respective OEMs (Original Equipment Manufacturer).The Substation-manager/Substation-Engineer/Asser manager should extend all possible help and support to the OEM's Erection Engineer

9 OLTC erection and checking;

iii. Connecting of Operating rod and the worm gear arrangement; While connecting Tap Driving Mechanism Box with tap-changer, care should be taken so that both DM (Driving Mechanism) Box and the Taps are on same tap number.

[Note: In the present scenario, the OEMs dispatch the transformer with completely assembled OLTC driving mechanism with the Diverter Switch/Switches.]

iv. Connect OSR (Oil Surge Relay) between the OLTC conservator and the diverter switch compartment. While connecting Oil Surge relays, it should be ensured that arrow given on relay is towards the OLTC conservator. (Refer Figure 15A and 15B)

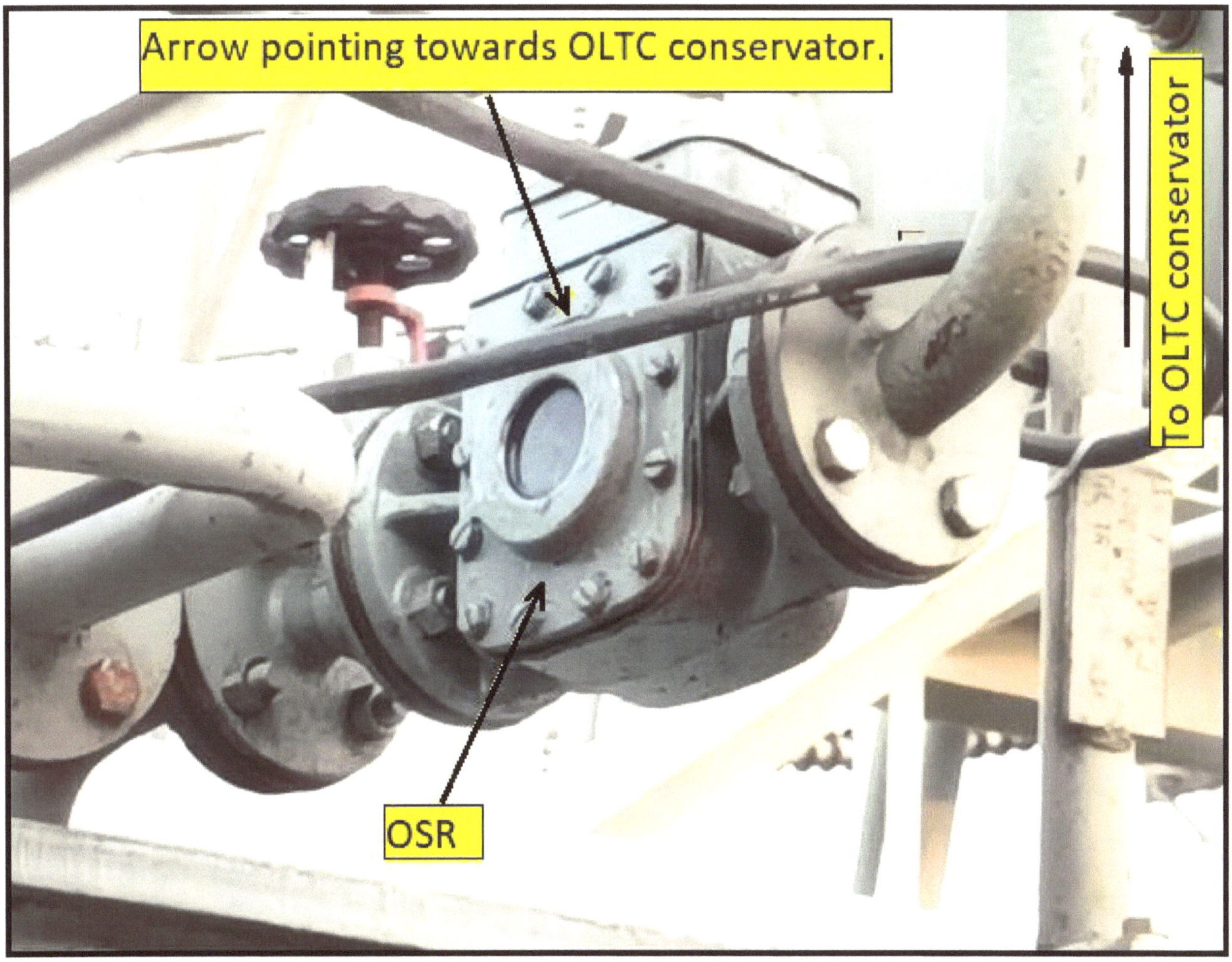

Figure15A: OSR.

Figure15B: OSR.

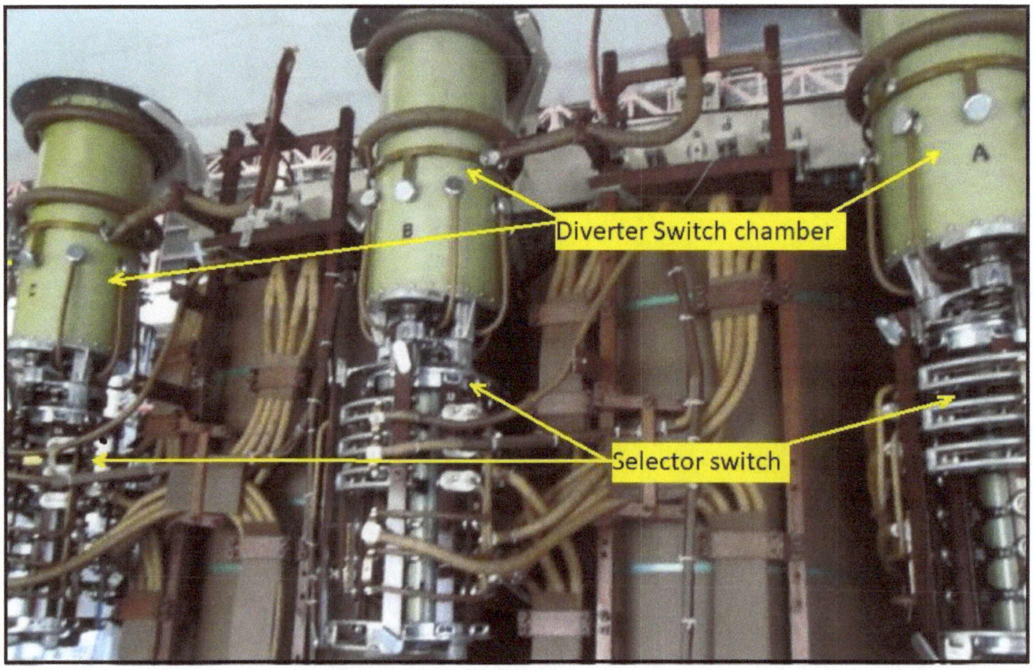

Figure 15C: In-tank type OLTC.

[Note: All EHV transformers are equipped with In-tank type OLTC. (Refer Figure15C]

v. Fill diverter switch chambers with new filtered mineral oil.
vi. Release air from air went from the diverter switch chambers.
vii. Check all the stop valves between oil conservator and driving unit of OLTC are open.
viii. Operate one full cycle of tap changer manually and check end limit switch stops.
ix. Operate one full cycle of tap changer electrically and ensure motor drive direction of rotation is correct.
x. Verify the operation of taps when operated from RTCC (Remote Tap Changer Control) panel.

[Note: The electrical and mechanical interlock should also be checked]

xi. Verify the OSR relay contacts, if closed must operate the Master trip relay module.

11.0 Neutral Grounding:

a. Copper- flats of suitable size must run through the Neutral CT where bushing CT is not provided in the neutral bushing (for REF protection) on the support insulators and connect them to the earth pit through 2 nos. ACSR/AAA conductors clamped with bimetallic pad to the copper flat as depicted in the figures 16A &16 B.

Figure 16A: External neutral CT for REF protection. connection through Neutral CT.

b. Neutral of HV and LV side of Star/star Transformers are grounded in separate Earth pits.
c. All the Earth Pits are connected in parallel and are in-turn connected to the earth-mat (Earth-grid) so that the combined Earth Resistance is always maintained below 1.0 Ω (preferably as low as 0.5Ω).

d. The individual and combined earth resistance is measured periodically, the Earth-Pits are maintained regularly and the electrodes are replaced if required.
e. The tank (body)should be permanently and effectively connected to earth by means of flat, flexible conductor of suitable size and materials (galvanized steel, copper, or acceptable equivalent) terminated on earthing terminals or pads provided at the bottom of the tank. (Refer figure 16C)

Figure16C: Transformer body earthing.

f. Cable box, marshalling box, OLTC operating box, etc. should also be earthed. (Refer figure 16D)

Figure16 D: Marshalling Box earthing.

g. Verify the integrity of the earthing system.

12.0 Delta Tertiary winding/bushings:3U, 3V&3W or Y1, Y2 Y3(as per IEEE std. C57.12.70-2000):

a. In a Transformer bank of three single phase units, ensure tertiary winding connection as per vector group YNa0d11 or YNa0d5 of the Auto-Transformer and connect them accordingly externally either on 33kV or 11kV bus drawn near the Auto transformer or through 33 kV or 11kV XLPE cables.

b. **Protection against transferred surges;**

i. Follow the instructions of the OEMs.

ii. In the absence of any instructions from the OEMs, in case of Auto transformers, earth one terminal of delta tertiary; either 3U or 3W (in UVW Nomenclature) or Y1 or Y3 (as per IEEE std. C57.12.70-2000) of delta tertiary bushing terminals.

iii. In case of 400/220/33kV Auto-Transformers and above, strictly follow the instructions of the OEMs.

iv. To protect the Tertiary winding against transferred surges, some OEMs recommend Surge Protection Capacitors in conjunction with LAs on each phase of Tertiary winding.

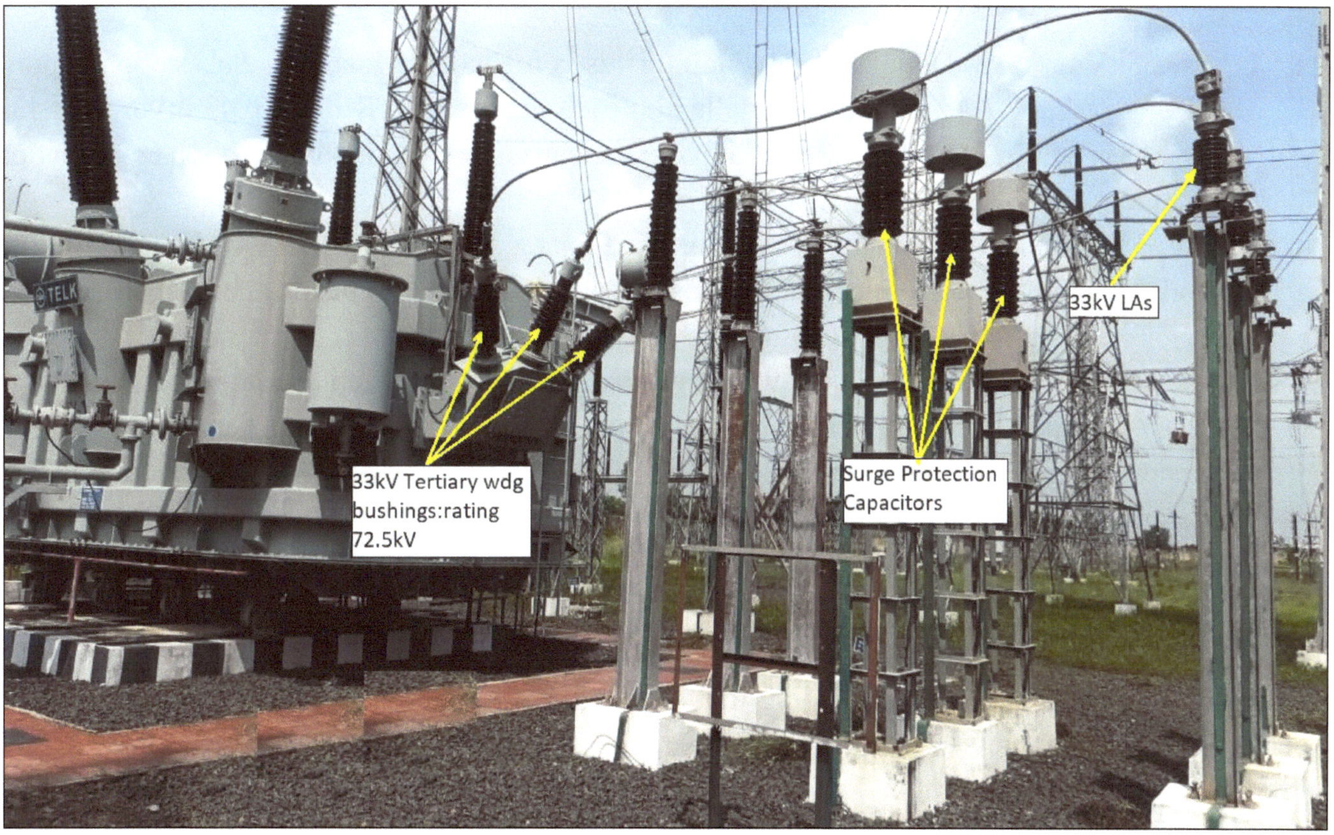

Figure 18A: Installation of Surge protection capacitors for protection of tertiary winding against transferred surges.

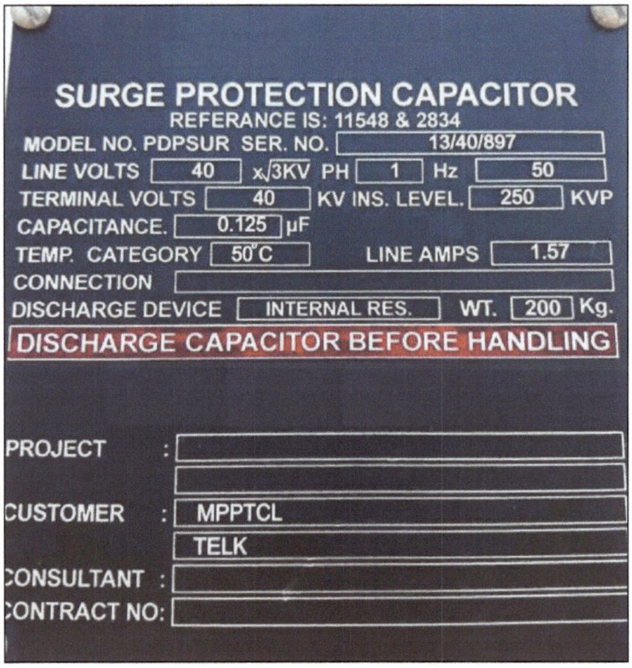

Figure18 B: Surge Protection capacitor name plate

c. **Tertiary winding utilized for feeding load;**

i. In case,33kV or 11kV Delta tertiary winding is used for feeding loads for Sub-station supply, 30kV LAs or 9 kV LAs should be installed on each phase. A DTR of 33/0.433kVor 11/0.433 kV, Dyn11 is required to be installed of appropriate capacity with suitable isolation & protection arrangement sin 33 kV or 11kV as well as on 0.433kV side.

ii. In case, 33kV or 11Delta tertiary winding is used for feeding loads to DISCOMs then Zig-Zag Neutral Grounding Transformer of appropriate capacity should also be installed with associated Relays, Circuit breakers, CTs, Isolators and LAs.

iii. In case, 33kVor 11kV Delta tertiary winding is used for feeding reactive compensation devices such as Capacitor banks or Reactor, necessary LAs, Isolators, CTs and Circuit breakers of adequate capacity should be installed. Some Utilities also use NDR Protection (Neutral displacement protection) using 3ph, 33kV RVTs or 12kV RVTs.

iv. Ensure that 1U,1V&1W (H1, H2, H3 as per IEEE std. C57.12.70.2000.) designated terminals of Transformer are matching with R, Y, B (R, S, T) phase configuration of buses respectively of switchyards on HV and 2U,2V & 2W (x1, x2& x3) on IV (in case of Auto Transformers /ICTs) and LV side (in case of two winding transformer) and 3U,3V & 3W (Y1, Y2 &Y3) in delta tertiary winding in case of Auto-Transformer (ICT.s)

[**Caution**: If Delta Tertiary is not used either for Reactive compensation equipment, Station supply or feeding 33kV loads, then one corner of Delta tertiary winding should be earthed.]`

13.0 Pre-Commissioning tests;

In some Utilities, the Job of relay settings and testing thereof are responsibilities of the Manager/ Engineer S/S, however in most of the Utilities at global-level, the Job responsibility lies with a separate wing / department / division which are regally designated as Protection & Testing Divisions, MRT divisions and Equipment & Relay protection Divisions etc.

13.1 Prior to pre commissioning Tests, the responsible authority should calculate and arrive at appropriate settings for each and every protective relay and shall arrange to test the protective relays accordingly.

13.2 Perform the pre-commissioning tests as recommended by the respective Utilities in particular; however, following Pre-commissioning tests are recommended and may be performed in agreement with the Commissioning Engineers of different OEMs.

i. Core Insulation Test.
ii. LT Magnetic balance test.
iii. LT Magnetizing current Test:
iv. Vector group test,
v. Short Circuit test.
vi. SFRA (Sweep Frequency Response Analysis) Test.
vii. Capacitance and Tanδ Transformer Windings.
viii. Capacitance and Tanδ Bushings.
ix. Insulation Resistance IR value measurement and PI.
x. Ratio Test.
xi. Winding Resistance Measurement.
xii. Through Fault Stability test.
xiii. REF Stability tests.

[Note:Carry-out all the Pre-commissioning tests mentioned above before energizing/ commissioning of the Transformer and verify the test results with the FAT results.

i. SFRA (Sweep Frequency Response Analysis) Tests should be carried out and the curves obtained during pre-commissioning test should essentially be verified with the signature curves of SFRA test taken during FAT.
ii. Ensure testing of the transformer with AC and thereafter DC tests should follow, otherwise the test results taken at the time of FAT would not tally.**]**
iii. Ensure that all the Protective relays, Circuit breakers, CTs, Isolators and LAs, etc are tested and are ready for charging.

[Note: In case of SPR Breaker (Single-Pole Reclosure breaker), Pole Discrepancy protection should be checked and the recommended setting is 300 ms to 500 ms]

14.0 OTI &WTI Settings: Sample procedure for setting OTI & WTI settings:

The Guaranteed temperature rise of Oil and Winding are embossed on the R&D plate of each Transformer.

Say Guaranteed Oil temperature raise is 45 ^{0}C and Guaranteed Winding temperature raise as 50 ^{0}C as mentioned on the R&D plate of the Transformer.

Assuming maximum ambient temperature as 45^{0}C in the tropical countries;

Settings ofOTI & WTI(Col.1)	Guaranteed Winding Temperature rise; deg. C(Col.2)	Ambient temperature;deg. C. (Col.3)	Calculation;(Col.2) +(Col.3). deg. C.(Col.4)	Setting recommendeddeg. C (Col.5)
Winding temperature trip.	50	45	95	100
Winding temperature Alarm.	50	45	May be set at	90
Oil temperature Alarm.	45	45	May be set at	80
Oil temperature trip,	45	45	May be set at	90
Fan start				60

Table 2.

Note: In case of any exigency, if overloading is required on the transformer, it can be done with higher settings of OTI & WTI, however overloading may be planned as per loading guide IEC 60076-7-2018.

14.1 LTI & WTI Settings;

In the present scenario, many utilities are switching over to Bio-degradable insulation liquid the ester fluid, having fire point above 300^{0}C and the other solid insulation materials with better thermal capability.

14.1.1 Proposed settings for LTI and WTI for Ester fluid filled transformers with Kraft paper and Thermally Up-graded Paper (TUP) as per IEC 60076-14:2013.

a. Effective Insulation thermal class for Kraft paper and TUP have been taken in to consideration.
b. Average winding temperature rise considered as 75^{0} C for the Kraft paper and 95 ^{0} C for the TUP. (Refer table C.3 of IEC 60076-14: 2013) [6]
c. Maximum temperature is taken as 40^{0} C.
d. WT trip setting is aligned to thermal index which is 10^{0}C lower than Effective insulation thermal class for kraft paper and for TUP.

	Kraft paper in deg. C	TUP in deg. C
Effective insulation thermalclass.	120	140
Average Winding-temperaturerise.	75	90
Winding temperature trip.	110	130
Winding-temperature Alarm.	105	125
Liquid temperature Alarm.	100	105
Liquid temperature trip.	110	115

Table 3.

15.0 All the annunciation indications should also be checked through simulating the conditions of each relay;

i. Oil level low,
ii. Buchholz alarm,
iii. Oil temperature alarm.
iv. WTI alarm.

Recommended limits for mineral insulating oils after filling in new electrical equipment prior to energization(Refer table3 of IEC 60422-2013)			
Property	**Highest voltage for equipment kV**		
kV level	**< 72,5**	**72,5 to 170**	**> 170**
Appearance	**Clear, free from sediment and suspended matter**		
Colour (on scale given in ISO 2049)	**Max. 2,0**	**Max. 2,0**	**Max. 2,0**
Breakdown voltage (kV)	**> 55**	**> 60**	**> 60**
Water content (mg/kg) [a]	**20 [b]**	**< 10**	**< 10**
Acidity (mg KOH/g)	**Max. 0,03**	**Max. 0,03**	**Max. 0,03**
Dielectric dissipation factor at 90 °C and 40 Hz to 60 Hz [c]	**Max. 0,015**	**Max. 0,015**	**Max. 0,010**
Resistivity at 90 °C (GΩ×m)	**Min. 60**	**Min. 60**	**Min. 60**
Corrosive sulphur	**Non-corrosive**		
DBDS content (mg/kg)	**< 5**		
Interfacial tension (mN/m)	**Min. 35**	**Min. 35**	**Min. 35**
Total PCB content (mg/kg)	**Not detectable (< 2 mg/kg total)**		
Particles	**--**	**--**	**See Table B.1 [d]**

[a] The values are not corrected for temperature since not enough time may have elapsed to reach anequilibrium between oil and cellulose insulation. [b] For use in transformers under 72,5 kV class, the maximum water content should be agreed betweensupplier and user depending upon local circumstances. [c] Higher dielectric dissipation factor values may indicate excessive contamination, or the misapplication of solid materials used in manufacture, and should be investigated. [d] A determination of particle size and quantity should be made as a baseline for future comparison intransformers >170 kV.

Table 4.

15.1 All other annunciations pertaining to the Transformer's external Protections areto be checked on the operator's desk for their correctness by manually operating or simulating the respective relays such as Differential relay, REF (restricted earth fault) relay, back up O/C & E/ F and Over-fluxing relay as required.

15.2 Trip test to be taken through the master-trip relay or module and ensure tripping of HV and LV breakers simultaneously through simulation of Master trip relay or Module for master trip.

16.0 Preparation before energizing the Transformer;

a. The OEM's personnel should be present at the time of 1st energizing the Transformer and should ensure readiness the NIFPS system.
b. Release air from all the ear-marked points including bushings before energizing.
c. Release air from each radiator or bank of radiators as the case may be.
d. Ensure that connectivity of all the tan-δ points and core-frame-tank terminals are earthed properly.
e. Before energizing of the Transformer release the air through the vent plug of the Diverter switch.
f. **Check the Oil parameters after filling the oil prior to charging as per the table of 3 of IEC 60422-2013.** (Refer table 4 of this article);
g. Check oil level in the conservator.
h. Check the Oil level in the Diverter switch.
i. Check the healthiness of Protection circuits.
j. Check the Aux DC system.
k. Check the pointers of all gauges for their free moment.
l. Check the Buchholz relay and readjust the float/ switches etc. if, necessary.
m. Check the healthiness of the air-cell by oil levels in Prismatic oil level indicator on the conservator tanks.
n. Ensure removal of the discharge-rods from HV, IV/LV bays wherever they were hung.

17.0 Energizing the transformer:

To be on safer side, charge the transformer at tap no. 2in case of the CFVV (Constant Flux Voltage Variation) system type of OLTCs.

[Note: In the CFVV, the value of HV voltage shall remain constant, thus the flux shall also remain constant in the Core irrespective of the tap position of the LV voltage]

a. At the time of charging of transformer, first switch-on the stuck breaker protection in/out switch to "in" position.
b. Put the NIFP system in service (on) position.
c. Energize the transformer and keep it without load for at least 24 hrs.
d. In the present scenario, since, numerical Differential relays are used for protection. Hence, curve of Inrush/charging current Vs time may also be captured.
e. After 24hrs of switching-off of the Transformer, release the air from all the earmarked points including bushings before taking load.
f. Switch-on the Transformer and load the Transformer.

18.0 Routine maintenance;

i. Earmark the working area of the transformer's HV & LV bays, including all the outdoor equipment with red-colored cloth strips about 1foot wide.

ii. Every precaution to safe-guard men and material at site should be ensured before and after taking shut-down on the Transformer.
iii. Carry out routine maintenance jobs as per the schedule defined by the Utility.
iv. **Transformers equipped with NIFPS;**

   a. After the transformer is switched off, following action /activities are to be performed to facilitate periodic maintenance and testing of the transformer protections etc.
   b. Auto/manual Switch of NIFPS to be turned to "Off" position on the control panel in the S/S Control room.
   c. Lock Nitrogen Injection solenoid valve and Oil drain solenoid valve by properly inserting the respective lock pins in the Control Cubicle in the Switch yard.

[**Caution**: If above activities are not done, there is every possibility that NIFPS may mal-operate during testing of protections of the Transformer resulting in to Nitrogen injection into the Transformer and simultaneously draining of 10% of oil thereof.]

18.1 While discharging the switched-off Transformer, care should be taken to connect the loose end of the earthing lead to be firmly tied / tightened on to a metallic structure after scraping the paint or rust thoroughly / earthing point FIRST, and then the earthing Clamp/ G-clamp of the discharge-rods should be connected to the HV- LV jumpers to discharge the Induced residual High voltage.

18.2 After the EHV Transformer is switched-off and discharged as mentioned in previous item, the phase jumpers should be removed first and there-after, the neutral jumper should be removed from the transformer bushings.

18.3 While normalizing the connections of the Transformer, first of all the neutral jumper/clamps should be connected and thereafter the phase jumpers to respective bushings.

19.0 Attend the leakage of the Bushings immediately, if found during operation, by taking a shut-down on the transformer.

a. The bushings should be cleaned with dry cotton cloth during each shut downs.
b. Special emphasis to be given to tan δ capacitance measurement and monitoring thereof of OIP/RIP bushings and Transformer Windings as stipulated in IEC 60137 or IEEE Std C57.19.01-2000-2003 as applicable.

19.0 Check the Oil in the Transformer and OLTC for Dielectric strength and Moisture content. Take suitable action for restoring the quality.

a. Check the oil level in the Oil-cup and ensure air passage is free in the breather. If oil is less, make-up the oil.
b. Check the oil for Acidity and sludge as per IS 1866-2017(IEC 60422-2013) in case of mineral Oil.
c. If any inspection cover is opened for some inspection or work it should be re- fixed with a new Gasket/O -ring
d. Check the WTI and OTI pockets and replenish the oil if found less in their respective pockets.

20.0 Check all the bearings and driving mechanism and lubricate them as per schedule.

21.0 Check the healthiness and readiness of NIFPS (Nitrogen Injection Fire- Prevention system).

22.0 a. Check the integrity of Grounding of HV & LV neutrals in their respective earth pits.

b. Check integrity of earthing of all the equipment in the Transformer bays with the help of Primary injection kit.
c. Check the earthing of Tertiary winding bushing.
d. Earth resistance to be checked periodically.

23.0 Thermo-vision survey/Infra-Red Scanning of all the Bushings, bushing clamps, Jumpers of the HV and LV bays should be carried out monthly or periodicity as defined by the Utility.

24.0 Check cooling fan's functions vis' a vis setting of temperatures thereof.

25.0 Check air cell in conservator, it should be in inflated position.

26.0 Trip/ appearance of Alarms;

a. Check the Transformer thoroughly after any alarm or protection operates.
b. Check the HV/IV/LV bays for damages of equipment, if any.
c. After each tripping of the Transformers, DGA should be carried out and the results should be compared with the results of earlier DGA report.

27.0 Condition Monitoring Tests;

Condition monitoring for transformers is the process of monitoring the parameters of conditions in transformer in order to identify a significant change which is indicative of a developing fault.

27.1 Monitoring tests, its frequency and acceptable values for Transformers and Reactors is tabulated in the table 5. [4]

<table>
<tr><th>Sl. No.</th><th>Test name</th><th colspan="2">Accepable Values</th><th>Frequency</th></tr>
<tr><td>1</td><td>Insulation Resistance Value of Winding Min</td><td colspan="2">Unless otherwise recommended by the manufacturer, 500 Mega - Ohm for 66kV and above voltage class</td><td>2 Yearly</td></tr>
<tr><td rowspan="7">2</td><td rowspan="7">Polarisation Index (Ratio of IR value at 10 min. to 1 min.)</td><td>Polarization Index</td><td>Insulation condition</td><td rowspan="7">SOS</td></tr>
<tr><td>Less than 1</td><td>Dangerous</td></tr>
<tr><td>1.0 - 1.1</td><td>Poor</td></tr>
<tr><td>1.1 - 1.25</td><td>Questionable</td></tr>
<tr><td>1.25 - 2.0</td><td>Fair</td></tr>
<tr><td>2.0 - 4.0</td><td>Good</td></tr>
<tr><td>Above 4.0</td><td>Excellent</td></tr>
<tr><td>3</td><td>Winding Resistance Measurement (Resistance converted to 75°C)</td><td colspan="2">± 5 % difference between phases or from Factory tests</td><td>SOS</td></tr>
<tr><td rowspan="4">4</td><td rowspan="4">TanDelta for bushing and winding</td><td colspan="2">0.007(Maximum) - for bushing</td><td>Yearly (bushing)</td></tr>
<tr><td colspan="2">0.005(Maximum)-for winding</td><td rowspan="3">4 Yearly (winding)</td></tr>
<tr><td colspan="2">Note:</td></tr>
<tr><td colspan="2">i. Tan delta of bushing should not go beyond 0.005 or should not increase more than 0.001 per annum w.r.t pre-commissioning values in the temperature range 10°C to 40°C.
ii. Winding tan delta should not go beyond 0.005 or should not increase
more than 0.001 per annum.</td></tr>
<tr><td>5</td><td>Capacitance for Bushing</td><td colspan="2">- 5 % to + 5 % Variation from Factory test results</td><td>Yearly</td></tr>
<tr><td>6</td><td>Capacitance for Winding</td><td colspan="2">- 5 % to + 10% Variation from Factory test results</td><td>4 Yearly</td></tr>
<tr><td>7</td><td>Oil parameters of use tank oil</td><td colspan="2">Refer IS:1866 -2017/IEC 60422</td><td>Yearly</td></tr>
</table>

<table>
<tr><td>8</td><td>DGA of tank oil And OLTC oil</td><td colspan="2">Refer IEC 60599-2007
Note: Refer to table 6 & 7 for DGA interpretations.</td><td>Half Yearly</td></tr>
<tr><td rowspan="7">9</td><td rowspan="7">Sweep Frequency response Analysis Tests (20 Hz to 5MHz)</td><td colspan="2">In general, changes of ± 3 db (or more) in following frequency range may</td><td rowspan="7">SOS</td></tr>
<tr><td>Frequency Range</td><td>Probable fault (assuming no other failure modes exist)</td></tr>
<tr><td>20 Hz - 10 kHz</td><td>Winding turn - to - turn short circuit (affected winding show the greatest changes), Core defects (OC test)</td></tr>
<tr><td>5 kHz - 100 kHz</td><td>Axial winding deformation, Bulk winding movement, Contact resistance, Open circuit winding</td></tr>
<tr><td>50 kHz - 1 MHz</td><td>Radial winding deformation</td></tr>
<tr><td>100 kHz - 500 kHz</td><td>Floating shield with local insulation carbonization(detectable response with changes in peaks and velleys)</td></tr>
<tr><td>1 MHz - 5 MHz</td><td>Winding looseness due to transportation(Difference will be greater for the most affected windings), Floating shield with local insulation carbonization (largest differences in peaks and valleys)</td></tr>
<tr><td>10</td><td>Oil parameters in use tank oil</td><td colspan="2">Refer IS:1866 -2017/IEC 60422</td><td>Yearly</td></tr>
</table>

11	Thermovision Scanning	Temperature rise above ambient(°C)/ criticality		Half Yearly
		0 - 10	Minor	
		11 - 39	Intermediate	
		40-75	Serious	
		>76	Critical	
		Note: Comparison to be made with similar joints/ items of the same transformer and values are to be nearly matching		
12	Furan analysis	For calculation of DP (Degree of Polymerization) using Chendong's formula; DP = Log (2FAL) − 1.51/ −0.0035, where 2FAL in ppm DP values for estimating remaining paper life;		Every 5 years, after the Transformer has served for 20 years.
		New insulation	1000 DP to 1400 DP	
		60 % to 66% life remaining	500 DP	
		30% life remaining	300 DP	
		0-Life remaining (end of life)	200 DP	
13	PD (Partial discharge)	Is recommended if there is increase in key gases in DGA analysis		As a supplementary test after DGA analysis.
14	FDS (Frequency Domain Spectrometry) of Bushings and windings	Supplemented by DGA is found to be a good method for condition assessment of bushings and windings. Measurement of C & tanδ over a frequency range of 1mHz to1kHz.		As a supplementary test after DGA and C & tan δ test.
	Note: SOS context means "As and when" required. Table 5			

Typical faults in Power Transformers are depicted in the table 6.

Typical faults in Power Transformers(refer IEC 60599-2007)		
Type	Fault	Examples
PD	**Partial discharges**	**Discharges in gas-filled cavities resulting from incomplete impregnation, high-humidity in paper, Oil super saturation or cavitation, and leading to X-wax formation.**

D1	**Discharges of low energy**	• **Sparking or arcing between bad connections of different or floating potential, from shielding rings, toroids, adjacent disks or conductors of winding, broken brazing or closed loops in the core.** • **Discharges between clamping parts, bushing and tank, high voltage and ground within windings, on tank walls.** • **Tracking in wooden blocks, glue of insulating beam, winding spacers,** **Breakdown of oil, selector breaking current.**
D2	**Discharges high energy**	**Flashover, tracking, or arcing or high local energy or with power follow-through**
		• **Short circuits between low voltage and ground, connectors, windings, bushings and tank, copper bus and tank, windings and core, in oil duct, turret. Closed loops between two adjacent conductors around the main magnetic flux, insulated bolts of core, metal rings holding core legs.**
T1	**Thermal fault** **t<300** °C	• **Overloading of the transformer in emergency situations** • **Blocked item restricting oil flow in windings** • **Stray flux in damping beams of yokes**
T2	**Thermal fault** **300** °C **< t< 700°C**	• **Defective contacts between bolted connections, gliding contacts, contacts within selector switch (pyrolitic carbon formation), connections from cable and draw-rod of bushings.**
		• **Circulating currents between yoke clamps and bolts, clamps and laminations. In ground wiring, defective welds or clamps in magnetic shields.** • **Abraded insulation between adjacent parallel conductors in windings.**
T3	**Thermal fault** **t > 700** °C	**Large circulating currents in tank and core Minor currents in tank walls created by a high uncompensated magnetic field Shorting links in core steel laminations.**

Table 6

27.3 Interpretation of DGA through ratios of key gas method.

DGA Interpretation Table(refer table 2 of IEC 60599-2007)

Case	Characteristic Fault	C_2H_2 / C_2H_4	CH_4 / H_2	C_2H_4 / C_2H_6
PD	Partial discharges	NS[a]	<0.1	<0.2
D1	Discharges of low energy	>1	0.1 – 0.5	>1
D2	Discharges of high energy	0.6 – 2.5	0.1 -1	>2
T1	Thermal fault T < 300°C	NS[a]	>1 but NS[a]	<1
T2	Thermal fault 300°C < 1 < 700°C	<0.1	>1	04-Jan
T3	Thermal fault > 700°C	<0.2 [b]	>1	>4

NOTE 1 – In some countries, the ratio C_2H_2/C_2H_6 is used, rather than the ratio CH_4/H_2. Also in some countries, slightly different ratio limits are used.

NOTE 2 – The above ratios are significant and should be calculated only if at least one of the gases is at a concentration and a rate of gas increase above typical values.

NOTE 3 – CH4/H2 <0.2 for partial discharges in instrument transformers. CH_4/H_2 <0.007 for partial discharges in bushings.

NOTE 4 – Gas decomposition patters similar to partial discharges have been reported as a result of stray gassing of oil.

a. NS = Non- significant whatever the value

b. An increasing value of the amount of C_2H_2 may indicate that the hot spot temperature is higher than 1000°C

Table 7

1.0 Don'ts;

1. Do not accept the consignment of the Transformer at site,
   a. If multiple earthing of core found while the transformer is in the loaded condition on the low bed trailer itself.
   b. If the nitrogen pressure not found having ve pressure.
   c. If the Impact (vibration) recorder shows some abnormalities and if the LT tests carried out while the transformer is on the low bed Trailer found unsatisfactory.

[Note: Do not forget to inform the OEM about the same as found status of the Transformer immediately]

2. While tightening/Loosening any bolt etc inside the Transformer tank through the inspection windows, ensure that the person carrying it out should not keep any article in his upper pocket.
3. Do not use low capacity lifting hydraulic-Jacks.
4. Do not forget to tie a cotton tape at one end of the spanner or any other tool and the other end of the cotton tape should be tied to any clamp or hook on with the Transformer top/body while tightening/ loosening operation of any bolt inside the Transformer through the Inspection window.
5. Do not store the Transformer/Reactor N2 gas filled at site for a long period >6months.
6. Do not commission the Transformer before conducting all the Pre-commissioning Tests.
7. Do not discard a new Transformer; If the PI value is found close to 1 as;

    a. The Polarization index method should not be used for accessing insulation condition in new Power Transformers.
    b. The polarization index for insulation liquid is always close to 1. Therefore, the Polarization index for Transformers with low conductivity liquids (e.g. new mineral oil) may be low despite of good insulation conditions.

8. Do not leave Delta tertiary winding terminals unprotected outside the tank. They should be connected to LAs/ Surge Protection capacitors in conjunction with LAs of appropriate rating.
9. Do not move the Transformer with Bushings mounted.
10. Do not parallel the Transformer without confirming that the transformer in question fulfills all the conditions/criteria such as Vector Group, Terminal Voltages, Voltage ratios, tap variations of OLTC and% Z.
11. Do not over load the Transformer beyond the limits as specified in IS:2026 (Part-7):2009(IEC equivalent IEC 60076-7:2005).

[Note:

a. Relative aging rate-If hot spot temperature exceeds from 80^{0}C. to 140^{0}C, as for every 6^{0}C temperature-rise the relative aging becomes double.
b. Loading beyond the rated Current (in pu) as described on the R&D Plate;

- Cyclic Over loading:1.3 pu keeping in view the WTI settings. (ii)Emergency loading (30 min): 1.5 pu keeping in view the WTI settings.]

12. Do not change settings of WTI and OTI, alarm and trip frequently.
13. Do not meddle with the protection circuits.
14. Do not allow conservator Oil to fall below; 35^0 C level.
15. Do not allow bushing Oil level to fall, it must be immediately topped-up.
16. Do not leave the marshalling box doors open. They should be locked.
17. Do not switch off heater in the Marshalling Box.
18. Do not leave ladder unlocked.
19. Do not allow any unauthorized persons to loiter near the Transformers.

20. Do not mix mineral oil to the existing mineral oil in an in-service transformer, unless it confirms to IS: 1866 -2017/IEC 60422-2013.
21. Do not mix ester-fluid to the existing Ester fluid filled in-service transformers, unless it confirms to IS: 16099 / IEC 61203.
22. Do not continue with deteriorated silica-gel it should be replaced immediately.
23. Do not energize the transformer without thoroughly investigating/observing and testing the Transformer after appearance of Buchholz relay Alarm or Trip indication and also tripping on Differential Protection, REF protection, OSR and PRV.
24. Do not energize the Transformer without thoroughly testing the Transformer for its healthiness and without analyzing the trapped Gas in the Buchholz relay and DGA of the Transformer oil.
25. Do not parallel the transformers without bringing them on appropriate tap position as per operating instructions.
26. Do not connect the Transformer on a live Bus-bar unless ensuring that the EHV shunt-capacitor bank is in 'OFF' condition.
27. Do not leave cap of tan delta measurement point on the bushings in open or loose condition.
28. Do not push the oil through drain valve.

29.0 Conclusion: The activities narrated in this Article may be treated as general guide-lines, if followed the life-expectancy of the transformer would be enhanced however, if there are other specific instructions of the utilities and the OEMs, they may be followed scrupulously.

1.0 Bibliography;

1. EEE Guide for Installation and Maintenance of Liquid Immersed Power Transformers Std C57.93-2019.
2. CBIP Manual on transformers No.:319-2013
3. Terminal marking UVW as per IS:2026-Part (iv)1977 reaffirmed in 2001 and IEEE std. C57.12.70-2000.
4. Central Electricity Authority (Govt. of India): Standard specifications and technical Parameters for Transformers and Reactors (66kv and above voltage class published in April2021.
5. Mineral oil fresh IS:335. (IEC equivalent IEC 60296 5th edition 2020).:
6. Unused Synthetic organic ester confirming to IS: 16081-June 2013IS / IEC 61099.
7. Power transformers –IEC 60076 Part 14: Liquid-immersed power transformers using high-temperature insulation materials
8. Loading guide for Oil immersed Power transformers IS:2026-part-7(IEC equivalent; 60076-7).
9. IEEE std C57.152-2013(IEEE Guide for fluid filled Power transformers, Regulators and Reactors).
10. IS: 16099 / IEC 61203 maintenance for used Synthetic organic ester fluid.

31.0 Authors;K.K.Murty, JJL KAPIL, Santosh DDubey & S.K.Chaturvedi

12. Transformers and Reactors Maintenance-I-Reasons and Need;

1.0 Introduction: The Power Transformer / reactors are the costliest equipments in a Transmission utility. In MPPTCL, there are as many as 323Nos EHV S/S (400kV S/s-9Nos.220kV S/s-70Nos.,132kV-243nos and 66kV-1 No)having around 730nos. of transformers.

Analyzing the factors affecting the life of these equipments and Types of faults has been dealt with in this lecture.

If proper design, up keep and operation as per IS: 6600 is carried out the life expectancy of the Transformer can be enhanced beyond the anticipated life of 35 years. One such example is there in 220kV S/S Jabalpur wherein the Mitsubishi make, 1+3X 40 MVA 220/132/33kV, Transformer has recently completed 52 years of successful service and are still in service.

2.0The reasons affecting the life of Transformers/Reactors are as follows:

2.1 Optimization of design, vis a vis the cost effectiveness: In the present scenario of cut–throat competition, the design of the Transformers are so developed that the physical size has reduced by reducing the clearances between the windings/core/tank etc in the process of optimization of design and reduction in the cost thereof, the transformer units, although withstand all the Acceptance/ Routine tests during pre-delivery inspections, but have failed in actual field conditions during LI stroke/ Switching Impulse and also during close-up faults. The manufacturers should improve in this respect and should not use sub-standard cheap material etc.

2.2 Lapses during Manufacturing Process.

1. Use of conductors having burrs etc.
2. Use of substandard insulating materials such as Press board for wedging the coils, Kraft paper/ cotton & PVC tapes, Perma-wood for end press rings etc.
3. Proper care to avoid any metal filling getting pressed in between the laminations of the core.

2.3 Cutting- short the commissioning Process:

Any kind of short cuts in the commissioning process should not be appreciated/allowed at all. It has been noticed that the Transformers are commissioned short cutting the drying out process, with the presumption that the Drying out is not essential since it is received Nitrogen filled. More so, the AE/EE generally resort to the short cuts in the Drying out process with the intention of getting appreciation / commendation from the management for commissioning of the transformer in record time of a few hours.

Every care as recommended by the manufacturer should be taken during erection / installation of the Transformer and its auxiliaries.

Care should be taken for filling Oil confirming to IS: 335 and particular care should be taken to fill the Transformer with an oil having moisture contain of about 5ppm.

Every care should be taken about Neutral Earthing.It should be done in the standard earth- pits made for the Transformer. The HV and the LV neutrals should be earthed separately.

As soon as the Transformer is received and is on the trailer itself, the core earthing should be checked. (Multiple earthing of Core may give rise to development of Hot spot).

2.4 Over Loading/ Running with high temperature: The loading pattern of the

Transformer should be such that it should be within the limits of loading specified in the IS: 6600. The insulation deteriorates at the rate of twice its life for every 6°C rise in Oil temperature above 80°C to 140°C

2.5 Fault Level/ Fault clearing Time: With the enhancement in generating capacity in the State, the Fault levels have gone up. The decision should be taken to shift the old Transformers which have already rendered their useful service life to the organization to remote places, instead of them new Transformers with higher BIL should be installed.

Note: Short circuit time is increased to 3 seconds by the erstwhile MPSEB from 2 seconds as per IS: 2026.

The fault clearing time of the downstream feeders/ Transformers has also to be examined critically with the E.E.(STM).

2.6 Effect of Short Circuit: As per the principle of the Transformer, EMF in developed in the opposite direction in the HV-LV windings. Whenever the fault appears in the downstream the tendency of the coils is to move away axially from one another, known as telescopic effect. This is pronounced when there is a close up 3-ph fault, the axial force is of such intensity that it can even break/damage the perma- wood laminated end pressure ring and the windings also dislocate and finally the Transformer fails.

Quality of the Perma-wood end Pressure plate should be such that it should sustain the telescopic effects due to the severe faults nearby.

2.7 Effect of additional core earthing: The Core can get earthed due to poor insulation of the laminations in addition to its original core-earthing and circulating current may develop giving rise to the Hot spot.(this should be checked during each maintenance of the Transformer.

2.7(a) Multicore earthing: It is not out of place to mention here that in the past MPEB had procured 40MVA 132/33kV Transformers- 4 Nos. from NGEF which were found to have multiple earthing and were getting hot. It is of prime importance that the core should be earthed at one point only. To ensure the same MPEB had included in the specs that the manufacturer should

bring 3 terminals onto bushings on the top cover, 1 – from the core, 1- from the end frame and 1- from the Transformer tank and earth them with a common strip on the transformer top cover.

2.8 Development of Hot spot inside the Transformer. Hot spot can be developed due to following reasons;

1. Circulating Stray currents may occur in the Core structure and/or tank.
2. Unintentional grounding of core at more than one point may cause hot spot.
3. Due to bad connection of the leads and joints.
4. Due to electrical arcing between winding and ground, between windings of different potentials.
5. Due to poor contacts of Tap changer.
6. Due to Over- fluxing of the Core. ($\varnothing \propto \frac{V}{f}$).

2.9 Moisture In the oil/winding: Bad oil preservation system shall give rise to increase in moisture content level. The transformer winding is highly hygroscopic. Any ingress of moisture can deteriorate the insulation of the oil as well as the windings, and subsequently lead to the failure of The Transformer. Thus, care should be taken to prevent entry of moisture inside the Transformer.

2.10 Bushing Failure: The vendors/ manufacturers give a call for inspections while the ear marked bushings have not reached the works for some reason or the other. The Transformers are tested with some other bushings(this is general practice of the vendors). The new bushings which are subsequently supplied had failed shortly after the Transformer was commissioned. This failure of bushing may be due to high tan δ value or due to failure of Stress control ring.

Suchoffers of inspections without the ear-marked bushings should not be entertained.

The tan δ and the capacitance of the bushings should be monitored on 1/2 -yearly basis with a 10kV kit. The tan δ should be less than the limiting value of 0.7% and the capacitance should not be more than 10% at any point of time from the commissioning value.

Increase in tan δ indicates increase in contamination in the insulation of the Bushing and increase in capacitance value indicates failure of any series capacitance of the OIP bushing.

Bushings have also failed due to improper connection/loose connection of the Thimble/lead.

2.9 Fluctuating Load and Effect of odd harmonics: Fluctuating Loads of like Rolling mills, Arc furnaces and also of Railway Traction, burden the Transformer with multiple odd harmonics, causing over heating of the Core. It should be ensured that such loads should be served from different Transformers as far as possible.

2.10 Voltage surges: As already mentioned, the voltage surges of Lightening strike or switching surges may damage the Transformer. The Lightening/surge arresters of adequate rating should be used and their earthing to be separately done in the ear marked earth pits.

2.11 Failure of Tap Changer: The failure of tap changers may also lead the Transformers to be defective and would be kept out of service. The Rating of the Tap changer should be properly selected by the procurement Deptt. and accordingly mention the same in the Sepcs. The sparking of the contacts and failure of limit switch ultimately lead the Transformer to failure.

2.12 Deterioration of Transformer Oil: As already mentioned that the oil should be free from moisture, it should also be kept away from Oxygen. Only oil preservation system like Air-cell and Silicagel breather together can perform this duty. Also thermo siphon should be kept in service with Activated alumina so as to extract water /moisture from the oil which has been produced due to breaking of high level alkenes.

These are a few causes of Transformer failure. The Transformer once failed and gutted with fire leaves no traces of cause of failure. In such cases only probability/anticipation is expressed against reasons for failure.

2.13. Ageing: This is a natural process. As the time passes with the age and the Transformer too becomes old as a living being. Over this period, the Transformer has undergone many shocks of faults and short circuits, Variation in Loads vis a vis seasons, variation in ambient Temperatures, Conditions of Over loads, Lightening over voltages/ switching Over Voltages etc. The insulation of the winding deteriorates and becomes brittle and also contains moisture. Such transformers may give further service if necessary precautions are taken as preventing moisture to enter through the silicagel breather. They should be saved against over loading and shocks of frequent and prolonged through faults occurring in the downstream feeders. The relay co-ordination with downstream feeders should be carried out with necessary dialogue with the E.E(STM) of the area.

13. Transformers and Reactors Maintenance-II-Techniques

1.0 Introduction: This Lecture is the complimentary to the previous lecture. The actions taken after the transformer has Tripped due to any reasons/ faults..

2.0 Visual checks: Following a tripping of a Transformer following visual checks are to carried out;

2.1 Relay indications and alarm fascia indications are to be observed /noted and reset thereafter.

Note:

a. If fault is involving any of the protections of the Transformer or due to the fault on HV or LV side of the Transformer, the tripping shall take place through the MT relay in case of Grid transformers.
b. If the fault is on to the LV side of the transformers feeding radial loads like 132kV /33kV Transformers., only LV breaker shall trip.
c. It has to be ensured that the tripping is not of spurious nature.

3.1 Check the transformer for;

a. Physical verification/ inspection going round the Transformer to see if the happenings are tallying with the relay indications.

- Splashing/ Gushing of Oil from the transformer.
- Failure of bushings etc.
- Buchholz relay.
- Oil surge relay.
- Low Oil level in the conservator.
- Winding Temperature and Oil temperature.
- PRD (Pressure Release Device) with splashing out of Oil.

3.2 Inspect the HV and LV bays thoroughly for any abnormal happening in the bays such as;

- Monkey/ cat or any other wild animal electrocuted and found dead.
- Snapping of any conductor.
- Failure/ bursting of any CT.
- Failure/ bursting of any Lightening Arrestor.

4.0 Preliminary Testing of Various components;

a. Operations of Buchholz relay by operation the lever manually.
b. Measurement of IR values and PI value out of the results of IR value tests.

c. Collection of Oil sample for Oil parameter tests.
d. Collection of Oil sample for Dissolved Gas analysis.
e. Magnetic balance Test with LT supply.
f. Magnetizing current test.
g. Winding resistance measurement.
h. tan δ of the winding.
i. tan δ test of all the bushings.
j. Ratio tests.
k. SFRA test.

5.0 Oil parameters and their significance:

The sample of oil is subjected to all the tests recommended in the IS: 1866 which is for the oil in service.

Sno	Particulars of tests	Values as per IS 335-1993(new oil	Values as per IS:1866(Oil in service	Interpretation
1	Density (gm/ cm^3) at 29°C	0.89(max)		
2.	IFT (n/m) at 27°C	0.04(Min)	0.018(min)	
3.	Flash Point	140°C (Min)	125°C (min)	Reduction in ignition temperature.
4	Pour Point	-6 °C (Max)	-------	
5	Tan delta at 90°C	0.002(max)	0.2(Max) for ≥ 45kV	Indicates presence of contaminated material
6	Resistivity (ohm-cm) At 90°C. At 27°C	$35x10^{12}$(min). $500x10^{12}$ (min)	$0.1x10^{12}$(min) $10x\ 10^{12}$(min)	It is detection of conducting impurities in the Oil
7.	Corrosive sulphur	No Corrosive		
8.	Water content (ppm)	50(max)	25(max) for >= 145KV 35(max) for < 145KV	It may reduce the electric strength, the resistivity and accelerate the deterioration of insulating paper.
9.	SK value (%)	4 to 8 (max)	-------	
10.	BDV (KV) rms a. (New oil Unfiltered). b.After filtration	------- 30(min) 60(min)	50 for >= 145KV 40 for 72 to 145KV 30 for<=72.5 kV	reveals extent of suspended Matter and advisable to carry out filtration
11.	Acidity(max)	0.03 mg KOH /gm		

5.0 DGA (Dissolved Gas Analysis):

While in service, particularly the EHV Transformers, are subjected to many adverse conditions, such as, Over-loading, System Over-voltages, Over-fluxing, Under- frequency, Switching Surges, Lightening Surges, Effects Due to Harmonics developed by non-Linear loads and severe faults on the LV side etc.The type of Operating conditions and conditions of some of the faults give rise to De-composition of the insulating oil inside the Transformer under service.

The Scientists and Engineers have studied the different conditions under which the mineral insulating oil decomposes and key-gasses are evolved which in-turn give significant indication in respect of the health of the Transformer.

6.0 Phenomena of Decomposition: Electrical & thermal faults give rise to session of some of the bonds of different hydrocarbon molecules which in-turn convert into free radicals or in ionic forms which combine rapidly through complex reaction to form the Methane(CH_4), Ethane(C_2H_6), Ethylene(C_2H4) Acetylene(C_2H_2).s which get dissolved in the Oil.

Low energy faults: Partial Discharge of cold plasma Type (Corona discharge) through ionization reactions and accumulation of Hydrogen as main recombination gas.

<500°C- Methane and Ethane are formed.

>500°C Ethylene is formed

800°Cto1200°C.-Acetylene is formed mainly in arcs, where temperature is thousands of degree Celsius.

Carbon Particles formed at 500°C to 800°C and are generally observed after arcing in the Oil or very hot spots.

Oil may oxidize with formation of small quantities of CO and $CO_{2,}$ which can accumulate over a long period of time.

7.0 Decomposition of Cellulosic Insulation:

Paper, Press-Board and Wood Blocks are in the category of Cellulosic Insulation which decompose at the temperature at >105° C with complete carbonization >300° C. Mostly Carbon Monoxide, Carbon Dioxide and water are formed in much larger quantities than by oxidition of oil at the same Temperature, together with minor amounts hydro carbon gases and furanic compounds.

CO and CO_2 formation increases not only with temperature but also with the Oxygen content of oil and the moisture content of paper.

8.0 Type of Faults:

On the basis of Data available of the internal inspections of large Nos. of faulty Transformers world-over, the reasons for development of faults are broadly classified as under;

- ➢ Partial discharge(PD) of cold plasma-wax deposition on paper insulation, or of the sparking type, including pin holes, Carbonized perforation(punctures) in the paper.
- ➢ Discharge of Low energy(D1), in Oil or /and paper, give rise to larger carbonized perforations(Punchers), Carbonization of the paper surface(Tacking) or Carbon Particles in oil(as found in OLTC Diverter switch).
- ➢ Discharge of High energy(D2) in oil or/and in paper, with power flow- through evidenced by excessive destruction and carbonization of paper, metal fusion, extensive carbonization in oil. In some cases Tripping of the Transformer is the evidence of excessive Power-flow through.
- ➢ Thermal Fault in Oil or/and in paper,

Below 300°C (T1) – If the paper turns Brownish.

Above 300°C (T2) – If it has carbonized.

- ➢ Thermal Fault Due to high temperatures (T3);

Above 700 °C – If there is strong evidence of Carbonization of oil.

Above 800 °C – If there is evidence of metal colorization.

Above 1000 °C – If there is appearance of Metal fusion.

8.1:Table of Abbreviations;

PD	Partial Discharge
D1	Discharge of Low energy.
D2	Discharge of high energy.
T1	Thermal fault, t<300°C
T2	Thermal fault, 300°C,<t<700 °C
T3	Thermal fault, t>700 °C

10.0 Typical Faults in the Power Transformers;

Typical faults related to the Transformers have been shown in the following table (Table- 1.A of IEC: 60599 is reproduced below).

Table A.1 – Typical faults in power transformers

Type	Fault	Examples
PD	Partial discharges	Discharges in gas-filled cavities resulting from incomplete impregnation, high-humidity in paper, oil supersaturation or cavitation, and leading to X-wax formation
D1	Discharges of low energy	Sparking or arcing between bad connections of different or floating potential, from shielding rings, toroïds, adjacent disks or conductors of winding, broken brazing or closed loops in the core Discharges between clamping parts, bushing and tank, high voltage and ground within windings, on tank walls Tracking in wooden blocks, glue of insulating beam, winding spacers. Breakdown of oil, selector breaking current
D2	Discharges of high energy	Flashover, tracking, or arcing of high local energy or with power follow-through Short circuits between low voltage and ground, connectors, windings, bushings and tank, copper bus and tank, windings and core, in oil duct, turret. Closed loops between two adjacent conductors around the main magnetic flux, insulated bolts of core, metal rings holding core legs
T1	Thermal fault $t < 300$ °C	Overloading of the transformer in emergency situations Blocked item restricting oil flow in windings Stray flux in damping beams of yokes
T2	Thermal fault 300 °C $< t <$ 700 °C	Defective contacts between bolted connections (particularly between aluminium busbar), gliding contacts, contacts within selector switch (pyrolitic carbon formation), connections from cable and draw-rod of bushings Circulating currents between yoke clamps and bolts, clamps and laminations, in ground wiring, defective welds or clamps in magnetic shields Abraded insulation between adjacent parallel conductors in windings
T3	Thermal fault $t >$ 700 °C	Large circulating currents in tank and core Minor currents in tank walls created by a high uncompensated magnetic field Shorting links in core steel laminations

A.1.3 Identification of faults by DGA

Table 2 (see 5.3) applies directly to all transformer sub-types, except those equipped with a communicating OLTC. In the latter cases, if C_2H_2/H_2 is higher than 2 to 3, there is contamination from the OLTC into the main tank, and table 2 does not apply or applies with care, after subtracting background contamination from the OLTC (see 5.6).

NOTE 1 Any gas formation below typical values of gas concentration and rates of gas increase should not be considered as an indication of "fault", but rather as "normal gas formation". Ratios are not significant in such a case (see note 2 of table 2).

NOTE 2 In the case of air-breathing power transformers, losses of gas occur very slowly with time by diffusion through the conservator or as a result of oil expansion cycles, with the result that the measured gas levels may be slightly less than the gas levels actually formed in the transformer. However, there is no agreement concerning the magnitude of this diffusion loss in service, some considering it as totally negligible, others as potentially significant, depending on the type of equipment considered. In case of doubt, to get an idea of the volume ventilated it may be expedient to measure the gas concentration in the conservator as well. Significant diffusion losses may affect gas ratios, typical values of gas concentrations and rates of gas increase (see 6.1).

11.0 Filter machine: FILTER MACHINE at, 220kV S/S Jabalpur,;

In previous year the concept of filter machine was rough, vacuum about - 680 Hg. And high temperature about 80 ºC, due to high temperature the transformer oil loses its property, now in present day filter machine used with high vacuum about one torr with low temperature about - 60 ºC. In this chapter we talk about high vacuum filter machine.

The Various Components of filter machine and their functions are as under:-

1. **PRELIMINARY FILTER: (Item No. 1)**Preliminary filter is at the oil inlet and is entering the gear pump. This perforated sheet prevents the particles having size above 1mm from entering provided with magnetic strainer and a perforated sheet. The magnetic strainer prevents magnetic particles from the gear pump. This is used for rough particles filtration and to arrest- Magnetic particles this protects inlet gear pump.
2. **OIL INLET PUMP: (Item No. 2)**This pump functions to force the oil through the filter at required pressure and flow rate. The input side oil is cool and more viscous there for it is preferable to use gear type pump.
3. **CENTRIFUGAL PUMP: (Item No.** 7)This is mechanical seal type oil discharge pump. This pump is provided at the outlet of degassing column and can draw the oil from degassing column and feed it to the transformer. The output side oil is hot and viscosity is low, so we prefer to use centrifugal type pump.
4. **HEATER TANK: (Item No. 3)** Heater tank is made up of m.s. welded construction and is thermally insulated with mineral wool. The electric heaters provided are of indirect type and their design is such that over heating or localized heating is eliminated. It is automatic in operation. Thermostats are provided to maintain the desired temperature. The total power is 48 K.W., divided in three groups. Heaters shall be interlocked with Gear Pump and shall not be in ON position, unless the Inlet Pump is working. This heater tank functions to heat the oil at required temperature.
5. **DEGASSING CHAMBER: SINGLE STAGE (Item No. 8)**A shower arrangement is provided in the degassing chamber. Oil comes out of the shower arrangement in the form of shower and falls on the cage filled with rasching rings. The purpose of rasching rings is to expose the incoming transformer oil to optimum surface area for efficient degassing & dehumidification. A Sight Glass with illuminating lamp shall be provided for observation of oil flow.
6. **EDGE FILTER: (Item No. 5)**This consist of two chambers one for incoming dirty oil and the other for outgoing clean oil. Several candles are held under spring compression in the dirty oil tank. The spring pressure can be adjusted by tightening or loosening the nuts. These candles consist of stacked ring type filter papers. Oil enters the candles in radial direction and flows towards clean oil chamber in axial direction. Filters shall be easily changeable.

This shall be useful for removal of sludge content in the used oil. This has got an oil inlet, outlet, drain and air pressure connections for clearing purpose. This is a very fine filter and it filters the oil to more than 1 micron size particles.

7. **AIR COMPRESSOR (Item No. 5.9)**It is provided for edge filter clearing by passing the compressed air. This compressor can also be used for spray painting and other purpose.
8. **IONIC REACTION COLUMN: (Item No. 11)**It content of activated alumina (20kgs). If the initial acidity of oil is higher as compared to IS 335, than this column can be taken in oil circulation to reduce the acidity of the oil.

9. **FLOAT SWITCH (HIGH LEVEL) (Item No. 8.2)**This is provided to prevent the rise of oil level in the degassing chamber to trip the inlet pump.
10. **FLOAT SWITCH (LOW LEVEL): (Item No. 6.3)**When the oil level in the degassing column (low level) falls to a specific level, the low level float switch opens the circuit and trips off the discharge pump thereby preventing it from dry running.
11. **# VACUUME PUMP (Item No. 8).**A rotary oil sealed vacuum pump WSVP 9060 is provided for evacuation of the plant.
12. OIL SAMPLING VALVE (Item No. 7.2) Processed oil sample can be collected from this valve.
13. THERMOSTATES: (Item No. 3.1,3.2,3.3,3.4)Four thermostats are provided on the heater tank to control the oil temperature. These are automatic in operation. One of them is used for temperature setting as a safety and the others are used for controlling the temperature, heating. The safety Thermostat shall be provided to take care of any accidental rise of temperature of oil and shall put off the Heaters in such eventuality. This Thermostat shall be set at slightly higher temperature than that of controlling thermostats. These thermostats are having a range of 30 to 110 ºC.
14. THERMO METERS: (Item No. 3.5,3.9.)Two nos of dial type thermometer are provided on the plant which are located at the inlet of heater tank to indicate the temperature of incoming oil and the other at the outlet of heater tank to indicate the temp. of outgoing oil.
15. VAPOUR CONDENSER (Item No. 8.4)One No. of vapor condenser (baffle type) is provided in the plant to protect the vacuum pump and to trap the vapours of transformer oil.A oil drain line is provided and connect to the degassing chamber to collect the trapped oil inside vapour trap.
16. SOLENOID VALVE AT INLET &OUTLET0Gne no. Soid Valve at Inlet and One no. at Outlet shall be provided. Valve at the Inlet & Outlet shall open automatically. The moment oil Inlet & Outlet Pump are switched ON. In case of power failure, these valves shall be capable of preventing the oil from entering into the Plant and thus avoiding the possibility of mixing processed oil with unprocessed oil.
17. OIL HOSES - 2 NOS. Two Nos. Nitrile Rubber Hoses each 15 Mtrs long with flanged end connection on both sides shall be provided. One for Oil Inlet & one for Oil Outlet. Oil Hoses shall be capable of handling the transformer oil at 100oC (max.) and vacuum.

    1. To summaries our specification requirement for filter plants are as under:
        a. The plant should be capable to remove solid impurities up to 99.9%
        b. The plant should be able to produce a vacuum of less than one torr.
        c. Filtration should be achieved is a maximum of 5 passes. The plant should produce a vacuum of around 10 torr in the first pass and less than one torr in the final pass.
        d. The plant should be capable to reduce water contender form 88 ppm to 5 ppm and gas content from 10% to 2%
        e. The above results should be achieved without raising the temp. of oil in the system beyond 70ºC

2. The following inter locking and important features must be available in the filter plant.
    a. Solenoid valves on the inlet as well as outlet side so that incase of power failure mixing of treated and untreated oil does not take place.
    b. Arrangement for control of flow of oil in the system.
    c. Arrangement for tripping of system incase of high level or low level of oil in the system.
    d. Arrangement to preserve aromatics present in The oil.
    e. Provision of separate motors for each pump and compressors from flexibility and efficiency point of view.
    f. Provision of thermostats for control of temperature.
    g. Pressure relief arrangement to release excessive pressure if any developed in the system.
    h. Indication to operating staff for changing filter system & Arrangement for oil Circulation.

HOW TO OPERATE THE PLANT

Ensure that the plant is in horizontal and leveled position.

Refer to valve operation chart for operation. It is understood that all valves should be closed initially. The valve operation chart functions like a check chart. Maintain the oil level in the vacuum pump. Change vacuum pump oil if it is turbidity. After checking all valve closed, connect inlet and outlet valves 1.1 and 7.5 to the respective hoses. Switch on circuits of various items, which will be involved in the processing operation.

DOS AND DO NOTS

DOS;

1. Close all necessary valves. Refer to valve operation chart.
2. Check direction of all pumps.
3. Check lubricating oil level for vacuum pump use only vacuum oil.
4. Keep the gas ballast - of vacuum pump open.
5. Store the oil in a separate tank. Completely drain the free water and the feed the oil into the plant.
6. for operation switch on vacuum pump. After achieving sufficient vacuum, switch on gear pump. When gear pump trips off (by float switch) switch on centrifugal pump. Keep necessary valves open as per valve operating chart.

DO NOTS.

1. Do not load the plant with oil having free water.
2. Do not run the plant with insufficient amount of oil.
3. Do not bad Lubricating oil for vacuum pump.

Do not run vacuum pump with gas ballast closed.

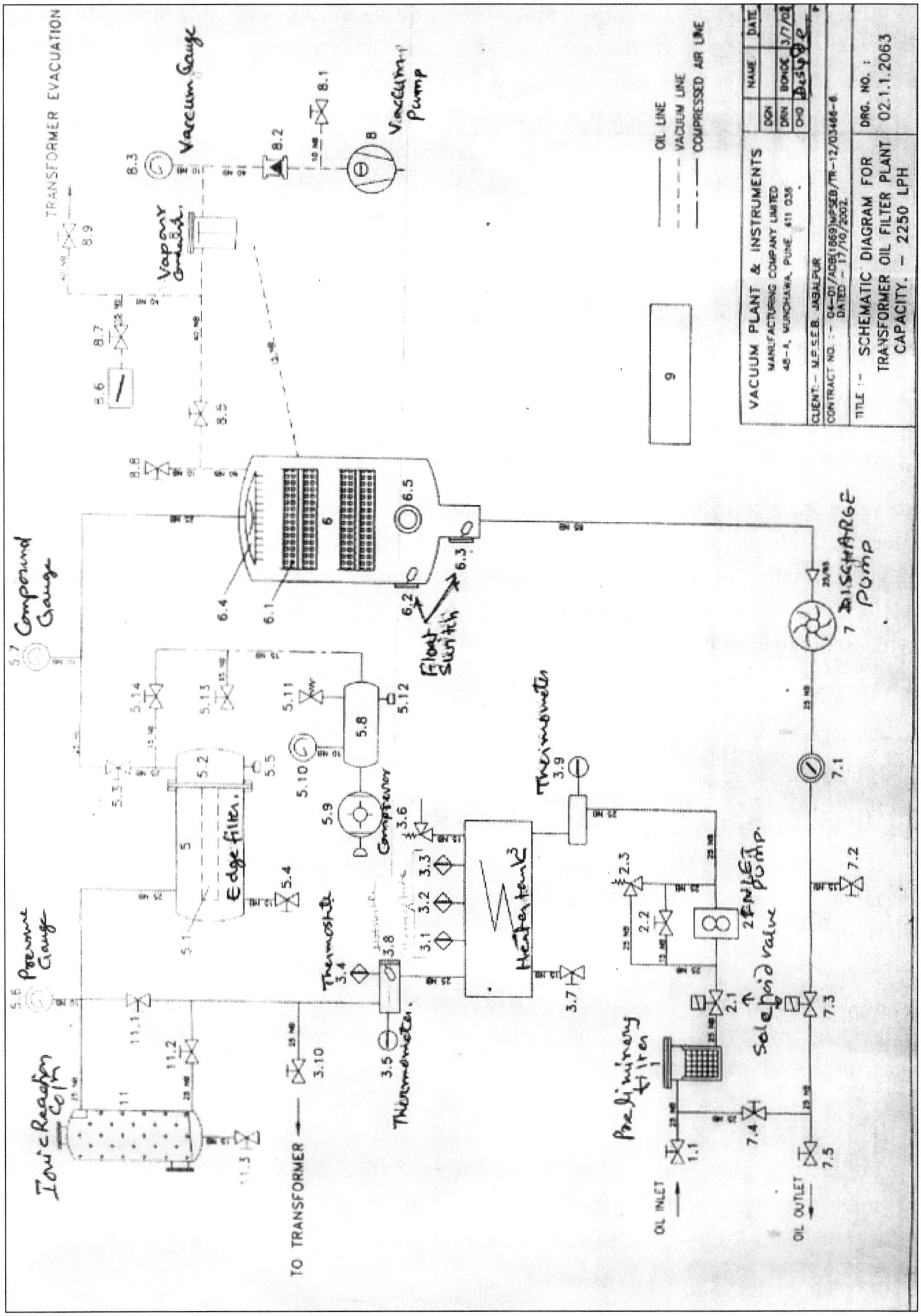
VACUUM PLANT & INSTRUMENTS
MANUFACTURING COMPANY LIMITED
CLIENT :- M.P.S.E.B. JABALPUR
TITLE :- SCHEMATIC DIAGRAM FOR TRANSFORMER OIL FILTER PLANT CAPACITY. - 2250 LPH
DRG. NO. : 02.1.1.2063
OIL LINE
VACUUM LINE
COMPRESSED AIR LINE
TRANSFORMER EVACUATION
TO TRANSFORMER
OIL INLET
OIL OUTLET
Vacuum Gauge
Vacuum Pump
Compound Gauge
Pressure Gauge
Edge filter
Heater tank
Thermometer
Compressor
Float Switch
DISCHARGE Pump
INLET Pump
Solenoid valve
Preliminary filter

14. "Role of Zig-Zag Transformer in Unearthed System- Its Concept & Usage with Citation of Exemplary Re-Use of Redundant 40 MVA 132/12kV, YNd11, Transformer"

1.0 ABSTRACT: The Unearthed System/ transformers are insulated from the earth. To provide a path for the Earth fault currents to return to such source, through Neutral, Earthing transformer is required. The Earthing Transformer of Zig-Zag type is the only answer to this. The author shall not only elaborate the concept of a Zig-Zag Transformer, its sizing but also cite an exemplary re-usage of a redundant Power transformer having 11kV LV winding as delta.

2.0 INTRODUCTION:

Question: Why do the Power Systems / Power equipments need earthing / grounding?

Answer: This is a statutory requirement as per the Clause 67 (1) of the IE rules-1956(with latest amendment in the year2000). The essence of relevant potions is reproduced below;

" 67. Connection to earth-(1)All non current carrying metal parts associated with HV/EHV installation shall be effectively earthed to a grounding system or mat which will:-

a. limit the touch and step potential to tolerable limit,
b. limit the ground potential rise to tolerable values.

In the clause (1A)(a) it has been emphasized therein that Neutral point of every Generator and Transformer shall be earthed by connecting it to the earthing System as defined in Rule-61(4)".

The clause (1A) (c) speaks about deriving neutral point of Delta connected system which is reproduced below;

"In case of Delta connected system the neutral point shall be obtained by insertion of Grounding Transformer and current limiting resistance or impedance where ever considered necessary at the commencement of such a system."

3.0 UNEARTHED SYSTEM:

Irrespective of the system whether it is a Delta connected or a star connected system if insulated from the earth(in case of a star connected system, if its neutral is not connected to earth) are known as unearthed systems. Following are the schematic diagrams and the corresponding phasor (vector) diagrams showing normal condition and the faulty condition depicting 1Phase to earth fault of unearthed system.

Note: The system is having ABC convention with counter clockwise rotation of sequence.

3.1 Normal condition;

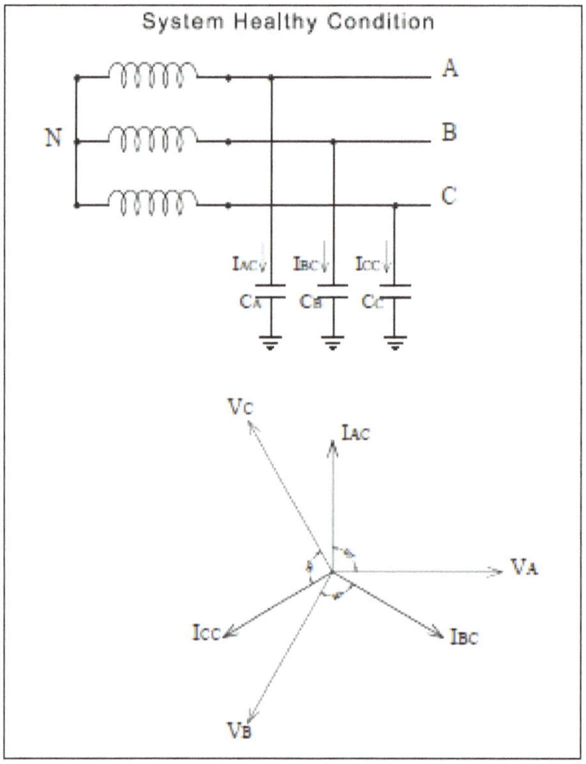

Figure 1: Unearthed system in healthy condition.

It is observed from the Figure1 that the voltages of the Conductors A, B & C are equal with respect to earth though insulated from the earth. Their equality is maintained due to the capacitive coupling through shunt capacitance between respective phases to earth and flow of capacitive current through them due to the potential difference between the Phases and the earth. It is assumed that capacitance of each conductor to earth is equal and the star point of these capacitive currents is formed in the Earth.

As per Kirchhoff's law IAc + IBc + ICc = 0. And as per the basic theory, the respective capacitive currents (charging currents) are leading their corresponding Phase to ground voltages by 90°.

The value of each capacitive current would be

$$Ic = \frac{Vph - n}{Xc},$$

3.2 SINGLE PHASE TO GROUND FAULT: Please refer to the schematic diagram as shown in the figure 2.wherein fault of "A phase to Ground" is depicted.

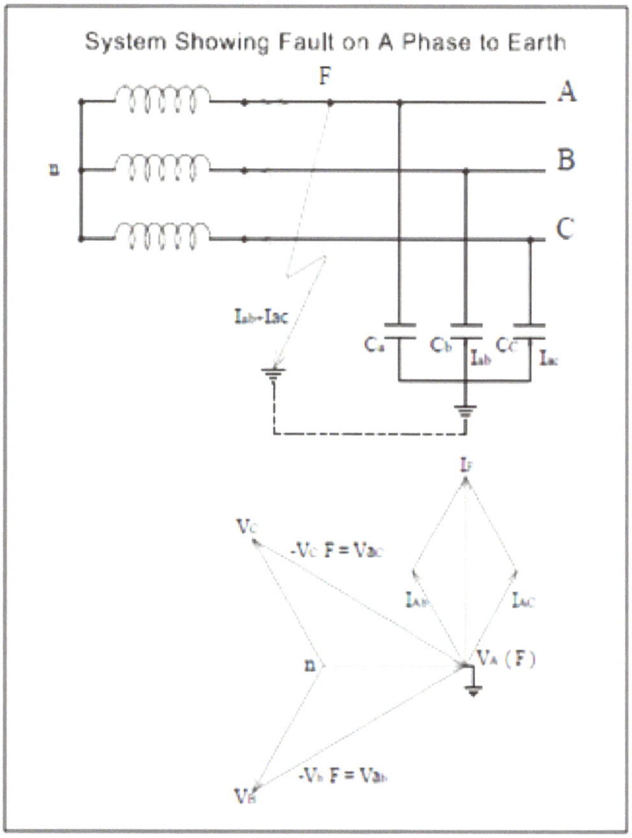

Figure2: Ph to earth fault on A phase.

Following Phenomena/conditions shall develop as soon as 'A' phase gets earthed at location F.

i. The A phase voltage shall assume the same voltage as Earth ie A phase voltage to earth will be '0'.
ii. The voltages of the healthy phases to A phase or earth shall become Line to line voltage ie $\sqrt{3}$ times the Phase to neutral voltage. This may damage the insulation of the System.
iii. Since the Capacitance of the faulty phase is shorted, the fault current shall pass through the mother earth and return to the circuit through remaining 2 healthy phases with the enhanced value of each capacitive current as, now the capacitive currents developed are due to the Potential difference between the enhanced C phase to earth and B phase to earth voltages respectively which have assumed the values as $\sqrt{3}$ times the phase to neutral values and th shall be developed through the enhanced voltage of each remaining Phase to ground ie

$$\textbf{Thus Icf} = \frac{\sqrt{3}\,.\text{Vph}-\text{n}}{\text{Xc}}$$

And Vector sum of Icb + Icc would be $\sqrt{3}\,.\text{Icf} = \sqrt{3}\ .\ \frac{\sqrt{3}\,.\text{Vph}-\text{n}}{\text{Xc}}$ = 3.Ic (i.e. 3 times the Capacitive current of healthy condition). (Refer figure1).

iv. While one phase is already grounded, if another phase develops E/F. It will result into serious Ph to Ph & E/F.

Note: It may be mentioned here, that unlike other Neutral grounding/ earthing systems, this system is advantageous for continuous process industries etc. In LT systems, it is the most essential requirement for the "Heart Surgeries" in particular.

v. On occurrence of an E/F capacitive current flows into the earth. While fault is hanging, if the closing command is given to the delta system, the capacitive fault current may develop an arc in the ionized path of the fault and may result in development of an arc. If the arc persists for some time it shall give rise to a condition known as " Arching Grounds" in which cyclic charging and discharging of the system capacitor occurs due to the fault, resulting in very high oscillations super imposed on the whole system, which in turn develop very high voltages to the tune of 5-6 times the normal voltages. These very high voltages may attribute to the cause of insulation break down of the system and also may damage the insulation of those loads like Transformers, HT/ LT motors etc which are connected to the system. Such phenomenon predominantly occurs in under-ground cable applications. The formation of capacitances in an underground cable is depicted in figure 3.

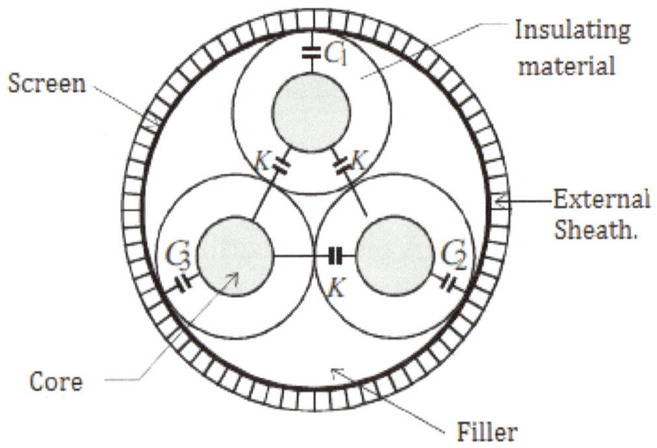

Underground screened cable

Figure 3: Formation of capacitances in screened 3- ph U/G cable.

4.0 ADDRESSING THE PROBLEM OF UN - EARTHED SYSTEM:

a. Since, it is useful in continuous- process industries, there should be some interlock to let it be continued till such time the process is not completed and the next similar process should not start till such time the ground fault in an ungrounded system continues.
b. Last but the most important answer to this problem is to create "an artificial neutral for allowing the Zero sequence current(E/F current) return to the source at the time of Earth faults only". There should not be any flow of currents in the normal service conditions.
c. The condition as mentioned in 'b)' above is possible only in Zig –Zag Transformers.

As a reserve Source of Supply: In case of failure of any YNyn0 Transformer, to meet this acute emergency, supply to the consumers is extended by using Zig-Zag Transformer on the 33kV Tertiary (Delta) winding of the 400/220 kV or 220/132 kV Auto –Transformers.

5.0 ZIG-ZAG TRANSFORMER:

The dictionary meaning of "Zig-Zag" is A line or pattern that looks like a Z or a row of Zs joined together.

Initially it was known as Inter -connected Star Earthing Transformer. There are 3 pairs (6 nos) of winding having 1:1 ratio. Each limb carries 2 windings as shown in the figure 4 (a). The connection and voltage available to each winding is shown in Figure 4 (b) And its vector diagram is shown in Figure 4(c). The ampere turn in each limb is cancelled as the windings are 1:1 and are wound oppositely. This arrangement gives a free path to the zero sequence current through the neutral in case of earth faults. Since the Zero sequence current is co-phasor, shall divide into 1/3 rd of the fault current through each limb.(Figure 6).

It may be observed from the figure 5 showing connections of inner and outer windings which resembles the inverted alphabet 'Z'. The Vector diagram in figure 4(C) also looks like inverted Zs connected together at one end and are at 120 ^{0} apart.

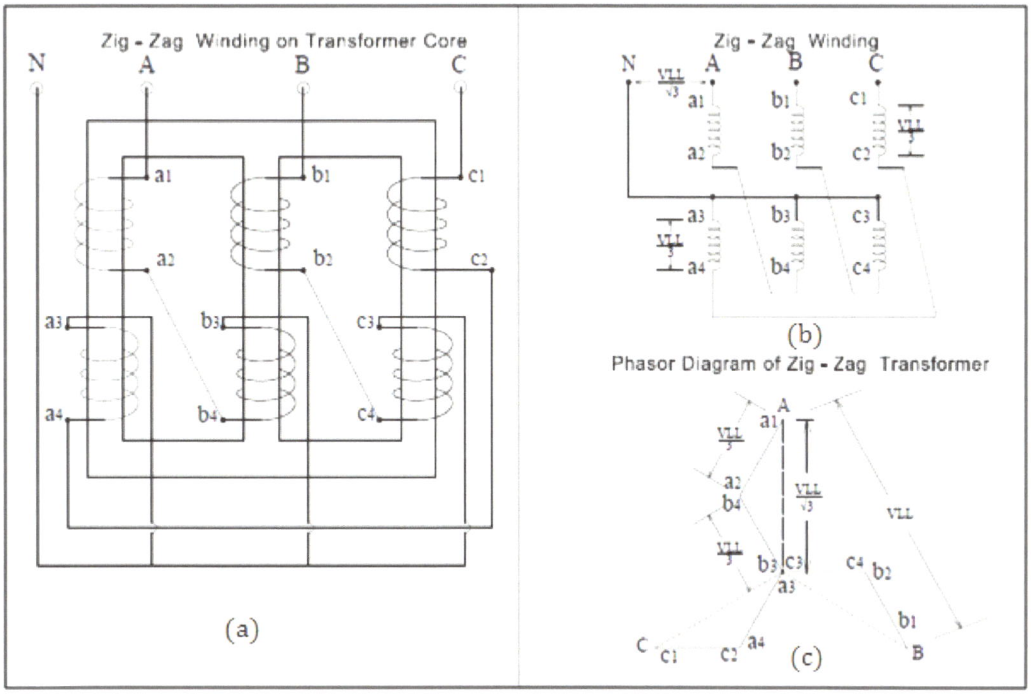

Figure.4: (a)Showing Zig Zag winding on the transformer core, (b) Voltage distribution on each coil as $V_{LL}/3$ volts and each phase to neutral = $\frac{Vll}{\sqrt{3}}$ volts. (c) Shows the vector diagram.

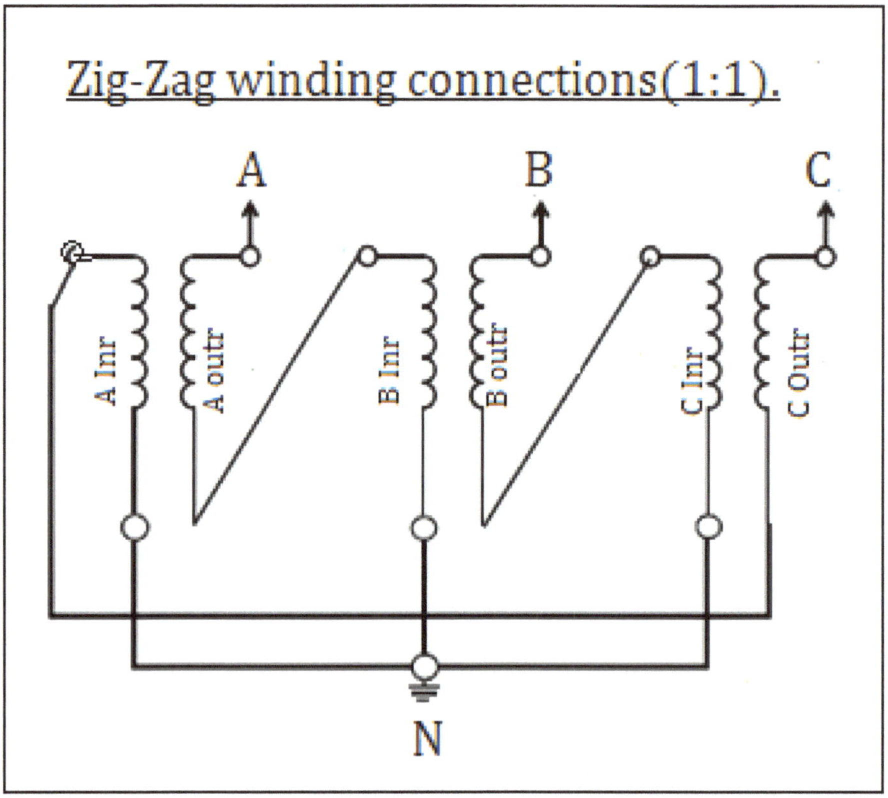

Figure 5: Zig-Zag Windings (1:1 ratio), showing actual connections of inner and outer windings over the core limbs.

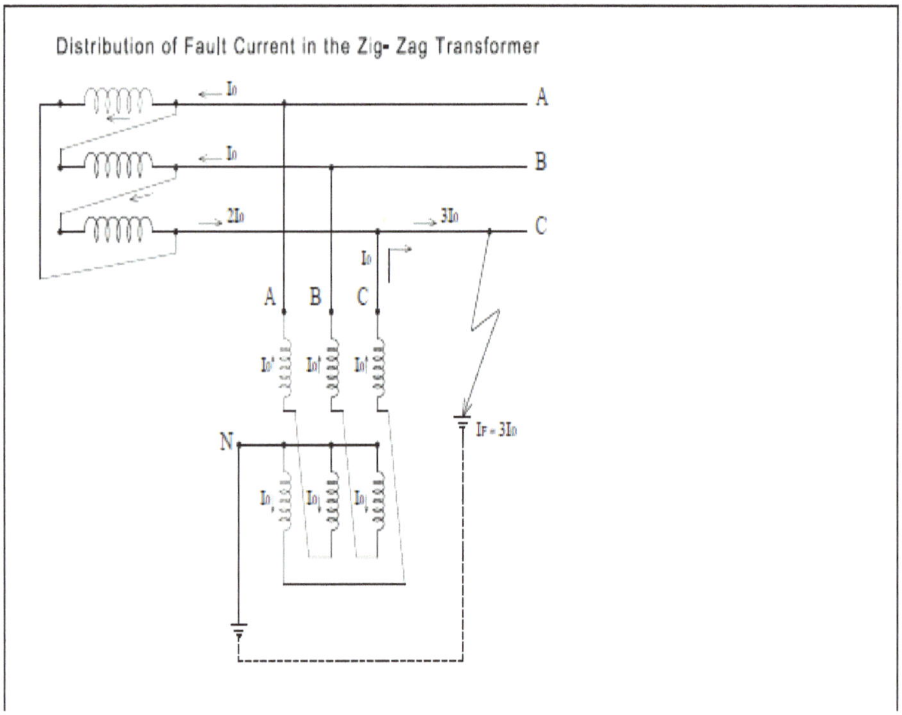

Figure 6: Fault current distribution, Fault on 'C' Ph to Earth. I_F=3 I_0.

6.0 SALIENT FEATURES OF ZIG- ZAG TRANSFORMER:

- Zig-Zag winding comprises of 2-windings in series, these windings are on different core limbs and have different induced voltages but since the current through the 2-in series is equal to the Ph to N current ie the current is the same in equivalent Y connected winding. The voltage across the each coil is 0.577 times the Phase to neutral voltage, Therefore the total length of conductor per phase in Zig-Zag winding is 2 X 0.577=1.155 times the conductor length of equivalent Y winding.
- The quantity of copper required in Zig- Zag Transformer is 33% more than the copper used in equivalent Y winding having same KVA rating.
- The Copper losses in Zig- Zag Transformer shall be 1.155 times more than the copper loss in equivalent Y winding having same KVA rating.

7.0 SIZING OF ZIG-ZAG TRANSFORMER: Since the earthing Transformer has to perform duty of allowing fault current through neutral for a very short time therefore as stipulated in IEEE std 32-1972, continuous rating of 3% for 10 sec. unit is to be adopted, ie 0.03 times the fault current for continuous operations.

The Power rating of Zig-Zag Transformer $P=V_{LG}.I_N/\sqrt{3}$. in KVA.

$V_{LG} = VLL/\sqrt{3}$. = 11/1.732=6.35kV, I_N= 13000A x 0.03=390Amp. Say, 13000 A is the Ph to Earth, fault current on 11kV side.

Therefore P= 6.35x390/1.723 = 1430 KVA.

8.0 CASE HISTORY:

RE-USE OF REDUNDANT 40MVA, 132/12 KV YND11 TRANSFORMER)(FIGURE:

To meet an immediate 11kV load demand to the tune of 3.5 MW near 220kV S/s, Damoh, a decision was taken to reuse a redundant 40MVA, 132/12 kV, YNd11 Transformer available on the plinth in the S/s. which appeared to be the earliest, easiest and cost effective solution to be used in conjunction with an Earthing Transformer (readily available in the nearby Area Store).

The said Transformer became redundant consequent to writing-off of the obsolete Dynamic ±30MVAR, 12kV Synchronous Condenser, which was used for MVAR load management in the EHV Grid System.

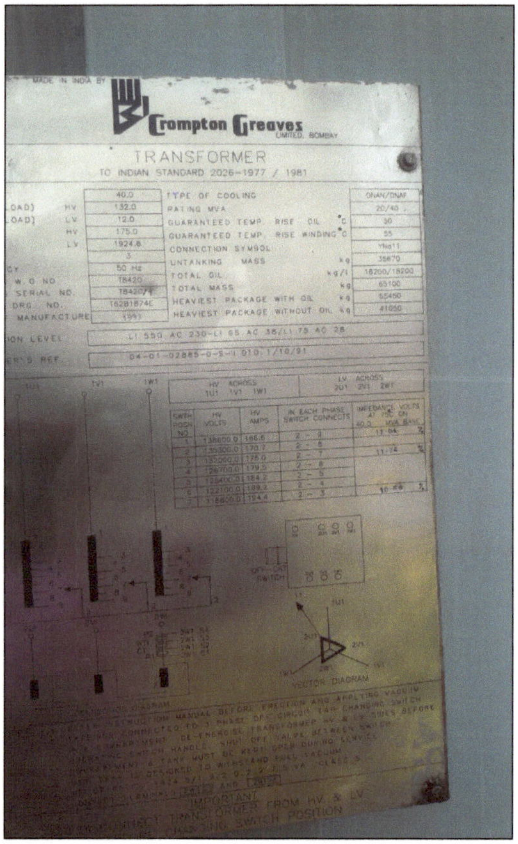

Figure 7: R&D plate of 40 MVA, 132/12kV, YNd11,

The existing System of 220kV Damoh is depicted through the single line diagram; (figure 8)

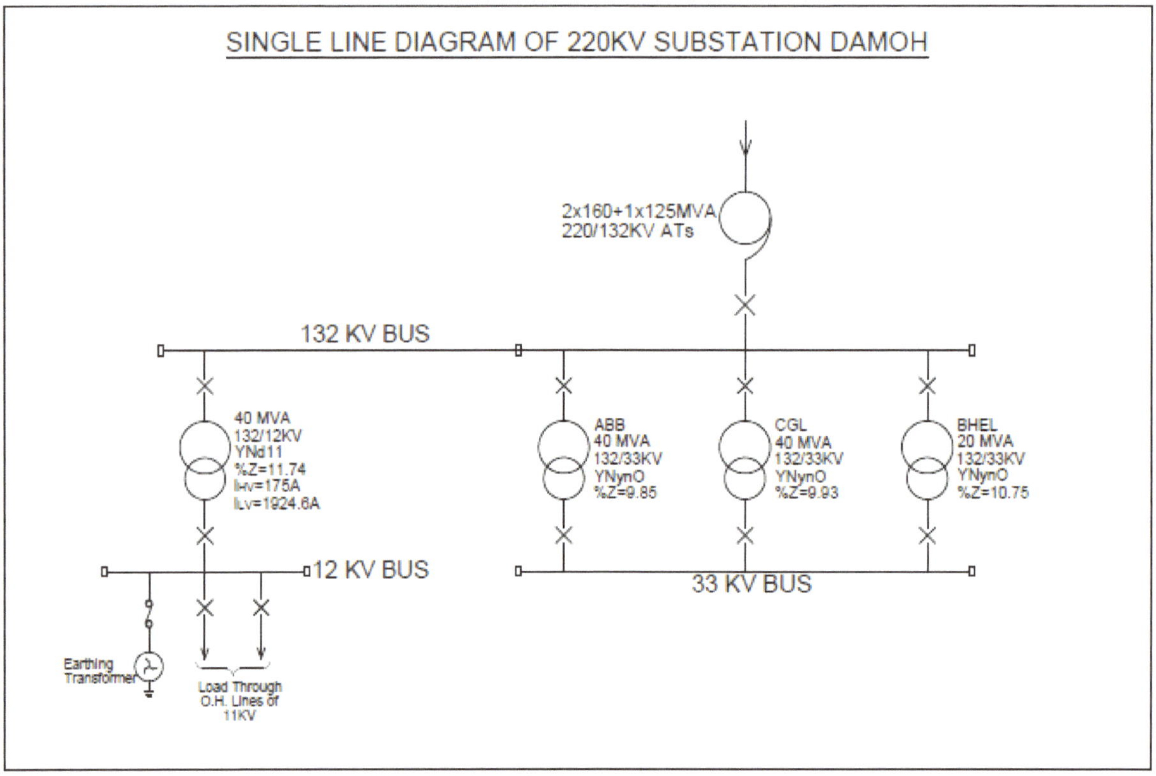

Figure 8: SLD of 220kV S/S Damoh.

11 kV System: For extending 11kV supply following arrangements were made in the substation switch yard;

a. Erection of 11kV bus with 2 feeder bays.
b. Connecting Earthing transformer onto the 11kV bus with fusing arrangement by directly grounding the neutral through a CT of adequate ratio.(refer figure 9)

Figure 9: Connection of installed Earthing Transformer onto the 11 kV bus.

The Earthing Transformer Details: (Figure 9)

a. Make: ETE Hydrabad.
b. Rating: 2750 KVA(Nom).
c. Volts: 11kV.
d. Zero seq Impedance: 27.003 Ω.
e. Neutral current limited to 919 A for 20 sec.
f. Neutral current limited to 1300 A for 10 sec.

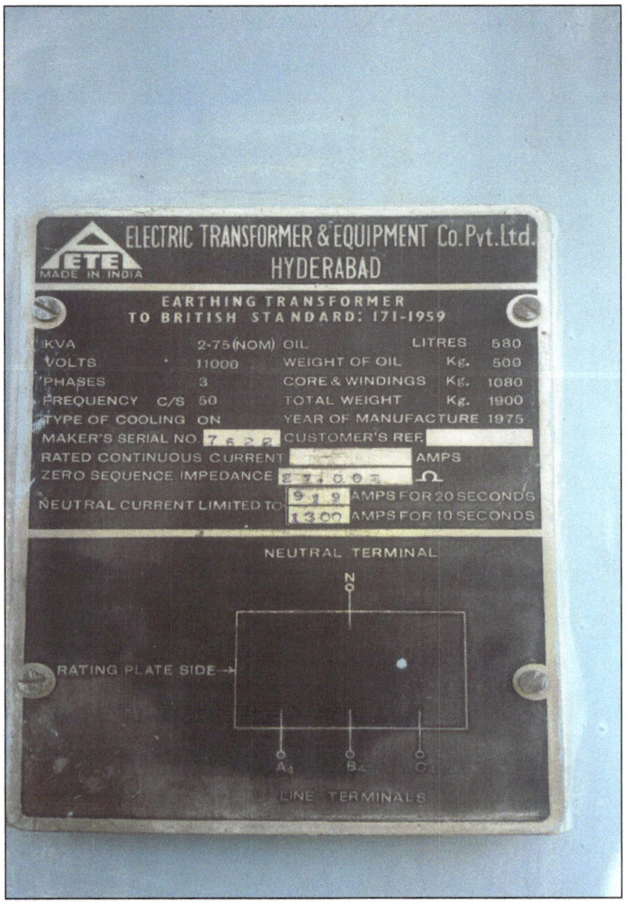

Figure 10:R&D plate of the Earthing transformer.

9.0 References:

i. IE rules 1956(amended up to 2000).
ii. Protective Relays Application Guide of GEC measurements.
iii. Power Transformers Principles and Application by J. J. Winders Jr.
iv. John R Boyle: Understanding Zig- Zag Grounding banks.
v. IEEE Std -32-1972(Reaffirmed 1990).
vi. IS: 2026-8-2009.

12 THE AUTHOR: K.K.Murty & M.Y.Mansoori

15. OIP bushings – Concept to Condition Monitoring".

A true case history of failure of 245kV Oil impregnated paper bushing of 160 MVA Transformer and a sample realistic cost benefit analysis, emphasizing the use of online tanδ monitoring system.

Figure 1: 245/145/72.5 kV, OIP (Oil Impregnated Paper) bushings on 220/132/33 kV 160MVA, transformer at 220 kV sub-station, in Madhya Pradesh (India).

Table of contents summary

This article focuses on the concept of condenser bushings (Oil impregnated Paper Bushings and Resin impregnated paper bushings) for extra high voltage (EHV) class transformers (132 kV and above), their selection, erection and monitoring, so as to avoid catastrophic failure of EHV transformers due to bushing failures. A brief description of on-line monitoring system highlighting its superiority over the conventional off-line monitoring practices have been depicted with a true case study and a specimen cost- benefit analysis pertaining to failure 160 MVA, 220/132/33kV Transformer due to a HV - OIP bushing failure and its-assumed down time of 72 Hrs. at a sub-station 150 Kms away from the HQ of the controlling authority of the Transmission Utility in the state of Madhya Pradesh(India).

Abstract;

The bushings provide insulation to one or more live leads of the power equipment from its metallic body. The bushings for the EHV (Extra High Voltage) power transformers (generally oil filled) depend upon the type of external insulation media where the Transformers are required to be installed. Following types of bushings are generally used;

- Oil-to-Air type - Used for installation in outdoor sub-stations/ switch yards.
- Oil-to-SF6 Gas type - Used for the Gas Insulated indoor sub-stations and in the *hybrid sub-stations.

***Note:** Hybrid Sub – Station: A sub-station having both indoor portion containing EHV equipment in SF6 gas enclosures, as well as outdoor portion of an EHV sub – station. This type of sub-station is constructed where space is a constraint for further expansion of the Sub-station.*

The condenser type bushings OIP (Oil impregnated Paper) & RIP(Resin impregnated Paper bushings) had proven their superiority over Synthetic Resin Bounded Paper (SRBP) bushings by meeting the stringent level of partial discharge (PD) requirement of maximum discharge quantity of 10pC, at Um (Power frequency maximum phase to Phase system voltage) [1].This criteria came into being since the year 1975, the year when manufacturing of 400 kV Transformers commenced.[2]. Now, the indigenously manufactured OIP bushings have become part and parcel of EHV Transformers, ie each and every EHV Power Transformer manufactured in INDIA is generally equipped with Oil Impregnated bushings only. (The price differential between the indigenously manufactured OIP bushings and the imported RIP bushings is approximately 1:2).

The important parameters for assessing the condition of the bushings are tan δ and capacitance. The conventional method of tan δ and capacitance measurement is through off-line portable measurement kits. Deterioration in the condition of the bushings in between the periods of two consecutive testing dates cannot be assessed. To obviate such bushing failures/catastrophic failures of transformers it is very essential to monitor the bushings on real-time basis with the "On-line tan δ monitoring systems which have proved to be superior to the conventional off – line testing methodology

2.0 Key words: CIGRE' (Consil International des Grands Re'seaux Electriques a'Haute tension), BIL (Basic Insulation level), kVp (kilo Volts peak), OIP bushings (Oil impregnated paper bushings) RIP bushings (Resin Impregnated Paper Bushings), tan δ (tangent of loss angle).UST. (Ungrounded Specimen Test), GSTg (Grounded Specimen Test with Guarding)

3.0 Conceptualizing capacitance graded condenser bushings.

3,1 Following example depicts the theory/concept of equi-potential gradient due to equal capacitors connected in series;

Example:

5 Nos Capacitors say C1, C2, C3, C4 & C5, each of 25μF are connected in series to a 250 volts supply.

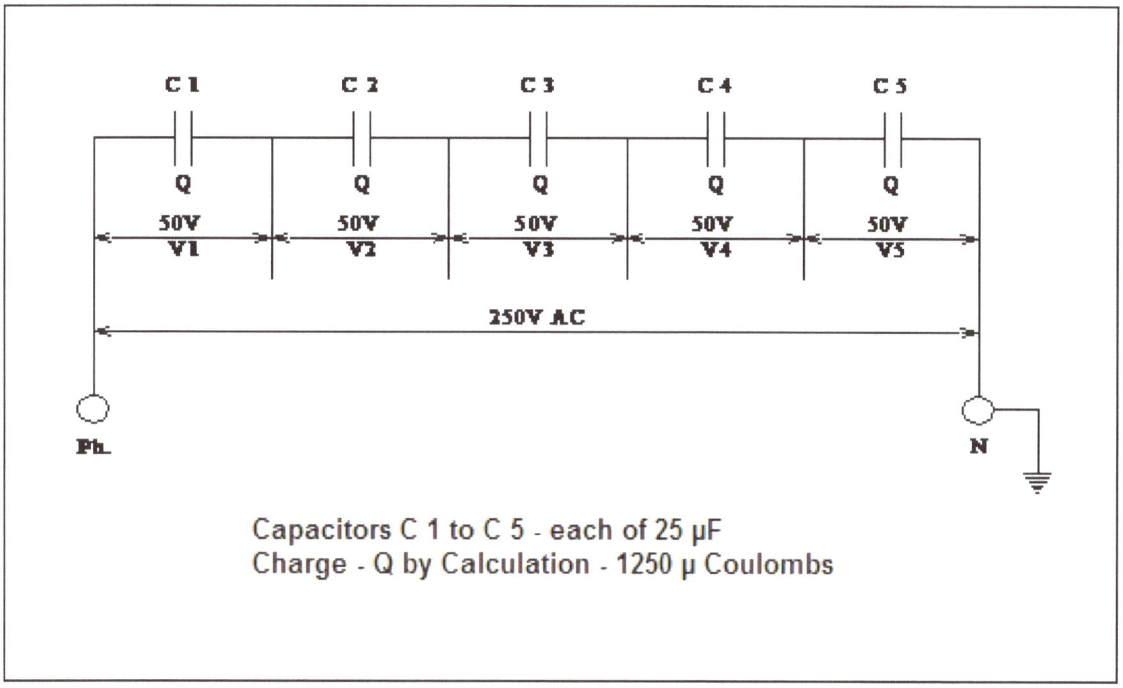

Figure.3: Specimen circuit of capacitors in series and to show the equipotential grading.

$$\frac{1}{Ct} = \frac{1}{C1} + \frac{1}{C2} + \frac{1}{C3} + \frac{1}{C4} + \frac{1}{C5}$$

$$\mathbf{Ct} = \frac{5}{1} = \mathbf{5^{\circ}}\,\text{F}.\ \mathbf{As\ C} = \frac{Q}{\text{Potential across Capacitors}} = \frac{Q}{V}$$

Total charge in the circuit = $Q = C_{t.} V = 5 \times 250 = 1250\mu$ Coulombs.

Now, $V1 = \frac{Q}{C1} = \frac{1250}{25} = 50V$, $V2 = \frac{Q}{C2} = \frac{1250}{25} = 50V$, $V3 = \frac{Q}{C3} = \frac{1250}{25} = 50V$, $V4 = \frac{Q}{C4} = \frac{1250}{25} = 50V$,

$V5 = \frac{Q}{C5} = \frac{1250}{25} = 50V$.

ie the equipotential gradient is maintained uniformly from 250 volts to the reference level(ground) with an uniform potential gradient of 50 volts each down to the earth potential.

3.2 Concept of Condenser bushings: The condenser bushing is conceptualized keeping in view the principle of uniform potential gradient/grading of electric field from the live EHV conductor to the fixing flange of the Transformer which is at the earth potential through series capacitors.

The Oil Impregnated Paper bushings / Resin Impregnated Paper bushings are manufactured with great precision maintaining equal capacitances between each cylindrical condenser, formed between two co-axial aluminum-foil cylinders encircling the bushing central pipe/rod over the layers of cellulose paper (Superior grade kraft paper) wrappings. The kraft paper is dried to the extent of water content of 1% maximum by heated cylinders. The condensers thus formed, are in series. The number of series condensers of equal capacitances is decided through the computer aided design, maintaining equal potential gradient from the central tube/rod at high voltage to the bushing flange at earth potential. The winding of the condenser core with superior grade kraft paper is carried out in a dust free chamber which is subjected to high degree of vacuum (0.005mm) and then impregnated with degassed transformer oil in case of oil impregnated paper bushings(OIP-bushings) or impregnated with cured epoxy resin in case of resin impregnated paper bushings(RIP bushings). They are finally assembled by encapsulating them under pressure in glazed hollow porcelain cylindrical bushings having appropriate creepage distance through the rain sheds/petty-coats with other necessary fittings and accessories.

The theory of equipotential gradient from the live conductor to flange at earth potential through cylindrical condensers formed as condensers in series, mentioned earlier is shown in Figure No.4A(a). [4-A] and also a formula showing calculation of the potential gradient **V** $_x$ at any distance **x** from the circumference of the central tube is shown. In figure No. 4A (b). [5]

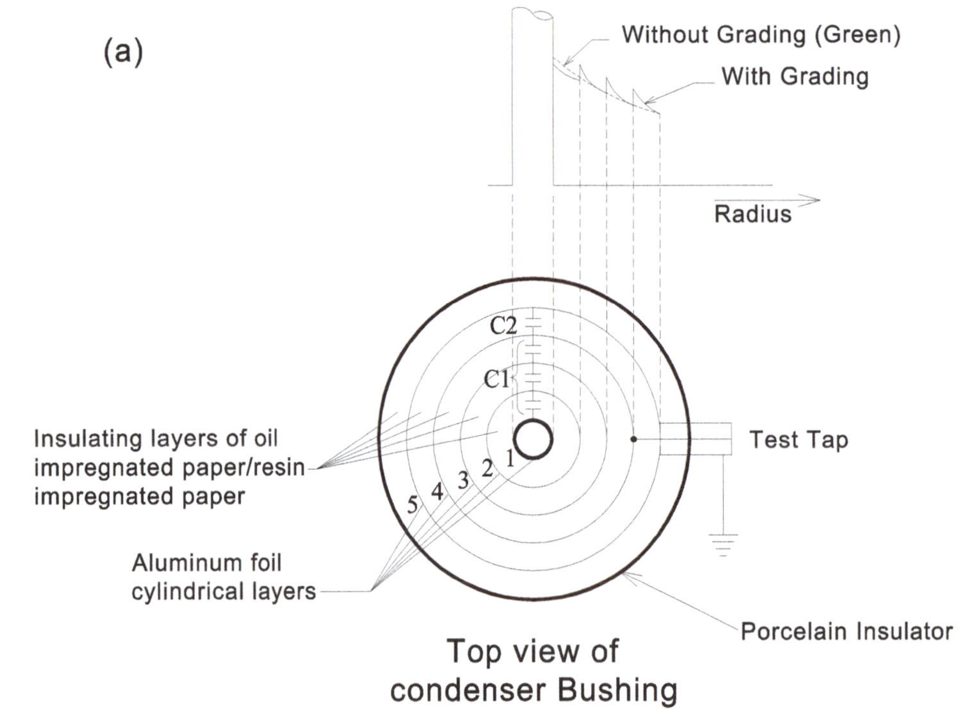

(b)

Potential gradient at any point x from central tube periphery

$$Vx = \frac{V}{x \log_e \frac{Rn}{R1}} \quad ....[5]$$

Where Rn is radious of nth circle from co-centric point.

Figure 4 A: (a) Concept of potential graded condenser bushings. [4-A]
(b) An expression for Potential gradient at a distance 'x 'from the circumference of the bushing. [5].

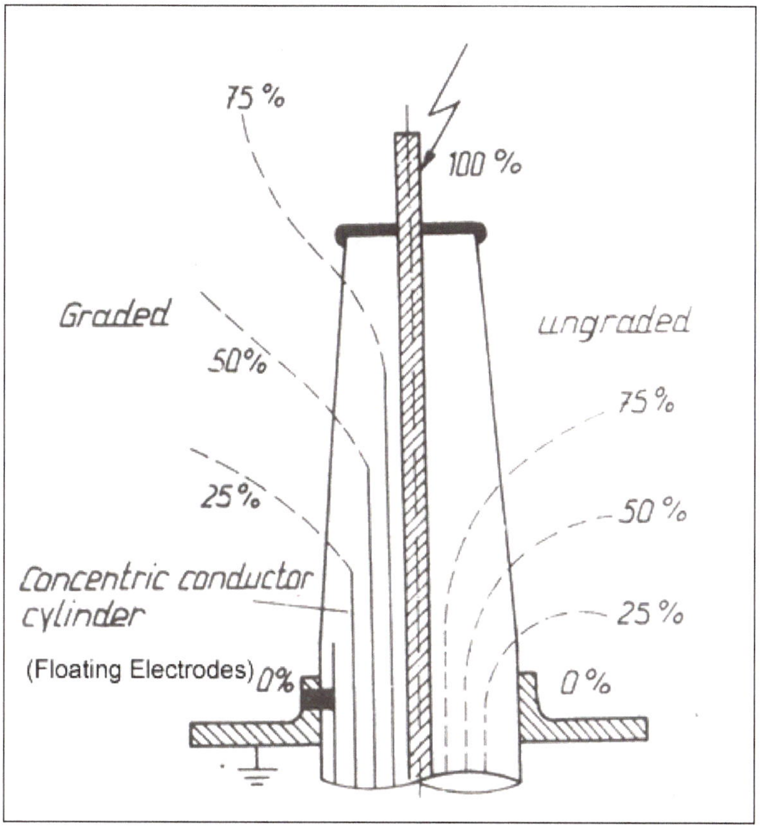

Figure 4.B: Comparison of Potential distribution with graded and ungraded Bushing. [4-B]

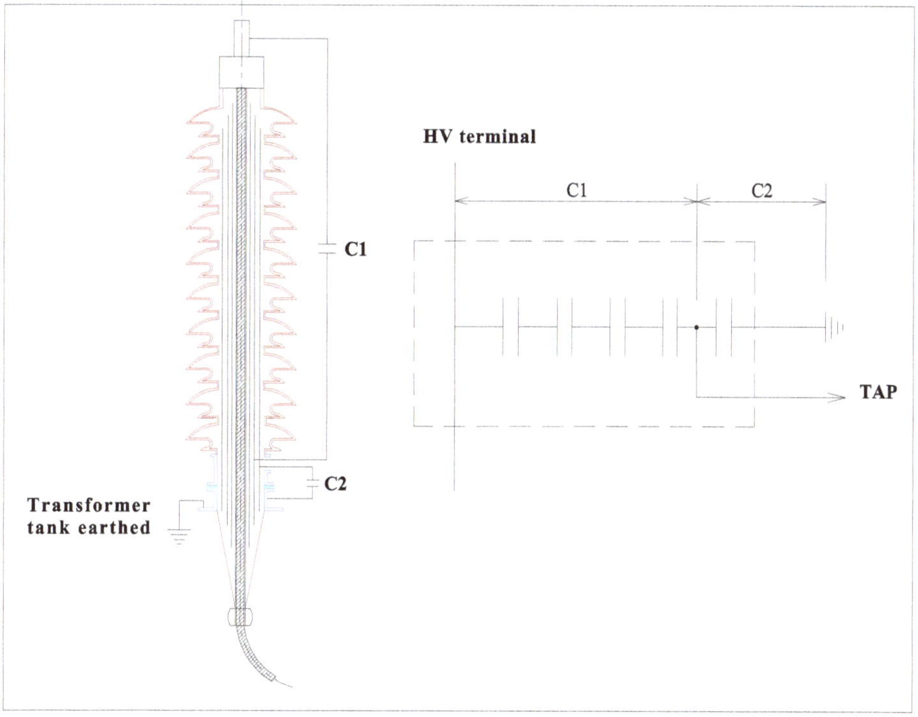

Figure 5: Locations of measurement of C1 & C2 in a condenser bushing with a symbolic series circuit showing capacitance C1 & C2 in series while test tap cap is closed.

Note:

i. C1 is the total capacitance of all the series capacitances formed from the Central conductor to the Test tap. The insulation provided is either Oil impregnated paper in case of OIP bushings and resin impregnated paper. The main insulation is between the central conductor of the bushing and the Test tap. It is described in terms of Dissipation factor(tan δ

ii. C2 is the Capacitance between Test tap to the flange of the bushing. The insulation between the test tap to the flange is also measured.

iii. When test tap cover is closed, C2 is connected in series with C1 to form total capacitance of the bushing from the central conductor to the flange which is kept at the earth potential.

4.0 Dissipation factor tan δ in context of Condenser bushings and its phasor representation;

The Dissipation factor is defined as the reciprocal of the ratio between the insulating material's capacitive reactance to its resistance at a specified frequency. It measures the inefficiency of an insulating material. In other words any increase in Dissipation factor tan δ is an indication of deterioration in its original insulating quality.The concept of OIP bushings having formed series capacitances with oil impregnated Kraft paper and cylindrical aluminum foils is as follows; With reference to the concept of construction of condenser bushings {refer paragraph 3(b) and the figures.4-A & 5}, the significance and the phasor representation of tan δ is explained here-under;

The angle δ is the loss angle and the tan δ is the loss factor (Dissipation factor).

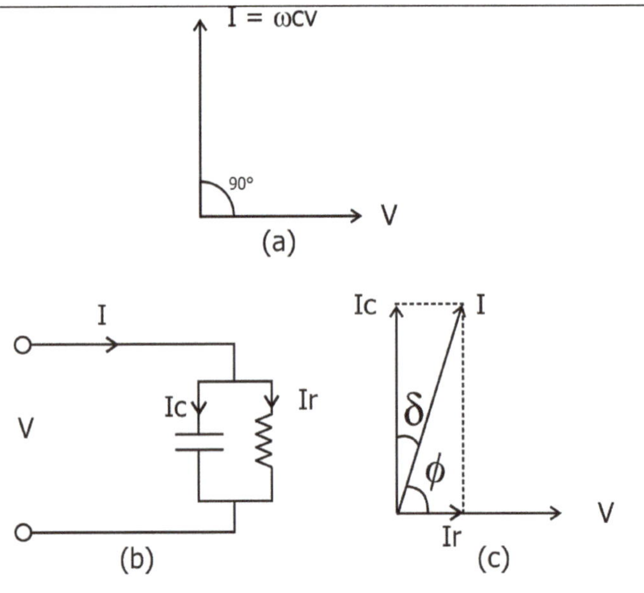

(a) Current leading voltage in a capacitor.
(b) Moisture and other immpurities offer parallel resistance and I gets two parallel paths and gets devided into Ic & Ir.
(c) Vector diagram of the circuit in (b) above wherein $\delta = 90 - \phi$ is called loss angle.

Fig. 2

Figure 6: Loss angle δ (90-∅ shown developed as the angle between the phasors Ic and I.

The AC voltage when applied to an ideal capacitor gives rise to a capacitive current which leads the applied voltage by 90° (ie Ic = ωCV)(Figure:6 (a). Impurities, if developed in the capacitor shall act as resistance in parallel. The current **I** shall now have two parallel paths, one through the capacitor and the other through the resistance (impurity). They are $\mathbf{I_C}$ and $\mathbf{I_R}$ respectively. {Figure: 6 (b)}

As per the phasor diagram, the current phasor I, no longer leads the voltage phasor V, by 90°, but now it is leading by ∅° =90°− δ {Figure: 6 (c)}

Therefore, Ic = I. cos δ, ωCV = I. cos δ, thus $I = \frac{\omega CV}{\cos\delta}$,

Dielectric loss = VI Cos ϕ, = VI sin δ, = $V.\frac{\omega CV}{\cos\delta}$. sin δ,

Dielectric loss = ωCV² tan δ, ---------------------------------- (A)

- $\tan\delta = \frac{IR}{IC}$ i.e. increase in tan δ value is an indication of deterioration of the insulation.
- Increase in C takes place when either moisture ingresses into the insulation or some of the condensers get shorted.
- Dielectric loss is directly proportional to tan δ, C and square of the voltage as ω =2πf, where f of a power system normally does not vary sharply therefore ω is considered to be constant.

5.0 Difference between Dissipation factor and Power factor; [6]

"Dissipation factor" tan δ is used in Europe and Asian countries including India whereas in US, Canada and North America etc., "Power Factor" Cos ∅ is used to specify the power losses in the insulation,

$\tan\delta = \frac{PF}{\sqrt{1+PF^2}}$ and $PF = \frac{\tan\delta}{\sqrt{1+\tan^2\delta}}$

Sample comparative calculation; $\tan\delta = \frac{PF}{\sqrt{1+PF^2}}$

i. For power factor = 0.4, $\tan\delta = \frac{0.4}{\sqrt{1+0.4^2}} = \frac{0.4}{\sqrt{1+0.16}} = \frac{0.4}{\sqrt{1.16}} = \frac{0.4}{1.077} = 0.371.$

ii. For power factor = 0.4, Using formula $PF = \frac{\tan\delta}{\sqrt{1+\tan^2\delta}}$ Putting the value of PF=0.4, the tanδ works out to be=0.436,

Note: Value of tanδ or PF of bushings are only for monitoring the condition of the Bushings.

6.0 Increase in total capacitance value if failure of any capacitor in series occur.(Specimen Calculation):

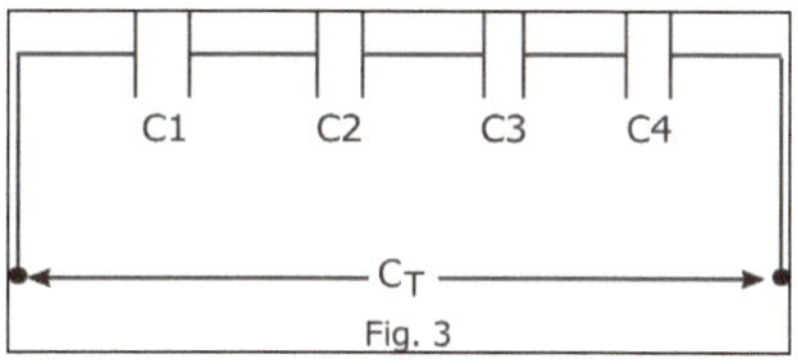

Figure7: Equal capacitances in series.

$$\frac{1}{CT} = \frac{1}{C1} + \frac{1}{C2} + \frac{1}{C3} + \frac{1}{C4},$$ Say C_1, C_2, C_3, C_4 are having capacitances of 100 pF each.

Therefore, $\frac{1}{CT} = \frac{1}{100} + \frac{1}{100} + \frac{1}{100} + \frac{1}{100} = \frac{4}{100}$, Thus $C_T = \frac{100}{4} = 25pF$

Suppose one of the capacitors is shorted, say C_3 is shorted,

$$\frac{1}{CT} = \frac{1}{C1} + \frac{1}{C2} + \frac{1}{C4} = \frac{3}{100},$$ Therefore $C_T = \frac{100}{3} = 33.33$ pF

It is, therefore established that total capacitance will increase if one or more of the capacitor in series are shorted.

Main parts/ components of OIP bushing. [7];

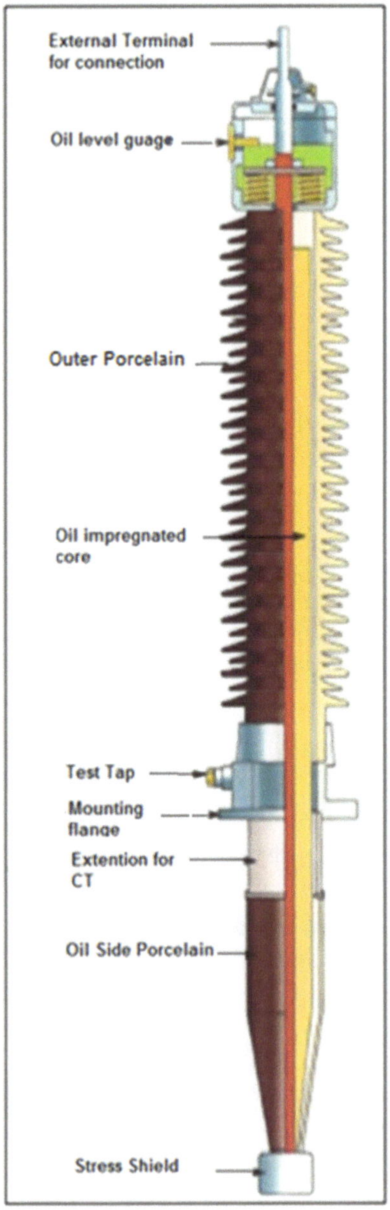

Figure: 8: Components of condenser bushing. [7]

7.0 Selection of bushings;

i. The voltage ratings of the bushings being manufactured are equivalent to the highest system voltages, such as HV bushings of 420kV voltage rating are used on the HV side and the bushings of 245kV voltage rating are used on the 220kV side of 400/220/33kV Transformers. The other voltage ratings of bushings used are 145kV for 132kV system, 36kV for 33kV system etc.. The HV bushings should also be selected such that they should have higher BIL(Basic Insulation level) as compared to the BIL of the Transformer winding. The BIL is the highest Lightening Impulse Voltage in kVp (kilo-Volts peak) which an equipment can withstand without any damage, For example the 400/220 kV class transformers are equipped with the HV bushings rated for 420kV with the BIL of 1425 kVp whereas the BIL rating of the transformer winding is 1300 kVp.[8] This is essentially required for maintaining insulation co-ordination between the insulation of the windings and that of the bushings against the ill-effects of peak lightening impulse voltages.

The Lightening impulse voltage wave rises from zero value to its peak value in 1.2 micro-seconds and then falls to 50% of the peak value in 50 micro-seconds. It is denoted as 1.2/50μ sec.(Figure:7). [1], [9]

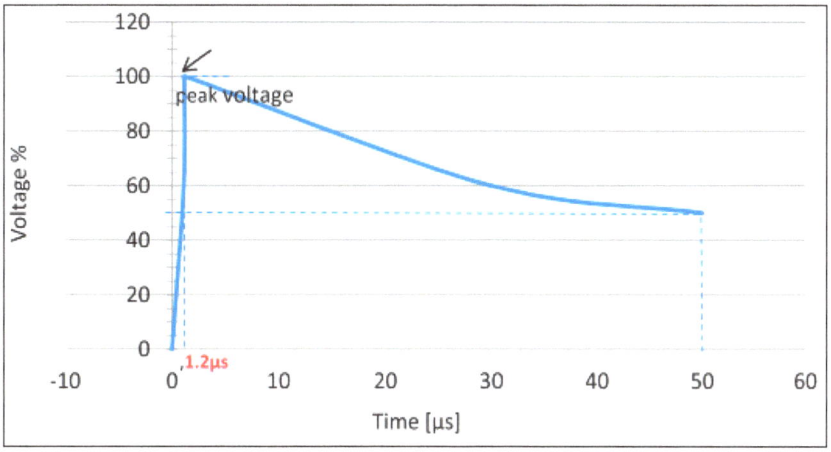

Figure 9: Curve showing Lightening Impulse Voltage profile.

ii. **Type of bushings with reference to insulation application**: It depends upon the location of installation of the Transformers.

   a. Oil-to-Air application bushings: For outdoor installations such as EHV substations/ Switchyards.
   b. Oil-to-SF6 application bushings: For GIS substations(Gas Insulated sub station).

iii. **Minimum nominal creepage distance:** The user may indicate the requirement keeping in view the pollution level of the location where the Power Transformers are planned to be installed. The requirements of creepage distance as per pollution levels are as follows. [1] & [9]

   a. **Lightly polluted. - 16mm/kV.**
   b. Medium polluted. - 20mm/kV.

Heavily polluted. - 25mm/kV.Very heavily polluted. - 31mm/kV.

The manufacturers generally install bushings suitable for heavily polluted atmosphere ie having a ceepage distance of 25mm /kV unless otherwise specific pollution level is mentioned by the customer.

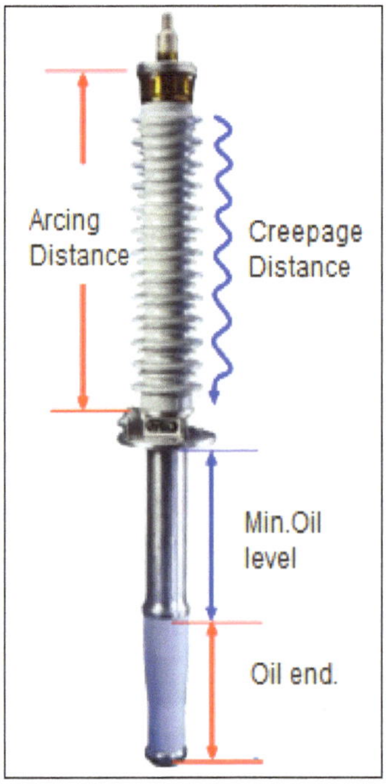

Figure 10: Arcing distance & creepage-distance of a bushing.[10]

iv. **Arcing distance**: If the altitude at which the Transformer is being procured for installation, is >1000 m, the arcing distance needs to be enhanced as the air density at high altitudes is lower than that at sea level, resulting in low dielectric strength of air. The arcing distance required to be increased by 1% for each 100 meters in excess of 1000 meters. [1], [9]

v. **Ambient Temperature:**

The maximum and minimum range of ambient temperature must be made known to the vendor/ manufacturer of the Transformers. They are as follows;

a. Ambient Temperature;
b. As per IEC 60137 for cold countries: (–) 40° C to + 40° C [1]

As per IS:2099 for Tropical countries like INDIA: (–) 05° C to + 50° C [9]

8.0 Acceptance and Type Tests at Manufacturer's works:

While accepting the bushings, Type and Acceptance tests (as the case may be), should be witnessed / carried out carefully at manufacturer's works in confirmation to IEC: 600137. [1] or IS:2099.[9].

a. 8.1 Acceptance tests: - They are same as Routine tests and are performed on the bushings offered for inspection. Following Tests are the Routine tests;
b. Measurement of dielectric dissipation factor tan delta and capacitance at ambient temperature.
c. Dry power frequency voltage withstand test.

d. Measurement of partial discharge quantity.
e. Tests on tap insulation.

Pressure test on liquid filled and liquid insulated bushings.

8.2 Type Tests:-

a. These tests are performed on one or two bushings which have already been tested for the Routine Tests successfully, so as to ensure their faithful performance with reference to their design and quality assurance. Following are the Type Tests;
b. Wet power frequency voltage withstand test.
c. Dry Lightning impulse voltage withstand test.
d. Dry or wet switching impulse voltage withstand test.
e. Thermal stability Test.
f. Temperature rise test.
g. Thermal short time current withstand test.
h. Dynamic current withstand test.
i. Cantilever load with stand test.

Tightness test on liquid filled and liquid insulated bushings.

Note: Before and after the thermal and dielectric tests, measurements of tan delta & capacitance and partial discharge quantity shall be carried out to check whether a puncture or any other significant damages have occurred.

9.0 Transportation, Storage, Erection/Installation of the bushings[11]; Following main points are to be strictly followed during transportation, storage, unpacking & handling and erection of the EHV bushings

A. **Transportation:**

a. Bushings can be transported in a wooden crate maintaining an angle of 6 to 8 degrees to the horizontal plane such that the oil end is at the lower level.
b. Its transportation should be jerk free.

B. **Storage:**

It can be stored outdoors with tarpaulin covering maintaining 6 to 8 degrees inclination with the horizontal plane keeping the oil end portion at the lower level.

C. **Unpacking & handling:**

a. While unpacking care should be taken such that the bushing should not experience any kind of jerks.
b. For lifting no steel rope or chain pulley block should be used. While lifting from the crate, manila rope /nylon rope should be used such that the rope holds the bushing at the top as well as bottom locations and should be lifted hoisting it horizontally.

c. Never rest the bottom stress-shield on the ground. (In the case of non detachable stress shields).

D. **Erection:**

a. The bushing should be thoroughly checked physically for any damages, cracks etc.
b. The bushings should be meggered for IR values also.No, workmen should be allowed to climb over the bushings for any work like connecting the leads & over head jumpers etc instead proper bucket truck/bucket crane should be used for the workmen to stand & pull the lead and make lead connections in draw - lead type bushings and also in the draw-rod connection type bushings.Similarly the jumpers from the overhead gantry to the bushing terminals should also be connected through this arrangement.
c. Before energizing the bushing, it must be kept in upright position or mounted on the transformer for a period of minimum 24 hrs. [11].
d. Non-metallic ropes should be essentially used for lifting and installing/ mounting of the bushings.(Refer Figure: 11)[11]

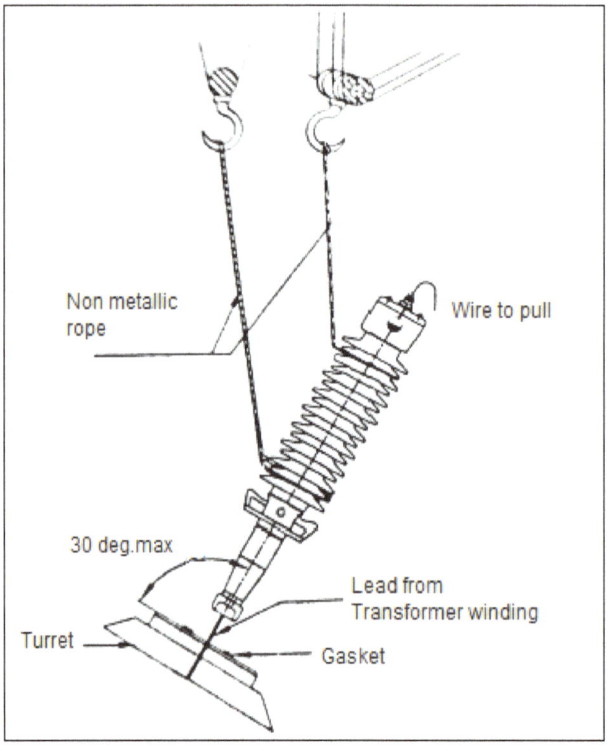

Figure 11: Installation of the bushing (Draw lead type). [11]

Procedure and the precautions should be strictly followed as recommended by the manufacturer while carrying out the erection of OIP bushings.

10.0 Causes of failure of the bushings.

- A few of the causes leading to bushing failure are enumerated below;
- High dielectric stress due switching surges and lightening surges.

- Ingress of moisture and other contaminants through deteriorated aged gaskets and hair line cracks in the porcelain.
- Oil leakage.
- High atmospheric temperature and humidity etc.

  a. The dielectric property also deteriorates due to:
  b. Rise in oil temperature consequent to the overloading of the transformers.

- Loose joint in the leads in the draw lead type bushings/draw rod type bushings, connections giving rise to excess temperature in the bushings.
- Failure due to improper earth connection of test tab.

Improper re-fixing of test tab cap after tan δ & capacitance measurements.

11.0 Need for condition monitoring of the bushings while in service:

About 17.16 % failures of the transformers are attributed to the bushing failures only, it is of prime importance to monitor the same, observe the trend and identify the defective bushings, whose tan δ values are increasing. If found increasing beyond .007, arrangements for replacement of the same have to be made.

Due to bushing failure, the dielectric distances shorten and flash over between the live EHV conductor to the Transformer body takes place resulting into hazardous fire/explosion in the Transformer and also damaging the nearby outdoor equipment. This phenomenon is generally observed in the 220kV, 400kV and 765kV voltage class Transformers.

i. The losses incurred are colossal losses as mentioned here under;
ii. Incurring loss of an amount by way of cost of the bushings.
iii. Incurring loss of a huge amount by way cost of the transformers (in case of Transformer failures) and associated activities.
iv. Incurring loss by way of system disturbances / black-outs affecting large number of Industries, railway tractions, collieries & mines and large number of the consumers in general.

Also incurring huge loss of revenue to the utilities (Generation, Transmission as well as Distribution utilities). Due to above mentioned reasons from (i) to (iii).

12.0. Off-line monitoring of the bushings:

This is carried out once or twice in a year as per practice followed by many Transmission utilities in India..

12.1 tan δ & Capacitance measurements;

The value of tan δ is measured off-line with 10kV, tan δ testing kit on UST mode (Ungrounded Specimen Test mode).(Refer figure.12,). The limiting value of tan δ for condenser bushings is 0.007 or 0.7**%** as per IEC-60137. [1], however, tan δ of new bushing is very low to the tune of 0.3 to 0.4%

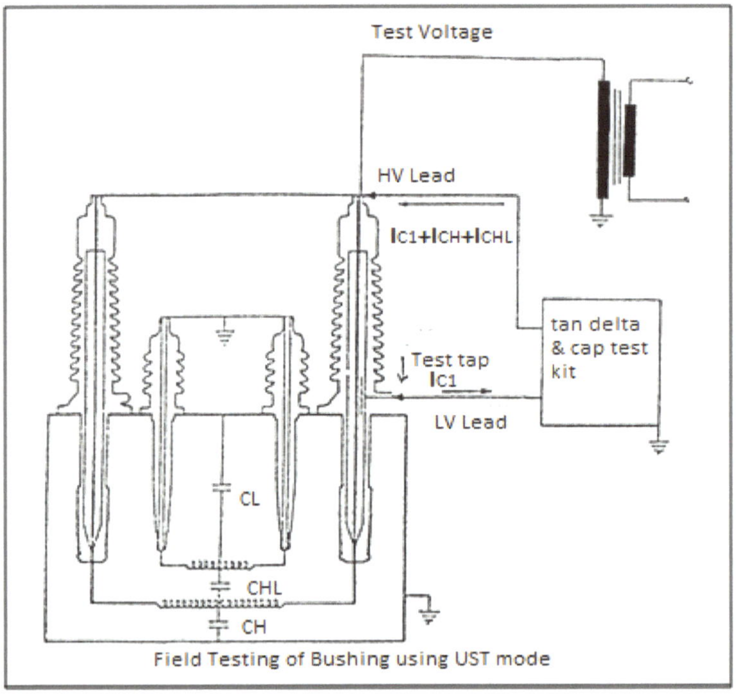

Figure 12: Off line testing arrangement/ schematic diagram for tan δ Measurement of the transformer bushing in UST mode.(Ungrounded Specimen Test mode).

12.2 Bushing tan δ and Capacitance - present trend / practice in vogue:

In contrary to IEC-60137 [1], the present practice as followed by ABB. is, that the tan δ of bushing central conductor to test tap, that is, C1, should not increase by +50 % with reference to the pre-commissioning value and the bushing should be replaced if the value increases up to +75 % and >0.4 %.(ABBs norms).[12]

i. Note: The present day practice by the Indian utilities is to specify the value of tan δ of C1 as less than 0.4 % at the time of procurement of the Transformer/ condenser bushings.
ii. **Insulation check of tan δ tap C2:**

The test voltage should be between 500 V to 2.5 kV only. The tan δ measurement is done in GSTg mode only. GSTg mode is defined as Grounded Specimen Test with guarding. Guarding lead is connected to the reference voltage point in this case, it is ground, resulting into avoiding any stray capacitances between tan δ tap and the ground.

- This value need not be converted to 20°C base for assessment comparison.
- The tan δ of the bushings test tap to flange (C2) insulation generally varies between 0.4 - 3%.[12]
- If the value of Capacitance C1 is found low as compared to the factory value, it is indicative of disruption due to transport damage, therefore it should not be installed.

iii. While in service if the value of C1 increases beyond the factory value by 3 %, it is indicative of partial puncture of the condenser of the bushing and the bushing should be replaced immediately.
iv. The rain sheds / petticoats and the outer surface of the ceramic or porcelain bushing should be cleaned thoroughly before tan δ measurement.

13.0 Interpretation of tan δ and capacitance values [6],[12],[13].

- Values of the Dissipation Factor (tan δ) and Capacitances are compared with one or more of the following;
- Rating plate/Name plate data.
- Results of the prior tests of the same bushing.

a. Results of similar tests on similar bushings.

The value of dissipation factor of modern condenser bushings is generally of the order of 0.5% after correction to 20° C. (However as specified in IEC: 60137 and IS:2099, the limiting value is 0.7% [1],[9]

b. ***[Note:** Limiting value of Power factor: As per IEEE C57.19.01-2000, the limiting value of power factor for OIP bushings is 0.5 %(+0.02/-0.04) and that for RIP bushings is 0.85 % (±0.04)].* [14]

c. Capacitances should be +/-5…..+/- 10% of name plate value, depending upon the total number of condenser layers.

d. Significance of test values of tan δ and capacitance of condenser bushings - Analysis of results/ interpretation thereof.[6], [13]

S.No.	tanδ and capacitance-Trend of test results.	Analysis/Interpretation.
1	Increase in Dissipation factor (between 0.7% and 1%) accompanied by marked increase of capacitance.	Indicative of exessive moisture in the insulation.
2	Very high increase in Dissipation factor alone. (more than1%).	Indicative of thermal deterioration, aging or contamination other than moisture.
3	Low tanδ.	Indicatie of weak potential connections.
4	Increased capacitance.	Indicative of possibility of short circuited condenser layers.
5	Decreased capacitance.	Indicative of possibility of floating ground sleeve, or open or poor test tap connection.
6	Very large variation in tan δ and capacitance values.	Indicative of no oil in the bushing.
7	Negative Dissipation factor accompanied with small reduction in capacitance.	May result from external surface leakages or internal leakages resulting from carbon tracking etc..

***Table1**: Significance of test values of tan δ and capacitance of condenser bushings - Analysis of results/ interpretation. [6], [13]*

14.0 Convertion of tan δ values to the base temperature of 20°C.

Since value of tan δ varies with the temperature, therefore tan δ recorded at different oil temperatures need to be converted to a common base temperature for comparisons. The base temperature is taken as 20°C for comparison of tan δ measured at different temperatures is tabulated in Table 2. [12]

i. The value of tan δ, the insulation dissipation factor increases with the temperature. Main tank oil temperature is measured while testing the tan δ of the bushings of the transformer. In order to compare the tan δ values taken at different oil temperatures, correction factor is applied to the reading so as to bring them to a common reference temperature which is universally accepted as 20° C (Correction factor table is shown in table No.2).[12]

Range of temp. in °C	Correction factor to 20° C		Range of temp in °C	Correction factor to 20° C	
	OIP bushing	RIP bushing		OIP bushing	RIP bushing
0-2	0.80	0.76	43-47	1.25	1.33
3-7	0.85	0.81	48-52	1.30	1.37
8-12	0.90	0.87	53-57	1.34	1.41
13-17	0.95	0.93	58-62	1.35	1.43
18-22	1.00	1.0	63-67	1.35	1.43
23-27	1.05	1.07	68-72	1.30	1.42
28-32	1.10	1.14	73-77	1.25	1.39
33-37	1.15	1.21	78-82	1.20	1.35
38-42	1.20	1.27	83-87	1.10	1.29

Table No.2: Tan δ and Power factor correction factors for OIP & RIP bushings to 20°C.

ii. **Formulae for recalculation/conversion at 20°C.** [6];

The temperature correction factors for tan δ or power factor of insulation is dependent upon the insulating material, material structure, the moisture contents etc; The following relationship holds good;

$$PF_{20} = \frac{\textbf{PF mt}}{\textbf{K}} \text{ or } \tan\delta_{20} = \frac{\textbf{tan}\boldsymbol{\delta}\textbf{ mt}}{\textbf{K}},$$

Where PF_{20}= Power factor at 20° C., PF mt = Power factor at measured test object temp.

$\tan\delta_{20}$ = tanδ at 20°C, tanδ mt = tanδ measured at test object temperature. K=Correction factor.

Example of calculation;

Say the tan δ measured was found as 0.0075 at oil temperature of 45° C;

- Type of bushing: OIP type.
- Conversion factor: From the table 2, K= 1.25(for temperature range 43 to 47°C)
- Using formula $\tan\delta_{20} = \frac{\textbf{tan}\boldsymbol{\delta}\textbf{ mt}}{\textbf{K}}$,
- $\tan\delta_{20} = \frac{\tan\delta 45}{K}, = \frac{0.0075}{1.25}, = 0.006$

15.0 Variation of Power factor with temperature and voltage.

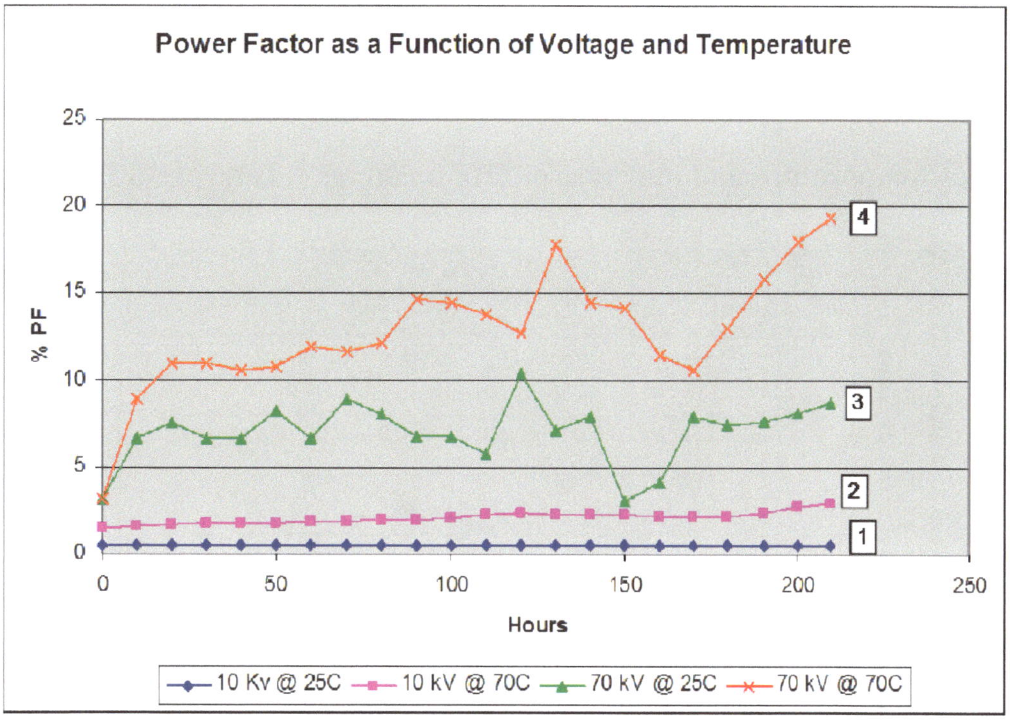

Figure 13: Curves indicating variation of % power factor with voltage and temperature over a period of 210 Hours. [15]

Since the value of tan δ and Power factor of insulation vary with the raise in temperature, the tan δ or pf of the bushings measured in off- line condition shall not give the true picture of the bushings, corresponding to the dynamic operating conditions.

Thus the researchers have studied the influence of variation of temperature and variation of the voltages on the power factor of the bushing and have come to the conclusion that the power factor of the bushing varies not only with variation of temperature but also with the variation of the voltages (Figure 13.). [15].

4-Nos curves were drawn for different conditions of temperature-voltages over a period of 200 Hours-Versus - Power Factor.

a. Curve No.1. Refers to response at 25°C – 10 kV voltage is applied i.e. offline monitoring. It can be seen that the pf is quite stable.
b. Curve No.2. Refers to response at 25°C – 70 kV voltage is applied on the same bushing, temp remaining constant, this indicates variation of pf with voltage. pf is on the higher side.
c. Curve No.3. Refers to response at 70°C – 10 kV voltage is applied. This depicts the variation of pf with temperature. It may be observed that the voltage applied is same as in case of the curve in (a) above i.e. 10kV.
d. Curve No.4. refers to response at 70°C – 70kV voltage is applied i.e. temp as well as voltage have been raised. The curve shows that the pf increases with the raise in temperature and the voltage, on the bushing insulation.

e. Since, power factor increases with higher voltage and higher temperature, need for online tan δ measurement is felt essential. Any abnormality observed, can be addressed to well in time by the utility and avert enormous revenue loss to the utility by way of down time /outage of transformer due to bushing failure or power transformer failure as well as dissatisfaction amongst the consumers at large.

16.0 tan δ versus temperature and moisture of OIP bushings – Interpretation of curves.[12]

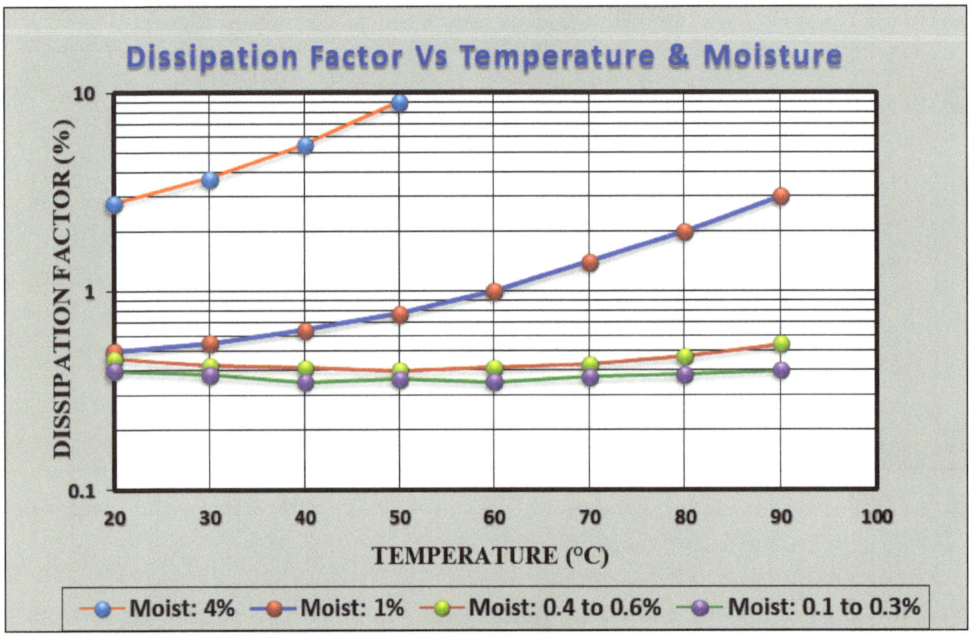

Figure 14: Tan δ as function of temperature and moisture in case of OIP bushings. [12]

It has already been mentioned earlier that tan δ of Condenser bushing varies with the temperature and voltage. These curves establish yet further conclusions which relate to the variation of dissipation factor with rising temperature and moisture contents of the OIP bushings. In the figure 14 following curves are drawn;

i. Curve No.1 is of moisture content of 4%.
ii. Curve No.2 is of moisture content of 1%
iii. Curve No.3 is of moisture content of 0.4 to 0.6%.
iv. Curve No.4 is of moisture content of 0.1 to 0.3%.

Interpretation:

a. It is observed from the curve No.3 & No.4 that the tanδ is almost constant from 20°C to 90°C.
b. In case of curve No.2 the tan δ is 0.7% at 50°C and with further rise in temperature; the tanδ has a rising trend from 0.7 to 3 % at 90°C.
c. In case of curve No.1 tanδ at 20°C is as high as 2.8 %, and abruptly rises to 9% at 50°C

With the above interpretation, if the tanδ is monitored continuously and the values of tanδ are such that at different temperatures they follow any of the mentioned curves, the moisture contents of the bushing can be easily established and faulty OIP bushing, if any can be replaced with a healthy one immediately.

17.0 Case study of failure of 245 kV bushing in one of the 220 kV substation at Narsinghpur in M.P. Power Transmission Co. Ltd. India.

Details:

- Transformer rating: 220/132 kV, 160 MVA, BHEL Transformer.
- Make of OIP Bushing: BHEL, year of Manufacturing 1993.
- Periodicity of off-line monitoring: 1 year.
- Period elapsed prior to failure after last monitoring of tan δ: 5 month
- *Tan δ & Cap Values last measured: tan δ – 0.52 %. Cap-393.9 p F.
- Load at the time of failure – 20 MW.
- Season/ weather condition: Rainy season, Date: 14th August 2013 / Cloudy.
- Relay Indications: Diff ABC(differential relay, ABC), Instantaneous. Differential, Instantaneous Over current, PRV(Pressure Relief Valve) and Buchholz Trip.
- The thermo-vision survey (in service and in loaded condition) was also carried out about 4 months prior to the failure. No hot points were observed in any of the bushings.
- Condition of the transformer: The Transformer was tested and found healthy, and then it was re-energized successfully after replacement of the failed bushing with a new one.

***Note**: The values of tan δ and capacitance taken from time to time with 5kV kit are tabulated in table No.3.

Sr.No.	Date of testing	Ambient temp.in °C	Capacitance in pF	tan δ in %
1	12.09.05	30	396.5	0.35
2	29.11.07	32	394.1	0.42
3	01.10.08	45	352.6	0.53
4	07.03.09	43	392.1	0.50
5	25.05.09	40	391.2	0.45
6	05.12.09	39	392.5	0.40
7	19.05.11	38	394.2	0.49
8	29.09.11	40	390.6	0.51
9	05.06.12	40	393.9	0.56

Table No.3: Tanδ delta and capacitance values as recorded.

18.0 Analysis;

On observing the above details and the photographs of the failed bushing. (Refer figure15 &16). it can be inferred as follows;

- The said bushing had failed between the successive periods of off- line monitoring.
- To assess the condition of the 160 MVA Transformer after failure of the 245kV bushing, it was tested at site and was found to be in healthy condition.

- Though the load recorded was only 20 MW but had met the load demand to the extent of 85-87 MW during irrigation season (irrigation season is – between November to February of following calendar year).
- Ingress of moisture must have taken place through;

    a. Cracked cementing which couples the porcelain portion of the bushing to the metallic dome, and/or
    b. through the deteriorated gasket of the oil level indicator.(Refer figure 15).

- This phenomenon is due to contraction of bushing core at low load periods and at cool night hours.
- (Location of ingress of moisture indicated encircled with yellow color. (Refer figure15).
- Non-bursting and non-shattering of the outer porcelain splinters of the OIP bushing at the air end side indicate that the failure had not taken place due to any violent conditions like heavy lightening impulse stroke / switching surges but had developed pressure inside it in a very slow pace, the bursting at the lower oil end side also confirms this.(refer figure 16).
- The development of excessive pressure inside the ill-fated bushing appears to have taken place due failure/shorting of some of capacitances caused by ingress of moisture and also increase in tan δ value and also occurrence of low intensity partial discharge, over the period of 5-months after the last off-line tan δ and capacitance measurement were made.
- Since, neither the Transformer nor any of other bushings or auxiliaries/components had any adverse effect; it is obvious that the fault was not of violent and of serious nature. It was confined to the failed bushing only.

Figure 15: Failed 245 kV, 1250 A OIP bushing (showing separated oil expansion metallic–chamber/dome from the porcelain insulator).

Figure 16: Failed 245 kV, 1250 A, OIP bushing.

18.0 Remedial Measures:

The practical remedial measure is to switch-over to "On line monitoring system of bushing-tan δ″ for high value transformers (the power utilities may weigh the advantages/gains vis' a vis cost investment) (Refer table 4 of Paragraph No.20.0.).

- It has been already been mentioned that the tan δ value of the insulation (in this case of bushings) increases with rise in temperature and moisture contents.
- The on-line monitoring system for tan δ, shall not only record the tan δ on real time basis on actual voltage, at different loading conditions, at different oil temperatures / winding temperatures but also it will identify the bushing getting red hot consequent to loose lead joints/connections, loose terminal-
- clamp connections, etc.
- Loose lead joints/connections and also loose terminal clamp connections etc. shall generate heat and get red hot which can be verified further with the Thermo-vision infra- red scanning.
- Further, it may also be mentioned here that the tan δ of the bushing is being monitored at the system voltage level and at actual temperature of the bushing (at dynamic real- time condition) unlike in off-load tan δ monitoring system wherein the voltage applied is typically 10kV only.

19.0 On Line monitoring of tan δ and capacitances of bushings-Theories / Methodologies;

- Sum of currents method.
- Absolute measurement - Voltage Transformer reference method.
- Dual transformer Comparison method.

The above methods are briefly described here under;

19.1 Sum of currents method: The voltages in a three-phase transformer are assumed to be almost equal in a grid system and having phase displacement of 120° between them sequentially. The bushings thereof are considered to have insulation level almost equal to each other. Since Insulation is of capacitive in nature, thus with the application of 3 phase balanced voltage, leakage capacitive currents shall flow between the respective phase & the ground through capacitances C1, leading the applied driving voltage by little less than 90° but maintaining phase displacement of 120° between them sequentially. The vector sum of all the 3 capacitive leakage currents shall be zero ideally, considering them equal in magnitude and angle of displacement as 120° between each other (basic

Electrical Engineering). However, in practice the sum of leakage capacitive current vector shall not be zero but shall be a very small value making a very small angle, say with the A-Ph voltage phasor It has been shown in red coloured.phasor. (Refer figure 16(a)}.

The deterioration of insulation means increase in the resistive component (i.e. contaminant/impurity) in the bushing, making the resistive current to flow in the parallel to capacitive current. (Refer paragraph No.4, figure 6). The resistive current component would be in phase with the driving voltage, the total resultant current through the capacitance of the bushing would be leading the driving voltage by ∠Ø less than 90^{0} (i.e. the resultant current would be lagging the original capacitive current vector by an ∠ δ. Thus tan δ is the dielectric dissipation factor of A – phase bushing now. Thus, the vector sum of capacitive currents of all the 3 phases shall deviate from the original value and will have small increased value (Red coloured increased vector).{Figure 16(B)}.

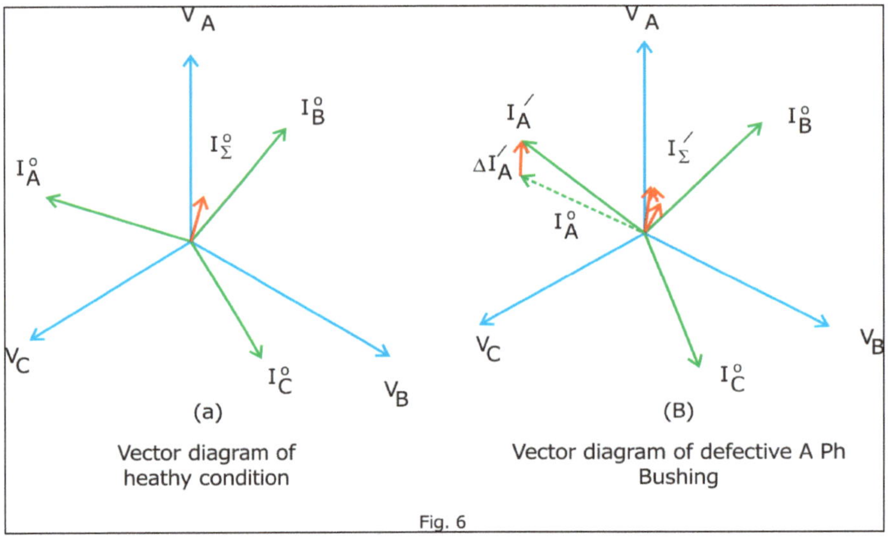

Figure 16: Phasor diagram showing drifted location of current vector with reference to the voltage vector of 'A" phase, indicating deterioration of insulation of A phase bushing. (Refer phasor diagram at figure16 (B).

The change in sum of the currents can approximately be estimated through the formula mentioned here under, with the assumption that only one bushing (A-Phase) has developed fault;

$$\Sigma I = \frac{\Delta I}{I_\circ} \approx \sqrt{(\Delta \tan\delta)^2 + (\Delta C / C_\circ)^2} \quad [15]$$

Where ΣI=parameter sum of currents.

Δtan δ= small change in tan δ.

ΔC/Co =Relative change in bushing capacitance.

Co = Initial Capacitance reading.

Io = Initial sum of Current value.

Using the above concept, on-line monitoring System has been developed by many of the manufacturers., wherein basic block diagram of bushing monitoring scheme is shown in Figure No.17

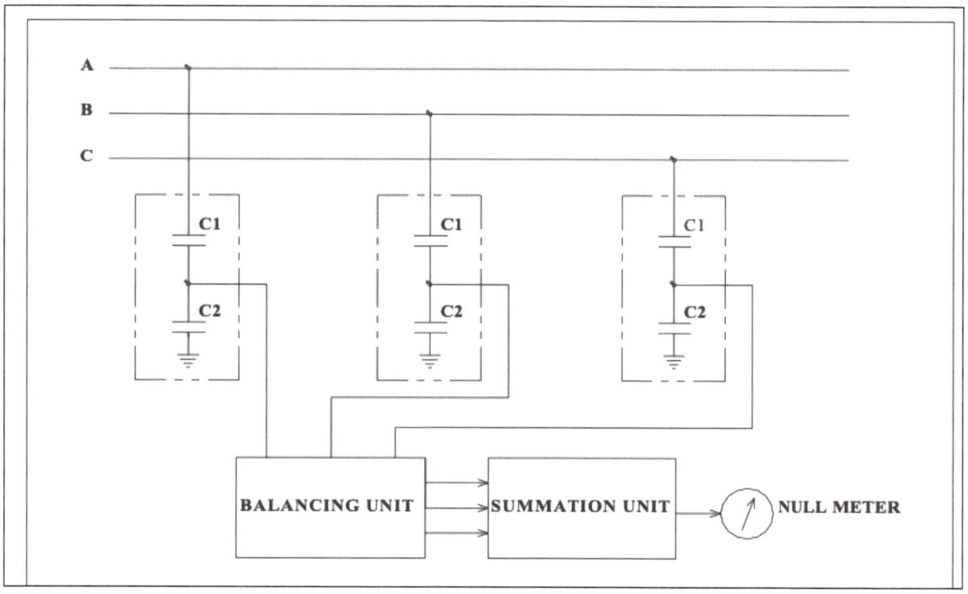

Figure 17: Schematic diagram of sum of current method.[15]

In this scheme, the sum of the bushing currents has been considered. In practice the capacitances of the bushings may not exactly be same. During commissioning the null meter is balanced to zero.

The purpose of balancing is to take into account the system voltage, phase fluctuations and bushing characteristics and thus fix an initial reference mark. With development of fault, the current and the phase angle change and the null meter shall no longer show null position.

With the change in amplitude and phase angle, the bushing losing its dielectric property can be identified.

19.1(a) Brief description of the features of the on line tan delta monitoring equipment on the principle of "Sum of currents method."[16]

The on-line tan δ monitoring devices are presently manufactured by the reputed/leading manufactures. They are being supplied to OEMs (Original Equipment Manufacturers) of the transformers for incorporating the same as per the requirement of the customers.

- **Monitoring of Bushings/CTs at a time:** Can monitor 2x3-Bushings/CTs
- **Monitoring of 1-ph units**: Monitoring of 4x1-phase transformers exists with 3 Nos. in service and 1 no. as spare. The spare unit gets grounded thru its sensor automatically in a kiosk provided.
- **Defective Bushing**: Specific software enables and depicts the imbalance current vector position on polar plot which in turns affects the Power factor and capacitances of the defective bushing.
- **Effect of Temperature/Voltage:** Can identify changes in tan δ and Capacitance with respect to temperature & voltage and identify deteriorating condition of such bushings.
- **False alerts:** Special algorithms eliminate false alert due to noise and other atmospheric conditions etc.
- **Communication protocol:** Can be locally accessible through RS 232 and remotely for SCADA etc. through Ethernet /RS 485 etc.

- **Data storage period:** 5 years on hourly basis.
- **Self-diagnostic feature.**
- **Change in PF and Capacitances:** Change in A phase (yellow) depicted on polar diagrams. (Figure 18).)

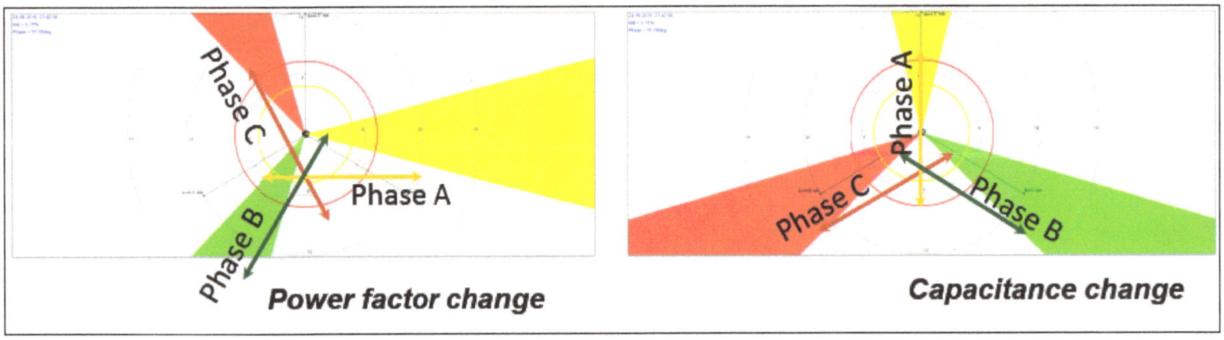

Figure 18: Phasor diagrams showing change in PF and Capacitance [16].

19.2 Absolute measurement-Voltage Transformers reference method. [17] [18]:

To overcome the variation in the tan δ values due to imbalance of system voltages, Absolute measurement (Voltage transformer reference) has been innovated. Imbalance voltages in the system occur due to unequal loading in the phases, System faults or by asymmetric induction from high currents in nearby over head lines. It has been observed that oil impregnated bushing having original tan δ value of 0.25% had increased to 0.5%.This could be misleading and alarming too.

For this Absolute measurement method, absolute values are taken from the Voltage Transformers of respective phases.The schematic diagram of connections is shown in the figure 19.

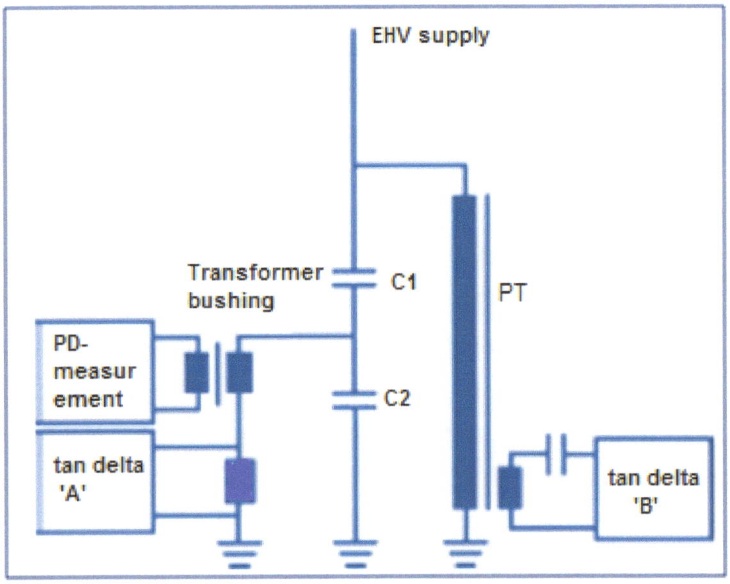

Figure 19: Schematic diagram of Absolute measurement-Voltage Transformer reference method.[17]

The equipment/setup for such a system is depicted in figure 20.[17],[18]

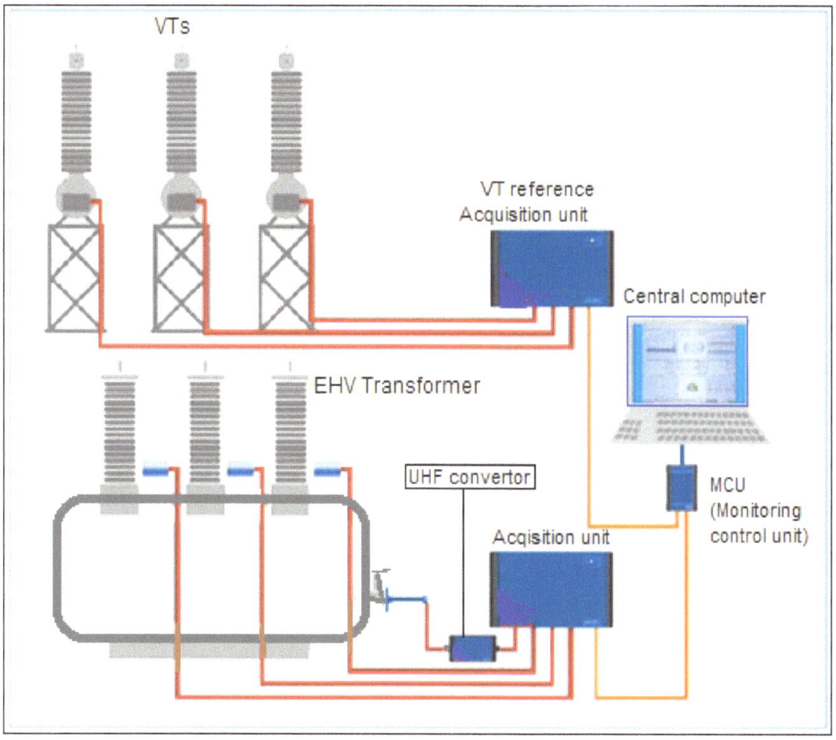

Figure 20: Absolute measurement-Voltage Transformer reference method.[17],[18]

Salient features and duties of main equipment of the scheme;[18]

a. **UHF sensor;**
   - Highly sensitive PD measurements in the winding insulation.
   - Wide frequency response.
   - Results can be correlated to the PD signals detected at the bushing.

b. **Acquisition unit / Transformer**:
   - Synchronous acquisition of data from bushing taps adopters and UHF sensors.
   - Advanced signal processing system for capacitance, dissipation/power factor, transient over voltages and Partial discharge etc,

c. **Acquisition unit / VT reference:**
   - Synchronously acquires reference signals from the VTs for absolute capacitance, dissipation/power factor measurement.
   - It can be used with three voltage transformers reference scheme or three power transformer bushings reference scheme of similar power transformer on the same bus.

d. **MCU Fiber optic bus controller:** It is used for connecting each acquisition unit to the central computer. Maintains galvanic isolation eliminating interference and ensures personal safety too.

e. **Central computer with compatible user friendly monitoring software:** It has following features.

   ➢ State of art data base system ensures long term data storage and retrieval.

- ➢ Processing of the data is performed intelligently to provide useful information of status/ condition of the bushings.
- ➢ Parameters are monitored to give signal, alarm and warning of limiting levels /defined thresholds.
- ➢ Accessible via web browser interface.

f. Specimen graphs/curves of Dissipation factor and Capacitance of the system;

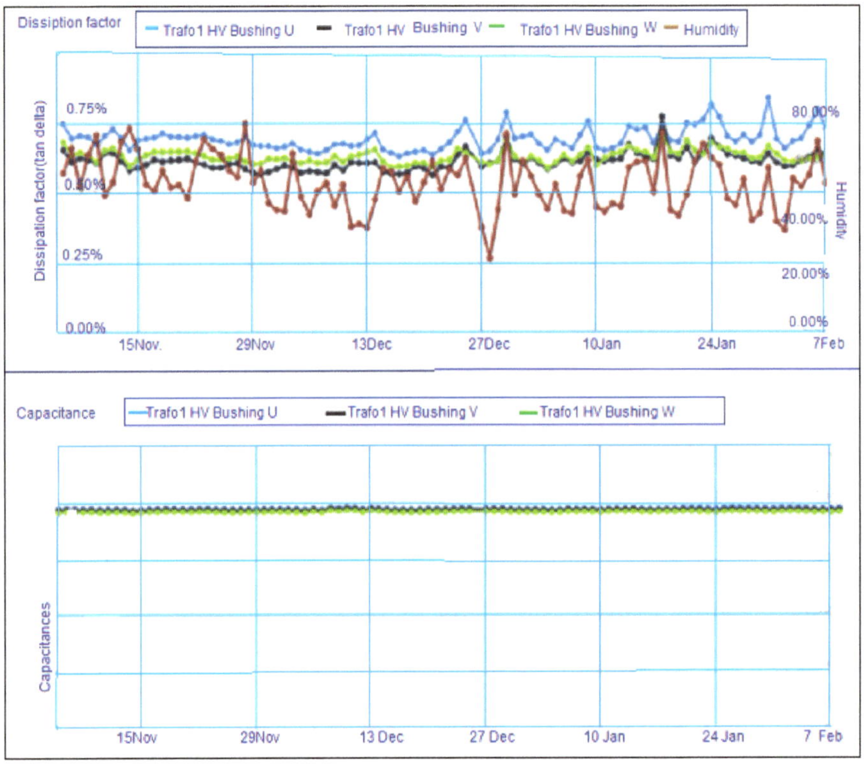

Figure 21: Specimen curves of Tan delta and Capacitances of all the 3 phase bushings.[18]

19.3 Dual Transformer Comparison Method[18]: The theory/concept of comparison is exactly similar to Absolute measurement (VT reference) method (refer paragraph19.2).Instead of three numbers of Voltage Transformers, the respective phase bushings of a Transformer having similar rating connected in parallel are used for reference /comparison. The test set up is similar to Absolute measurement (VT reference) method.

NOTE: Cost-benefits analysis mentioned in the original paper is scratched, it may be treated as cancelled and in place thereof following cost benefit analysis is replaced which is narrated below; **Revised Sample cost- benefits analysis incorporating values in US Dollars is mentioned below.**

20.0 A sample cost- benefits analysis of an "On-line tan δ monitoring System for the bushings";

This analysis is highlighted to show the benefits of installing an "On-line tanδ monitoring system for the bushings" of the Power Transformers. This system shall monitor the condition of the condenser bushings while the transformer is in live condition. It is a sample case of failure of a 245 kV, oil impregnated paper bushing of a 220/132kV, 160 MVA Transformer installed at a Sub-station 150 Km away.

Assumptions,

a. Season: Peak irrigation season for the agricultural fields. (Nov 2015).
b. Date of Failure: 15.11.2015.
c. Load: 164 MW fed by 2 x160 MVA, 220/132kV transformers having equal % impedances and tap position variations.
d. Load on the affected Transformer prior to failure: 82 MW.
e. Location of Site: The 220KV Substation is 150 Kms. away from the head quarters of the Executive Engineer, Testing Division & the utility's material & equipment stores.
f. Extent of Failure: Failure of 1 No.245kV 1250Amp, OIP bushing.
g. Value of parameters found during the last off-line testing of the bushings prior to its failure: tan δ: 0.0053 & Capacitance: 394 pF.
h. Period of failure elapsed after last off-line tan δ measurement: 5 months after its last tan δ measurement with a portable 10kV - kit.
i. Periodicity of testing of tan δ and capacitance of bushings: 6 monthly. (As per practice in M.P. Power Transmission Co. Ltd., India).
j. Duration of down time after failure of the bushing: 72 Hrs.(Assumed).

Required activities & Costing;

k. Availability of heavy duty crane at site: Not available.
l. Personnel required:
   a. For removal of the failed bushing and testing the Transformer.
   b. At the time of installation testing and energizing of the transformer.

Note: Since crane was not available at site, the above works were carried out manually.

Sr No.	Pariculars	Rate per engineer /man-days in USD	Engineer/ man- days	Amount in USD(USD Exchange rate as on 15.11.2015:1USD = 66.1918 INR)
i	Maintenance Engineers + Testing Engrs.	4500.00 INR= 67.9842	3 x 3 = 9	611.8578
ii	Maintenance Technicians/Testing Personnel(Skilled)	2500.00 INR= 37.7690	5x 3 = 15	566.5350
iii	Un-skilled personnel	500.00 INR= 7.5538	8 x 3 = 24	181.2912
			Total	1359.684

Table No.4: Expenditure towards Engineers, Skilled and unskilled personnel involved in carrying out works at the SubStation.

m. Testing of the Transformer, HV & LV bushings after the incidence (As found). and the new 245kV bushing & Pre – commissioning Tests and energization of the Transformer.; charges.= 1,50,000.00 INR = 266.1417 USD
n. To & fro journey charges by Testing Engineers & Testing Staff from Headquarters – 2 times to the site (220kV S/S) 150 km away from their HQ @ Rs.15.00 INR per Km = 0.2266 USD per Km (average) by 2 Nos. Light motor vehicles.
o. Charges towards transportation of Testing Staff:= 2 x 2 x2 x150 x0.2266 = 271.92 USD
p. Cost of the 245kV, 1250 Amp, OIP bushing: 6,00,000 INR = 9064.5669 USD.
q. Cost towards inventory & other miscellaneous charges towards storing the bushing @ 25% of 9064.5669 USD(the cost of the bushing): 2266.1417 USD
r. Expenditure towards transportation of Bushing:

- Through road transport(truck) –freight; 18,000 INR = 271.9370 USD
- Transit Insurance @ 5% of the cost of the bushing(5% of 9064.5669 USD); 453.2283 USD
- Loading/unloading charges: 8000 INR = 120.8608 USD

Total: 846.0261 USD

s. Loss of revenue due to down time of the transformer.

(Note: As per tariff order 2015-16 of M.P. Electricity Regulatory Commission, India)

Calculation of Loss of revenue; Loss of revenue assuming outage of 50.00 MW during the down-time of 72 hrs. Price of unit at 132kV level = 5.29 INR/unit and assuming Load factor of 85%, 50,000 x 0.85 x72 x 5.29 INR = 16187400.00 INR = 244552.951 USD.

Note:

- Since there was one more similar Transformer running in parallel, thus outage of 50 MW is considered for calculation of revenue loss.
- The other portion of load has been shared by nearby 132kV substations.
- Load factor of 0.85 has been taken for calculation.
- A down time of 72 hrs (assumed).
- Average cost of unit = 5.29 INR/unit

t. The cost/expenditure incurred in replacing the failed 245kV OIP bushing with a new one is shown in table No.4 (Above realistic assumptions have been considered while preparing this statement).

SN.	Particulars	Investment in USD.	Investment in USD.	Expenditure incurred/ Cost to the Utility in USD (Exchange rate as on 15.11.2015: 1 USD=66.1918 INR).
1	Cost of On-line tan δ monitoring equipment based on sum of currents method at site after loading, all taxes, duties, export–import formalities, shipment and transit insurance etc including erection & commissioning at site(approximately).	30215.223		
2	Cost of On-line tan δ monitoring equipment and accessories based on Absolute measurement (Voltage Transformer reference) method at site after loading, all taxes, duties, export –import formalities, shipment and transit insurance etc including erection & commissioning charges at site(approximately).		77909.086	
3.	Cost of OIP bushing.(refer: p).			9064.5669.
4.	Cost towards inventory & other Misc towards storing the bushing etc. @ 25%.(refer: q)			2266.1417.
5	Transportation of the bushing to the site.(refer: r).			846.0261
6	Charges towards testing of the Transformer, HV & LV bushings as found, after the incidence and the new 245kV bushing & pre – commissioning tests prior to re-energization of the Transformer.			2266.1417
7	Expenditure towards to & fro journeys to site by the Testing & Commissioning Engineers and Testing Team.(refer: n)			271.92

8	Charges towards works at site by Engineers, testing Technicians & the maintenance staff.(refer: 1)			1359.684
9	Estimated loss of revenue due to down time.(refer: s)			244552.951
	Total(in USD)	30215.223.	77909.086.	250429.431.

Table No.5: Cost benefits analysis.

It is observed from the above cost/expenditure statement that the cost of replacement of 245kV OIP bushing- 01 No. including loss of revenue to the utility consequent to this failure is;

a. more than (250429.431/30215.223).= 8.28 times, the approximate cost of the on-line tan δ monitoring system based on sum of the currents method including its commissioning at site.
b. more than (250429.431/77909.086)=3.21 times, the approximate cost of the on-line tan δ monitoring system based on Absolute measurement(Voltage Transformer reference) Method, including its commissioning at site.

The above example has been cited as a sample case of failure of 245kV OIP bushing - 01No.only.Had there been catastrophic failure of the Transformer, the Power utility shall incur enormous / colossal loss and the down-time would be approximately 4- 6 weeks.

The cost of a new 220/132/33kV, 160 MVA Transformer is approximately 5.13 crore INR= 775020.471 USD (Exchange rate of 66.1918 INR to 1USD as on 15.11.2015).

21.0 **Dos ad Do nots of the OIP bushings;**

21.1 **Dos;**

i. Check and verify if the bushing has been transported in a in crate such a way that the oil end is at the lower level and the bushing is inclined making an angle of 6-8° with the horizon.
ii. Check and verify if there is any damage to the crate externally before unpacking the bushing.
iii. Unpack the bushing with care so as to avoid any kind of damage to the bushing or the porcelain of the bushing.
iv. Verify the name plate details.
v. Handle the bushings with non- metallic nylon ropes while removing from the crate, hoisting and fixing/ assembling on to the transformer.
vi. While storing the bushing in a crate, care should be taken such that the bushing is inclined making an angle of 6-8° with the horizontal plane with its oil end at the bottom.
vii. The bushing should be covered with tarpaulin and stored in a shed to safe guard it against rains and atmospheric moisture.
viii. Clean the bushing insulator thoroughly before testing it for IR value, tanδ & capacitance.

ix. Check for Oil level and oil leakages in a bushing.
x. If spare bushing is to be tested, it should be fitted in up-right position in an oil chamber filled with Transformer oil and then tested.
xi. Before installation/ erection of the bushing, it should be kept at up- right position for more than 24 hrs.
xii. Verify that the Test tap cover is electrically connected to the Transformer body (at the same potential).
xiii. Verify that the Test tap cover is properly fixed on to the Test tap.
xiv. Proper safety measures to be taken so as to safe guard the life of men and material/ instruments from high voltage induction.
xv. Proper record of the bushings should be maintained for routine maintenance such as IR value, tanδ & capacitance measurements and Thermo-vision survey reports.
xvi. Care should be taken while testing for tanδ & capacitance of C2 that is Test Tap to flange, the test voltage range should be 500 volts to 2 kV only.
xvii. Immediately replace the bushing whose tanδ & capacitance values are deteriorating from the standard values.

21.2 Do nots;

i. Do not accept the bushings without Test certificates of the tests witnessed by the authorized representative of the customer.
ii. Do not unpack the bushing from the crate unless required for erection/ assembling.
iii. Do not, at any case or circumstances use any metallic rope for lifting, hoisting and mounting on the transformer.
iv. Do not test the bushing if the weather is very moist and raining.
v. Care should be taken not to test the tanδ & capacitance of C2 that is Test Tap to flange with any voltage beyond 2kV
vi. Do not try to dismantle any bushing at site without approval of the competent authorities/ manufacturers.
vii. Do not climb on the bushings for cleaning, making jumpers/connecting clamps etc.

22.0 Conclusions;

Right from the basic concept of Capacitive graded bushings to the condition monitoring thereof have been elaborated in this article, with a special emphasis on the on-line monitoring system for the condenser bushings, designating them as a superior methodologies for monitoring of condenser bushings at dynamic / live conditions on real time basis as compared to the off-line monitoring system.

Such on-line monitoring system which is cost effective but, shall give a caution signal well before the utility may suffer catastrophic failure of the transformer. (Refer paragraph 20,0 table 5 for cost -benefits analysis).

To sum up, in view of above and considering the sample cost benefit analysis, the Central as well as State Transmission /Generation Utilities and also the Private Entrepreneurs may resort to online tan δ monitoring system for condenser bushings for their valuable assets like the 220 kV, 400 kV and 765 kV Transformers and the Reactors.

24.0 Bibliography/References:

[1] IEC - 60137 edition 6.0 2008-07:International Electro- technical Commission Standard for Insulated bushings for Alternating Voltages above 1000 V.

[2] P. Ramachandran, ABB's article on "Evolution of Transformer Technology in India" in IEEMA Journal, Aug,, 2005.

[3] S.Tenbohlen, J Jagers & others: Presented a paper titled as"Development and results of a world wide Transformer reliability survey" at CIGRE A2 COLLOQUIUM 2015, in China between 20-25 Sept,2015 on behalf of CIGRE W G A 2.37.

[4-A]Marcos E.G Alves, TreetechSistem as DigitaisLtda. and Jorge C. da Silva, Cemig S.A. on" Online Capacitance & Tangent Delta Monitoring for Condensive Bushings" in CIGRE Conf. Brasil-20-24 Dec 2007.

[4-B]Lecture notes of NPTEL- http://nptel.ac.in/courses/108104048/lecture3/slide3.htm

[5] E. W. Golding's Text book "Electrical Measurements and measuring instruments" 4th edition-1955(reprinted in 1962), Sub-title "Potential Gradient" in Chapter IV "Capacitors, Capacitance and Dielectrics".

[6] HAEFLEY TEST AG's Midas 228x -Operating instructions.

[7] THVS OIP Condenser Bushing for High Voltage Transformers, Type-OTA.72.5 kV to1200 kV up to 5000 A.

[8] CBIP's (Central Board of Irrigation and Power – an autonomous body under the Govt. of India) "Manual on Transformers, Publication No.317, April 2013".

[9] IS: 2099 - 1986. (Re affirmed 2003). Indian Standard Specification for Bushings for Alternating Voltage above 1000 volts.

(Oil to Air)(24kV…….420kV).

[10] ABB's publication No; 2750 515-142 en, Rev.3. on Bushing diagnostics and conditioning.

[11] BHEL's Presentation on Bushings-6645329.

[12] IEEE C57.19.01-2000(R2005): IEEE Standard Performance Characteristics & Dimensions for outdoor apparatus Bushings.

[13] Claude Kane & Alexander Golubev, Dynamic Ratings, USA paper on " Bushing monitoring Methods Case studied." In PdMSA conference. Sept 2010.

[14] Dynamic rating: Online bushing monitoring and comparison of off-line testing.

[15] Product brochure of Omricon.

[16] Krueger M., Ranninger U., Badicu L.V.: Monitoring, diagnosis and fault findings on a power transformer, Transformers Magazine, Volume2, Issue 3, 2015.

25.0: Authors:K.K.Murty & Dr.Monalisa Biswal,

16. A Unique Non-Invasive (non-destructive) Test; SFRA Used for Verifying Mechanical and Geometrical Integrity of Transformers.

1.0 Preamble; SFRA is the acronym of Sweep Frequency Response Analysis. This particular test was invented by the Chinese Engineers/ Scientists in the year 2004 and a Standard was issued to this effect by the People's Republic of China as DL/T 911-2004.

The Transformers, inherently have resistance, inductance and capacitances, therefore, they constitute a unique RLC circuit of each one of them. Since, the Inductance and the Capacitance are frequency dependent parameter, therefore the total impedance of the transformer shall vary if it is subjected to an AC supply with variable frequencies. SFRA has emerged as a powerful and sensitive tool for evaluating the mechanical and geometrical integrity of the core, windings and clamping structure of the Power transformers by measuring their electrical magnitude in dBs over the frequency range from 20 Hz to 2 M Hz. The health-status can be assessed without opening or dismantling the transformer. Thus, this test can be designated as "non-invasive test (non-destructive test)". This test is invariably performed at the OEM's works and also at site to verify its health-status of the Transformers. Further, this test is also performed on the transformer after major-trippings and severe seismic conditions for verifying the health-status of the Transformers. The maiden frequency response curves are called as the signature curves or thumb-impression curves for each individual Transformer.

The Authors have narrated the basic theory of the SFRA test and have narrated" Where to look-into, over the frequency range for each type of suspected faults? "

2.0 Key words; SFRA, RLC Circuit. Non- invasive test.

3.0 Theory/ Concept; The transformer has RLC circuit which is shown in the figure 1.

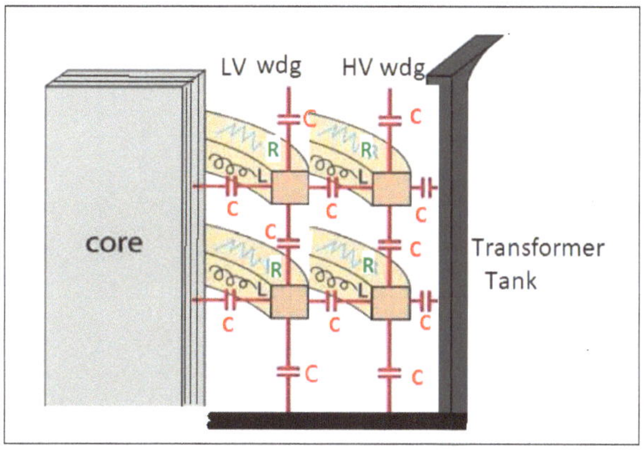

Figure1: Parameters RLC shown.

As the Power Transformer has winding which is inherently a resistance of the conductor, since it is wound in a circular fashion, the inductance is developed due to the AC system. The construction of Power Transformer causes development of stray capacitances between turn-to-turn, winding to winding, LV winding to core and windings to tank. Therefore, it is a complex piece of equipment having two frequency-based components viz Inductance ($X_L = 2\pi fL$) and capacitance ($\mathbf{X_C} = \frac{1}{2\pi fC}$) in addition to the resistance. So, the whole equipment has RLC circuit of its own.

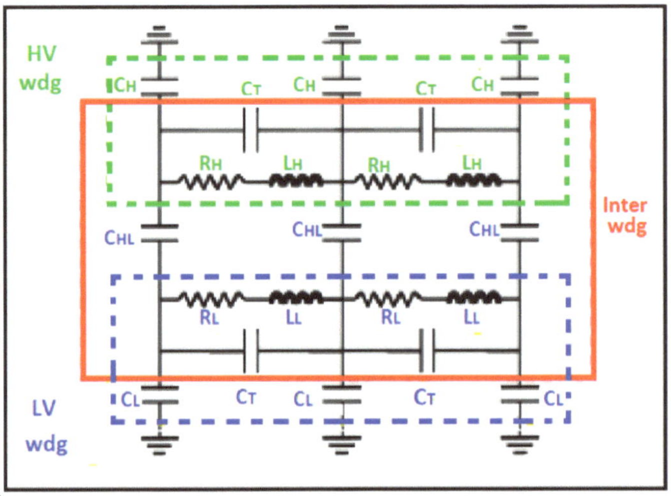

Figure 2. RLC Circuit of a healthy Transformer.[5]

If jerks and jolts are experienced by the Transformer, post - transportation on a low bed trailer would dislocate the core of the Transformer, thus the capacitances would vary considerably as shown in the figure 3 with pink colored capacitances resulting into change in the Impedance of the transformer.

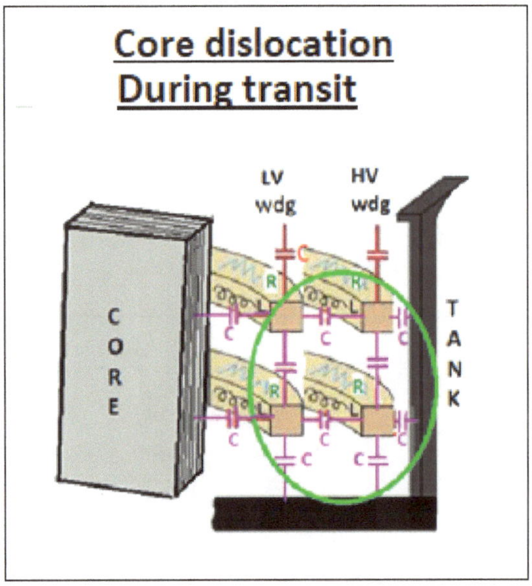

Figure.3: Dislocated core resulting into change in capacitances shown in pink color.

Note: Post transportation condition due to heavy jerks and jolts depicted in the figure 3. The values of the capacitances changed as shown in pink colour as compared to the values shown in the figure

1 in red, resulting into change in the frequency response curves as compared to the SFRA signature curves taken prior to commencement of the transportation of the Transformer.

3.1 Brief description of process for plotting SFRA traces;

Test set up;

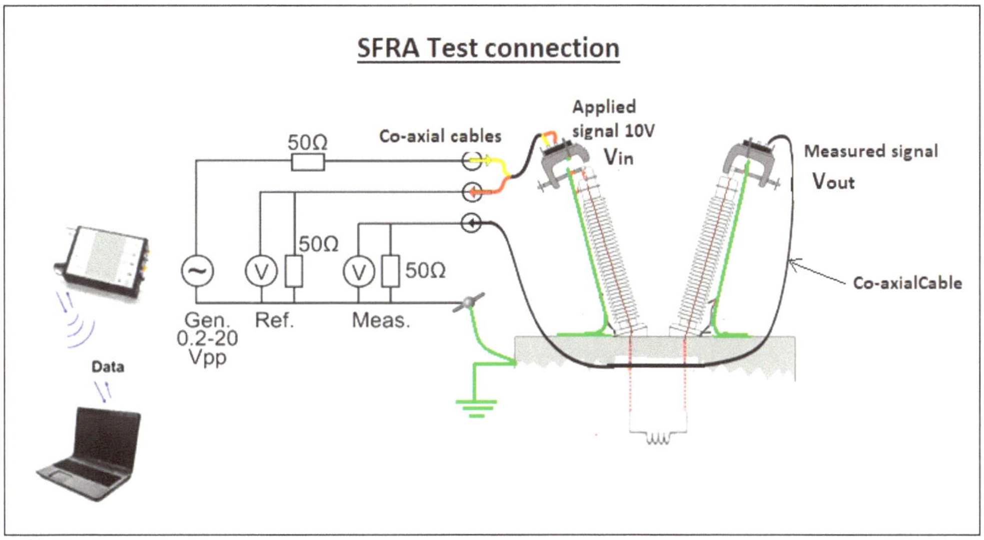

Figure.4: SFRA kit connections for plotting the frequency response of a winding keeping other terminals floating.[5]

Note:

i. Cables from kit to the bushings are Shielded co-axial cables of same length and they are to be perfectly clamped with the C-clamps.
ii. Bushing clamps to the transformer tank are copper braided strips (green in colour) firmly connected onto the Transformer tank. The transformer tank is in turn properly grounded.
iii. The SFRA Curve can be up-loaded in the lap top for any further references.

The Magnitude(dB)=20 x $\log_{10}\left(\frac{\text{Voutput}}{\text{Vinput}}\right)$ [4]

The SFRA is typically performed in a frequency range from 20 Hz to 2 MHz at an output voltage of 10 volts from the kit. The SFRA test is the most reliable non-invasive technique. The SFRA test set-up is shown in the figure.4.

4.0 When to subject the transformer with SFRA test? [2]

Each transformer has distinct response to frequency depending upon its RLC circuit. The Sweep frequency response trace of each transformer would be unique by itself and therefore, the maiden Sweep Frequency Response Curves are called as signature-curves or thumb- impression curves. There are several occasions/events or incidents after which generation of as found SFRA traces or curves are essential for evaluating the health-status of a transformer by comparing them with the signature-curves. The list of occasions/events or incidents are enlisted here-under;

a. For verification that no damage has occurred during Short Circuit test at OEM's works.
b. After receipt of the consignment at site for checking the mechanical and geometrical integrity of the transformer. Please refer figure.3 wherein the transformer has sustained dislocation of core due to heavy jerks and jolts, causing variation in original capacitances resulting into variation in SFRA curve as compared to its signature curve,

Figure 5: consignment of the transformer arrived at site.

Figure 6: Test bushings for Testing the Transformer on receipt.

Note: Test bushings are provided of 12 kV insulation level to facilitate SFRA and other relevant tests on receipt of the transformer at site. It the SFRA test is taken just prior to dispatch of the Transformer from OEM's works then on receipt of the transformer at site becomes essential to test SFRA test using these test bushings.

c. Relocation from one site to other site and also relocation at the same site due to say augmentation of capacity or shifting to another bay etc.
d. Post incident verification: Lightening, external through-fault, internal – short circuits, failure of vital component like "the bushings" and severe seismic conditions.
e. After Transformer Trip/Alarm;

- ➢ Differential or REF protection trip.
- ➢ PRV trip
- ➢ Buchholz alarm/trip.
- ➢ DGA report indicating thermal faults above 300°C indicating T-2 or T-3 conditions as depicted in the IEC 60599.-2007.

f. Condition assessment of old transformers.

5.0 Region-wise frequency response; [4]

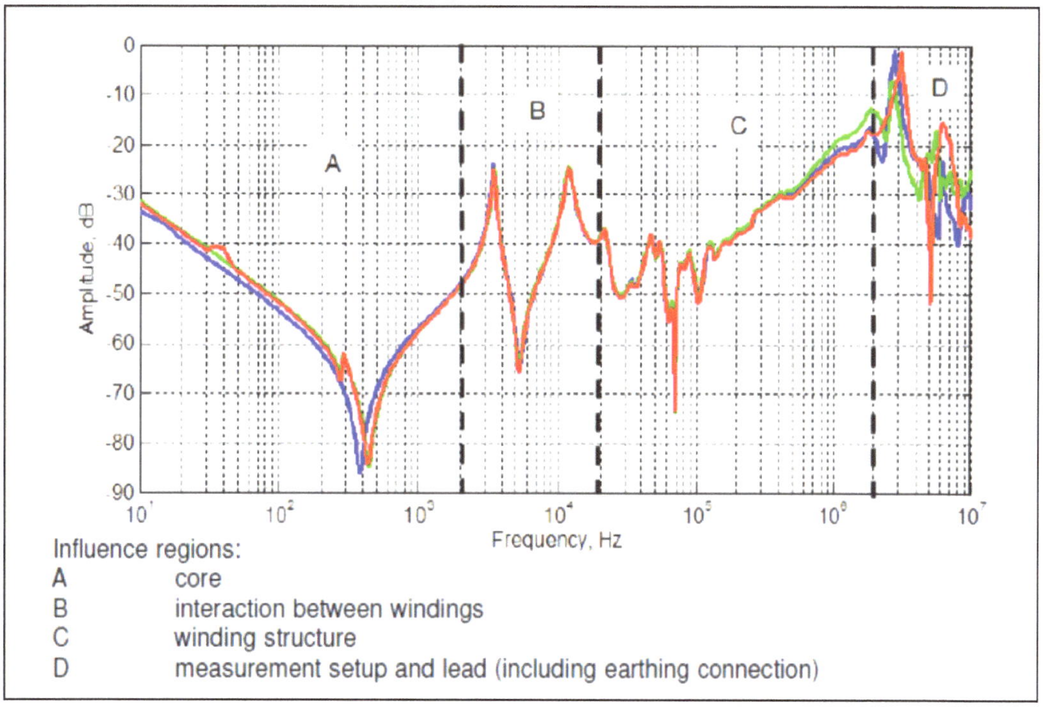

Figure.7 frequency response region-wise.[4]

5.1 Region A - Core Effect; [2]

At low frequency (below 2 kHz), the frequency response begins with the decreasing magnitude of -20dB/decade. This is due to the magnetizing inductance of the core. This is then followed by a minimum which occurs due to the series resonance between the magnetizing inductance of the core and the total capacitance of the transformer. In this example, the B phase (in blue color) is clearly different from the other two phases due to the different magnetic reluctance path of the middle leg of a three-limb core.

5.2 Region B - Interaction Between Windings;

Between 2 kHz –20 kHz, the frequency response is influenced by the interaction between the windings as well as by how the windings and neutrals are connected and terminated (open/closed delta; floating/ grounded).

5.3 Region C - Effect of Winding Structure of the Winding under Test;

From 20 kHz up to 1 MHz, the winding structure will heavily influence the frequency response. In this example, the FRA plot for this frequency range has **less resonances** and anti-resonance and a mainly capacitive raising trend, due to the high series capacitance of the HV winding.

5.4 Region D - Effect of Leads of Taps and Earthing leads

Beyond the frequency of 1MHz, the trend of the frequency response is irregular and complex, influenced by the tap leads and the measurement earthing leads.

5.4 Summary in a tabular form;

Frequency region	Frequency	Influencing factors
A-Low	< 2 kHz	Core
B- intermediate	2-20 kHz	Inter action between windings.
C-High	20 kHz	Individual winding structure
D- Very high	>1MHz(for ratings >72.5 kV). >2MHz(for rating ≤ 72.5 kV	Measurement setup and lead (including earthing connection)

Table1: Affected components/area in diferent Frequency regions.[7]

5.5.1 Note: The field engineers often raise a question that they have noticed that in a Sweep Frequency Response plot of any transformer, the curve droops to the lowest possible level then rises up. As could be seen from the figure 8, the curve droops from -38dB to -97dB. As the transformer has Inductance and the capacitance which have opposite characteristics to each other ie Inductance reactance XL= $2\pi f L$ and capacitance reactance $XC = \frac{1}{2\pi C}$, ie XL increases with increase of frequency whereas XC reduces with increase of frequency. Therefore, as the frequency increases XC reduces and XL increases and then at the lowest point of say 'V' like curve, it is the resonance frequency at which XL= Xc.

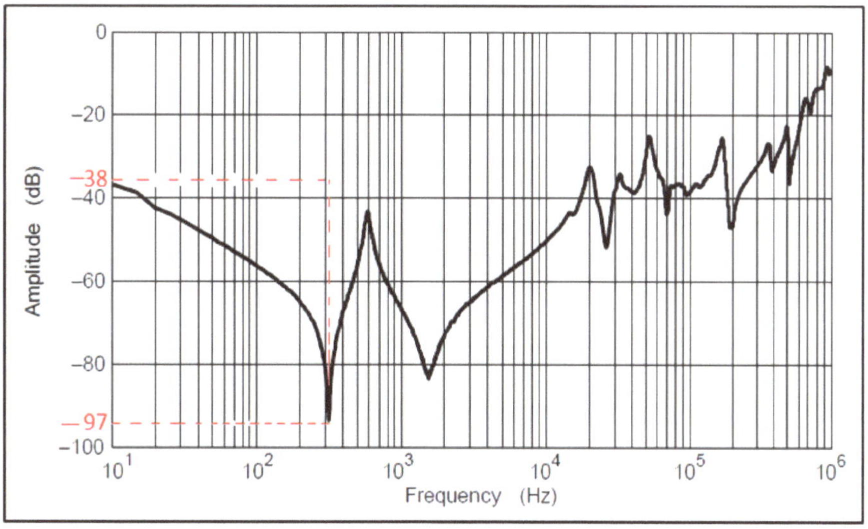

Figure 8:Graph showing resonanse frequency in Sweep Frequrncy response curve.

5.2 Concept of resonance frequency;

XL= Xc.

$$2\pi f\, L = \frac{1}{2\pi fC}$$

frso $= \frac{1}{2\pi\sqrt{LC}}$, where frso is the resonance frequency.(figure 9)

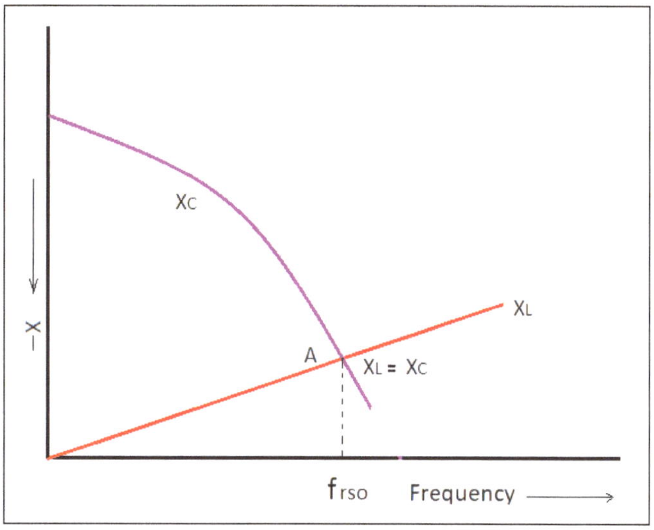

Figure 9: Concept of Resonance frequency.

6.0 Reference Conditions for Status Evaluation;[5]

6.1 Time-based comparison;

Such comparison is essential to assess the status of the transformer over a period of time as the transformer is subjected to many stresses and strains during its operation such as mal-operations by the operating staff, over-loads, over-voltages/under frequency conditions causing over-fluxing, trippings due to external faults and also during seismic disturbance conditions etc.(figure10)

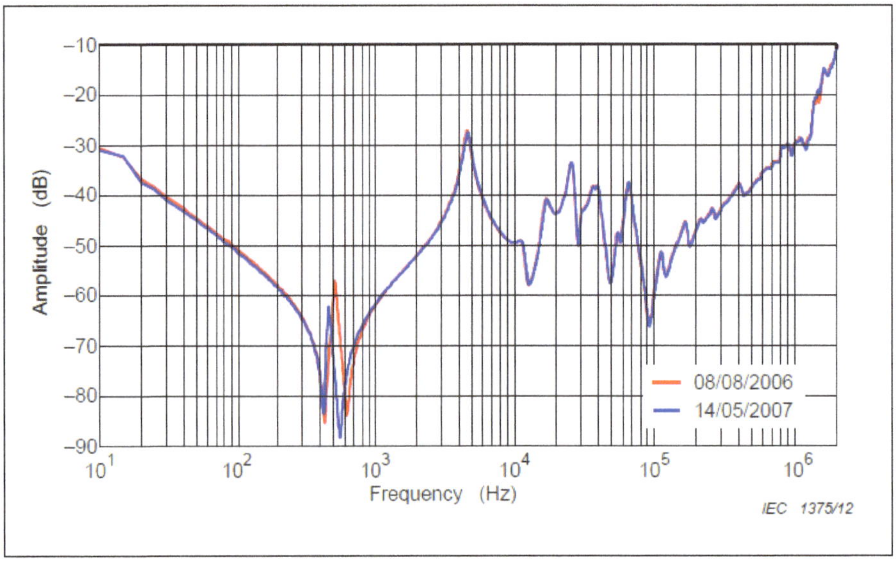

Figure 10: Time based comparison.[4]

6.2 Type or design-based comparison;

Such comparison is possible if a similar designed transformer manufactured in the same lot, with same series of serial numbers may be called as a sister transformer, if available at the site. However, some variation may be observed in the traces in the core region due to say minor variation in capacitances between the core and the LV winding attributed to variation in thickness of insulation and axial cooling ducts but the trend will be similar as depicted in Figure.11.

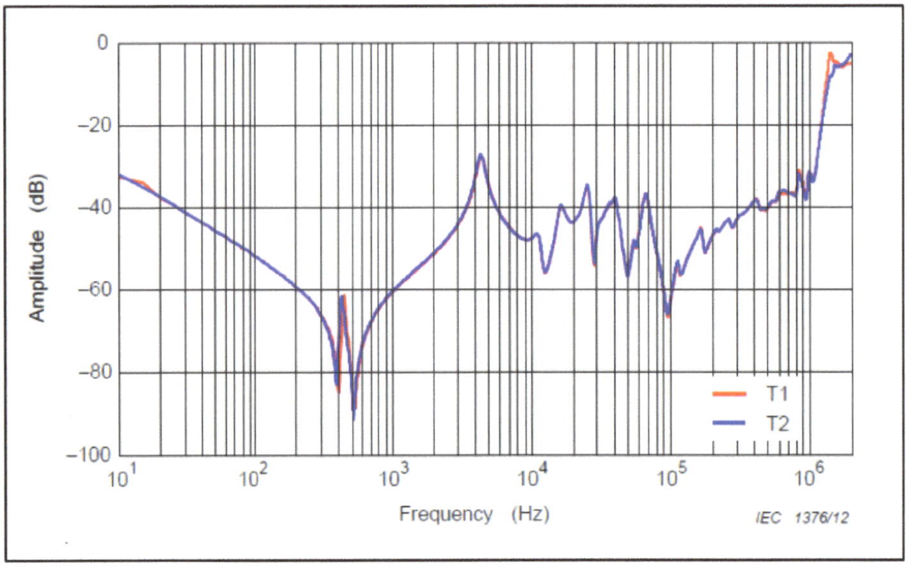

Figure.11: Design based comparison with sister transformer.[4]

6.3 Phase-based comparison;

When no other transformer is available at site then the only option is Phase based comparison. Since, the magnetizing current of the middle limb is less than the outer limbs as magnetic reluctance path is less than the outer limbs. Therefore, the magnitude thereof is less than the outer limbs. The frequency response curve of outer limbs may overlap each other. If some deviation is found in SFRA trace, it will confirm that the Transformer has developed some defect. Please refer figure12.

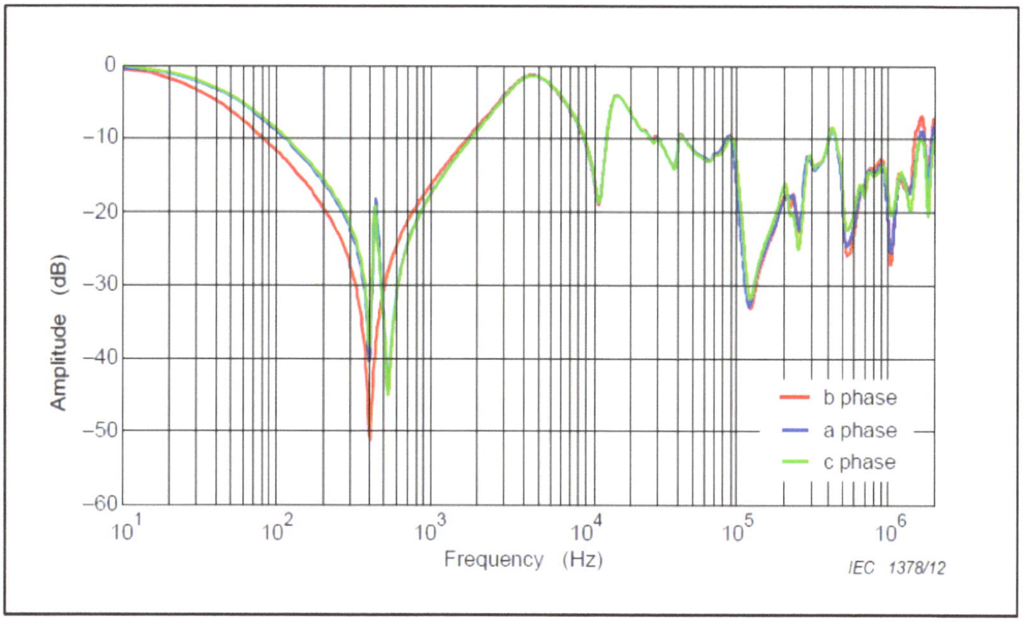

Figure.12: Phase comparison method.[4]

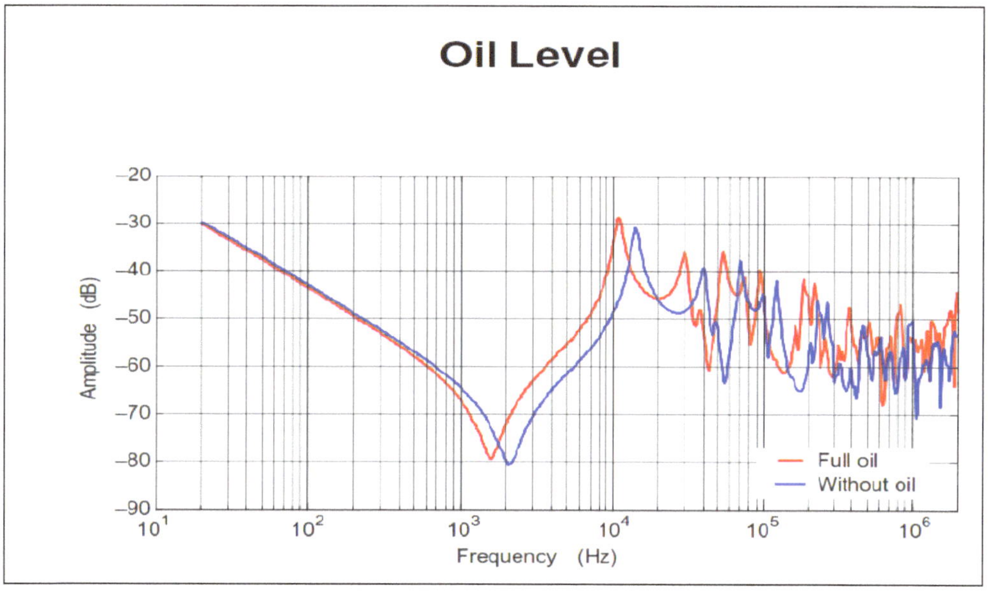

Figure.13; Deviation in curve of frequency response between full Oil and with-out oil. [4]

7.2 Comparison between before DC or after DC test;

Influence of DC prior to SFRA test.; Since winding resistance test is a DC test, thereof in such conditions the core gets magnetized and carries residual flux. The deviation in frequency response in core area within 2 kHz takes place, which has been depicted in the figure14.

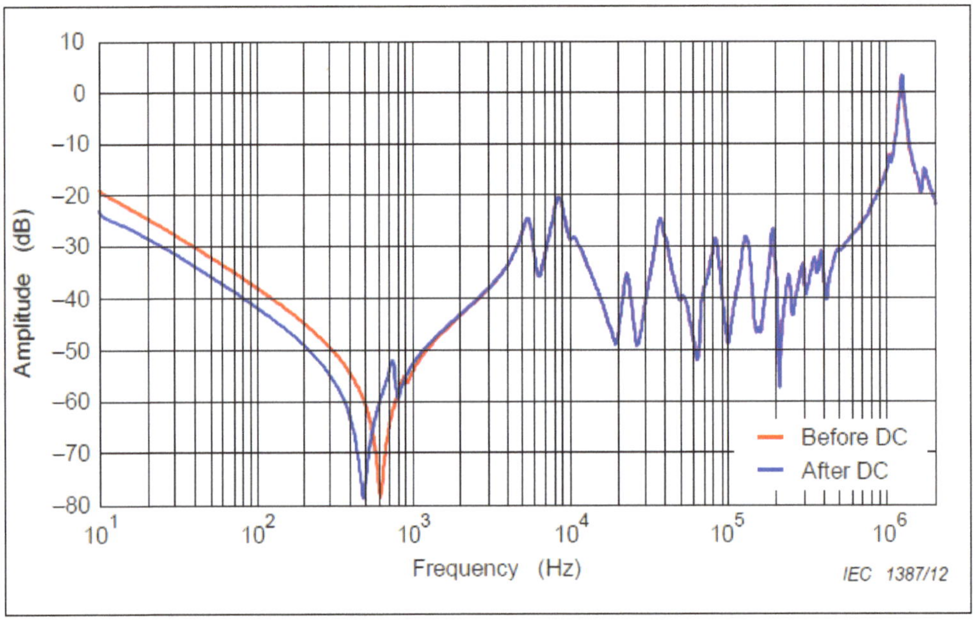

Figure14: SFRA curve before DC and after DC test such as winding resistance test.[4]

7.3 Connections inter-changed;

If the connections are inter-changed inadvertently i.e. connecting the lead of Vin to the neutral terminal and Vout to the phase terminal in case of open circuit measurement. The response curve shall show discrepancy in higher frequencies as shown in the figure 15.

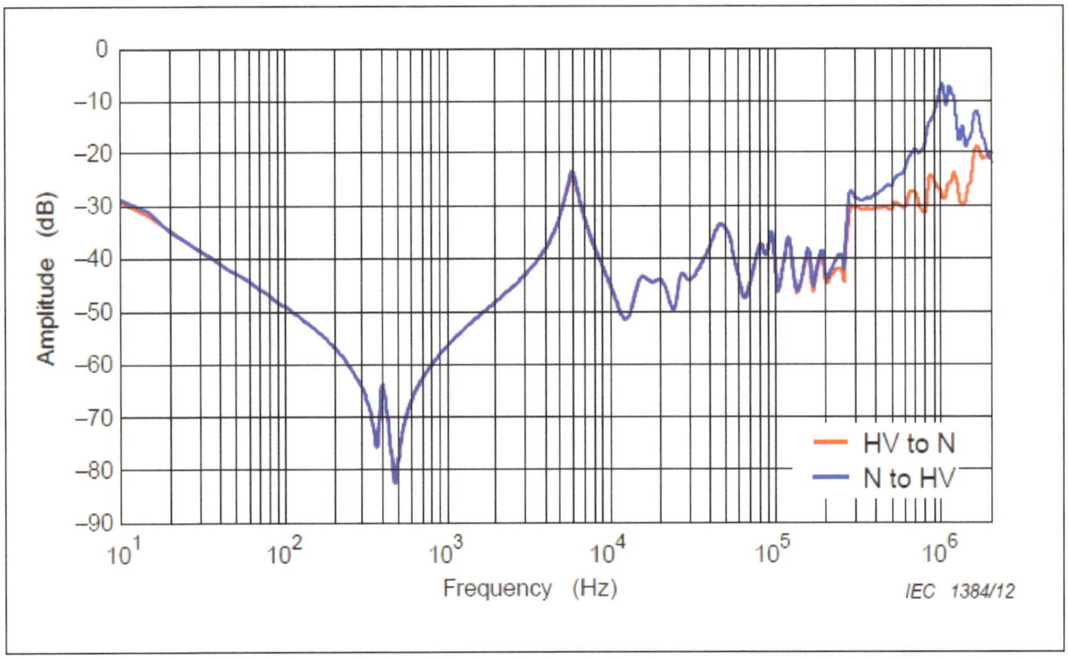

Figure 15: Discrepancy in FR curve in higher frequencies if terminals are interchanged.[4]

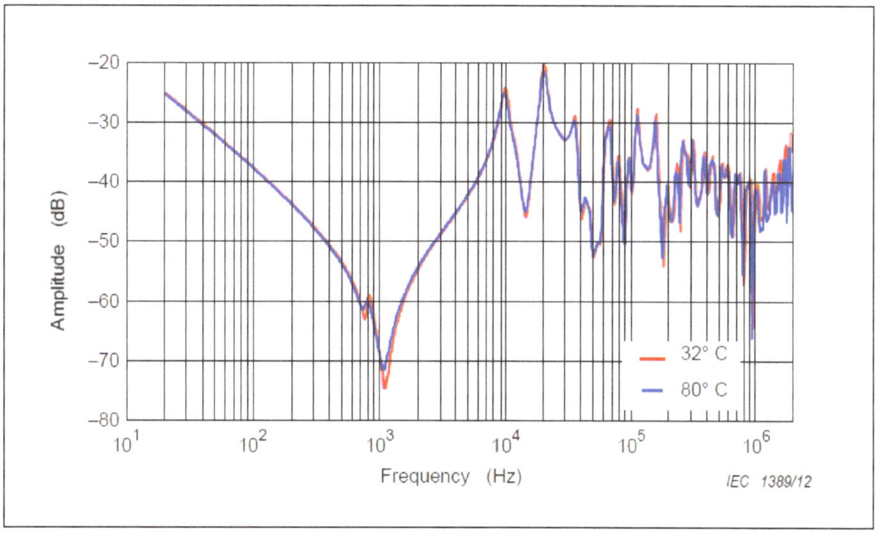

Figure 16: Effect of temperature beyond 50° C. [4]

7.5 Effect due to Delta tertiary winding Earthed or unearthed;

The delta tertiary winding is provided specially in Auto transformer beyond the rating of 100 MVA. As practiced in most of the Transmission Utilities; one terminal of the tertiary winding is solidly earthed if not in use. In such cases significant changes in the in the Frequency response curve would take place due to change in capacitive coupling of Delta tertiary winding with the series winding and also with the earth between 2kHz to 20kHZ as shown in the figure 17.

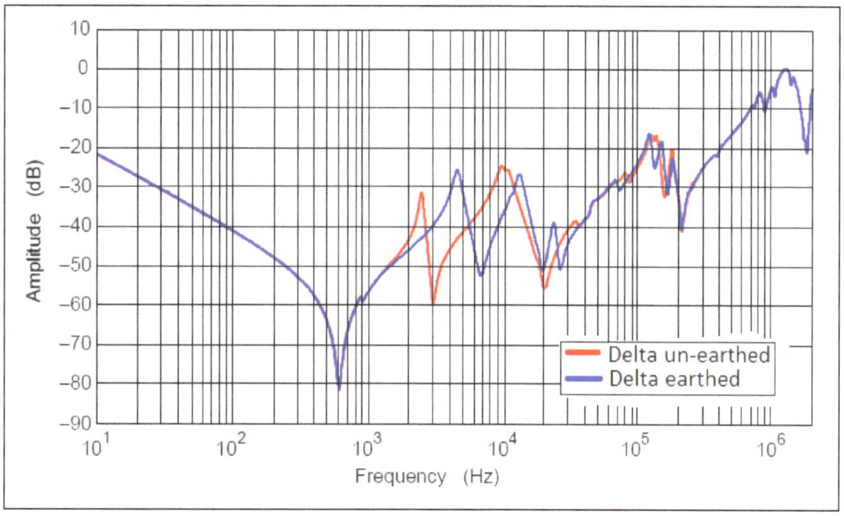

Figure 17: Frequency response curve when supply connected to series winding of an auto transformer keeping delta winding in earthed and un-earthed conditions.[4]

Note: In the OEM's works the delta tertiary winding is not earthed, however, as stated above one of the terminals of the delta tertiary winding is solidly earthed if the delta tertiary is not used to feed any load in most of the transmission Utilities.

8.0 Different Circuits for plotting Sweep Frequency Response traces [2]:

8.1 Open circuit measurement.

An open circuit measurement is made from one end of the winding to other with all other terminal floating open circuit test can be performed on single Phase and three phase transformers. This test can be performed either from HV side or LV side,

Open circuit tests are primarily influenced by core properties at or around fundamental power frequency. The open circuit tests can be used in conjunction with exciting current tests in determining failure modes that affect the exciting current circuit of the transformer.

Connection diagram;

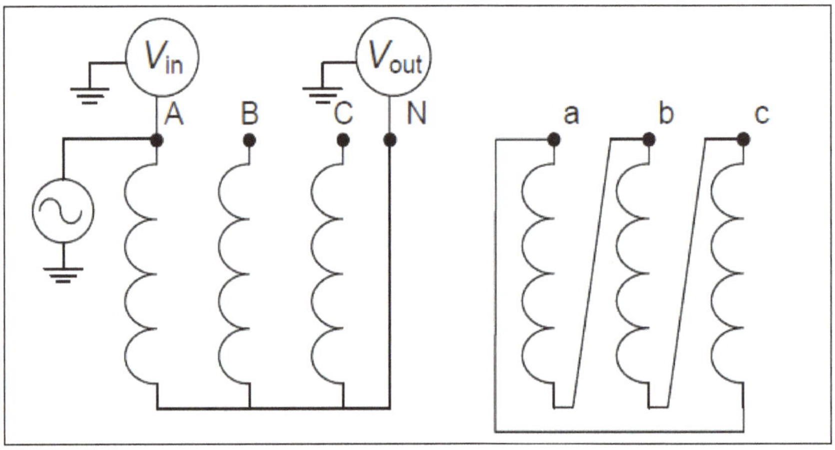

Figure.18: Open circuit measurement.

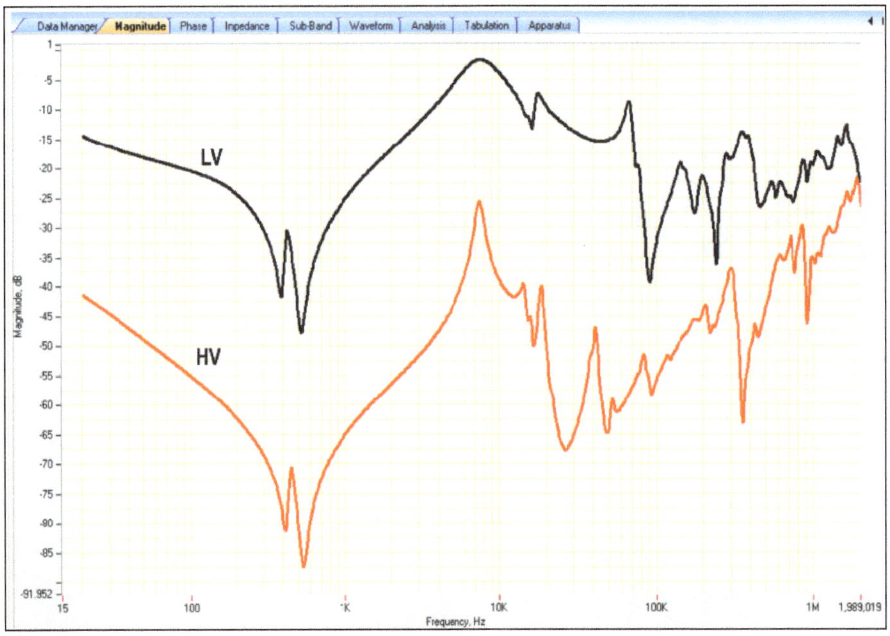

Figure. 19: Typical open circuit frequency response measurements of HV and LV windings.[3]

i. The LT windings are kept floating (open)when testing from HV side and vice-versa if testing from LV side.
ii. Test signal Vin is applied between 1U and N i.e. both the ends of the 1U Ph winding and the transmitted signal is measured on Vout.

iii. Open circuit tests are primarily influenced by core properties at or around fundamental power frequency.
iv. It may be noticed from the Figure 7 that the frequency below 2 kHz is the response zone depicting the core.
v. This test is analogues to excitation current test.

Note. [3];

a. (a) LV winding responses typically have higher amplitude than HV winding responses, especially at low and medium frequencies. This is caused by smaller number of turns, lower impedance and shorter length of turns.
b. (b) HV winding have pronounced increasing trend in the 100-1000 kHz range due its location being in outer side thus creating higher series to shunt capacitance.

Connection diagram;

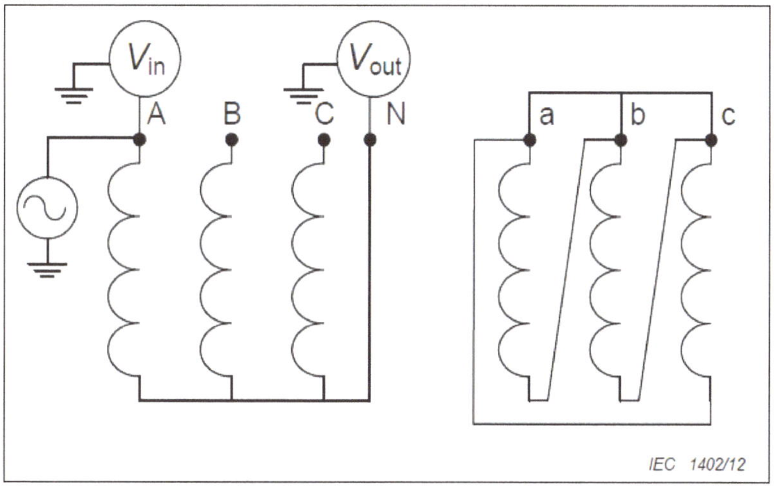

Figure.20: Short circuit measurement.[4]

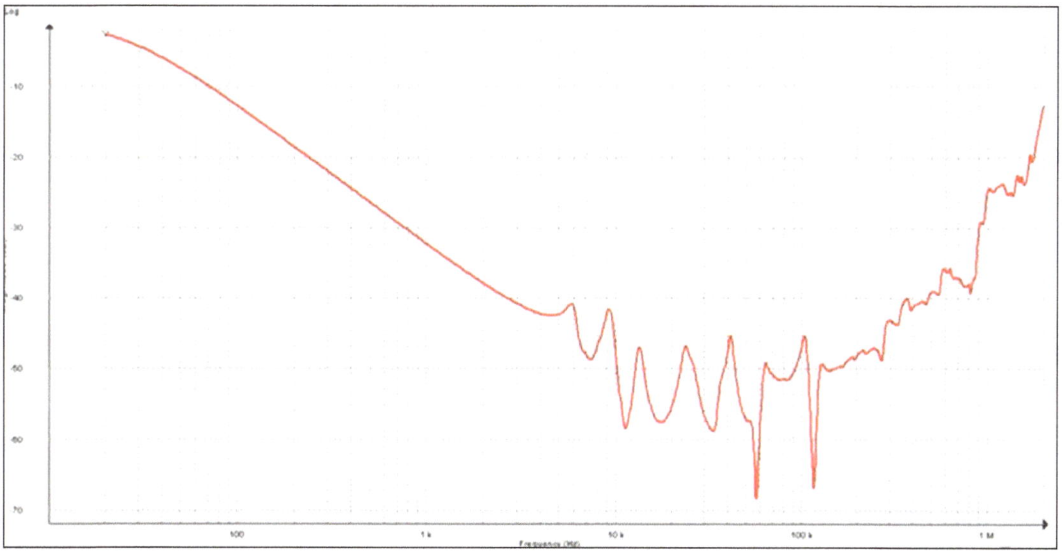

Figure21.Sample short-circuit measurement response curve.[5]

i. All the LV windings are short-circuited.
ii. Signal Vin is applied between1U and N and the transmitted signal is received in Vout at the neutral end.
iii. The short- circuit test isolates the winding impedance from the core properties. The response thereof can be seen beyond the 20 kHz.
iv. This test is analogues to leakage reactance test.
v. Note: Open circuit measurements and Short- Circuit Measurements are the vital measurements for diagnosing the health-status of a Transformer.
vi. **8.3 Capacitive inter-winding measurement** [2];
vii. The capacitive inter-winding measurement is the measurement between two electrically isolated windings. For capacitive inter-winding measurement Vin is made from one end of the HV winding and measuring the signal Vout through one of the terminals of LV winding or the Tertiary winding in case of an autotransformer with all other terminals floating. Since, the measurements are capacitive, thus exhibit a high impedance at low frequencies(<100Hz); the impedance generally decreases as frequency increases. This would include for example;

- In a 2-winding transformer: measure Vin on1U and Vout at 2U terminal.
- In an Auto -transformer with tertiary: measure Vin on1U and Vout at 3U terminal.

Note: (i) U to 2U on an Autotransformer is not inter-winding measurement but an open-circuit measurement on the series winding.

(ii) for winding-configuration of an Autotransformer please refer figure 2, of case-study 1.

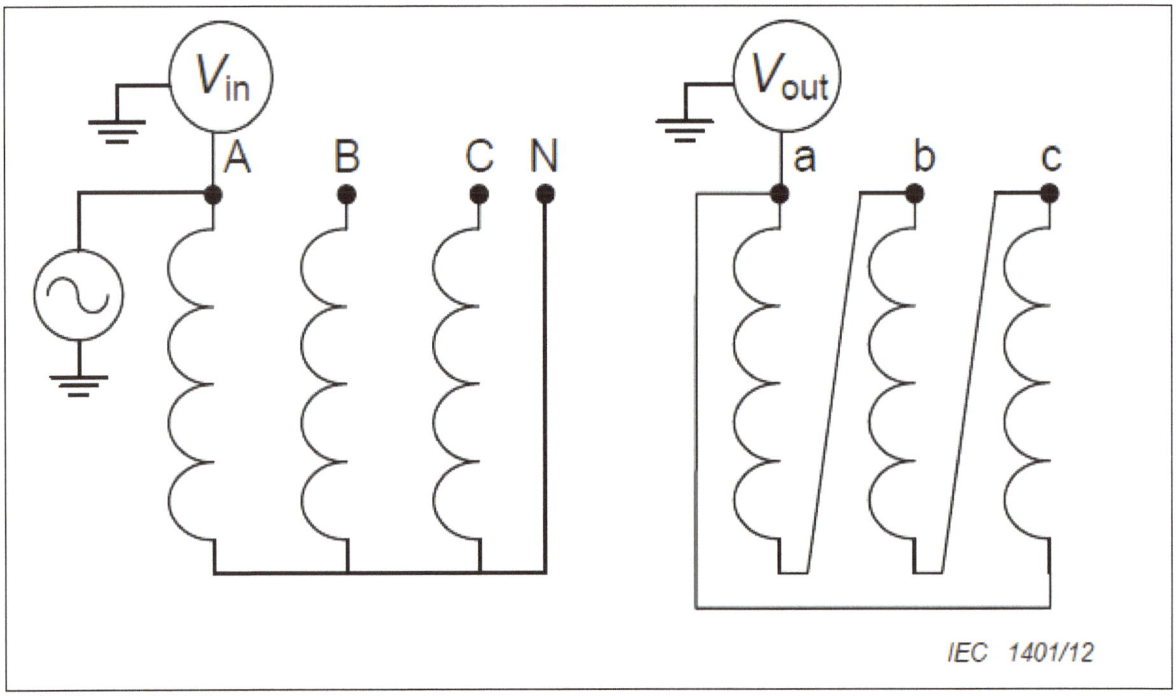

Figure 22: Capacitive inter-winding measurement.[4]

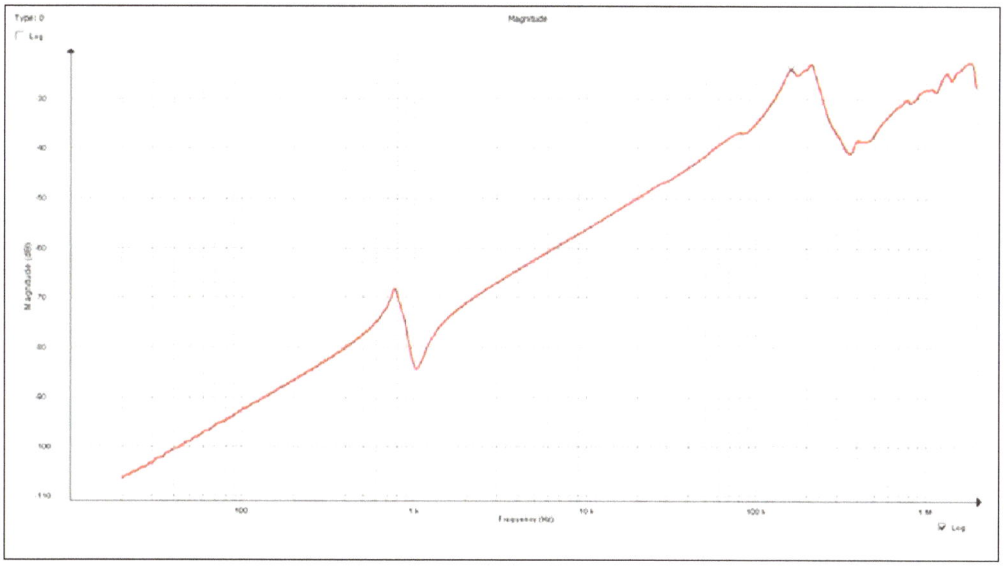

Figure 23: Sample response curve of inter-winding capacitive measurement.[5]

i. The signal Vin is applied to a HV winding and the response Vout is measured at one end of LV winding.
ii. In case of auto transformers with tertiary winding, the signal Vin is applied to HV winding and the response Vout is measured in the one end of the tertiary winding.
iii. The response using this configuration is dominated at low frequencies by the inter – winding capacitance.
iv. This test identifies, deviation if any, in the capacitance value.

8.4 Inductive inter-winding measurement;

Inductive inter-winding measurement, is also known as transfer voltage measurement, is performed between two windings with one end of each winding grounded. All other winding terminals not under test kept floating. The inductive inter-winding measurement most closely resembles to conventional turn-ration test at or around Power frequency.[2]

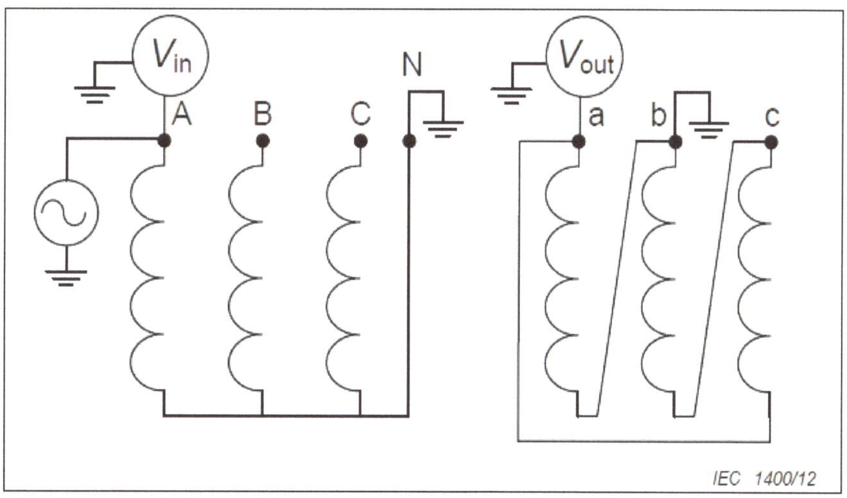

Figure 24: Inter winding inductance measurement.[4]

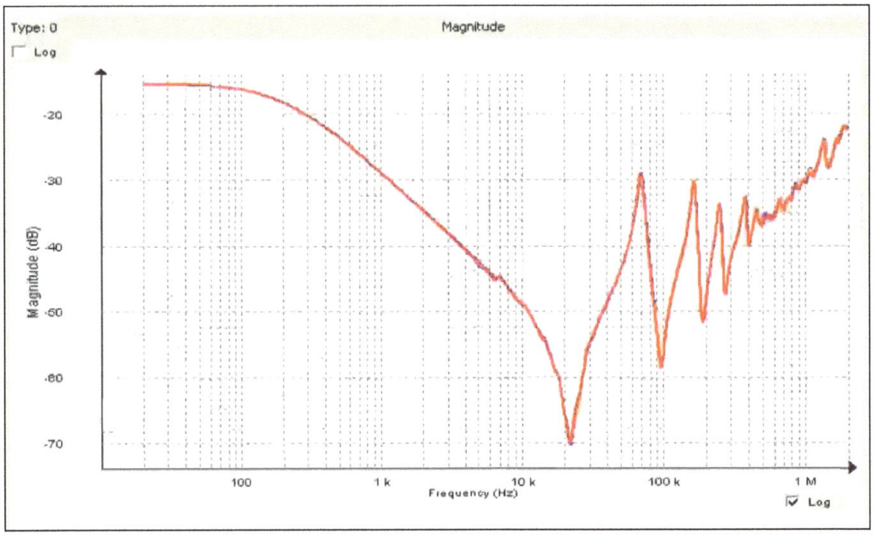

Figure 25: Sample response curve of inter-winding inductance.[5]

i. The signal Vin is applied to a terminal on HV side and response Vout is measured on the corresponding Terminal on LV side with the other end of HV as well as LV windings being grounded.
ii. This test resembles TTR test at and around Power frequency.
iii. This test identifies, any deviation in Inductance value.

8.5 Summary of SFRA tests;

The test connections described from para 8.1 to 8.4 are for 3 -phase transformers, however, all these connections have been shown in respect of single-phase transformers in table 2, indicating type of each test, looking at each parameter and also their similarity with the tests performed with the rated frequency at 50 Hz.

Summary of SFRA tests at a glance.

Type of Test	Transformer Characteristic	@ 50 Hz
1.) **OPEN** Circuit Self Admittance (Vin, Vout, OPEN)	Looks at winding and core characteristics	similar to excitation test
2.) **SHORT** Circuit Self Admittance (Vin, Vout, SHORT)	Looks at winding	similar to leakage reactance test
3.) **CAPACITVE** Inter-Winding (Vin, Vout)	Looks at capacitance between winding	capacitance
4.) **INDUCTIVE** Inter-Winding (Vin, Vout)	Looks at Inductance of both windings	similar toTTR

Table2: Summary of SFRA test on Single phase Transformer at a glance.[5]

9.0 Repeatability and its importance;[6]

In order to replicate the conditions of first SFRA Signature or thumb impression curves/traces, it is very important to create similar conditions which can be referred to, in case as and when required to evaluate the health status of a Transformer. Following precautions should be essentially taken care of;

- All the connections from the transformer except tank -grounding should be removed.
- The contact of bushings should be cleaned and the connection clamps have to be tightened firmly to ensure reliable electrical contact.
- Three shielded high frequency cables (usually Co-axial cables) of exactly same length should be used.
- Ground extension of test leads must be of low inductance (broad braided copper strips).
- Ground extensions to the base of bushing (reference potential is Transformer tank) shall be as short as possible with smallest achievable loop.
- It is very important to ensure reliable contact between ground extension to tank.
- Detailed information about the test set-up should be stored together with test data. Detailed Photographs are recommended.
- A very noisy curve is an indication of bad connections as depicted in Figure 26.[4]

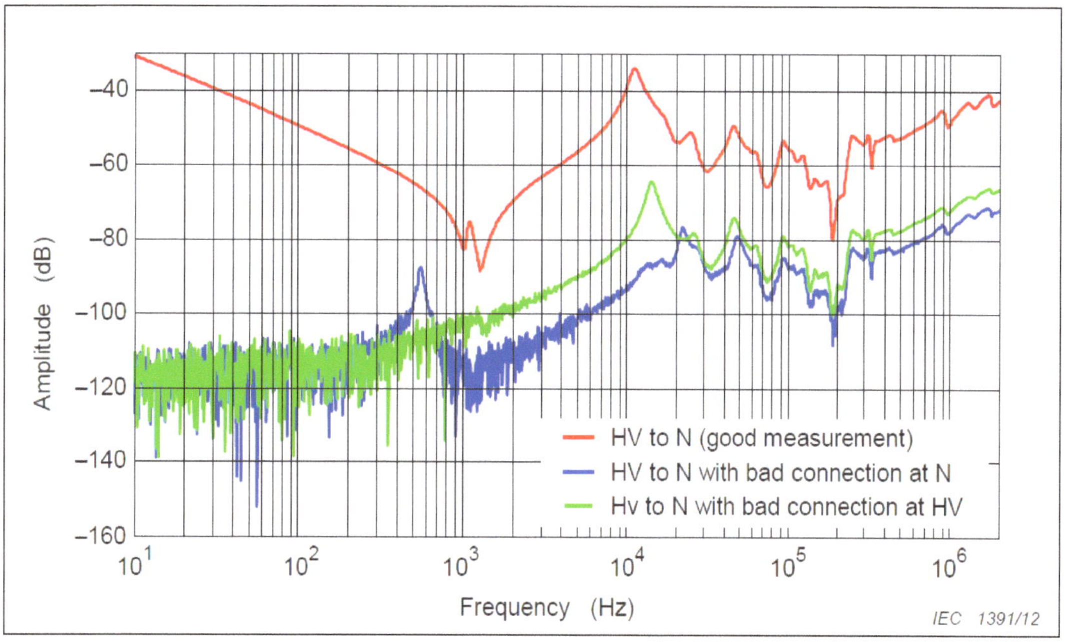

Figure 26: Noisy curves due to bad connections.[4]

9.1 Effect of Induction;

The SFRA curve shown in the figure 27 is a true Frequency response curve taken of HV W-ph and neutral-Open circuit on a new 315 MVA,400/220/33kV transformer in a 400kV S/s in an STU. The curve found was noisy in nature in the frequency range between 20 Hz to 2 kHz. It was due to the effect of the induction attributed by an over-head 400kV gantry to gantry stringing. After making 400kV Gantry to Gantry supply dead the SFRA test was again repeated and the SFRA curve obtained tallied exactly with the signature curve taken at OEM's works during FAT. The SFRA curve, thus obtained is shown in the figure 28.

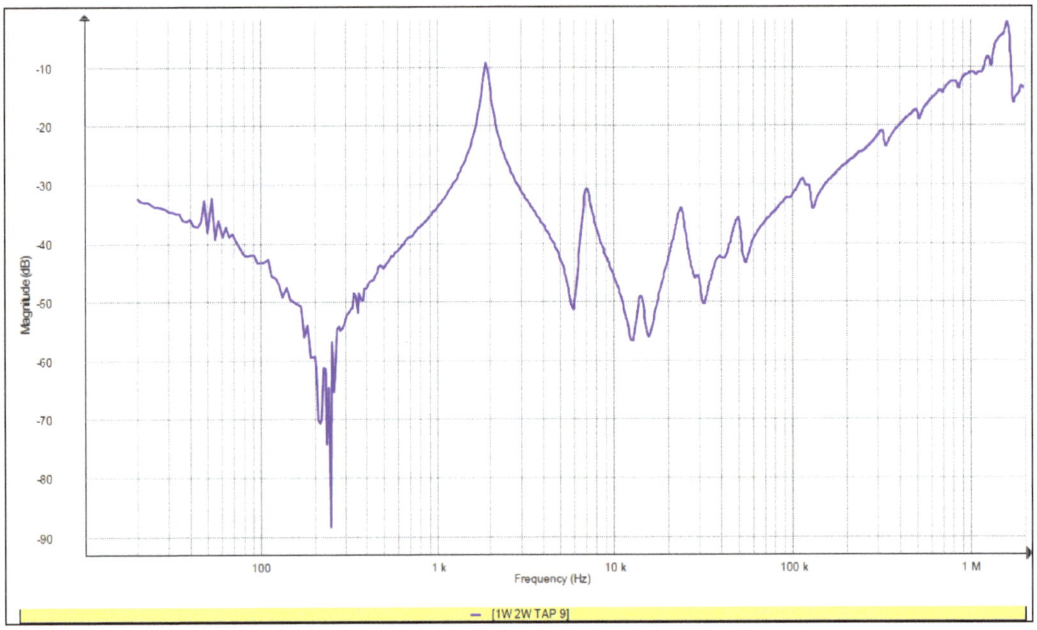

Figure 27: True noisy curve is due to induction in a 400kV Switch-yard.

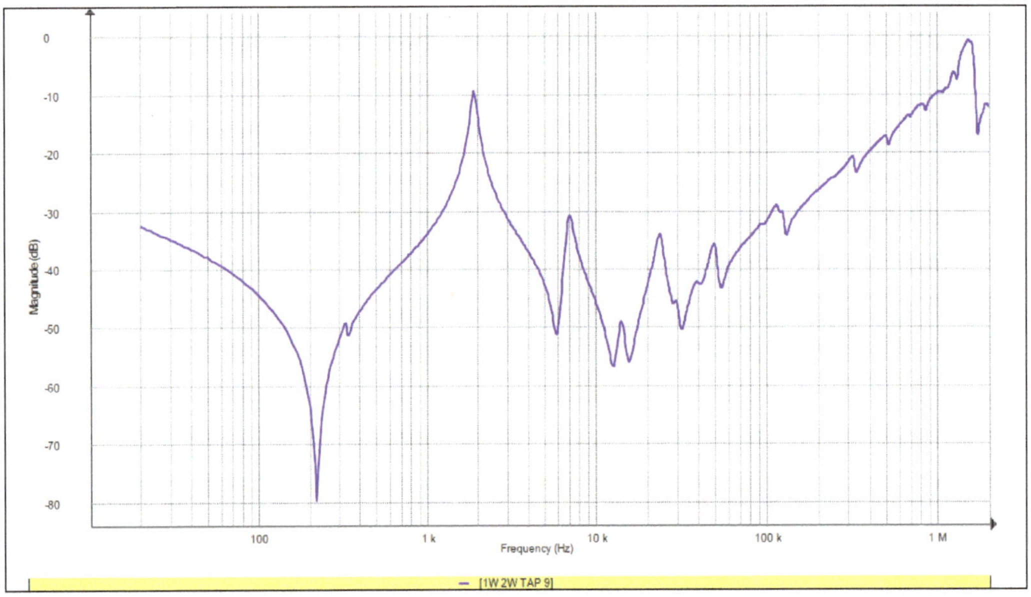

Figure 28: Curve after curbing the effect of induction.

1.2 Effect of bad earthing of the SFRA kit;

In one 400kV S/S in an STU, while testing a 315 MVA 400/220/33kV auto-transformer for IV-N; open circuit test, the curve observed was full of noise as shown in the figure 29. On thoroughly checking the connections of the SFRA kit, it was found that the earthing connection of the SFRA kit was not firmly tightened. After it was attended to the SFRA curve was found as shown in the figure 30.

Figure29: Effect of bad earthing of SFRA kit.

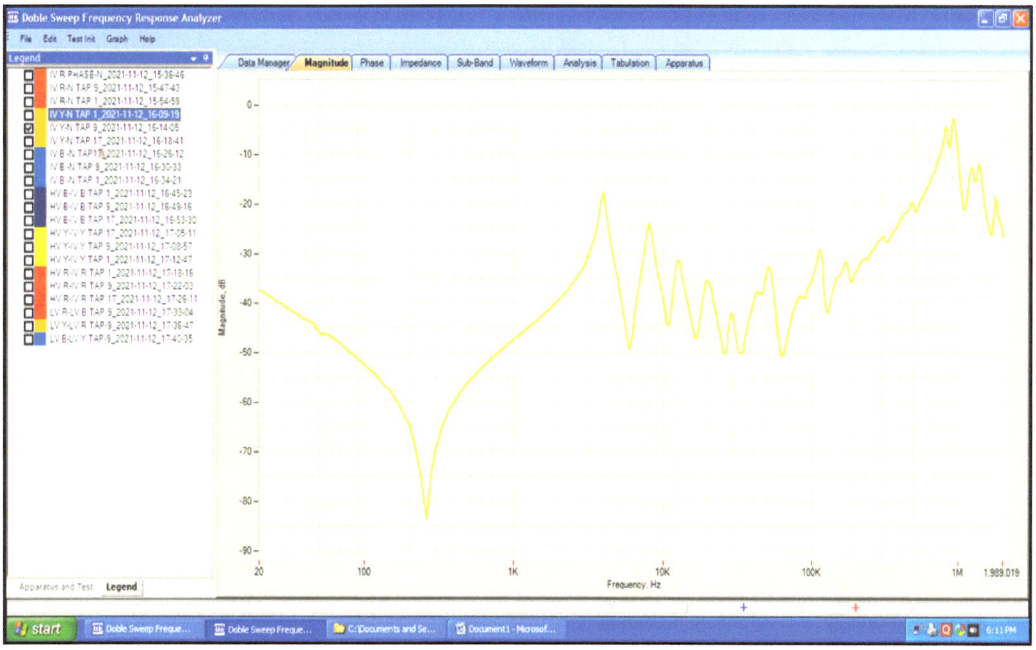

Figure30: Curve of IV-N; open circuit test post-attending to the earthing connection of

the SFRA kit.

10.0 CASE STUDY;

SFRA test analysis;

i. **Capacity:**315 MVA, 400/220/33kV.Auto transformer.

ii. **Tripping Indications and alarms:**

- Differential trip A-ph, High set differential,
- PRV Operated,

- ➢ Buchholz alarm & trip appeared.
- ➢ High set R ph E/F operated.

iii. **Incidence/episode:** While normalizing after maintenance, the transformer was charged from 400kV side, by inadvertently not removing the earth blades from "on" condition. Heavy flash over was observed.

iv. **SFRA Curves and analysis:**

a. **Figure.1**
Sweep frequency response Curves of Individual HV of each phase-N, keeping all other terminals floating (open circuited).

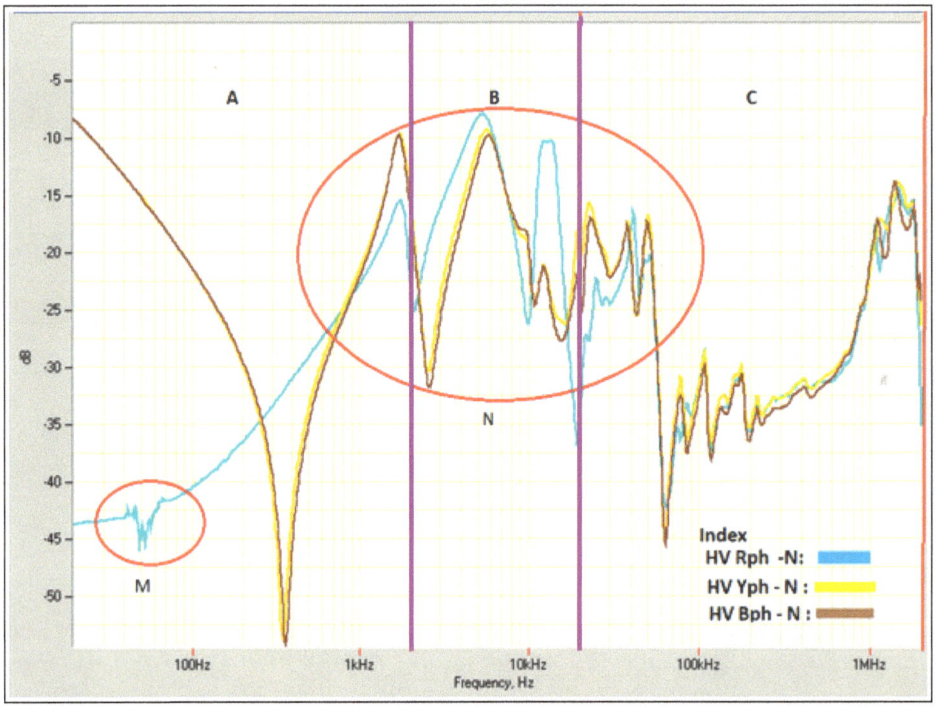

Figure 1: Graph showing frequency response Curves of Individual HV of each phase-N, keeping all other terminals floating (open circuited).

Analysis;

i. The R ph (Blue colored) amplitude raises from - 44dB to -15dB ie the frequency response curve is raising in amplitude, contrary to the characteristic of response curve drooping from higher amplitude to lower amplitude.

ii. The irregular distortion in amplitude of R ph in "A" region between 30 Hz to 60Hz frequency range shown under "M" indicating that the core got affected. The core laminations could be seen in the Photograph in the figure 5.

iii. The R ph response curve is irregular and distorted in the frequency range zone "B" and also extending up to 60 kHz in the zone "C" indicates that the winding is badly affected involving *common winding.

[Note: * Winding configuration in an Auto transformer is shown in the figure 2.].

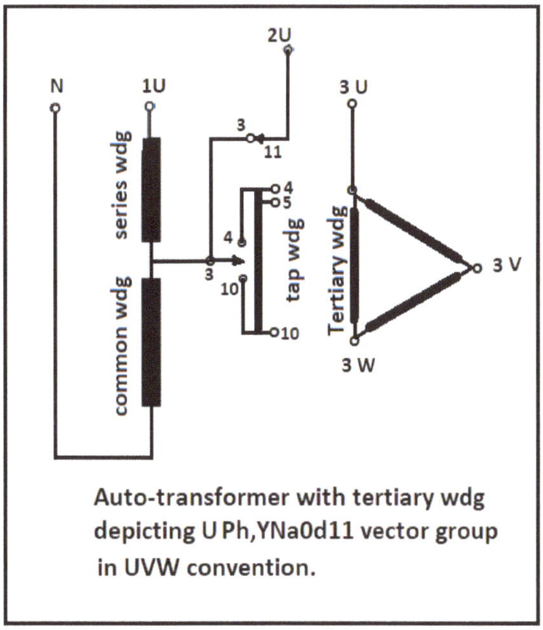

Figure 2: Auto transformer showing series, common, tap & Tertiary windings.

iv. The Y phase & B phase curves are congruent show their status as healthy.

b. Figure.2 Sweep frequency response curves of Individual IV of each phase-N, keeping all other terminals floating (open Circuited).

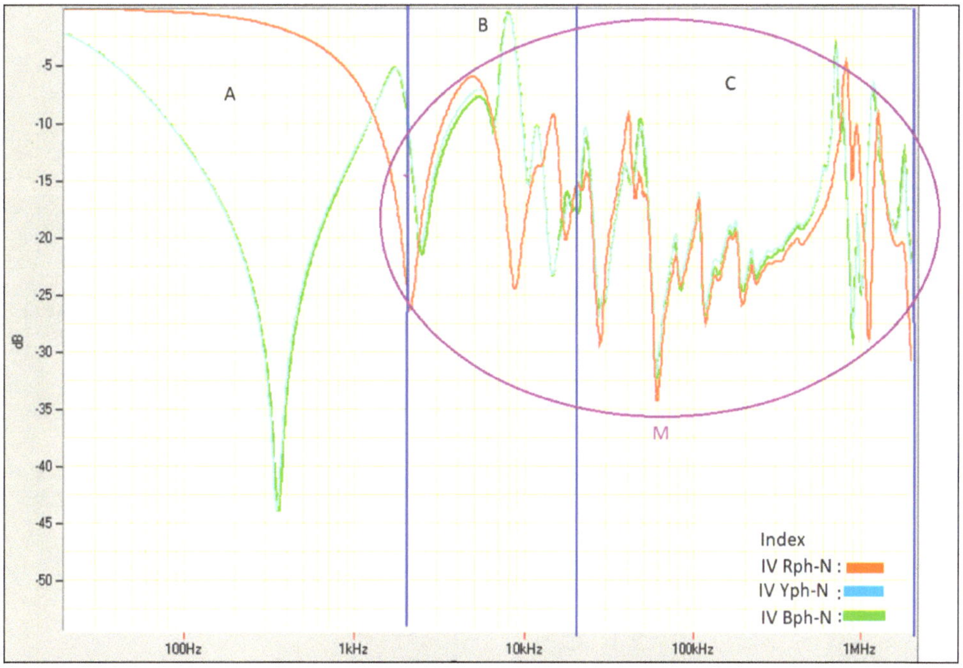

Figure 3: Graph showing frequency response Curves of Individual IV of each phase-N, keeping all other terminals floating (open Circuited).

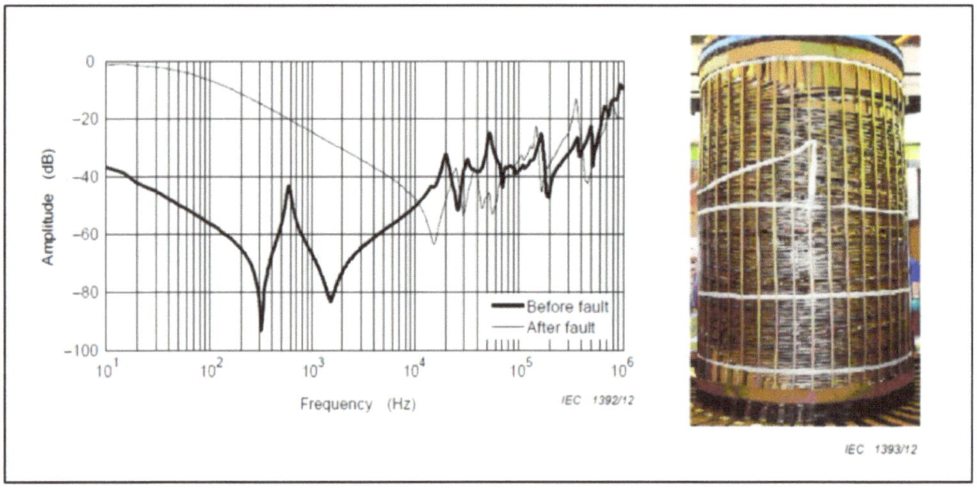

Figure 4: Sample frequency response of tap winding of an Auto transformer before and after fault showing axial collapse and inter-turn short circuit as depicted in the photograph along-side.[4]

Analysis;

i. The trace of R ph curve in the zone" A", droops in parabolic shape from -1dB to -27dB. Reference may be made to the sample graph.4. which is from the standard IEC 60076-18-2012.[4], it could be noticed that the R ph trace in the figure 3 resembles that in the figure 4 in the frequency range of zone "A". Looking to this resemblance it could be opined that turn to turn short in the tap-winding must have taken place in R phase between the IV terminal to neutral. The heavily carbonized lead and in the vicinity thereof with burning of wrapped paper insulation confirms as shown in figure 5.

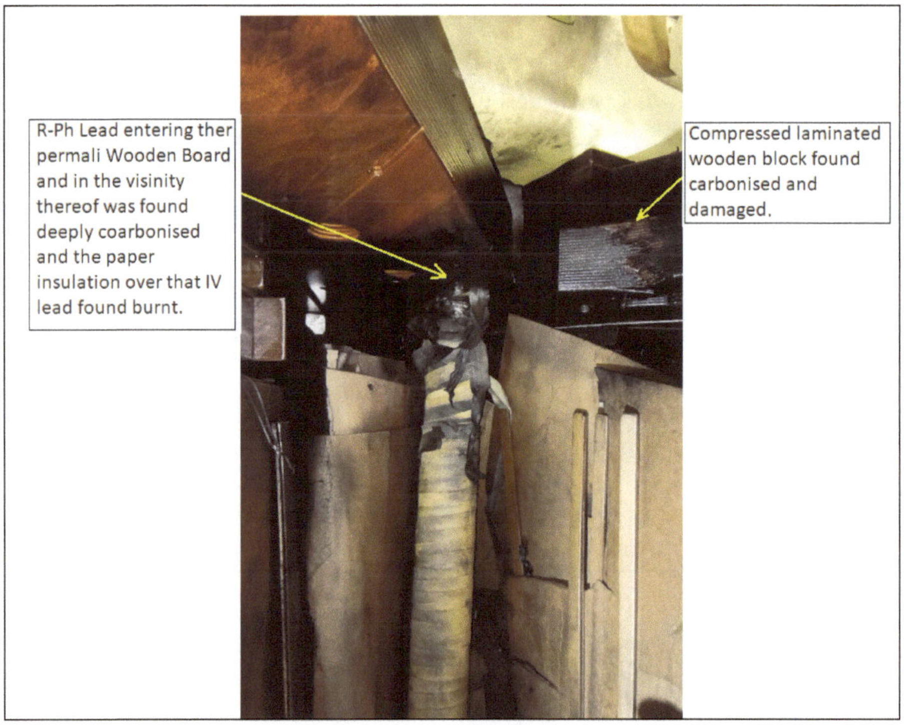

Figure 5: Lead of IV -R ph and the portion in its vicinity was seen heavily carbonized paper insulation burnt and portion of wooden block appears to have damaged.

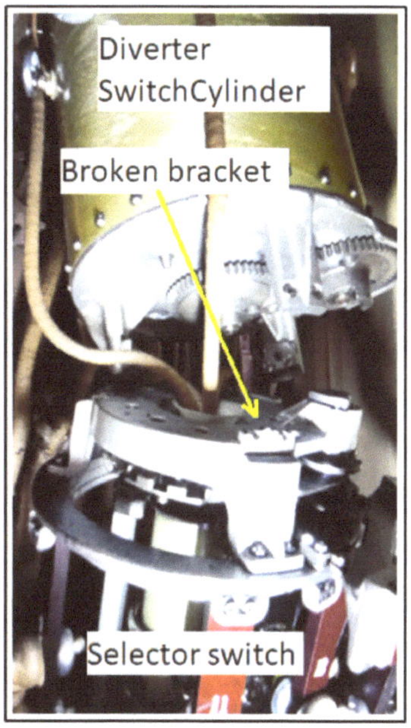

Figure 6: The coupling between the selector switch and the diverter switch of R ph OLTC is separated due to breaking of the fastening bracket.

- The Yph & Bph curves are congruent throughout the sweep-frequency range. Which confirm that Y and B phase windings are healthy.

11.0 Conclusion:

i. The SFRA is a powerful non-invasive technique for assessing the health-status of the transformer. It is a single test to evaluate the mechanical and geometrical integrity of the transformer. This Test is adopted in a big way all over the world, subsequent to issuance of International Standards and CIGRE reports viz: IEEE C.57 149-2012, IEC 60076-18-2012 and reports of CIGRE working Group A2 report Nos. 342 -April 2008 and 812-Sept 2020.

ii. The authors have narrated all the four measurement-circuits from para 8.1 to 8.4 for academic interest only; however, open circuit (para-8.1) and close circuit (para-8.2) measurements give clues leading to proper diagnosis of the suspected faults.

12.0 References;

1. CIGRE report by Group A, No.342-April 2008.
2. IEEE Std.C57.149-2012.
3. CIGRE report by Group A2, No.812-Sept, 2020.
4. IEC 60076-18-2012
5. Webinars by Megger.
6. Alexander Kraetge of Omicron Electronics, Klaus, Austria and associates: Experience with the practical application of SFRA on Power Transformers.
7. Tapan Kumar Saha (Editor), Prithwiraj Purkait (Editor) for the book titled as “Transformer Ageing: Monitoring and Estimation Techniques” (IEEE Press) 1st Edition, Kindle Edition

11.0 Authors;1) K.K.Murty2) Sunil Saini **3)** Smt. Nutan Sharma.**4)** Anurag Pant.

17. In-House Retrofitting of a Draw-rod type Bushing Replacing Deteriorating Tan δ Value Draw -Lead Type of a 420 OIP Bushing of 315 MVA Transformer- A Challenge

ABSTRACT

On the observation of rapid deterioration in tan δ value to 1.02% ie surpassing the limiting value of 0.7% as defined in the IEC Std.60137 and limiting value of 0.5% IEEE Std.C57.19.01-2000 of a 420kV OIP bushing of 315 BHEL Transformer at Katni, a decision was taken by the MPPTCL management to replace the bushing as early as possible.

In the absence of spare bushing and BHEL's denial to either supply a spare bushing or to assist MPPTCL in this matter, there was no option but to use a spare draw-rod OIP Bushing of TELK made at site, looking to the threat of catastrophic failure for the transformer, which would have resulted in-to Grid-system Disturbance causing colossal loss to the Utilities and dissatisfaction amongst the consumers at large.

The challenge was accepted by the MPPTCL engineers and executed the work successfully with limited resources and with minimum down-time and the transformer was put back in service.

Deteriorating in the value of tan δ, DGA results of 420kV, OIP, BHEL Bushing and difference in measurements in the physical dimensions of the Bushing of TELK as compared to BHEL Bushing and the mythology used to retrofit a Draw-Rod type TELK bushing in place of draw lead type bushing has been narrated in this paper.

Key Words: Bushings, Draw-rod type, Draw-lead type, DGA &Tan δ

ACRONYMS: MPPTCL (M.P. Power Transmission Co. ltd.), BHEL (Bharat Electrical Ltd), TELK (Transformers and Electrical Ltd. Kerala), OIP (Oil impregnated Paper), RIP (Resin impregnated Paper), DGA (Dissolved gas Ltd. WRLDC (Western Region Load Despatch Centre), SLDC (State Load Despatch centre), OEM (Original Equipment Manufacturer)

1.0 Decision for replacement of bushing: As per Maintenance schedule laid down my MPPTCL each transformer must be taken out of service and tested /maintained every year, however, as far as 400/220 kV 315 MVA Transformers are concern, the responsivity of allowing shut down on these inter-connecting transformers lies with the WRLDC/SLDC. During the scheduled maintenance period of 315 MVA BHEL Transformer, shut-down was not permitted by the WRLDC/SLDC in the year 2019.

The value of tan δ showed increasing trend abruptly from the year 2016 to 2018 and was observed as 1.02% in the year 2020 (Skipping in the year 2029 as shut down was not permitted by the SLDC). (Refer table1).

Date of Measurement	Test voltage	Tan δ	Capacitance	Oil temp.	Remarks
25 March 2010	5Kv	0.20%	423.78pF	40.0°C	During commissioning
13 March2016	10kV	0.15%	246.3 pF	42.2°C	Routine maintenance
15 Sept2016	10kV	0.23%	421.71pF	39.9°C	Routine maintenance
20 Sept.2018	10kV	0.42%	420.87pF	32.6°C	Routine maintenance
*2019	N/A	N/A	N/A	N/A	N /A
15 May2020	10kV	1.02	421.71	35.0°C	Routine maintenance

Table 1.

*Shut down not given by SLDC.

1.1 DGA report by POWER GRID Oil Test Lab At ITARSI:

DGA test report of the DGA testing carried out on date 120th May 2018 by the POWERGRID, Oil testing lab at Itarsi of the OIL sample of the 315 MVA BHEL Transformer. The test report is furnished here under;

Dissolved Gas Analysis Test Method: IEC 60567 Reference Standard: IEEE-C57.104-2008			
PARAMETER	UoM	Measured Value	Violation Limits, max
Total Gas Content	%	8.57	---
Nitrogen (N2)	%	6.46	---
Oxygen (O2)	%	1.36	---
Hydrogen (H2)	ppm	24	100
Methane (CH4)	ppm	77	120
Ethylene (C2H4)	ppm	265	50
Ethane (C2H6)	ppm	29	65
Acetylene (C2H2)	ppm	2.3	1
Carbon Monoxide (CO)	ppm	630	350
Carbon Dioxide (CO2)	ppm	6392	2500

Table2.

OIL Test lab of POWERGRID's Concluding remarks;

- As per IEEE-C57.104-1991 (Rogers Ratio): Thermal Fault > 700 deg. C
- IEC-60599-1999 ANALYSIS: Thermal Fault of >700 deg. C (T3).

Further, high level of carbon oxides (CO:630 & CO2:6392) indicate that the thermal fault in the paper is causing *pyrolysis.

Note: *Pyrolysis is thermo-chemical decomposition of organic matters into non condensable gases in the absence of oxygen. The amount of non-condensable gases (CO_2, CO, H_2 & CH_4) increases with temperature.

1.2 Decision of MPPTCL Authorities: Keeping in view, the MPPTCL Authorities took decision to replace the busing.

1.3 The Pictorial Confirmation commensurate with the DGA report.

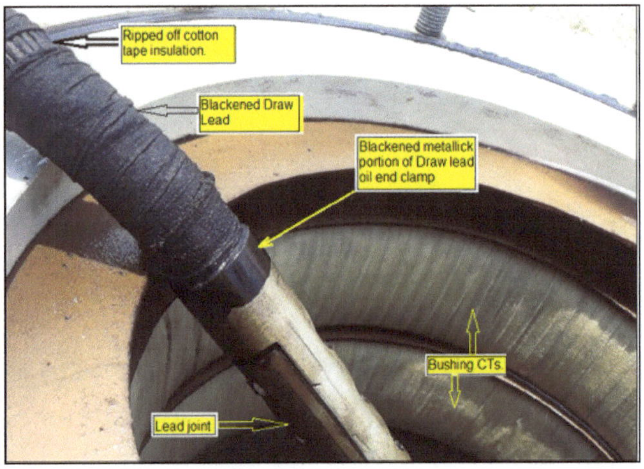

Figure 1

Figure 2

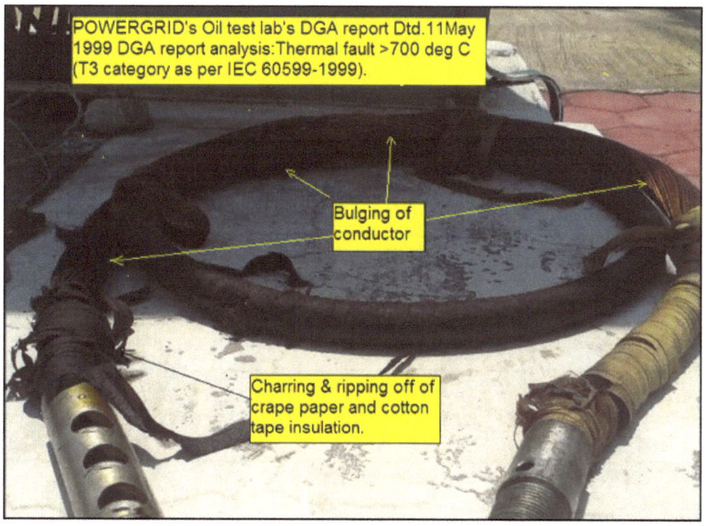

Figure 3

After removal of draw-lead type conductor of the deteriorated 420 kV BHEL bushing careful examination was carried out. Thermal heating is confirmed on following grounds.

a. Thermal fault of > 700 °C (T3), IEC: 60599 as per analysis of DGA by POWERGRID's oil test lab, Itarsi, report dated 11 May 2018, substantiates over heating of the draw lead.
b. Looking at the Figure 1, 2 and 3, the root-cause appears to be heating of bulged peripheral strands due to skin effect over a period of time. It appears that bulging of strands must have taken place during erection / installation of the 420 kV OIP bushing of 1V phase of BHEL transformer.

2.0 Retrofitting activities;

2.1 OIP Bushing Dimensions as defined in IS:12676-1989.

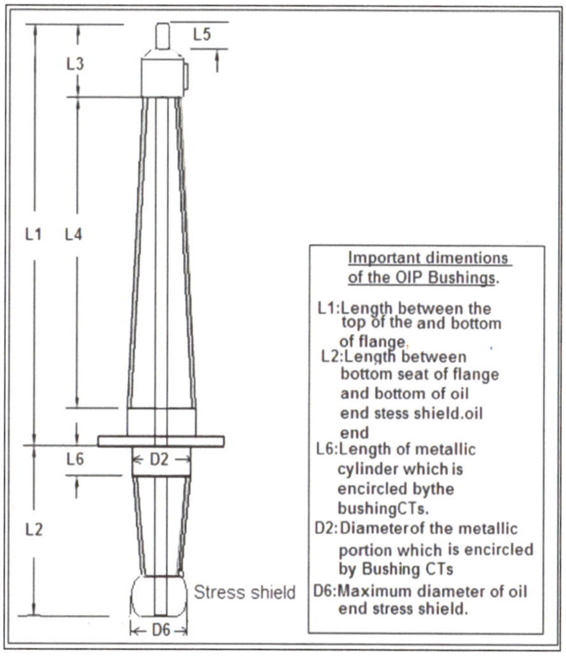

Figure 4 Figure of OIP bushing with dimensions as pet IS: IS:12676-1989.

Particular	BHEL	TELK
Type of bushing	Draw-lead type	Draw-rod type
L1	4050 ± 10 mm	4195 ± mm
L2	1640 ± 5mm	1640 ± 10mm
Flange diameter	770mm	770mm
PDC*	660 mm	660 mm
PCD diameter D2**	340 mm	350 mm
Space for BCD (length)L6	400 mm	400 mm
Stress shield physical size	Small	Larged
Stress shield physical shape	Oval	Cylindrical

Table 3.

Note:

* PCD is the acronym of Pitch centre Diameter.

** BCT is the acronym of Bushing Current-Transformer.

2.2 Challenge: How to use TELK bushing with difference in the dimensions as mentioned in the table 3, it is evident that there are some differences in the dimensions, such as in BCT diameter D2, physical size and physical shape.

Difference in dimensions in the Bushings;

Make	BHEL	TELK
BCT diameter D2	340mm	350mm
Stress shield Physical size	Small	Larger
Stress shield Physical Shape	Ovel	Cylindrical

Table 4.

Looking to the differences in vital dimensions, it was a challenge to use the 420 kV OIP bushing (draw-rod type bushing) of TELK make in place of 420 kV OIP bushing (draw lead type bushing) of BHEL make. The challenge was accepted by the engineers of (testing) division Katni without the help or support of either the transformer OEM or the bushing OEMs.

2.3 Draw rod of Draw-rod type TELK made bushing;

Figure 5: Draw rod of TELK make bushing.

Figure 6.: Stress shield of TELK made busing.

i. The draw-rod was thoroughly cleaned with acetone.
ii. Thorough cleaning of existing TELK make spare bushing was done and was held in the vertical position for a period of 24 hours with the support of a locally fabricated structure facilitating the settling down of oil in the bushing. Then the measurements of tan δ, capacitance and IR values were carried out and found to be in order.

iii. To replace BHEL make draw-lead type bushing by with TELK make draw-rod type bushing a 10,000-litre oil storage tank, oil filtration plant, 2 hydra cranes (with long booms) and all T&P were arranged.
iv. All gate valves of cooling banks header pipes and conservator were closed and oil about 1 kL drained up to the turret level in the storage tank.
v. 420 kV OIP BHEL make, draw-lead type bushing was removed with the help of long-boom hydra crane.
vi. Then TELK bushing lifted vertically over the turret and draw-rod thimble round nut opened with the help of locally fabricated pin-spanner and lowered downwards to establish joint of draw-rod to 420kV winding lead of the Transformer. Draw-rod of TELK bushing was successfully joined with the transformer lead by retrofitting the mismatch of the half-lap joint.
vii.

Figure 7: Lowering down of the TELK make bushing with utmost precision with the help of 2 nos. Long- boom Hydra canes.

Then an appropriate angle of bushing was established with the second hydra-crane and it was gradually lowered down carefully through the turret bushing CT, with utmost precession keeping all precautions in view;

a. that the portion of the bushing CT area diameter of the TELK bushing is greater than the BHEL bushing by 10.0 mm.
b. there was hardly any play between the inner diameter of the bushing CT (installed in the groove of the turret) and the diameter of the metallic portion of the bushing (diameter D2 in Figure 4), which is encircled by the bushing CTs.

The bushing was lowered down through the turret with utmost precision, maintaining an accurate angle with a successfully fitted stress shield.

viii. Contrary to the TELK's procedure of lowering down the bushing without a stress shield and fixing the stress shield by entering the transformer tank after decanting all of the oil, we performed the activity / procedure as narrated in (vii). Our procedure not only saved time by way of decanting complete oil and then entering inside the transformer and fixing the stress shied after lowering the bushing without the stress-shield. We have also averted any possibility of injury to the person entering the soiled / slippery transformer tank after the oil filtration has been done in a storage tank and the transformer oil was poured back into the transformer.

ix. As per TELK's guide lines the settle-down time after the bushing has been installed is 72.00 hrs. Accordingly. the 420 kV OIP draw rod type bushing was installed at around 16.00 hours on 18 May 2020 and it was successfully energised at 16.55 hrs on 21 May 2020.

3.0 Total down time;

Transformer switched off	11May 2020	09.22 hrs
Charged on	21May 2020	16.50 hrs
Load taken on	22 May 2020	11.30 hrs
Total down Time		11days, 2hrs,8 min.

Table 5.

4.0 Benefit to Transmission utility and the Grid network;

1. Averted unaware catastrophic failure of the transformer resulting into collapsing Grid partially.
2. Savings in revenue since the procedure was carried out with available resources without any support from the OEM.
3. Shorter down time, facilitating grid security.

5.0 Recommendation;

1. The Utilities should procure spare bushings with the Purchase order for the Transformers to the OEMs.
2. RIP bushings should be used in place of OIP bushings.
3. Standardization of size and dimensions should be defined by the BIS (Bureau of Indian Standards) such that inter-changeability of bushings shall become possible.
4. Testing of such Transformers should be planned in the period of load in the Grid.
5. DGA of transformers oil should be carried out on a regular interval of say 6-months.
6. For Tan δ value Test, test set of 10kV should be used.

6.0 References;

1. IEC 60137 and IS 2099-1986, Edition 6.0, 2008-07, Standard: Insulated bushings for alternating voltage above 1000 V
2. IEEE Std. C57.19.01-2000- Performance, Characteristics and Dimensions for Outdoor Apparatus Bushings.

3. IEC 60599, Edition 2.1, 2007-05, Standard: Mineral oil-impregnated electrical equipment - Guide to the interpretation of dissolved and free gases analysis.
4. Report of DGA by POWERGRID Oil test lab, Itarsi, of date 11 May 2018.
5. IS 12676, 1989 (reaffirmed in 2001)– Indian standard for oil-impregnated paper condenser bushings – Dimensions and requirements.
6. Instruction manual of TELK OIP condenser type bushings up to 400 kV, publication No. T 9007 D.

Authors: K.K.Murty, Rajeev agrawal, S.K Chaturvedi

Typical arrangement of Filtration

18. On –Line Refurbishment of Old EHV Transformers in the State of MP (India) – Selection of Transformers Through % of Relative Saturation of Water in oil.

Abstract;

This article focuses on refurbishment/reconditioning for extending the life expectancy of old transformers which were continuously in service for more than 20 years in a large fleet of transformers of around 600 nos..in the M.P. State Transmission Utility in India The salient feature of this Article is the selection criterion adopted for refurbishment. The selection criterion was based on % Relative saturation of water in oil which is different from the conventional practice of observing trend of deterioration of IR & PI values and DGA etc. The reconditioning of all such old transformers was taken up on-line in a big way avoiding long down-times and frequent load sheddings and was completed over a period of 5 years.

1.0 Keywords: Moisture content, ppm (parts per million= mg/kg), Karl Fischer Titration,%Relative Saturation (%RS). Oil Filter plant/machine. OEM(Original Equipment Manufacturer)

2.0 Introduction: The expected life span of the Transformer has been mentioned as 25 years as per Ministry of Power, Govt. of India's notification No.S.O.266 (E) Dtd.29th March; 1994[1].However, Transmission utilities should endeavor to extend life expectancy of the Transformer to 35-40 years, adopting good practices of maintenance of the Transformers, as they are the most expensive and vital equipment in a Power Utility. The old Power transformers were identified adopting the % Relative saturation of water in the transformer oil. All such old Transformers were identified for On-line refurbishment based on the following guidelines /criteria

a. which had rendered services beyond 20 years continuously,
b. which had >15% Relative saturation of water in the transformer oil.[2]
c. deterioration trend in IR and PI values.

This process not only enhanced the life-expectancy of the old transformers but also gave tangible advantages to the utility as well as to the consumers at large by way of avoiding; long down-times, frequent load-sheddings, loss of revenue, investment for replacement of transformers and consumer dissatisfaction at large etc. As many as 205 nos. of old EHV transformers of the total fleet of around 600 nos. of EHV Transformers were refurbished successfully over a period of 5 years in the Transmission Utility in the state of M.P.in (India) The authors have narrated the procedure adopted for identifying the old EHV transformers which needed refurbishment /reconditioning through % RS which is different from the conventional method of deteriorating IR&PI values, DGA values etc. and carried out online refurbishment /reconditioning project successfully over a period of 5 years.

3.0 Causes of Deterioration of insulation; Deterioration of insulation is attributed to the following;

i. Increase of moisture contents in the Oil beyond 20 ppm (mg/kg).
ii. Increase in water contents in the winding beyond 4%.[2]
iii. Continuous over loading of Transformers beyond 80 $^{\circ}$C to 140 $^{\circ}$ C leads to loss of life to half for every 6° C beyond 80 $^{\circ}$C as stipulated in the IS: 2026 (Part-7):2009 (IEC equivalent IEC 60076-7:2005).[3]. Resulting into fall in DP value nearing 200.

4.0 **Water, the deadly enemy of Transformer:** The most dangerous enemy of the Transformer is the moisture / water in excess in the Transformer oil and in the winding. The moisture content of the oil in the Transformer in service is governed by the IS: 1866-2000./IEC 60422/IEEE std. C57.106 -2013.[4]

4.1 **How does water get into the Transformer?**:

i. Inadequacy in drying out process at works of OEM (Original Equipment Manufacturer).
ii. Leakage/ seepage from gaskets and welding spots at various locations.
iii. Inhaling atmospheric air while oil volume shrinks during cold & moist nights and also during off peak hours of the day.

4.2 **Reason for increase in water contents in oil:** Most of the practicing Sub-station Engineers/ Managers wonder how moisture content of the oil increases while every precaution against entry of atmospheric moisture into the Transformer is taken care-of as narrated in paragraph 4.1.

Earlier only silica gel breathers were provided for the duty as mentioned in Paragraph 4.1(iii), but in the present scenario Air-cells are provided which act as the barrier between the atmospheric air and the Transformer oil.

5.0 Solid Insulation in the transformer. The paper is the major solid dielectric material, used for conductor wrappings, as barrier Boards, spacers and clamps (in compressed or resin bonded forms. The major constituents of the paper are Cellulose (90%), Lignin (about 6%) and reminder (about 4%). [5]

The structural formula of Cellulose is as shown in the Figure.1

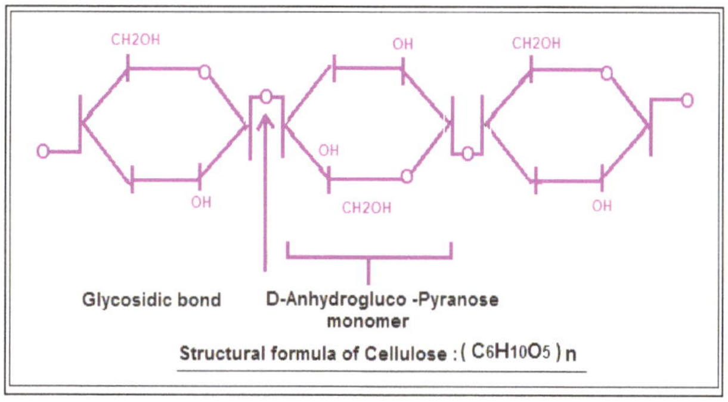

Figure 1: structural formula of Cellulose ($C_6H_{10}O_5$)n

The three most common degradation factors of cellulose have been identified which are thermal, oxidative, and hydrolytic. When cellulose is subjected to a temperature of 200 ^{o}C, the beta linkages (glycosidic bonds) tend to break and open the glucose molecule rings and thereby lose mechanical strength. The by-products of this reaction;

a. free glucose molecules,
b. Moisture,
c. CO & CO2
d. Organic Acids.

Moisture is also a byproduct of oxidation of hydroxyl ion or group.

The moisture produced as explained above, is also transferred to the oil.

6.0 Water content in paper insulation of winding;

In any mineral oil filled Power transformer, the amount of paper insulation is considered to be approximately 4% of the total weight of the winding with paper insulation Since, paper is highly hygroscopic, it is bound to contain some water from its manufacturing stage unless otherwise it is properly dried at the OEM's works.

Guidelines for interpretation of % moisture by dry weight of paper as per IEEE 62-1995[6] is shown in the table.1

% moisture content by dry weight in paper	Condition
0-2	Dry paper
2-*4	Wet paper
>4.5 Excessively wet paper	

Table1

6.1 Weight of water in paper insulation of winding (Sample calculation).[7]:

A transformer of 20 MVA,132/33kV EMCO made had failed after rendering service for >20 years was written-off for auctioning in the MP State Transmission Utility(India). The calculation of weight of water the paper insulation must have contained is calculated in the table 1(a).

Sr no.	20MVA,132/33kV, Transformer(written-off for auction)	Calculation	Weight
1	Wt of core and winding		18100kg
2	WT of copper winding with insulation 40% of wt of Core-coil assembly	18100 x 0.4	7240 kg.
3	Wt of paper insulation@ 4% of Wt of copper winding with insulation	7240 x 0.04	289.6 kg
4	Wt of water present in the paper considering as *5% as per table 1	289.6 x 0.05	14.48kg

Table.1 (a)

7.0 Phenomena of Water migration (Water dynamics): There is always some moisture present in the Transformer as the. paper is very hygroscopic and has greater affinity to water. Most of the moisture present would be in paper only.

Water migrates between the paper and liquid insulation in a transformer with changes in load and therefore temperature. Consequently, the concentration of water-in-oil alone expressed in parts per million does not provide sufficient information to arrive at for adequate evaluation of dryness of insulation system. Relative saturation provides a better evaluation under wide range of operating conditions and temperatures.

7.1 Percentage saturation of water in mineral oil [2]: The quantity of dissolved and dispersed water in mineral oil is significant for two reasons (i) presence of polar water molecules in the, mineral oil adversely affect the dielectric properties of the Mineral oil and (ii)The amount of moisture in the oil can be reflective of amount of moisture in the paper insulation. The solubility in mineral oil is temperature dependent. Therefore a statement as mg/kg of water in the oil without temperature information is insignificant. The calculation of % of water saturation, (mg/kg water/mg/kg of water at saturation) x100, has greater significance as it indicates the possibility of free water formation in the oil. Free water in oil results into poor the dielectric strength of oil.

Table 2 gives general guide lines for Interpretation of data expressed in % saturation.[7]

% Water saturation of oil	Condition
0-5	Dry insulation
6-*20	Moderate to wet; Lower nos indicate fairly dry to moderate levels of water in the insulation whereas Values towards upper limit indicate moderately wet insulation.
21-30	Wet
>30	Extremely wet

Table.2

7.2 Formula for Solubility of water in Oil and Specimen calculation [8]

$Log_{10}S_o = (-\frac{1567}{K}) + 7.0895$..(1) Where

S_o... is the solubility of water in Mineral oil.

K......is the absolute temperature in Kelvins. (^oC + 273)

^oC is the Oil temperature in Celsius at the time of sampling.

% Saturation= 100(mg/kg of water)/S_o

7.2(a) Specimen calculation;

Water Content =16 mg/kg,

Temperature of Top oil=30^oC

Temperature in Kelvin=30^oC+273=303 K

$Log_{10}S_o = (-\frac{1567}{303}) + \mathbf{7.0895} = -5.1716 + 7.0895 = 1.9197$

$S_o = 82.775$

Therefore % Saturation $= \left(\frac{16}{82.775}\right) 100 = 19.32\ \%$

7.3 Table-3, shows Solubility (S_0) of water as derived from the equation No1 in the paragraph 7.2. at temperature interval of 10^oC.

Oil Temperature in ^{o} C	Water Content (S_0) in oil, ppm
0 ^{o}	22
10 ^{o}	36
20 ^{o}	55
30 ^{o}	83
40 ^{o}	121
50 ^{o}	173
60 ^{o}	242
70 ^{o}	331
80 ^{o}	446
90 ^{o}	592
100 ^{o}	772

Table- 3: Water content in Oil at different Temperatures.

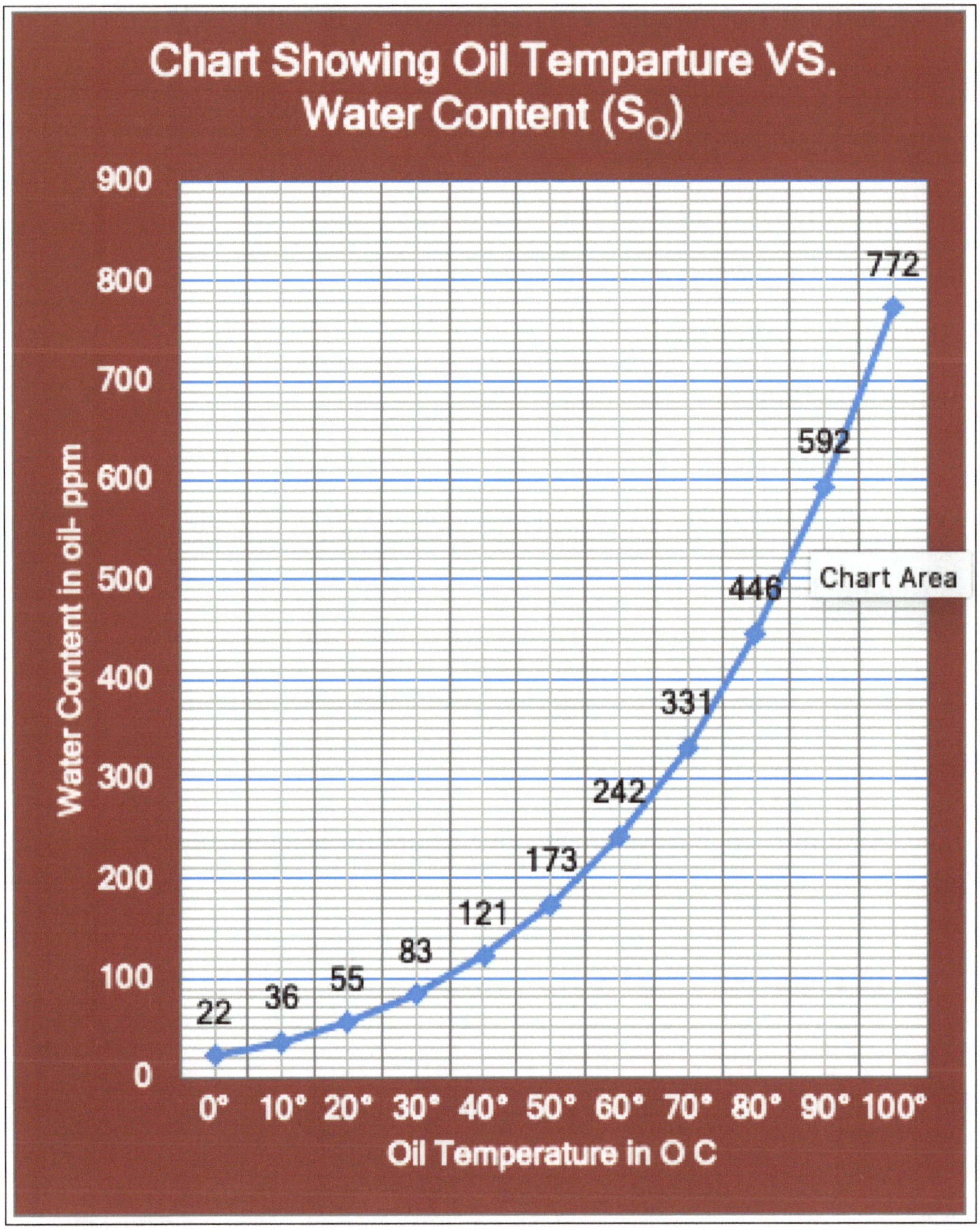

Figure-2: Graph showing solubility of water in the transformer oil at different temperature.

8.0 Criteria for selection of the old Transformers for refurbishment:

i. The transformers which had served more than 20 years.
ii. Deterioration in IR and PI values.
iii. Water saturation in oil if found >*15%.[2]

*Note: this figure is selected looking to the table.2. Indicating guidelines for interpretation of % saturation of water in oil.

8.1 Selection of Transformers for Refurbishment:

A sample list of a few old Transformers which have rendered services>20 years with the calculated RS% is shown in the Table.4

Sl. No.	Name of Substation	Name of Transformer	TR. Sl. No.	W.C. (PPM)	Temp. °C	Calculated R.S.%	Remarks
1	132KV S/s Patan	40MVA 132/33KV BBL	473372	16	46	9.24	CAL at 50° C
2	132KV S/s Patan	20MVA 132/33KV GEC	1328571	18	38	14.87	CAL at 40° C
3	132KV S/s VFJ	40MVA 132/33KV BBL	4733/21	15	52	8.67	CAL at 50 °C
4	132KV S/s VFJ	5MVA 33/11KV Kirloskar	50008304	36	36	29.75	CAL at 40 °C
5	132KV S/s VFJ	5MVA 33/11KV Accurate	18343/25	30	54	17.34	CAL at 50 °C
6	132KV S/s Kymore	16MVA 132/33KV NGEF	2800034714	21	50	12.13	CAL at 50 °C
7	132KV S/s Kymore	20MVA 132/33KV BHEL	6004548	27	50	15.60	CAL at 50 °C
8	132KV S/s Kymore	5MVA 33/11KV UE	183775	38	48	21.95	CAL at 50 °C
9	220KV S/s Jabalpur	40MVA 220/132KV BHEL	6004131	20	50	11.56	CAL at 50 °C
10	220KV S/s Jabalpur	40MVA 220/132KV BHEL	6004129	24	52	13.67	CAL at 50 °C

Table - 4: Sample list showing transformer which have served more than 20 years. with moisture contents of the Oil at the sampling temperature.

8.2 Calculation of % saturation of water in oil with true values:

A sample case of calculation of % saturation of water in oil in respect of 20 MVA, 132/33 kV BHEL Transformer bearing Sr. no. 6004548 at 132kV S/s Kymore appearing at Sr. No 7 in the table-4.is depicted here-under;

The Water content of Oil at 50° C as reported by Oil testing laboratory is 27ppm. from the curve at figure-2, the Solubility of Water in oil at 50° C is 173 ppm.

a. Therefore % Saturation of water in oil = (27/173) x 100 = 15.60%.
b. Since *15.60 > 15%, this transformer is identified for refurbishment.

*Note: 15% saturation of water in oil was taken as the bench mark for selection of the old transformers for refurbishment.

9.0 Filter plant with brief description of Salient feature thereof:

a. Before going into detailed method/technique adopted for on-line filtration process, it is imperative to know about the filter Machine. The flow diagram of a conventional filter machine of 500 GPH is depicted in the figure No (3).

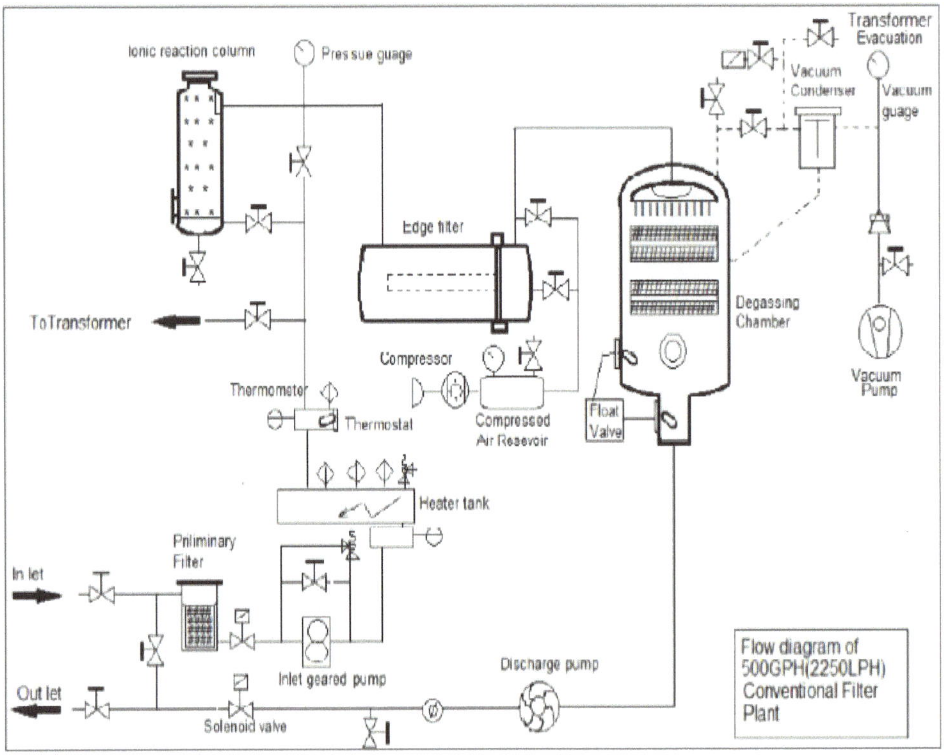

Figure 3: Flow diagram of 500GPH(2250LPH) conventional Filter Machine.

b. **Various Components of filter machine/plant and their functions:-**

1. **Preliminary Filter:**

Preliminary filter is at the oil inlet and is provided with magnetic strainer and a perforated sheet. The magnetic strainer prevents magnetic particles from entering the gear pump. This perforated sheet prevents the particles having size above 1mm from entering the gear pump. This is used for rough particles filtration and to arrest- Magnetic particles this protects inlet gear pump.

2. **Oil inlet Pump:**

This is positive displacement type of gear pump. This pump functions to force the oil through the filter at required pressure and flow rate. The input side oil is cool and more viscous there for it is preferable to use gear type pump.

3. **Heater Tank:**

Heater tank is made up of m.s. welded construction and is thermally insulated with mineral/glass wool. The electric heaters provided are of indirect type and their design is such that over heating or localized heating is eliminated.

Thermostats are provided to prevent overheating. It is automatic in operation. Thermostats are provided to maintain the desired temperature. The total power is 48/200 kW., divided in three groups. Heaters groups are interlocked with Inlet Pump and shall not be in ON position, unless the Inlet Pump is working. This heater tank functions to heat the oil at required temperature.

4. **Centrifugal Pump:**

This is mechanical seal type oil discharge pump. This pump is provided at the outlet of degassing column and can draw the oil from degassing chamber held under vacuum and feed it to the transformer. The output side oil is hot and viscosity is low, so we prefer to use centrifugal type pump.

5. **Degassing chamber:**

A shower arrangement is provided in the degassing chamber. Oil comes out of the shower arrangement in the form of shower and falls on the cage filled with rasching rings. The purpose of rasching rings is to expose the incoming transformer oil to optimum surface area for efficient degassing & dehumidification. A Sight Glass with Illuminating Lamp shall be provided for observation of oil flow.

6. **Edge Filter:**

This consist of two chambers one for incoming dirty oil and the other for outgoing clean oil. Several candles of filter packs are held under spring compression in the dirty oil tank. The spring pressure can be adjusted by tightening or loosening the nuts. These candles consist of stacked ring type filter papers. Oil enters the candles in radial direction and flows towards clean oil chamber in axial direction. Filters shall be easily changeable.

This shall be useful for removal of sludge content in the used oil. This has got an oil inlet, outlet, drain and air pressure connections for clearing purpose. This is a very fine filter and it filters the oil to more than 1 micron size particles.

7. **Air Compressor:**

It is provided for edge filter clearing by passing the compressed air. This compressor can also be used for spray painting and other purpose.

8. **Ionic Reaction Column:**

It content of activated alumina (20/200kgs). If the initial acidity of oil is higher as compared to IS: 1866-2000 [9] or IEC 60422-2013[10], then this column can be taken in oil circulation to reduce the acidity of the oil.

9. **Float Switch (High Level):**

This is provided to prevent the rise of oil level in the degassing chamber to trip the inlet pump.

10. **Float Switch (Low Level):**

When the oil level in the degassing column (low level) falls to a specific level, the low level float switch opens the circuit and trips off the discharge pump thereby preventing it from dry running.

11. **Vacuum Pump:**

A rotary oil sealed vacuum pump WSVP 9060 (for 2250 LPH)/SR 2500 B (for 6000 LPH) is provided for evacuation of the plant.For 6000 LPH machine two vacuum pumps are installed. First pump to create vacuum in first stage and second is for second stage of degassing column and transformer.

12. **Oil Sampling Valve:**

Processed oil sample can be collected from this valve.

13. **Thermostats:**

Four thermostats are provided on the heater tank to control the oil temperature. These are automatic in operation. One of them is used for temperature setting as a safety and the others are used for controlling the temperature, heating. The safety Thermostat shall be provided to take care of any accidental rise of temperature of oil and shall put off the Heaters in such eventuality. This Thermostat shall be set at slightly higher temperature than that of controlling thermostats. These thermostats are having a range of 30 to 110 °C.

14. **Thermometers:**

Two nos of dial type thermometer are provided on the plant which are located at the inlet of heater tank to indicate the temperature of incoming oil and the other at the outlet of heater tank to indicate the temperature of outgoing oil.

15. **Vapour Condenser:**

One No. of vapor condenser (baffle type) is provided in the plant to protect the vacuum pump and to trap the vapours of transformer oil. An oil drain line is provided and connect to the degassing chamber to collect the trapped oil inside vapour trap.

16. **Solenoid Valve at inlet & outlet**:

One no. Solenoid Valve at Inlet and One no. at Outlet is provided. Valve at the Inlet & Outlet shall open automatically. The moment oil Inlet & Outlet Pump are switched ON. In case of power failure, these valves shall be capable of preventing the oil from entering into the Plant and thus avoiding the possibility of mixing processed oil with unprocessed oil.

17. **Oil Hoses - 2 nos.:**

Two nos. Nitrile Rubber Hoses each 15/20 Meters long with flanged end connection on both sides are provided. One for the oil-inlet & other for the oil-outlet.Oil-hoses are capable of handling the transformer oil up to 100ºC (max.) and full vacuum.

Note: In 6000 LPH Filter plant two vacuum pumps and two roots pumps are installed. Vacuum pumps are to create vacuum and root pumps are to provided to create fine vacuum in the second stage. One water circulating (Cooling) pump is also installed to cool the the vacuum and root pumps.

10.0 REMOVAL OF WATER FROM WINDINGS

a. **Traditional hot oil circulation;** Traditional technique of circulation of hot oil through the transformer is widely used in many countries around the world. This technique has some limitations;
   i. The transformer needs to be taken out of service, as many cycles of oil circulation has to be carried out.
   ii. It removes moisture mainly from oil but removal of moisture from paper is very little.
   iii. Once the transformer is put back in service, the heat generated in the winding due to increase in load results in migration of moisture from paper to oil and thus moisture contents in oil increases.

b. **Process of On-line Filtration;** With the constraint as narrated in the paragraph 10.a), the only option is to resort to the "ON-LINE Filtration". Filter plant of adequate capacity is connected to the loaded transformer. At higher temperature moisture released by winding migrates to oil which is further removed by filter plant. As moisture is released from winding is at a very slow rate, there is no possibility of substantial improvement in breakdown voltage of oil during the process. This process may take few weeks to months depending upon quantity of water in the transformer and its temperature during load conditions. As a precautionary measure, heaters of filter plant are kept off during whole process.

c. **Technique/ Methodology adopted;**
   i. Complete oil from the transformer was decanted (drained-out in some *vessel/tank) cleaning of sludge at the bottom is done. After cleaning, vacuum of* 740 to 760 mm Hg.(1torr) applied on the transformer and maintained for 24 hrs.
   ii. In the mean while filtration of oil was taken up in the oil tank wherein the transformer oil was decanted.
   iii. After 24 hrs of vacuum, the filter machine is connected in such a manner that the oil-outlet (delivery) hose pipe of the Filter machine is connected to the top filter valve of the transformer and the Oil-inlet (suction) hose pipe is connected to the bottom filter valve of the transformer.
   iv. Oil is filled through the top filter valve of the Transformer..
   v. Air is released from all the pockets provided on the top cover and the Buchholz relay.
   vi. The transformer is switched on and observed for any abnormality without load.
   vii. After observing that no abnormality has taken place, the transformer is subjected to load.
   viii. On-load filtration of the transformer is carried out till moisture content come down to almost 10 ppm.
   ix. After 15 days of filtration a sample of oil is collected and tested for moisture content. If an increment in ppm is noticed in oil sample, then on-line filtration of transformer is carried

out until moisture content in the oil reduces to 10 ppm or very near 10 ppm. It is to mention here that on-line filtration of transformer eliminates water particles from the winding.

x. After a lapse of 3-months of the activity of on-line filtration is over, sample is taken for moisture contents test.

11.0 Sample test results; Results of a 20 MVA 132/33kV transformer Sr.no. 6004548 at 132kV S/s Kymore are as follows;

a I.R &PI values of 20 MVA 132/33kV BHEL Transformer Sr.No.6004548 before filtration in MΩ. (O.T. 46°C & M.F. 0.43)is shown in the table no.5.[11]

	R 15	**R 60**	**Converted**	**P.I**
HV-E	**111**	**132**	**56.76**	**1.18**
LV-E	**104**	**126**	**54.18**	**1.21**
HV-LV	**145**	**171**	**73.53**	**1.17**

Table 5

I.R &PI values of 20 MVA 132/33kV BHEL Transformer (sr.No.6004548) after online filtration in MΩ. (O.T. 50°C & M.F. 0.560) is shown in the table no.5.[11]

	R 15	R 60	Converted	P.I
HV-E	184	236	132.16	1.28
LV-E	181	254	142.24	1.4
HV-LV	270	372	208.32	1.37

Table. 6

Moisture contents and BDV;

Parameter	Before filtration	After filtration
PPM	23	13
BDV	60	76

Table. 7

12.0 Measurement of tanδ and capacitance of the bushings: tanδ and capacitance of bushings are measured half yearly. If tan δ of any transformer bushing is found approaching to 0.7 %, we consider it to be an alarming stage and process for replacement thereof is initiated. In case of capacitance of the bushings, if the variation is found to be > 5% action for replacement of the bushing is taken up.

13.0 Acknowledgement: The authors extend their sincere gratitude to the Managing Director and the Chief Engineer (Testing & Communication), M.P, Power Transmission Co. Ltd. for granting permission to use the data from the establishments of M.P. Power Transmission Co. Ltd and to permit serving officer of M.P, Power Transmission Co. Ltd., to be a co-Author, with them.

14.0 References;

1. Ministry of Power's, Govt. of India, notification No.S.O.266 (E) Dtd.29th March; 1994.
2. IEEE std.62-1995(Cl.6.3.12).

3. IS: 2026 (Part-7):2009(IEC equivalent IEC 60076-7:2005).Loading guide for Power Transformers.
4. IS: 1866-2000./IEC 60422/IEEE std. C57.106 -2013.
5. IEEE Std C57.152-2013; Guide for diagnostic field testing of fluid filled Power Transformers, Regulators and Reactors. (Annex-E).
6. IEEE Std 62-1995(Cl. 6.3.12).
7. Schedule "A "of Survey Report of 20MVA failed Transformer in 400kV S/s, M PPower Transmission Co. Ltd Katni.(Document Enclosed)
8. IEEE Std C57.106-2002)-Guide for Acceptance and Maintenance of insulating oil in Equipment. (Cl. 4.5)
9. 9) IS: 1866 -2000-Code of practice for electrical maintenance and supervision of mineral insulating oil in equipment
10. IEC 60422-2013 –Mineral Insulating oil in Electrical equipment supervision maintenance
11. IR value conversion factor chart followed in M.P. Power Transmission Co. Ltd.(Document Enclosed)

15.0 The Authors: K.K.Murty, 2) D.C.Jain, 3) R.K. Agrawal4) J.J.L Kapil.

19. "HOT SPOT" Temperature as Diagnostic Tool in Conjunction with DGA

1.0 Synopsis:

"HOT SPOT Temperature" as defined in IS: 2026-2 in Annex B-2 is reproduced here under;

"The maximum temperature occurring in any part of the winding insulation system is called the hot-spot temperature. This parameter is assumed to represent the thermal limitation of loading of the transformer".

If the same is ignored over a period of time, the paper insulation of the windings become brittle and the Transformer's life-expectancy shall reduce in accordance with IS: 2026-7/ IEC600-7.Contrary to this theory, there has been a case in MPPTCL wherein, a 63 MVA, 132/33kV Transformer's the DGA was showing deterioration in condition within a short period of 2 years but winding temperature indicator did not give any indication thereof. Perhaps, had there been Temperature sensors inside the Transformer as narrated in IS 2026-Annex B-4, the end user MPPTCL would have been alerted in the initial stages only. The Authors have narrated this case study indicating the methodology of accessing the Hot spot Temperature in the conventional practice as well as Optical fibre probe method.

2.0 OTI / WTI:

The Oil temperature is always indicated directly on an OTI meter, whether the Transformer is in service or off the service, however it is not the case with the WTI (winding Temperature indicator). The Winding temperature is proportional to the square of the current flowing through it (the replica thereof fed to the resistance through WTI-CTs), thickness of Kraft paper insulation wrapping and conductor-area for dissipation of heat into the Oil.

While in service at no point of time the winding temperature is actually measured. The winding temperature is measured only & only while Temperature - Rise test of a Transformer is being carried out at the Manufacturer's works. Since the Temperature-Rise Test is a Type -Test as per IS: 2026-1 & IEC 60076-1, generally only 1No - unit of the offered Lot is subjected to the same, the other units are not subjected to Temperature-Rise Test at all.

The winding temperature is based on the readings of resistance of the Winding taken at regular intervals of time, as per the test procedure laid - out in the IS: 2026 part-II / IEC: 60076-part 2.The resistance of the winding at the time of shutting-down the supply, is arrived at by extrapolation of the resistance to time zero, on Time vs Resistance curve. (ref.fig:1).

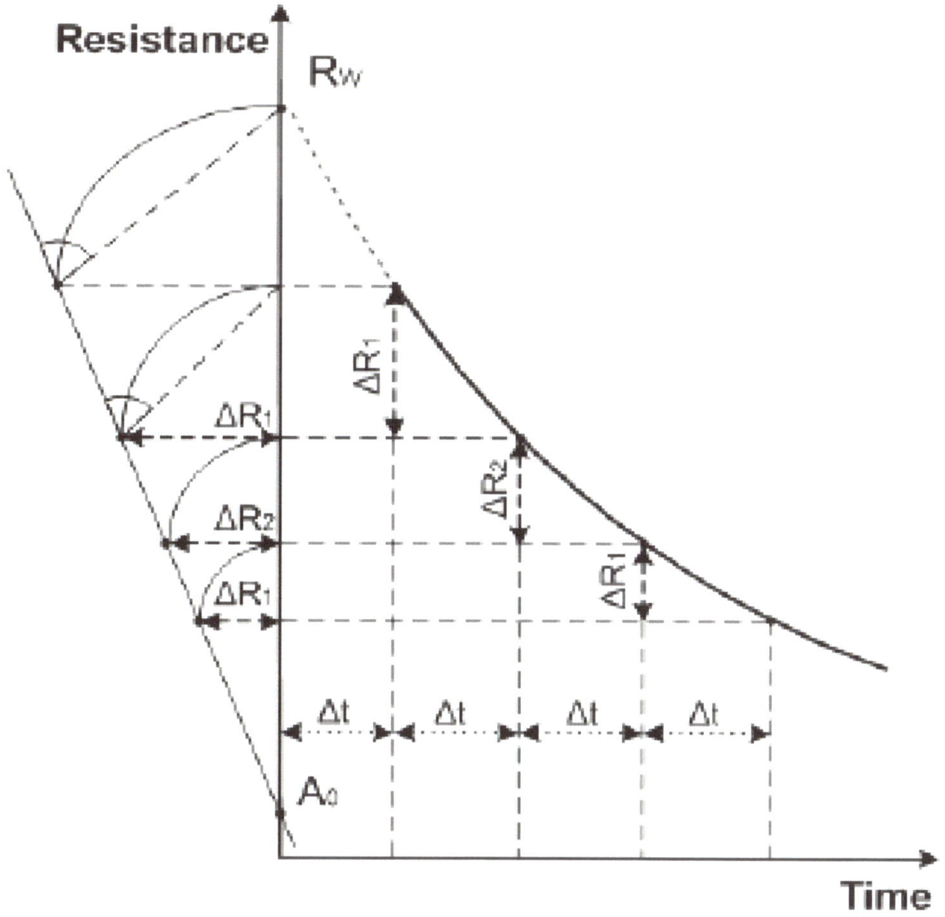

Fig. 5. Fall of the resistance at the instant of shutdown

Fig1: Winding resistance at time 0 through extrapolation

3.0 HOT-SPOT temperature: This is the maximum temperature at any point/part of the winding. Since it cannot be directly measured through conventional method, the same is arrived at through a formula which is based on thermal imaging.

3.1. Why Hot-Spot Factor?: Towards the upper end of the winding, there is usually a concentration of eddy current losses, therefore the windings in this portion are provided with extra electrical insulation which increases the thermal insulation. The actual local temperature between conductor and oil is therefore assumed to be higher by the 'Hot-Spot factor". The factor is assumed 1.1 for Distribution Transformers & 1.3 for Medium size Power Transformers.

The steady state Hot-Spot temperature rise above External cooling medium temp. (air) = [Top Oil temp. rise above cooling medium temp.+ {(Hot-Spot factor) x (Average temp. Difference, Winding to Oil)}].

3.2 Hot spot gradient & Hot spot temp. The following figure (Figure 2) depicts the Hot spot gradient & Hot spot temp., considering Hot spot factor as 1.3.

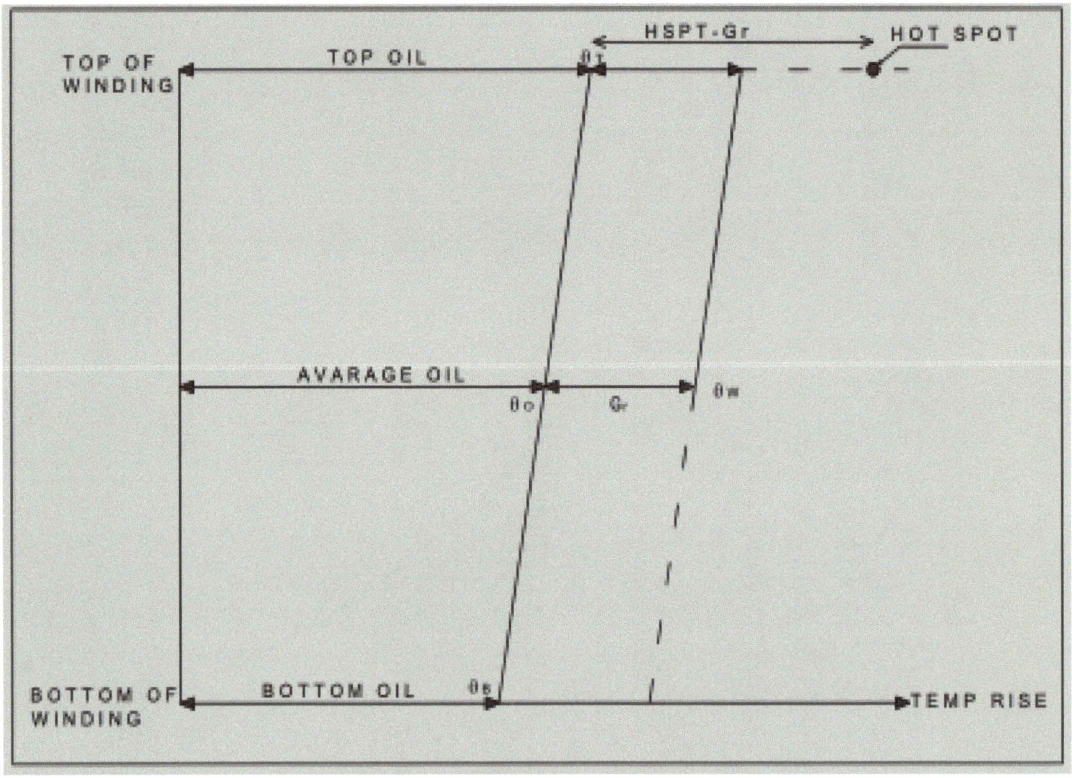

Figure 2: Hot-spot temperature gradient

As could be seen from the diagram above that the;

- Average oil temp$\theta 0 = \theta T + \theta B/2$. (Where T suffix is for Top & B is for bottom).

Average winding Temperature θw is derived from resistance measurement when the winding was carrying Rated current rated current just before disconnection of power supply.

Average winding gradient= $\theta w - \theta 0$

This is the Temperature- gradient between the Winding conductor and the transformer Oil surrounding the said conductor.

Typical average Winding Gradient is assumed as 10°C to 20°C.

- Hot -Spot gradient, Hspt gr. = 1.3 x Average winding gradient.

WTI is calibrated based on following equation;

$$\theta wti = \theta T + \text{Hspt gr}\left(\frac{I}{Iratd}\right)^{n}$$

Where,

- θwti = Hot Spot temperature as indicated by WTI.
- I = actual primary current measured by the CT.
- $I_{ratd.}$ = Rated full load current.
- n=exponent between1.6 to 2.

From the above, it is seen that the WTI reading is the assumption-based Temperature of the Winding.

4.0 Assumed Hot-Spot temp. vis' a vis Actual Hot- spot temperature.

The life expectancy of a Transformer depends on the deterioration of the cellulose insulation (Kraft-paper, press-board, wooden-block insulation etc), which is inversely proportional to rise in temperature (ref.IS:6600/IEC: 60076-7).

It is now, felt very essential to monitor the Hot-spot temperature online on real- time basis.

The fact, to the effect that the Hot-spot temperature obtained through conventional method is considerably lower than the actual temperature of the winding obtained through direct measurement. Such situations may shorten the life of the Transformers, un-noticed.

It has been seen through practical experiment carried out by Korean Engineers, using both conventional method and Direct measurement method by using Fiber Optic Sensors, simultaneously on a 154kV, 15 MVA Power Transformer, wherein the temperature recorded by Fiber Optic sensor was found to be 13.3°C higher than the conventional thermal imaging method at an ambient Temp of 24.7°C. (Reference:3).

5.0 Brief Description of the "Fluoroptic-Technology" & the Optical fiber sensor;

5.1. Fluoroptic-Technology: The advantage of the inherent optical properties of Phosphorescent materials has been taken-in, in developing the Fluoroptic Technology.

i. In the Fluoroptic Technology, a **sensor** (A phosphor, similar to the phosphor used in TV screen) is connected to the **instrument*** through a fiber-optic cable.
ii. The **instrument*** sends **pulses of light** to the **sensor** through this fiber-optic cable.
iii. The pulses of this light **"Excites**" the **phosphor** causing it to **'emit light'** at **different colours (long wavelength).**
iv. The **"emitted light"** Couples back into fiber-optic and **travels back** to the **Instrument*.**
v. Since, the **excitation light and** the **Emitted light** are of **different colours,** the **instrument*** can **distinguish between the two.**
vi. **The instrument*, determines the temperature of the sensor by measuring the decay time of the emitted light.**

It is a **persistent property** of the **sensor** that its **decay time varies precisely with temperature.**

Generally speaking, the **colder the sensor, the longer the decay time** of the **Phosphor's emitted light.**

Index:- *instrument*= The Optic Monitor/ Controller, Which, in the case of Transformer is mounted on the Transformer Wall.*

1.2 Fiber Optic Sensor Device: There are mainly three components;

i. Fiber Optic sensor. They are 2 types;

a. Probe type.

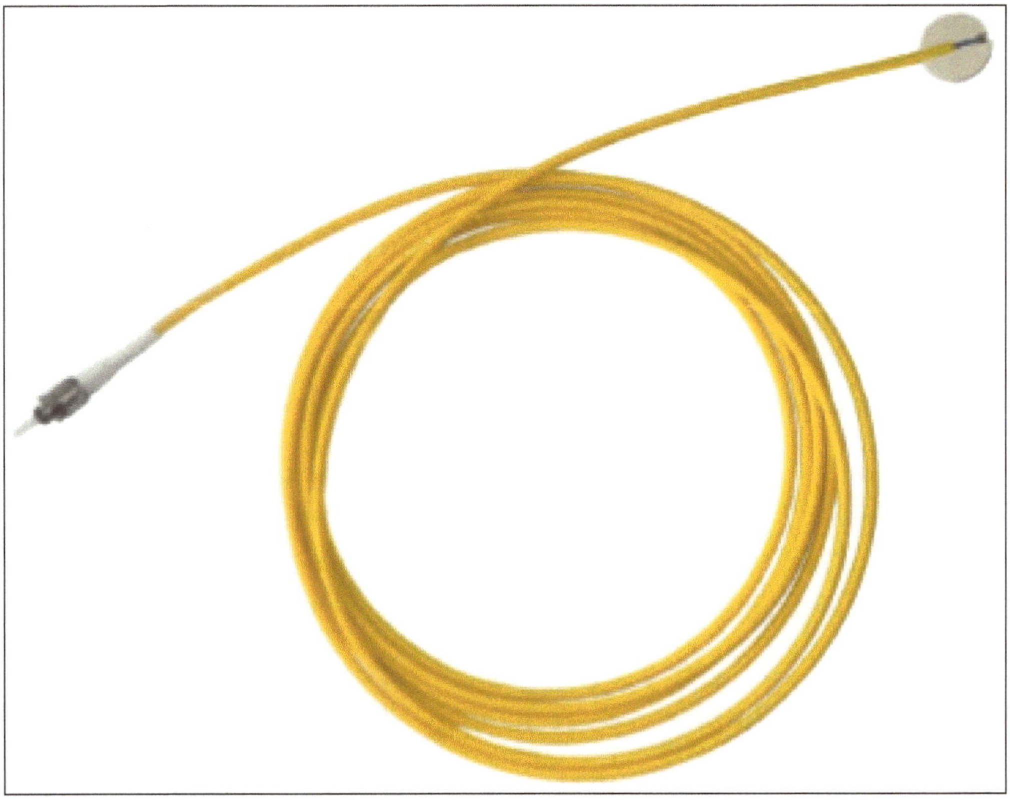

Fig3(a): Probe Type FO sensor device

b. Embedded in the Spacers

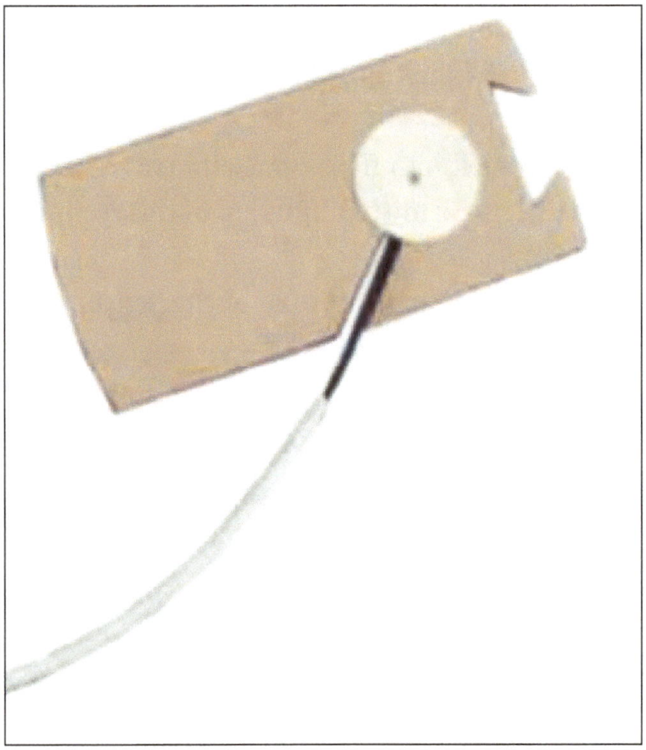

Fig3(b): Embedded in the Spacers type sensor

ii. Transfomer Tank-wall Plate assembly;

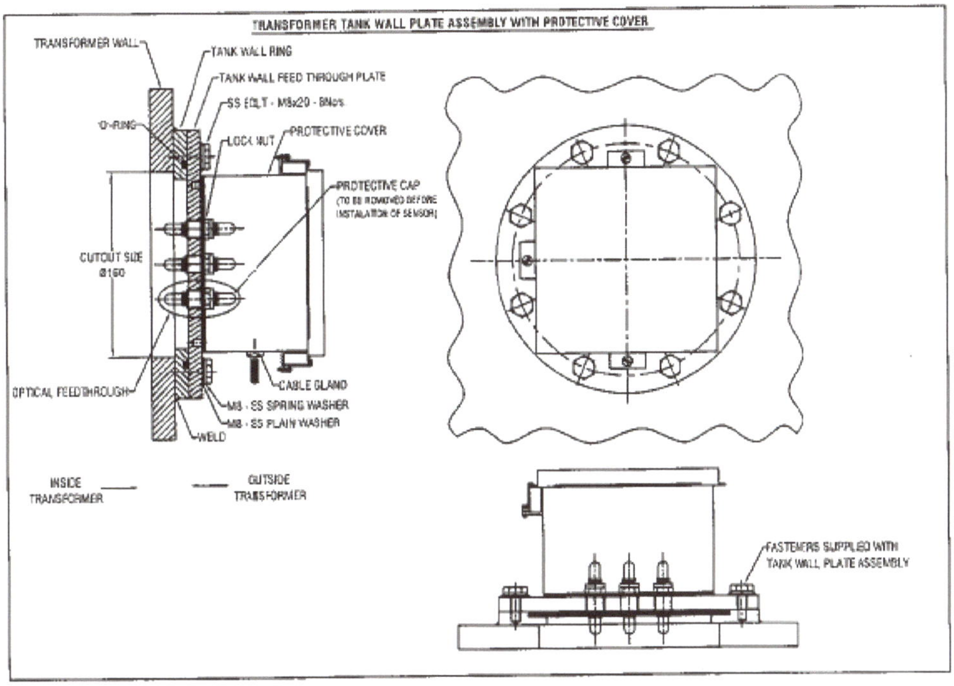

Fig4: Transformer Tank-wall Plate assembly;

6. Placement of Sensor probes: The sensors are inserted / placed in the top portion of the HV & LV winding, this portion is considerably hotter due to **spurious eddy currents which are at their highest value. The figure bellow shows locations of Losses.**

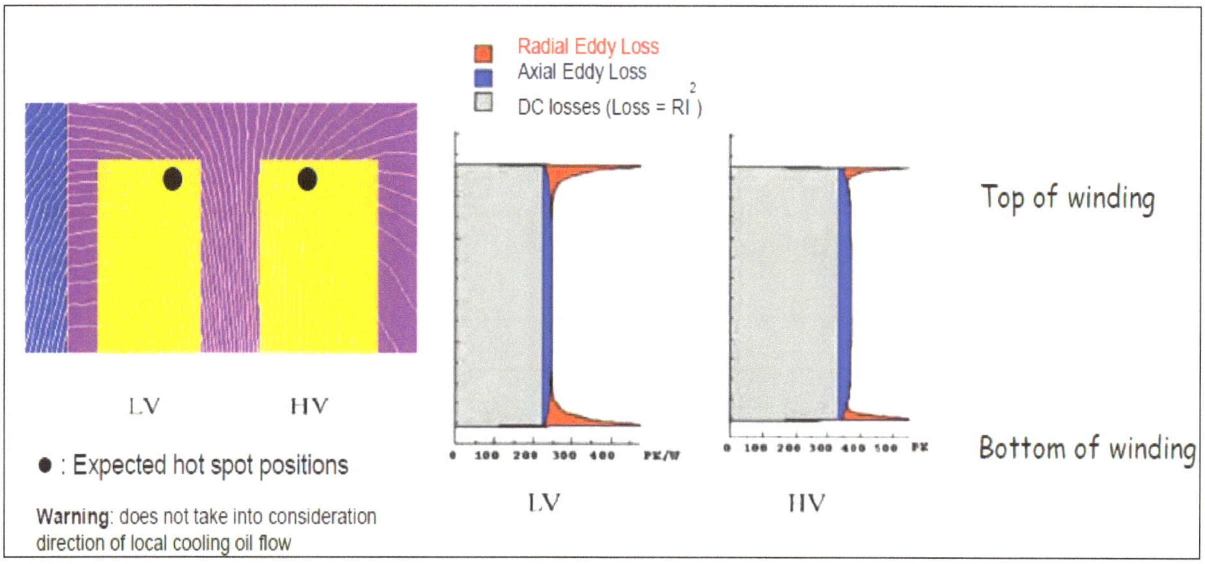

Fig5: Location of losses

The recommended Fiber-Optic Probes is 8-10 nos, in 3- Ph Transformer windings and 4-6 Nos.in Single Phase Transformers.In the fig below, the red spots indicate the placement of Fiber Optic probes in HV and LV. windings.

The recommended Fiber-Optic Probes is 8-10 nos, in 3- Ph Transformer windings and 4-6 Nos.in Single Phase Transformers.

In the fig below, the red spots indicate the placement of Fiber Optic probes in HV and LV. windings.

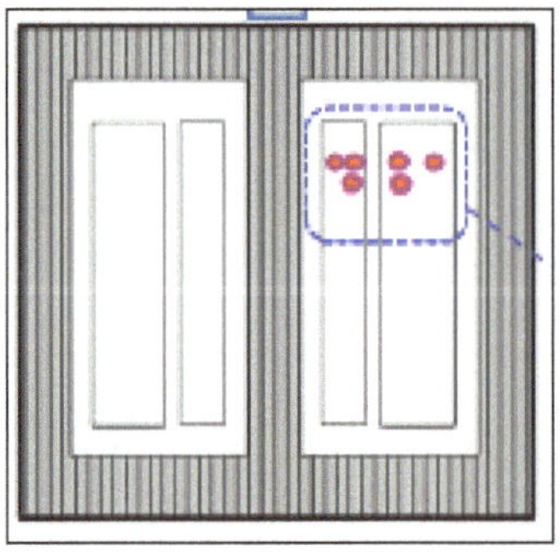

Figure 6: placement of fibre optic probe(red spots).

Figure 7: Insertion of Fibre- Optic Probes

iii. Multi-channel Fibre Optic Monitor/Controller;

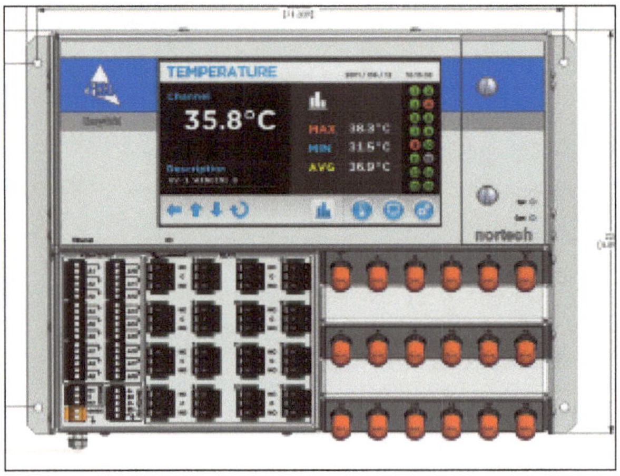

Figure 8: Multi-channel Fibre Optic Monitor/Controller;

The measurement of Hot-Spot Temperature shall be displayed on and also recorded on real-time basis in the fiber-optic multi- channel Monitor/controller. The output of this monitor/Controller can be directly sent for further remote SCADA applications, through Ethernet/RS232-RS 485 & compatible protocols like Modbus & IEC 61850 protocols etc.

Note: The figures shown above are only indicative of the main components of fiber-Optic Hot- spot measurement device. This is to confirm that Make/Manufacturer's names etc have not been shown / displayed on any of the figures / drawings anywhere, therein. The display of the above figures / drawings, is in no-way connected to any advertisement.

6.0 DGA (Dissolved Gas Analysis):

While in service, particularly the EHV Transformers, are subjected to many adverse conditions, such as, Over-loading, System Over-voltages, Over-fluxing, Under- frequency, Switching Surges, Lightening Surges, Effects Due to Harmonics developed by non-Linear loads and severe faults on the LV side etc. The type of Operating conditions and conditions of some of the faults give rise to De-composition of the insulating oil inside the Transformer under service.

The Scientists and Engineers have studied the different conditions under which the mineral insulating oil decomposes and key-gasses are evolved which in-turn give significant indication in respect of the health of the Transformer.

7.0 Phenomena of Decomposition: Electrical & thermal faults give rise to session of some of the bonds of different hydrocarbon molecules which in-turn convert into free radicals or in ionic forms which combine rapidly through complex reaction to form the Methane(CH_4), Ethane(C_2H_6), Ethylene(C_2H4) Acetylene(C_2H_2).s which get dissolved in the Oil.

Low energy faults: Partial Discharge of cold plasma Type (Corona discharge) through ionization reactions and accumulation of Hydrogen as main recombination gas.

Methane and Ethane are formed below 500°C while Ethylene is formed above 500°C. Acetylene requires Temp. of 800°C to 1200 °C. Acetylene is formed mainly in arcs, where temperature is thousands of degree Celsius.

Carbon Particles form at 500°C to 800°C and are generally observed after arcing in the Oil or very hot spots.

Oil may oxidize with formation of small quantities of CO and $CO_{2,}$ which can accumulate over a long period of time.

8.0 Decomposition of Cellulosic Insulation:

Paper, Press-Board and Wood Blocks are in the category of Cellulosic Insulation which decompose at the temperature at >105° C with complete carbonization >300° C. Mostly Carbon Monoxide, Carbon Dioxide and water are formed in much larger quantities than by oxidition of oil at the same Temperature, together with minor amounts hydro carbon gases and furanic compounds.

CO and CO_2 formation increases not only with temperature but also with the Oxygen content of oil and the moisture content of paper.

9.0 Category of Faults:

On the basis of Data available of the internal inspections of large Nos. of faulty Transformers world-over, the reasons for development of faults are broadly classified as under;

- Partial discharge(PD) of cold plasma-wax deposition on paper insulation, or of the sparking type, including pin holes, Carbonized perforation(punctures) in the paper.
- Discharge of Low energy(D1), in Oil or /and paper, give rise to larger carbonized perforations(Punchers), Carbonization of the paper surface(Tacking) or Carbon Particles in oil(as found in OLTC Diverter switch).
- Discharge of High energy(D2) in oil or/and in paper, with power flow- through evidenced by excessive destruction and carbonization of paper, metal fusion, extensive carbonization in oil. In some cases Tripping of the Transformer is the evidence of excessive Power-flow through.

  - Thermal Fault in Oil or/and in paper,
    Below 300°C (T1) – If the paper turns Brownish.
    Above 300°C (T2) – If it has carbonized.
  - Thermal Fault Due to high temperatures (T3);
    Above 700 °C – If there is strong evidence of Carbonization of oil.
    Above 800 °C – If there is evidence of metal colorization.
    Above 1000 °C – If there is appearance of Metal fusion.

Table of Abbreviations

PD	Partial Discharge
D1	Discharge of Low energy.
D2	Discharge of high energy.
T1	Thermal fault, t<300°C
T2	Thermal fault, 300°C,<t<700 °C
T3TT T3	Thermal fault, t>700 °C

Table1

10.0 Typical Faults in the Power Transformers;

Typical faults related to the Transformers have been shown in the following table (Table- 1.A of IEC: 60599 is reproduced below).

Table A.1 – Typical faults in power transformers

Type	Fault	Examples
PD	Partial discharges	Discharges in gas-filled cavities resulting from incomplete impregnation, high-humidity in paper, oil supersaturation or cavitation, and leading to X-wax formation
D1	Discharges of low energy	Sparking or arcing between bad connections of different or floating potential, from shielding rings, toroïds, adjacent disks or conductors of winding, broken brazing or closed loops in the core Discharges between clamping parts, bushing and tank, high voltage and ground within windings, on tank walls Tracking in wooden blocks, glue of insulating beam, winding spacers. Breakdown of oil, selector breaking current
D2	Discharges of high energy	Flashover, tracking, or arcing of high local energy or with power follow-through Short circuits between low voltage and ground, connectors, windings, bushings and tank, copper bus and tank, windings and core, in oil duct, turret. Closed loops between two adjacent conductors around the main magnetic flux, insulated bolts of core, metal rings holding core legs
T1	Thermal fault $t < 300$ °C	Overloading of the transformer in emergency situations Blocked item restricting oil flow in windings Stray flux in damping beams of yokes
T2	Thermal fault 300 °C $< t <$ 700 °C	Defective contacts between bolted connections (particularly between aluminium busbar), gliding contacts, contacts within selector switch (pyrolitic carbon formation), connections from cable and draw-rod of bushings Circulating currents between yoke clamps and bolts, clamps and laminations, in ground wiring, defective welds or clamps in magnetic shields Abraded insulation between adjacent parallel conductors in windings
T3	Thermal fault $t > 700$ °C	Large circulating currents in tank and core Minor currents in tank walls created by a high uncompensated magnetic field Shorting links in core steel laminations

Table 2

11.0 HISTORY/ CASE STUDY: There has been a case of a 132/33 kV, 63 MVA Transformer in MPPTCL, wherein DGA carried out at some intervals, revealed that there appeared to be development of fault which came true, when the transformer was got internally inspected after evacuating the complete oil from the Transformer at site.

12.0 The Details pertaining to The Transformer are as follows;

- **Capacity;** 132/33kV, 63 MVA, Yyn0 Transformer,
- **Date of Commissioning**: 08-08-2011.
- **Date of removing from Service:** 22.12.2014.
- **Max Load recorded**: on 132kV side; 226 A as against 272.55A.

- **Max Temp.Recorded**;OT-63°C, WT-68,68,69 °C
- **Failure of 33kV side bushings:** (i) B phase, 72kV bushing on 13.06.14 & R ph 72kV bushing on10.08.14.

13.0 Observation from the reports of DGA;

DGA was taken on regular intervals. It was found that the key gasses generally showed increasing trend with every subsequent DGA.also the formation of the CO and CO_2 found in increasing trend.

Date	09.02.12/ 17.02.12	16.11.12	05.08.13	02.05.14	10.07.14	11.11.14	27.11.14	05.12.14
Methane	1.133	0.27	0.95	5.27	10.52	18.73	23.7	20.51
Ethane	0	0	7.11	1.46	0.81	4.31	1.88	2.22
Ethylene	0	0	0.24	23.8	16.43	48.45	37.72	36.13
Acetylene	0	0	-	0.036	16.046	37.88	109	132.42
CO	18.43	62.5	43.19	246.04	0.26	11.50	107.19	95.53
CO2	91.99	37.16	59.73	572.73	138.66	260.76	73.49	47.26

Table 3

Photographs taken during internal Inspection at site are attached here-with as annex-1.

14.0. Analysis:-

14.1 Probable causes of formation of Dissolved Gasses:-

- As could be seen from the Photographs that Carbonization/burning/ pitting of HV windings "U & W" phases have taken place in a very short period after commissioning (3 yrs, 4 months approximately), It is a probability that the Kraft paper used by the manufacturer does not appear to be of good quality (as against 25 years of Life span the Transformer has developed such condition in only 3 yrs & 4 months).Once Partial discharge takes place the same would gradually spread and Carbonization/ burning/pitting shall take place.
- Increase in Key- gasses is indicative of Discharge through High energy, causing Tracking, arcing etc.
- Failure of LV bushings may have made heavy current to flow through the windings, categorizing in Discharge of high energy, giving rise to pitting of the winding conductor and burning of Paper insulation.

Note: Bushing failure taking place of 72 kV class cannot be due to power frequency over voltage on 33kV side, LI voltages on LV side or due to severe short-circuit condition in LV side feeders

Since similar U phase bushing has failed within a short period of failure of W phase bushing, the lower porcelain of the bushings have found damaged in both the cases, the bushings were also of the same batch of manufacturing, It is indicative of some manufacturing snags/deficiency.

- The increase of Methane and Ethane gasses indicate fault of about 300°C. and increase in Ethylene indicates that the fault must have developed giving rise to a temperature of more than 500° C. Increase in Acetylene shows that the fault must have developed which has given rise to temperature about 800°C or beyond. Increase in CO and CO_2 is indicative of decomposition of Cellulosic insulation. ie decomposition of paper insulation due to increase in temperature.
- The hot spot must have developed which the conventional WTI could not have detected. Had there been Fiber-Optic Hot-spot measurement device, the actual temperature on real-time basis could have been known & remedial measures could have been taken.(Pl ref to para 4.0 wherein experience of difference in Conventional method and Fiber-optic methods to the tune of 13.3°C has been experienced by the Korean Engineers) in a 15 MVA Transformer)

15.0 Conclusion: Irrespective of the age of the transformers, old or new, need for scrupulous monitoring of DGA and also HOT-SPOT Temperatures are essential. The Hot- spot temperature differs considerably between the On-line fiber-optic measurement device and the conventional measurement method which is purely on assumption basis. The utilities must think of procuring Transformers with Fiber Optic measurement devices for Hot spot measurements in-line with as described in IS: 2026 Annex B-4.

16.0 References:

1. IEC: 60076 -2, IEC:60076-7 & IEC:60599.
2. IS: 2026 part 2.& IS:6600.
3. Article "A study on the Hot Spot Temperature in 154 kV Power Transformers " by Korean Engineers Mr.Dong-Jin Kweon & his fellow engineers. Published in Journal of Electrical Engineering & Technology Vol 7, No.3.

17.0 The authors: Er.K.K.Murty & Er.R.K.Agrawal

18.0 Enclosed: 4 Nos Photographs in the Annex- 1.

Annex-1

Photographs taken during internal inspection of the132kV/33kV, 63 MVA Transformer at site without dismantling (Only HV windings could be inspected) Carbonized/burnt marks found on the U ph and W ph windings

SECTION-3

ARTICLES ON RELAYS & PROTECTIONS

1. Basic Concept of Reading Schematic Electrical Drawings in Conjunction with Some of the Device Numbers as Per ANSI.

1.0 Introduction; It is the general consensus that any drawing must speak by itself.

As far as Schematic Electrical Drawings are concerned, they are prepared as if;

- All the isolators or disconnecting Switch are in Open position.
- Circuit breakers are in opened position.
- Spring of Spring-charging motors are in discharged position.
- All relays are in de-energized /inoperative position.
- All the contractors are in de-energized position.

So, while reading such Electrical schematic drawings, one must read the drawing by imagining that all the above-mentioned items ie Beakers and isolators are in closed position and protective relay contact are in closed position.

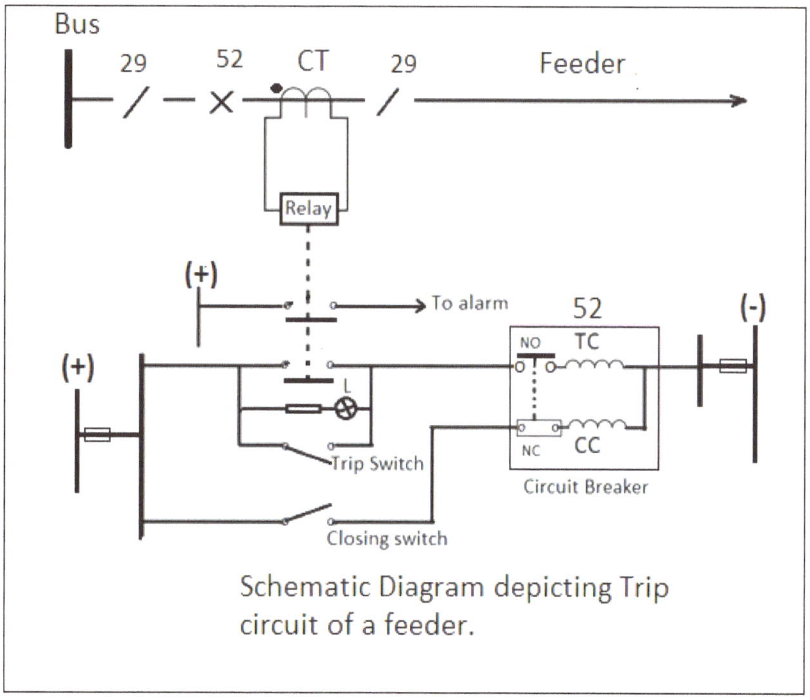

Figure 1: Schematic Diagram of a feeder.

In the figure 1,

- Device nos.;
  a. 29: isolator.
  b. 52: Circuit breaker.
- TC: Trip coil & CC: Closing coil.
- Isolators are in open.
- Circuit breaker is in off position.

Aux. contacts: NO (Normally Open) in Trip Circuit.

NC (Normally Closed) in closing circuit.

Note: NO or NC are the position;

- of the Auxiliary contacts when the Breaker is in off position.
- Contactors are in de-energized position.
- Relays are in de-energized position.

- L: Trip circuit healthy lamp.
- 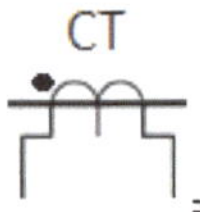

= CT with polarity marked with a dot.

2.0 Type of contacts;

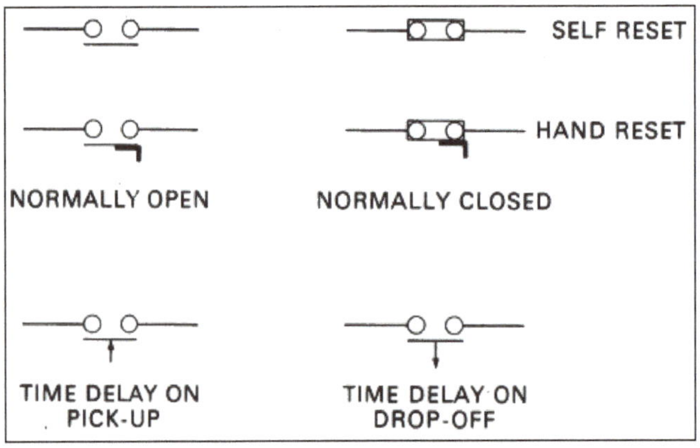

Figure 2

3.0 Relay Tripping Circuits;

Auxiliary contactors can be used to supplement Protective relays;

a. Series sealing.
b. Shunt sealing.
c. Shunt reinforcement with sealing.

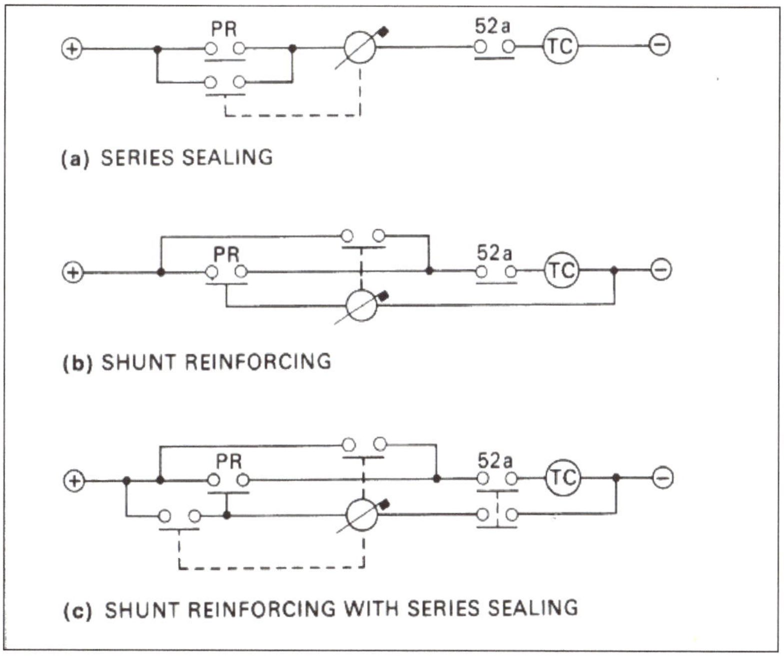

Figure 3: Sealing arrangements.

4.0 Trip circuit Supervision: It is very essential to supervise the healthiness of the trip Circuit. In order to supervise the healthiness of the Trip circuit there are 3 ways generally used as depicted in the figure 4

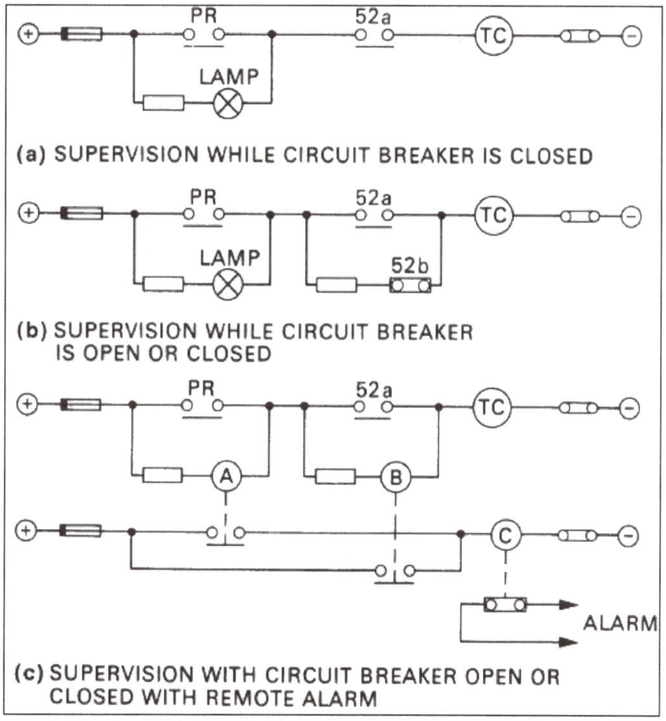

Figure 4: Trip-circuit supervision.

a. In the circuit shown in (a)a lamp known as Trip circuit healthy lamp connected in parallel to the Protection Contact with an adequate resistance.
Constraint: This shall glow only when the Circuit breaker is in on condition.

b. The advantage of this circuit is that the status of trip circuit can be verified even when the circuit breaker is in off condition. This is achieved by connecting the NC 52b (normally closed contact) in parallel to NO 52a (normally open contact), contact of the Circuit breaker auxiliary contact.
c. This has an advantage as in case of disruption in the Trip circuit healthy, an audible alarm can be heard. It is equipped with an alarm circuit through relay C's NC contact.
   Analysis; When the breaker is in closed position, the auxiliary contact 52a will close and also Relay A will operate to close the contact in series with relay C. This in turn will energies relay C and open its NC contact, thus, the alarm circuit is disrupted.

   - As soon as any disruption occurs in the Trip- circuit healthy circuit, relay A will get de-energized opening its NO contact, de-energizing relay C resulting into completing the alarm circuit through its NC contact. Then an audible alarm can be heard by the operator on duly.
   - Likewise, during breaker off position Relay B will be in actuated condition closing its NO contact in the alarm Circuit actuating the C relay and keeping its NC contact open, so long the trip circuit is healthy. If there is any disruption in the Trip healthy in the circuit breaker Off condition, the NO contact of relay B will open de-energizing relay C resulting in closing the alarm circuit through its NC contact. Then an audible alarm can be heard by the operator on duly.

5.0 Sample Schematic control Circuit:

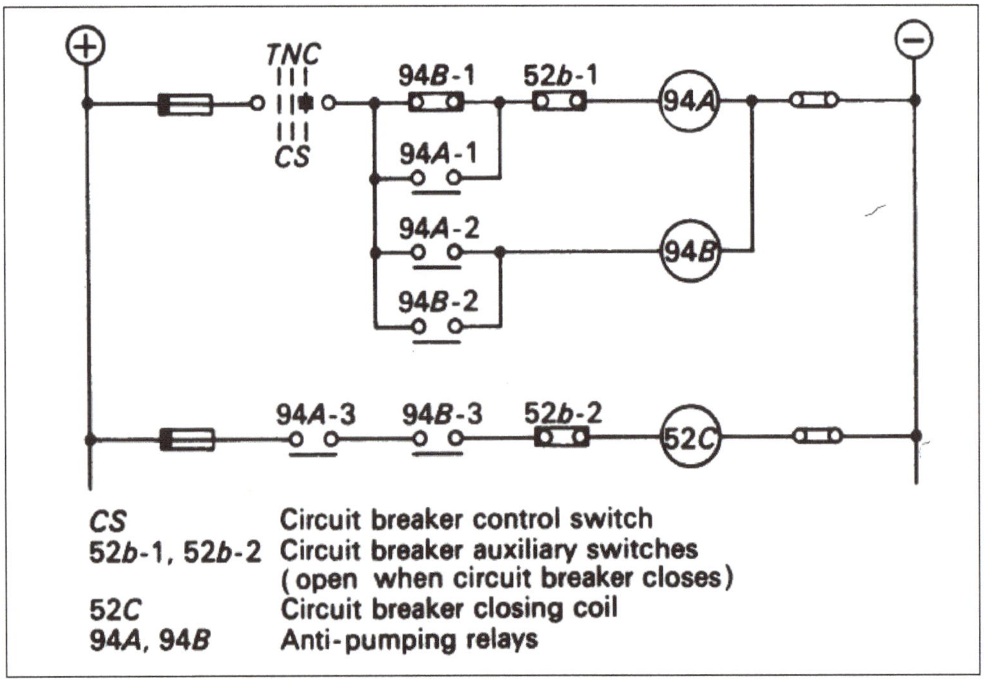

Figure 5: Anti-pumping relay control circuit.

Figure 5 shows a typical Circuit breaker anti-pumping circuit A manual control switch CS is shown for giving the closing pulse; in an auto-reclose relay initiation would be by means of the appropriate timer contact.

1. With circuit breaker open, operation of the CS (control switch) to the closing position would pick up relay 94A, which;
    a. seals it through contact 94AS-1,
    b. and through 94A2 energizes 94B relay.
2. 94B relay opens the NC contact 94B-1 and seals itself in via 94B-2
3. Since both relays 094A & 94B are energized, contact 94A-3 and 94B-3 are closed.
4. Therefore, closure of contact 94A-3 and 94B-3, energize the closing coil 52 C and the circuit breaker will close.
5. As soon as the circuit breaker closes, the auxiliary contact 52B opens and de-energizes relay 94A and closing coil 52C.
6. If the breaker stays in, on release of release of the control switch CS, relay 94B will be de-energized and the scheme will return to its original position. With control switch
7. However, if the circuit breaker trips out, as protective has operated with control switch CS still in the closed position, the breaker will not be able to reclose.
8. Reason: As relay 94A is de-energized and 94A-3 in the closing coil circuit is open.
9. The circuit breaker will remain open as long as the control switch CS is kept in close position and can be reclosed only if the control switch is first released and then returned to the close position by the operator.

6.0 Important Device Numbers as per ANSI; Some of the device nos. generally used are enlisted hereunder;

- 2 – Time delay Relay.
- 21 – Distance Relay.
- 25 – Synchronizing or Synchronize-Check Device.
- 27 – Under voltage Relay.
- 29 – Isolating Contactor or Switch.
- 50 – Instantaneous Overcurrent Relay.
- 51 – AC Inverse Time Overcurrent Relay.
- 52 – AC Circuit Breaker.
- 52a – Circuit breaker Auxiliary switch “Normally open” (‘a’ contact).
- 52b-Circuit breaker Auxiliary switch “Normally closed” (‘b’ contact).
- 59 – Overvoltage Relay.
- 64 – Ground Detector Relay (Earth fault relay).
- 64R – Restricted earth fault Relay.
- 67 – AC Directional Overcurrent Relay.
- 79 – Auto reclose relay.
- 86 – Lockout Relay/Master Trip.
- 87 – Differential Protective Relay.
- 94 – Anti pumping relay.
- 95 – Trip circuit supervision relay.

- 99 – Over-flux relay.
- 186A – Auto reclose lockout relay.
- 186B – Auto reclose lockout relay.

7.0 Alphabets depicting different circuits in a drawing as well as ferrule on the wires in Control & relay Panel;

J Series	**D.C Incoming**	**J1, J2, etc.**
K Series	**Control - Closing, Tripping, etc.**	**K1, K2, K3 etc.**
L Series	**Alarms, indications and annunciations**	**L1, L2, L3, etc.**
E Series	**Potential transformer secondaries**	**E1, E2, E3, etc.**
H Series	**LT A.C Supply**	**H1, H2, H3, etc.**
A Series	**C.T secondary for special protection**	**A1, A2, A3, etc.**
B Series	**Bus bar protection**	**B1, B2, B3, etc.**
C Series	**Protection Circuits**	**C1, C2, C3, etc.**
D Series	**Metering Circuits**	**D1, D2, D3, etc.**

7.1 STANDARD LEAD/FERULE NUMBERS;

Certain lead numbers are standardized with ferrules at terminations of leads;

J1 – Main DC Positive from the Battery-bank.

J2 – Main DC Negative from the Battery-bank.

K1 – DC Positive for Protection function circuits.

K2 – DC for Negative Protection function circuits.

L1 – DC Positive for Alarm/annunciation function circuits.

L2 – DC Negative for Alarm/annunciation function circuits.

K15R – Remote close.

K15L – Local close.

K 5R – Remote trip.

K5L – Local Trip

2. Numerical Relays - For Over Current (O/C) & Earth Fault (E/F) Application

1.0 INTRODUCTION;

1.1. RELAYS: The relay in the power system is analogous to human brain, It is a very vital secondary device in the power system which takes decisions if the quantity of parameters (received from the instrument transformers-CT, PT) exceed the pre-set value and gives command to the circuit breaker to trip which in turn isolates the faulty equipment/feeder etc from the system, resulting into saving the healthy circuit.

1.2. Initially electro- mechanical/ electro-magnetic relays were used, however with the technological advancement the solid state static relays came in into existence and in the present scenario Microprocessor based Numerical relays have replaced them.

2.0 Operation of Microprocessor based Relay:

A current signal from CT is converted into proportional voltage signal using I to V converter.

The AC voltage proportional to load current is converted into DC using precision rectifier and is given to multiplexer (MUX) which accepts more than one input and gives one output. Microprocessor sends command signal to the multiplexer to switch on desired channel to accept rectified voltage proportional to current in a desired circuit.

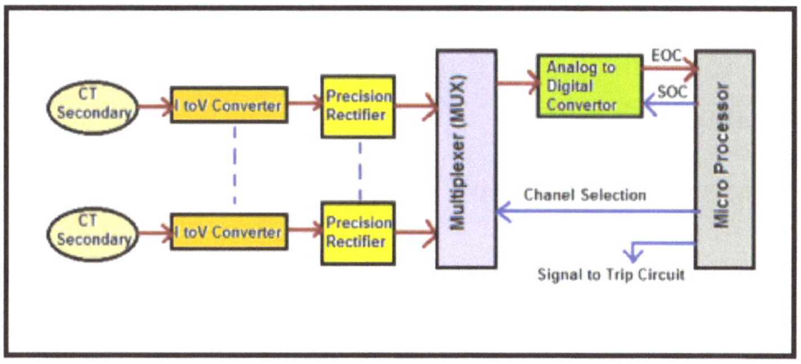

Figure 1 Microprocessor Relay - Operation diagram

Output of Multiplexer is fed to analog to digital converter (ADC) to obtain signal in digital form. Microprocessor then sends a signal ADC for start of conversion (SOC), examines whether the conversion is completed and on receipt of end of conversion (EOC) from ADC, receives the data in digital form.

The microprocessor then compares the data with pick-up value. If the input is greater than pick-up value the microprocessor send a trip signal to circuit breaker of the desired circuit.

In case of instantaneous over current relay there is no intentional time delay and circuit breaker trips instantly. In case of normal inverse, very inverse, extremely inverse and long inverse over current relay the inverse current-time characteristics are stored in the memory of microprocessor in tabular form called as look-up table.

2.2. Numerical Relays, the multi-function devices: As these are microprocessor-based relays, unlike electromechanical and static relays which offer single function and single characteristics, the Numerical relays offer multi-functions like protection, control, monitoring, measuring and also the communication capability facilitating remote control, monitoring and data transfer etc.

3.0 Numerical Over Current and E/F relay Type MJT-315: With the flexibility of microprocessor technology the Numerical relay Type MJT-315 of ERL- make(Pl Ref fig:) has wide range of protection elements and also features of metering, communication and fault recording etc. The M.P.Poorve Kshetra Vidyut Vitataran Co.Ltd.is taking steps to replace all the electromechanical relays of the 33/11kV Power Transformers and 11kV feeders with this relay.

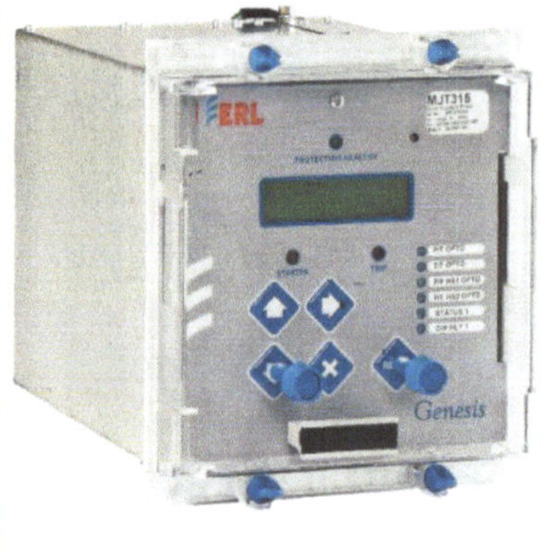

Numerical Overcurrent & Earthfault Protection - MJT315

Figure 2

3.1 Available Protection functions: This Relay offers inverse time and definite time over current protection for phase faults and earth fault.

4.0 Design: This relay consists of following Modules;

- Input module.
- Power Supply and output relay module.
- Measuring module and communication inter face.
- Front fascia.
- Status Input Module.

Above mentioned 4-modules namely (i) input module, (ii) Power supply and Output module, (iii) Measuring module and communication interface and (iv) Status Input Module are plugged into Front fascia which contains switches and LED/LCDs for human machine interface. All the PCBs are well protected with best shielding from each other as well as from the external environment and have electromagnetic compatibility.They are mounted on to a chassis which is draw out type.

The facility to short CT connection at the relay terminal while Relay is being drawn out is provided.

4.1 Relay Rating; The relay has the facility to configure either 1 amp or 5 amp current inputs. This selection is provided at relay terminals..

5.0 Applications;

- Distribution feeders.
- Transmission Lines.
- AC Machines.
- Capacitor Banks.
- Transformers.

6.0. Key Features;

- Trip circuit super vision.
- Primary fault current shall be available on fault recorder menu.
- Non volatile memory for trip indication.
- True RMS measurement.
- Self Diagnostic feature.
- Storage up to records of 5 faults.
- Internal real time records.
- 26 spontaneous events.
- Time stamped of spontaneous events for status for inputs and output relays configurable pick up and drop off.
- 8 configurable status inputs.
- 6 configurable output relays.
- 6 configurable output LEDs.
- Output relays can be configured for self, hand or electrical reset (through status input).
- Built in CB fail Protection.
- Status input can be configured to operate to O/P contacts or to block any protection function.
- Any output relay can be configured from any protection function.
- LEDs can be set for self or hand reset.
- One fixed output relay for protection healthy.
- CT ratio setting facility up to 10,000/1or 5 A.
- 1A or 5A rating can be selected at site.
- Wide range of current setting.
- Fault data can be retrieved for remote communication usingIEC-608705-103 protocol.
- RS 285 communication electrical terminals at the rear of the relay.
- Out puts can be directly driven from remote communication.
- Low AC and DC burden.

7.0 Metering;

- Status of digital (Status) input and digital output(relays)can be known by general interrogation.
- Currents- Ir, Iy, Ib, Ie & $I_{R,}$ $I_{Y,}$ I_B & $I_{E.}$

8.1 User Interface;

This relay is user friendly.

- Readable Human /Machine backlit LED display.
- Configurable LED indication.
- Push button for viewing/editing the settings and/or resetting.

9.0 Technical Information:

9.1 CT Rating: 1A/5A, Frequency- 50 Hzs.

In our 33/11kV S/S the CTs used are having 5 amp. secondary.

9.2 Aux Supply:

Nominal Voltage	Operating Range
30, 48, 110V DC	**24V** to135V DC

In our case, our 33/11kV S/Ss are having DC auxiliary supply as 30V DC.

10.0 Protection Settings:

Settings for O/C and E/F can be set as per the system requirement. The relay can be set for definite time operation or it can be set for inverse characteristic as per the formula given by the manufacturer. This relay has got the facility of highest element with setting provision of 0.5 x In to 30 x In (in the steps of 0.5In) for both O/C and E/F protection.

11.0 Inverse time Characteristic;

The time of operation of the relay can be calculated from the formula mentioned here under which shall follow the inverse time operating characteristic.

$$\mathbf{t} = \frac{\mathbf{K}}{\left[\frac{\mathbf{I}}{\mathbf{Is}}\right]^{\alpha-1}} \times T_{\mathbf{m},}$$

Where I=Fault Current, In=Current setting and Tm=time Multiplier setting.

SI 3 - K=0.14, $\alpha = 0.02$.(This is the standard Inverse curve known as 3 sec. curve ie for fault current of 10xIn the time taken is 3 sec for Tms =1)

SI 1 - K =0.0613, $\alpha = 0.02$. (This is standard Inverse curve known as 1. 3 sec. curve ie for fault current of 10xIn the time taken is 1.3 sec for Tms =1).

VI - K =13.5, α = 1.0 (Very inverse).

EI - K =0.80. α = 2. (Extremely inverse).

LTI - K = 120. α = 1.0. (Long time Inverse).

11.1 For IDMT Characteristics Please see the curve (Current in multiples of set current Vs Time in Sec with Time multiplier setting of 1 Sec.) (Ref. Figure 2)

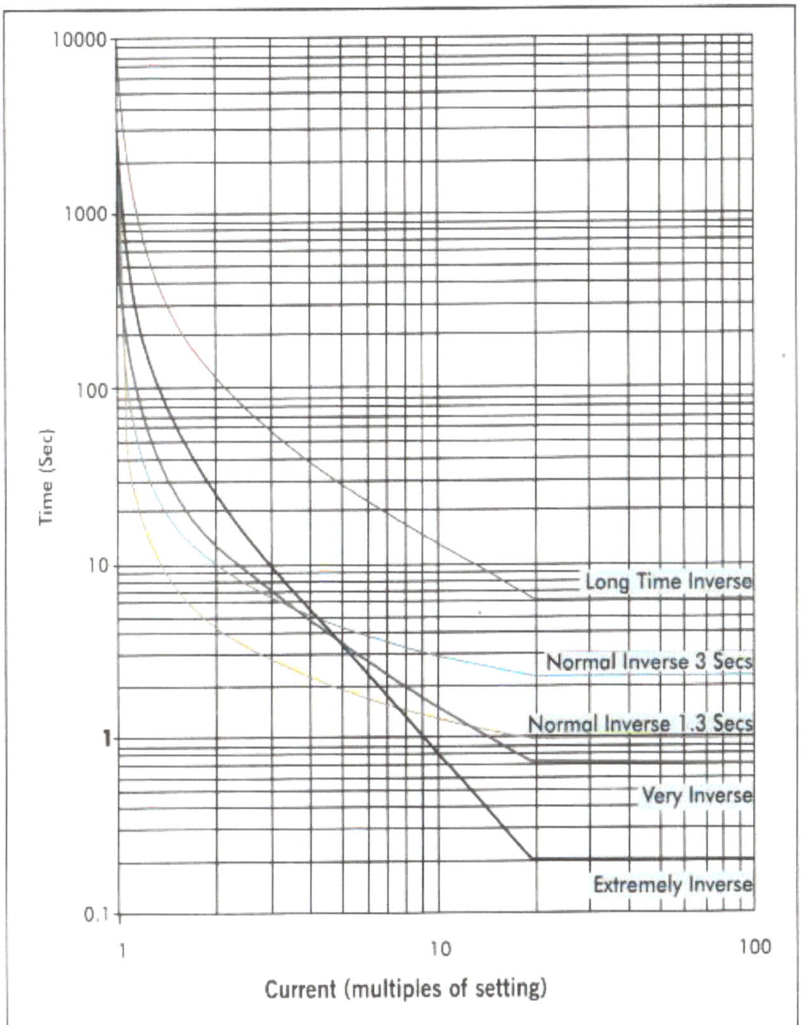

Figure 1 - IDMTL Curves (Time Multiplier = 1.0)

(Current Vs Inverse time Curve)

Figure 3

Note: (i) The Curves depict time taken by the relay to operate when Time multiplier setting is 1 sec. depending up on their characteristics.

(ii) In the relays installed in the system in MPPKVVCL, the IDMT Curve(Charecteristic-SI-1) having 1.3 secs ie the curve is such that at fault current of 10 times the set current shall take 1.3 secs to trip if Time multiplier setting is at 1sec has been adopted.

12.0 Principle of Protection:

The Relay shall operate for O/C or E/F conditions of the Line or the Transformer as the case may be.

The Schematic single line diagram of Trip Circuit is depicted in the fig.4

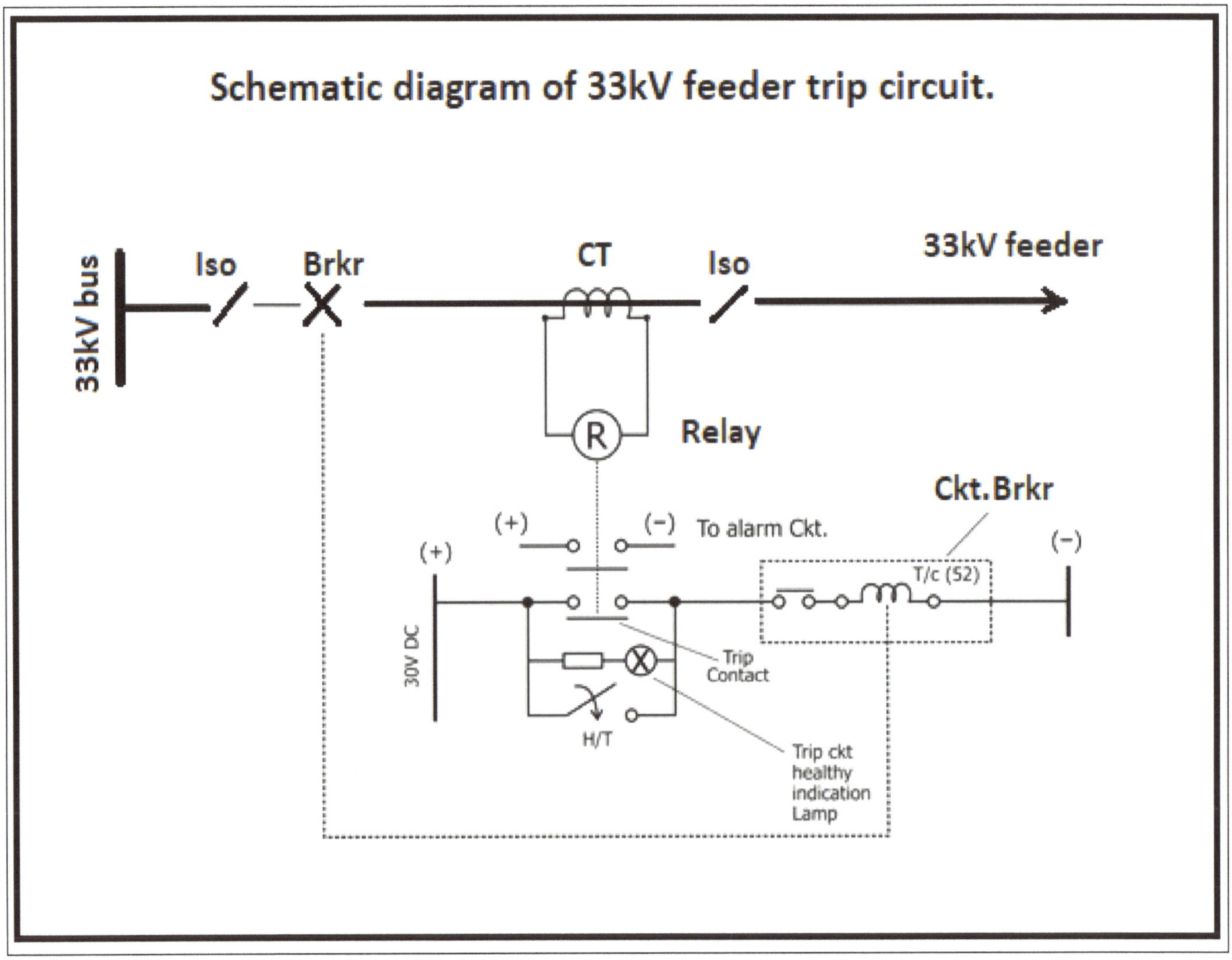

Figure 4

The O/C condition shall be of two types (i) Over Loading of the feeder and (ii) Phase faults ie. Phase To Phase fault involving phases and Three Phase fault involves all the three phases is if dead 3 phase short takes place..

The E/F condition shall occur when any phase conductor /jumper is touching the ground (Earth) or some tree branch falls on the line creating E/F.

Basic protection scheme shown in the Figure 5.

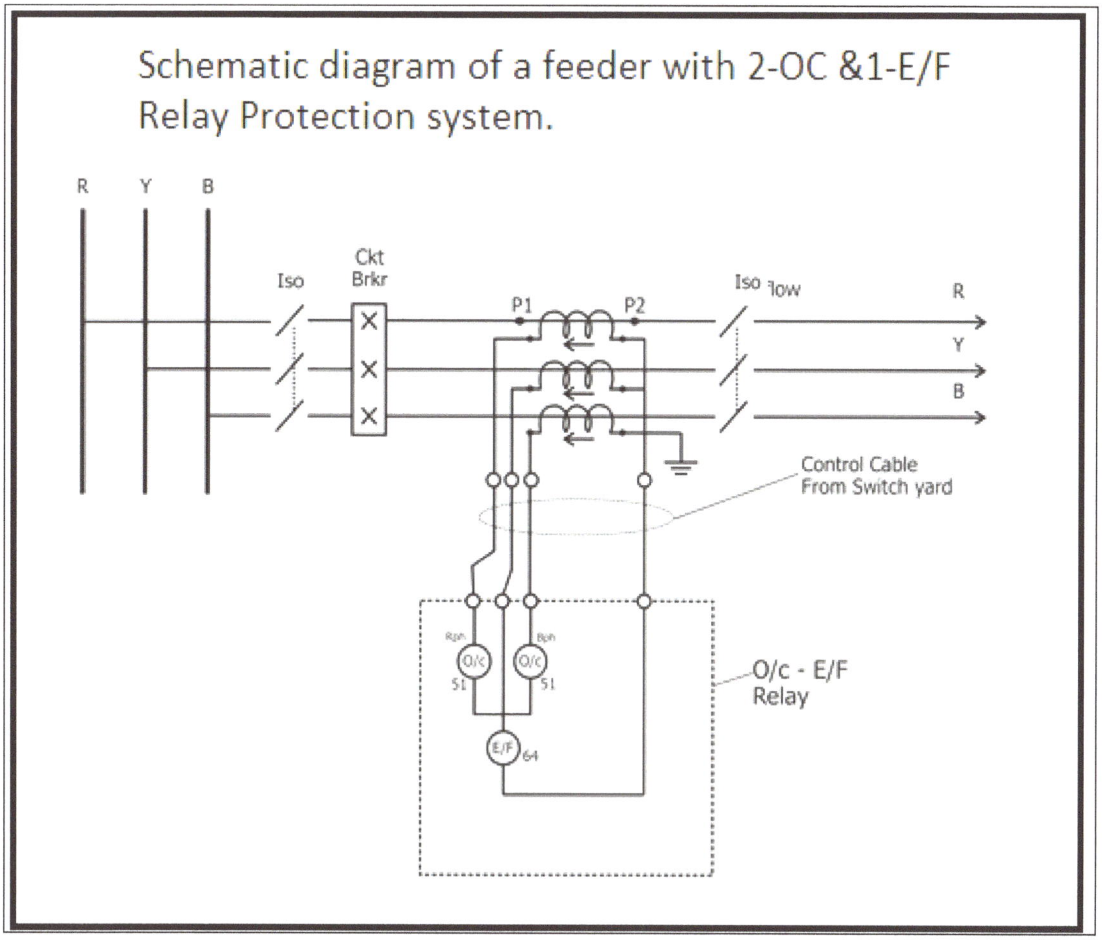

Figure 5

13.0 Settings Provided (As Specimen); In the 33/11 kV Ranital S/S at Jabalpur the settings provided (as specimen) on the numerical relay Type MJT-315 ERL make. are tabulated in the enclosed sheet. (Pl ref Table-1).

Table: 1 Numerical Relay Type MJT-315, ERL make, relay settings as provided at 33/11kV S/S Ranital, Jabalpur

Sr No.	Partic-ulars	Trans/ Fdr	CT-A	Ph Flt Curr-ent	E/F Flt Curr-ent	O/C setti-ng-%	E/F sett-ing-%	O/C TMS	E/F TMS	Charect. Curve	Op-Time Ph	Op-time E/F
1	33kV panel	Power Trans-frmr	150	1200	1200	60	20	0.15	o.25	1.3	0.1730.	0.200
2	11kV Panel	Power Trans-frmr	300	3600	1200	80	20	0.12	0.12	1.3	0.132	0.119
3	11kV Fdr	Golbazar	200	3600	1200	100	15	0.06	0.04	1.3	0.062	0.032

14.0 Wiring Diagram: Please ref. to the wiring diagram. The CT leads coming from the 33/11kV Sub-Station yard are having 5Amp secondary. Therefore care should be taken to connect the CT wires on to the ear-marked terminals for 5Amps at the rear of the MJT 315 relay.

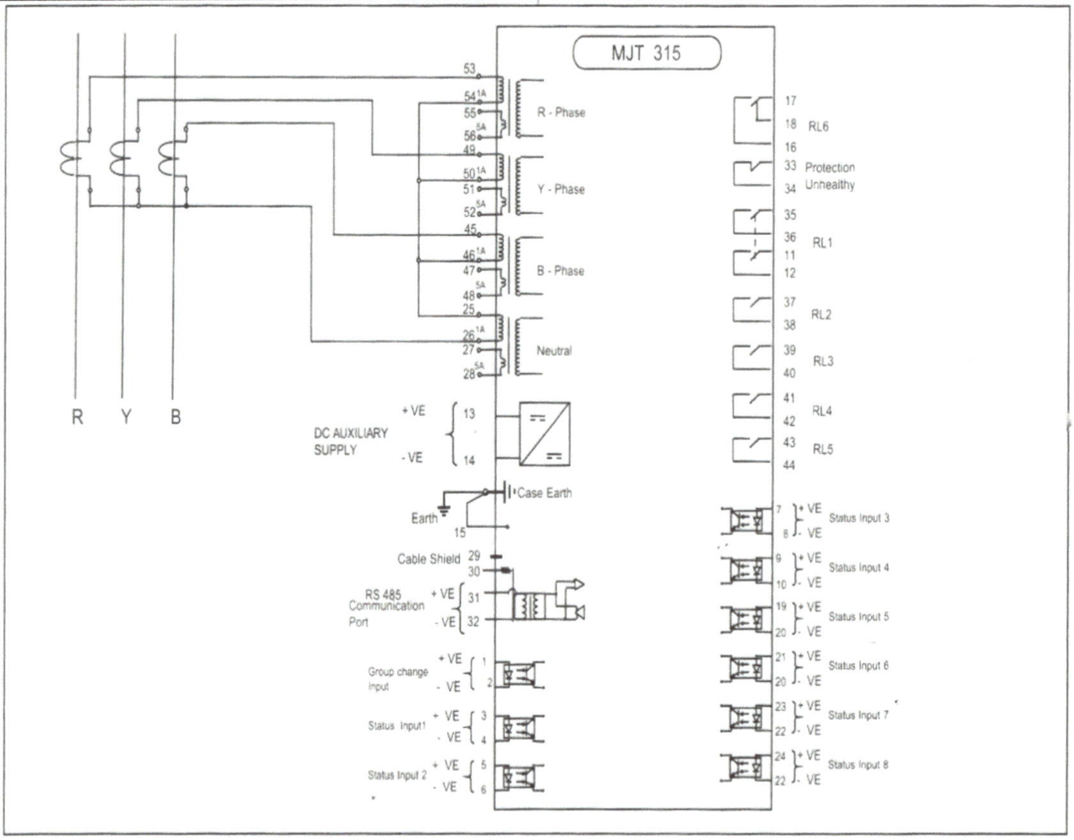

Typical Wiring Diagram for MJT315

Figure 6

15.0 Output Contacts: The Relay has got following output contacts;

- Protection Healthy.- 1N/C contact
- Change Over Contacts.-1 No (Configurable)
- Normally open Contacts-5Nos. (Configurable).

15.0 Indication:

- Green LED -Protection Healthy.
- Orange LED- Starter.
- Red LED – Trip.
- 16x2 Alpha- numeric LCD Display for setting, reading and fault data.

16.0 Burden:

By virtue of being microprocessor based numerical relay the AC Burden is very negligible and shall not over burden the CTs. The burden for 5 Amp Relay is <0.4VA.

3. Need for Directional Relays & Concept of operation of Directional Over Current and Earth Fault Relays.

1.0 Need for directional relay;

Following are the main reasons why directional relays are essential;

1. To increase reliability of power supplies to critical loads through parallel feeders.
2. More selective as compared to non-directional relays.
3. Plays a supervisory role as a back-up to Distance protection Relays.
4. Reliability of supply in Ring main Systems.

The reasons narrated in 1) & 2) in the list above are explained through figures;

Case 1: Load being fed through 2 parallel feeders with non-directional relays.

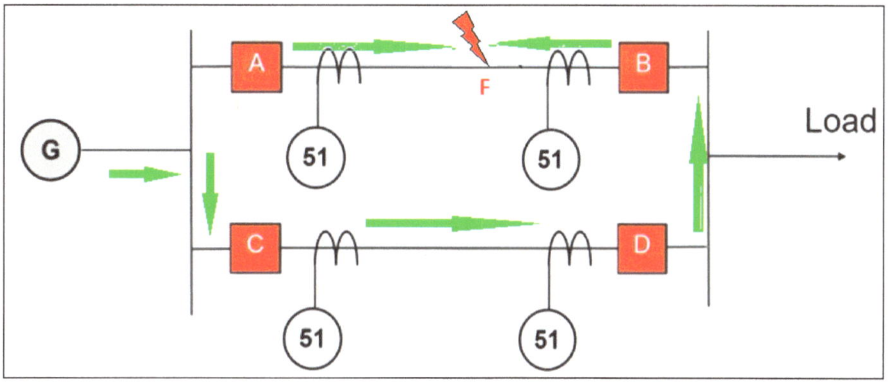

Figure 1: Parallel feeders feeding load with non-directional relays.

Description:

a. Load is fed through 2 parallel feeders having similar lengths and R/X ratio such that they share the load equally.
b. ABC & D are the circuit breakers.
c. 51is the nomenclature for the non-Directional over current relays.
d. F is the location of fault.
e. Green arrows indicate the path of fault current being fed from the generator G

Result: Both the feeders shall trip depriving the power to the load.

Case no2: Load being fed through 2 parallel feeders with non-directional relays breaker A & C and Directional Over current relays at the breaker

Location B&D.

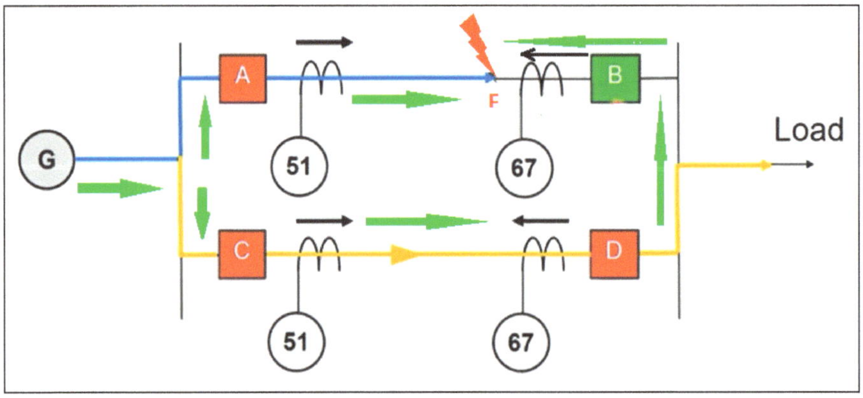

Figure 2: Parallel feeders feeding load with non-directional relays and directional relays.

Description:

a. Load is fed with 2 parallel feeders having similar length of feeders and their R/X ratio such that they share the load equally.
b. ABC & D are the circuit breakers.
c. 51 is the nomenclature for the non-Directional over current relays.
d. 67 is the nomenclature for the Directional over current relays.
e. 67 Directional relays see the faults in the forward direction as shown with black arrows
f. F is the location of fault.
g. Green arrows indicate the path of fault current being fed from the generator G.

Result;

a. Fault is still hanging despite the operation the directional relay (67) sees the fault in its forward direction operating the circuit breaker B.
b. Non-directional relay (51) at A shall operate after some time lag as per time setting provided maintain time discrimination adopted.
c. After operation of the relays mentioned in item no. a) & b) the fault is cleared.
d. However, the power to the load was not disrupted as the entire load was now shifted to the feeder having circuit breaker C&D.
e. Reason for non-tripping of circuit breaker D is as follows; the Directional relay (67) sees the fault in the forward direction as shown by the arrow, thus has not operated.

2.0 Directional Relay Concept with reference to modern relaying;

These relays have been conceptualized on the basic theory of Electro mechanical directional relays wherein direction of operation depends on the inter-action of flux φ developed from fault current and the φ developed by the current drawn by potential coil.

In the electromechanical and static relays, the required phase displacements are generally obtained by connecting the input quantities to the relay. However, in a digital or numerical relay, the phase displacements are obtained by software (different algorithms).

2.1.1 Type of connections for Directional O/C & E/F;

Directional decision for Directional O/C relays;

i. Self-polarization.
ii. Cross polarization.

Directional decision for Directional E/F relays;

i. Zero seq. Polarization;
   a. Zero seq current polarization.
   b. Zero Seq voltage polarization.

2.1.2 Polarization:

Polarization is nothing but the voltage given to the Directional O/C & E/F relays to achieve correct direction.

2.1.3 Definition of RCA (Rely Characteristic Angle);

Relay Maximum Torque Angle (MTA); It is the same as RCA.it is defined as the angle by which the current applied to the Relay must be displaced from the voltage applied to the relay to produce maximum torque.

RCA example; The angle by which the current applied to the relay must be displaced from voltage applied to the relay to produce maximum Operational sensitivity eg 45° (figure1)

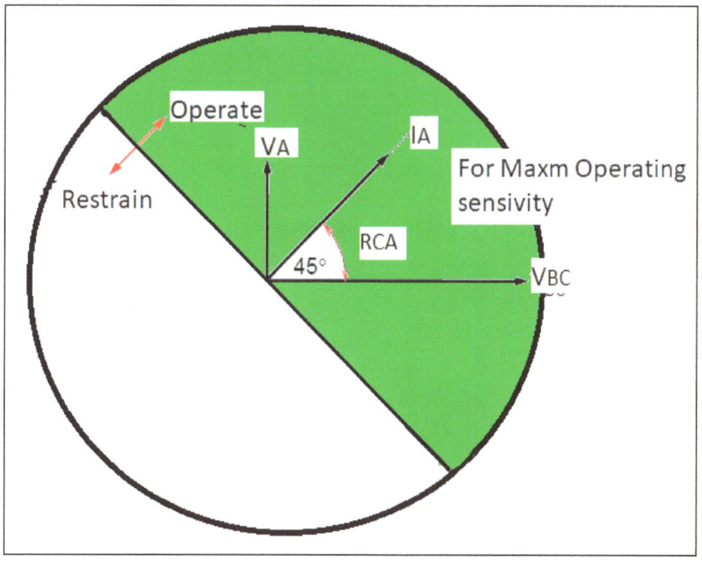

Figure 1: RCA Concept.

2.2 Direction Selectivity; Polarization plays a vital role in direction selectivity.

2.2.1 Self-polarization; This is not adopted by most of the numerical Directional relays as the voltage collapses at the point of fault, thus it cannot find the correct direction. Therefore, it is not adopted by most of the manufacturers.

2.2.2 Cross polarization:

An example highlighting usage of Cross polarization in correct direction selectivity:

Say, there is a fault in the R phase on a 11kV system. The secondary voltage through PT is 110V and the fault current through CT in secondary Amps be 2A.

In this case the fault is in R phase I_R and the polarization voltage will be V_Y-V_B.

(Note; All the calculations are following Polar Coordinate system)

Currents: R ph = 2A ∠ 90°, Y ph= 0.A ∠ 330° and B ph= 0.A ∠ 210°

Voltages: VR = 50 ∠ 90°, VY= 63.51 ∠ 330° **and** VB = 63.51 ∠ 210°.

RCA=45°.

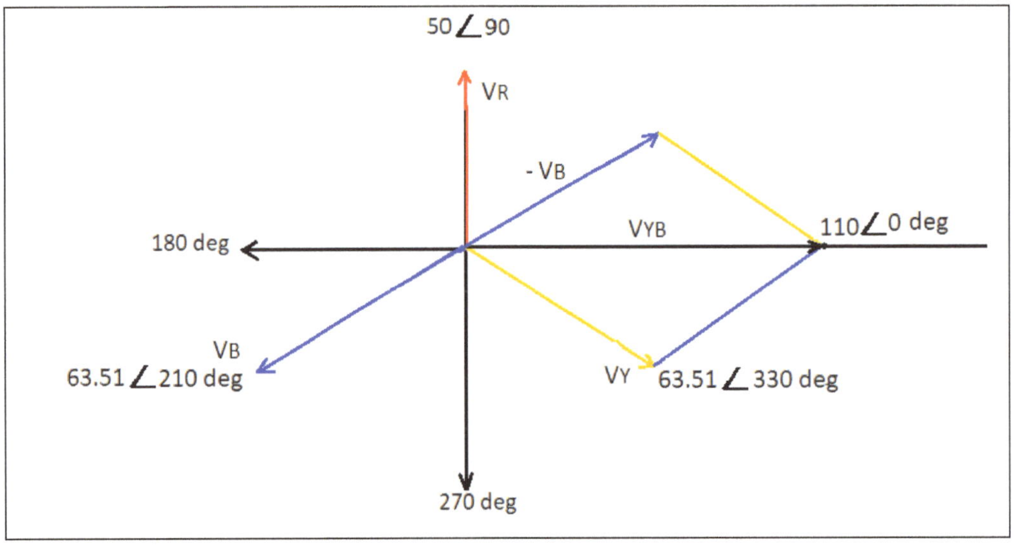

Figure 2: Vector Diagram authenticating VYB.

Calculation for V_{YB} voltage in Polar vector;

$V_{YB} = V_Y - V_B$;

= 63.51 ∠330 - 63.51∠210.

= 63.51(cos 330 +j Sin330) - 63.51(cos 210 +j Sin210),

= 63.51[(0.866 -j 0.5) - (-0.866 – j0.5),

= 63.51(2x0.866)

$$= 63.51 \times 2 \frac{\sqrt{3}}{2} = 110\angle 0^{\circ}, 0.866 \text{ has been replaced by } \frac{\sqrt{3}}{2}$$

$V_{YB} = 110\angle 0^{o}$

Therefore $V_{YB} = 110\angle 0^{o}$ is the un-faulted reference voltage for direction decision.

Note: $V_{YB} = 110\angle 0^{o}$ is also authenticated Vectorially as shown in the figure 2.

Add the RCA angle to reference angle;

Therefore $=110\angle\ 0^{o} + 45^{o} = 110\angle 5^{o}$

To find out the Operating zone add the RCA to $+90^{o}$ to -90°,

Therefore, operating boundary will be as follows;

$$45^{o} + 90^{o} = 135^{o},$$

$$45^{o} - 90^{o} = -45^{o}$$

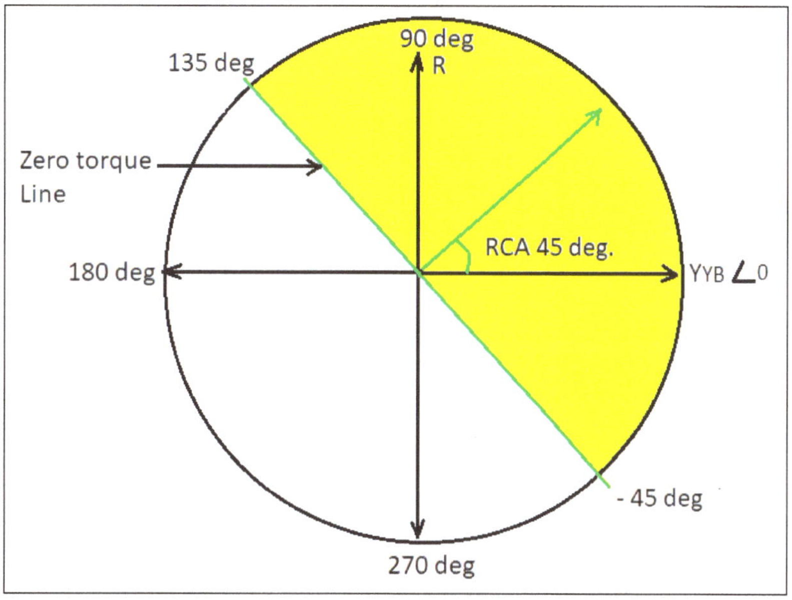

Figure 3: Operating boundary for R Phase fault.

Note:

1. Above Direction selectivity for R phase fault has been described.
2. For Y phase fault the polarizing voltage polarizing voltage shall be VBR.
3. For B phase fault the polarizing voltage polarizing voltage shall be VRY.

3.0 Directional Earth fault Protection;

Majority of faults on the transmission lines are Earth-faults. It is thus essential to use an earth fault relay with directional sensitivity.

The methodology used for the same is mentioned hereunder;

i. It is based on Zero sequence Voltage & Zero sequence Current in case of solidly earthed System.
ii. Negative Sequence polarization is used for ungrounded system.

Note:

1. Before we go further please look back on Symertical componets.
2. Zero Sequence current flows from the point of fault to the neutral of the source Transformer.
3. As most of the transmission system is solidly Grounded, thus Directional earth fault relays based on Zero Sequence current & Zero sequence voltage are used.
4. Operating signal:
   Zero Seq.Current:
   3Io= [IR + IY + IB].
   This is obtained from residual connection of line CTs.
5. Polarising Signal;
   It is the Residual Voltage taken from broken delta.

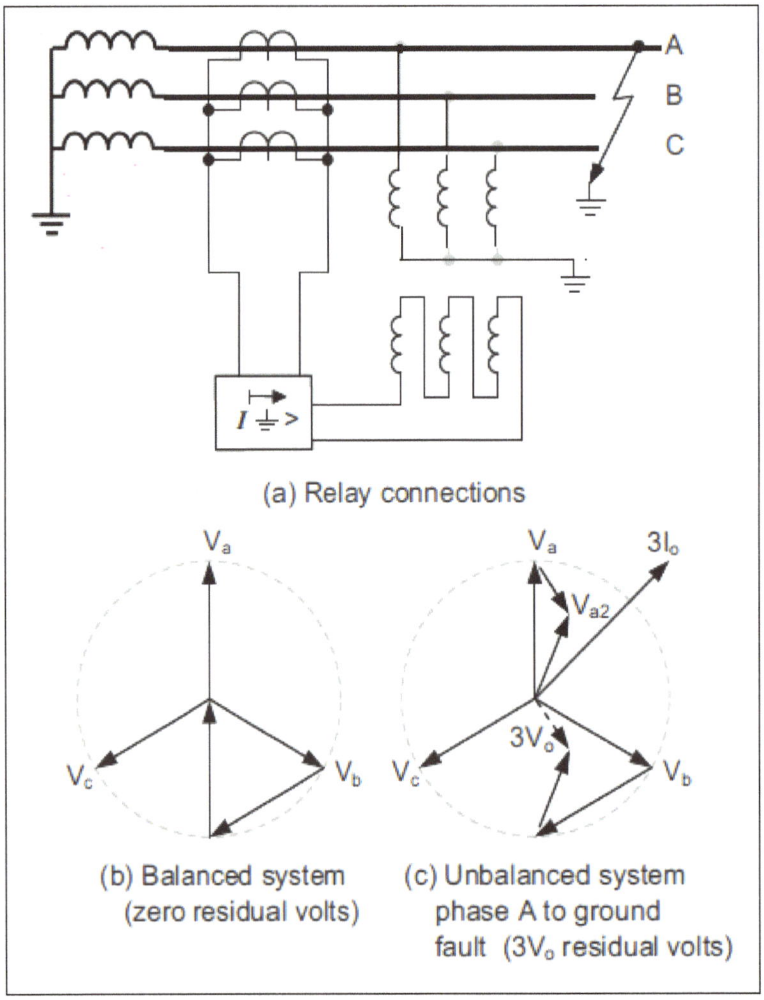

Figure 4: Residual voltage polarised directional E/F Relay.

3V0 = VR+ VY+VB

Notes:

1. PT primary must be earthed.
2. PT must be of the '5 limb' construction type or 3 x single phase units.

3.1 Direction selectivity for R Phase to earth fault;

An example highlighting Direction selectivity for R Phase to earth fault.

Operating quantity; Zero sequence current,

Polarising quantity; Inverted Residual voltage.

Let,

$IR = 0.5\angle 90^{O}$, $IY = 0.0\angle 330^{o}$, $IB = 0.0\angle 210^{O}$
$VR = 10\angle 90^{O,}$ $VY = 63.51\angle 330^{O}$, $VB = 63.51\angle 210^{O}$,
$RCA = 45^{O}$,

To find the zero Seq.Current;

$3Io = [IR + IY + IB]$.
$Io = 1/3\ [0.5\angle 90^{O} + 0.0\angle 330^{o} + 0.0\angle 210^{O}]$
$= 1/3\ [0.5\angle 90^{O}]$
$I_o = 0.166A\angle 90^{o}$,
$I_o = j0.166$

To find out Zero seq.voltage(Residual voltage);

$Vo = 1/3\ [\ VR + VY + YB]$,
$= 1/3\ [\ 10\angle 90^{O} + 6(3.51\angle 330^{O} + 63.51\angle 210^{O}\ ]$,
$= 1/3[(j10) + (55 - j31.755) + (-55 - j31.755)]$,
$= 1/3[j10 - 63.51]$
$Vo = 1/3[-j53.51]$,
$Vo = -j17.83 = 17.83\angle 270^{O}$,

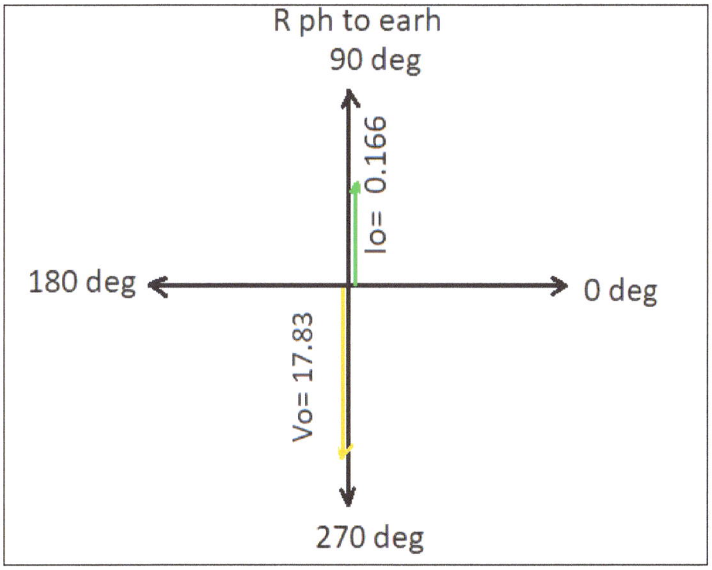

Figure 5: Zero seq.voltage is in opposction of the Zero seq.current

Vo & Io are in opposite direction.(Figure 5)

Inverting the zero sequence voltage gives +j17.83(=17.83∠90°)

ie Vo = + j17.83(17.83∠ 90°)

Both are in same direction.(Figure6)

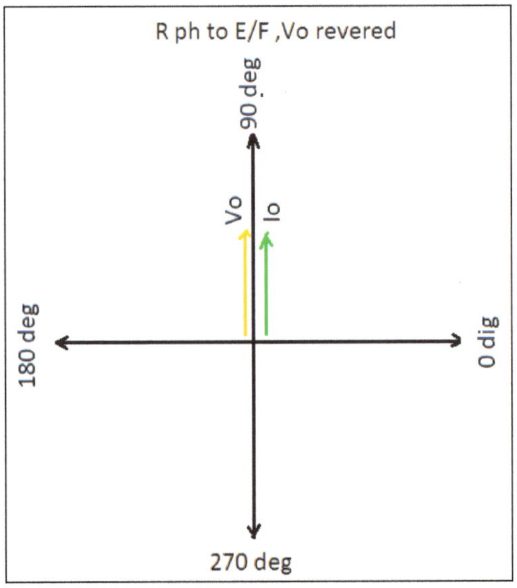

Figure 6: Inverted the polarity of zero seq.voltage.

3.2 To find out the Operating zone add the RCA to +90° to -90°, Therefore, operating boundary will be as follows;

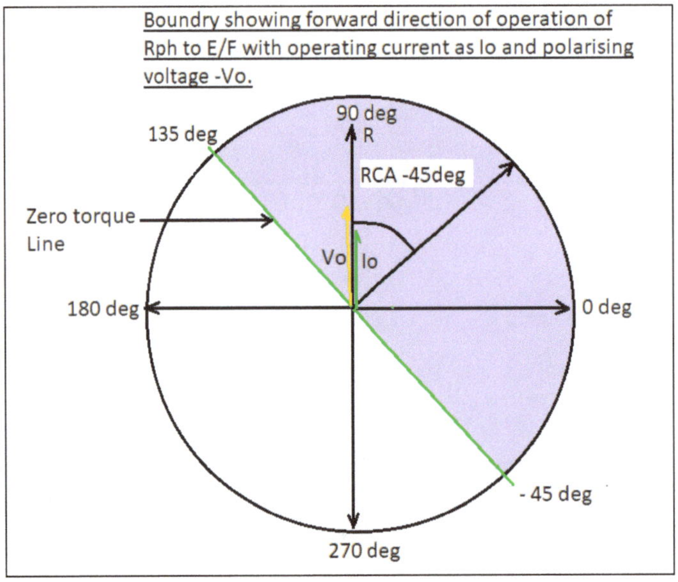

Figure: Operating boundary Rph -E/F.

Note: Like wise the boundary of operation in the forward direction can be calculated and drawn;

a. Y phase to E/F: Io & inverted Vo are required to be calculated and placed in an angle of 330°.
b. B phase to E/F: Io & inverted Vo are required to be calculated and placed in an angle of 210°.

4. Distance Protection- Concept, Characteristics and Setting Philosophy

1.0 Principle: Since the impedance of transmission line is proportional to its length for distance measurement that's why a relay has been innovated as distance relay.

The basic principle of measurement involves the comparison of fault current 'seen' by the relay with voltage at relaying point; by comparing these two quantities it is possible to measure the Impedance of the lineup to the point of fault.

Following figure 1 may be referred to using balanced beam type relay.

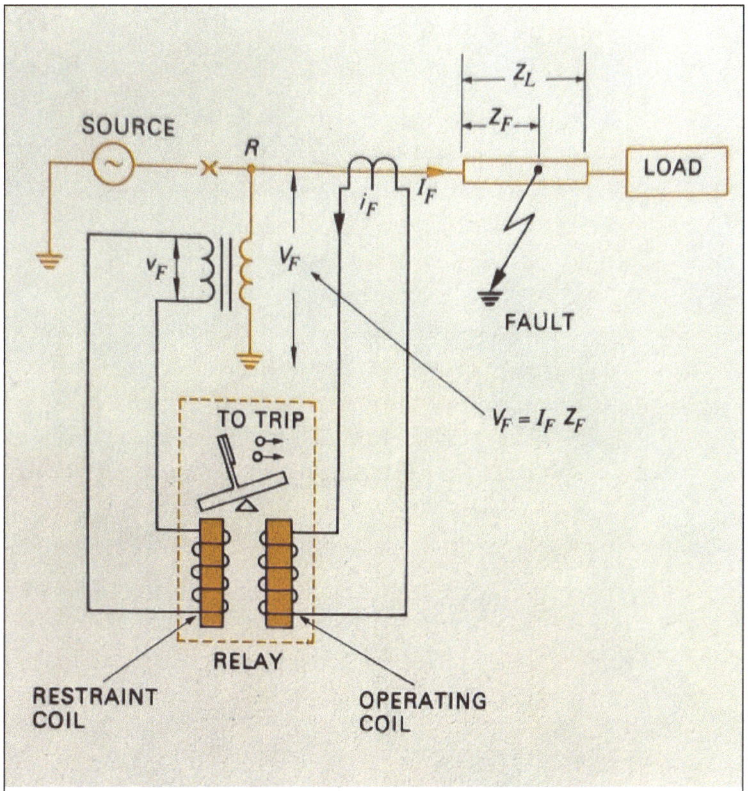

Concept of Balanced beam typr DPR

Figure 1: Concept of distance relay through balance beam relay.

It could be seen from the fig.1 that $V_F = I_F \times Z_F$, Therefore, $Z_F = \frac{VF}{IF}$, ie whenever fault is there, I_F increases thus Z_F decreases. The basic principle of distance protection involves the division of the voltage at the relaying point by the measured current. The apparent impedance so calculated is compared with the reach point impedance. If the measured impedance is less than the reach point Impedance, it is assumed that a fault exists on the line between the relay and the reach point.

2.0 Plain Impedance Characteristic; Characteristic on R/X diagram. This characteristic takes no account of the phase angle between the current and the voltage applied to it; for this reason its impedance characteristic when plotted on an R/X diagram is a circle with its centre at the origin of the co-ordinates and of radius equal to its setting in ohms. Operation occurs for all impedance values less than the setting, that is, for all points within the circle. The relay characteristic, shown in Figure is therefore non-directional, and in this form would operate for all faults along the vector AL and also for all faults behind the bus-bars up to an impedance AM. It is to be noted that A is the relaying point and RAB is the angle by which the fault current lags the relay voltage for a fault on the line AB and RAC is the equivalent leading angle for a fault on line C. Vector AB represents the impedance in front of the relay between the relaying point A and the end of line AB. Vector AC represents the impedance of line AC behind the relaying point. AL represents the reach of instantaneous Zone 1 protection, set to cover 80% to 85% of the protected line.

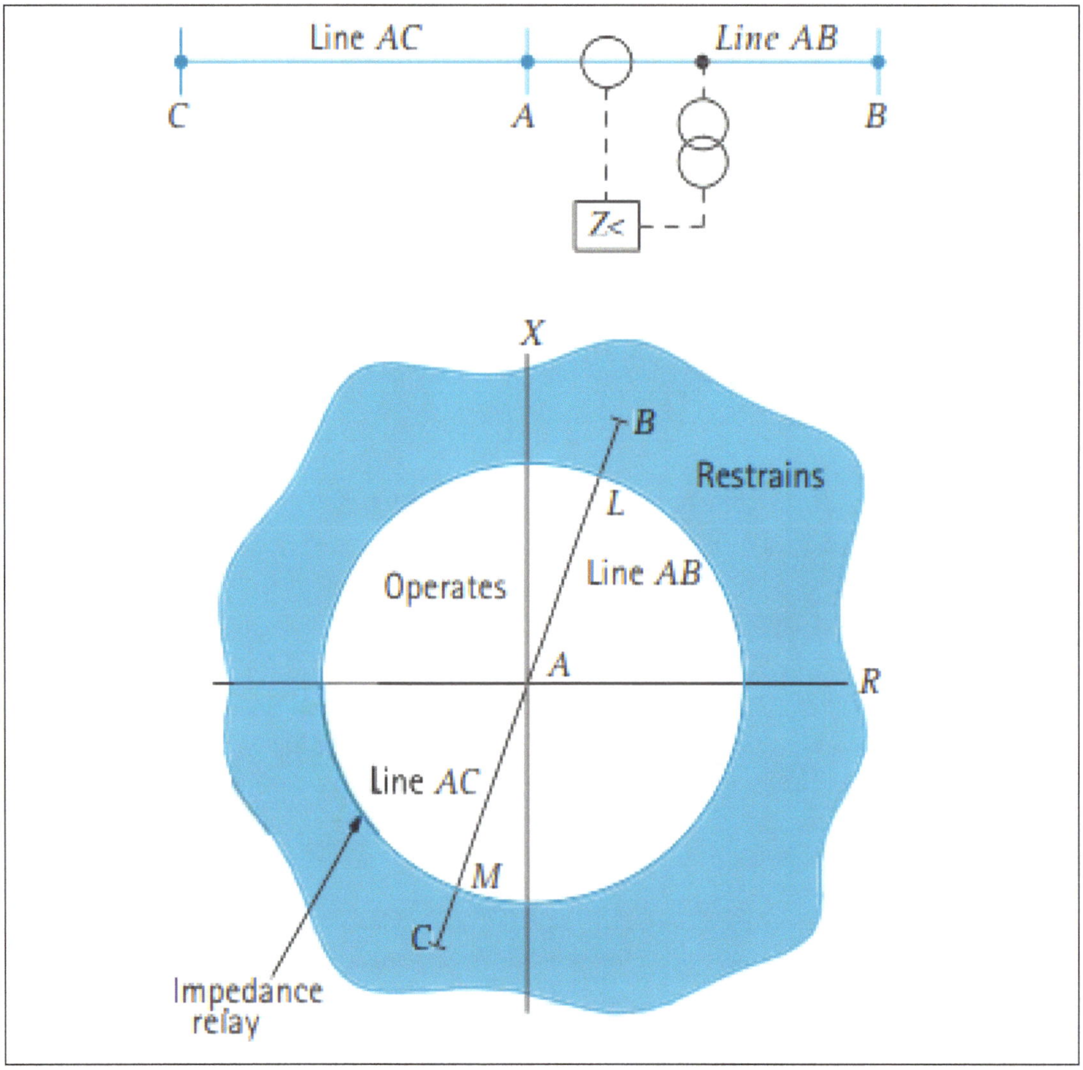

Figure 2: Characteristic on R/X Diagram of Z relay.

When it is plotted on R/X diagram it is found to be non-directional it is non-directional.

3.1 Impedance relay with directional feature;

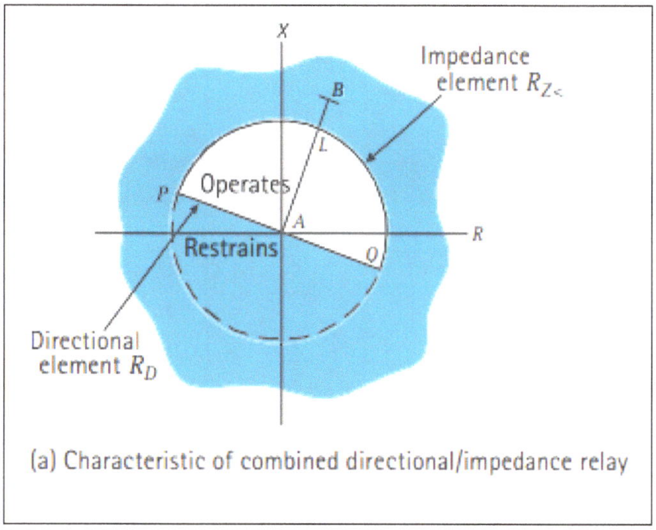

Figure 3: Characteristic of Impedance relay with directional feature.

4.0 Characteristic of MHO relay;

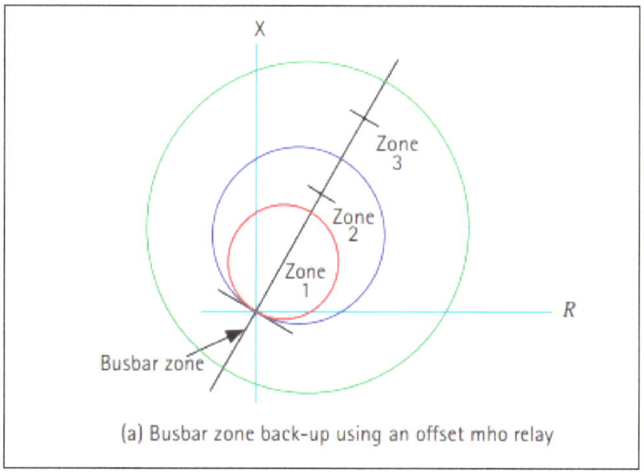

Figure 4: Characteristic of MHO relay.

MHO Relay is inherently a directional relay on R/X diagram in addition to the 3rd Zone covers the bus zone too.

5.0 Application of lenticular characteristic;

There is a danger that the offset mho relay shown in Figure 6 may operate under maximum load transfer conditions if Zone 3 of the relay has a large reach setting. A large Zone 3 reach may be required to provide remote back-up protection for faults on the adjacent feeder.

To avoid this, a shaped type of characteristic may be used, where the resistive coverage is restricted. With a 'lenticular' characteristic, the aspect ratio of the lens is (a/b) adjustable, enabling it to be set to provide the maximum fault Resistance coverage consistent with non-operation under maximum load transfer conditions. Figure 5 shows how the lenticular characteristic can tolerate much higher degrees of line loading than offset mho and plain impedance characteristics. Reduction of load impedance from ZD3 to ZD1 will correspond to an equivalent increase in load current.

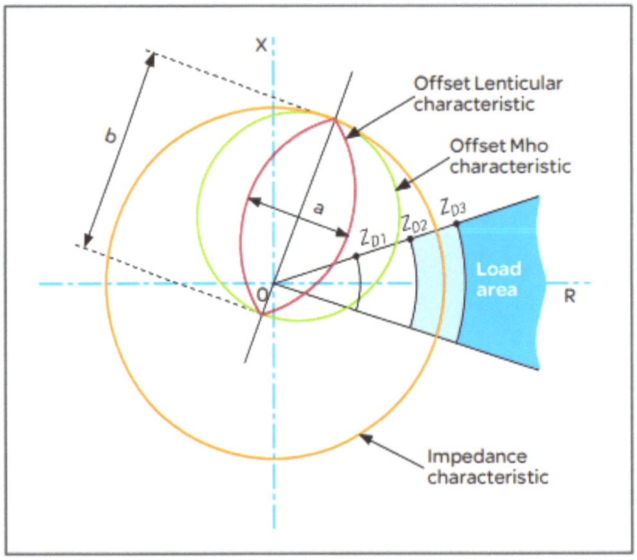

Figure 5: Minimum load impedance permitted with lenticular, offset mho & impedance relays.

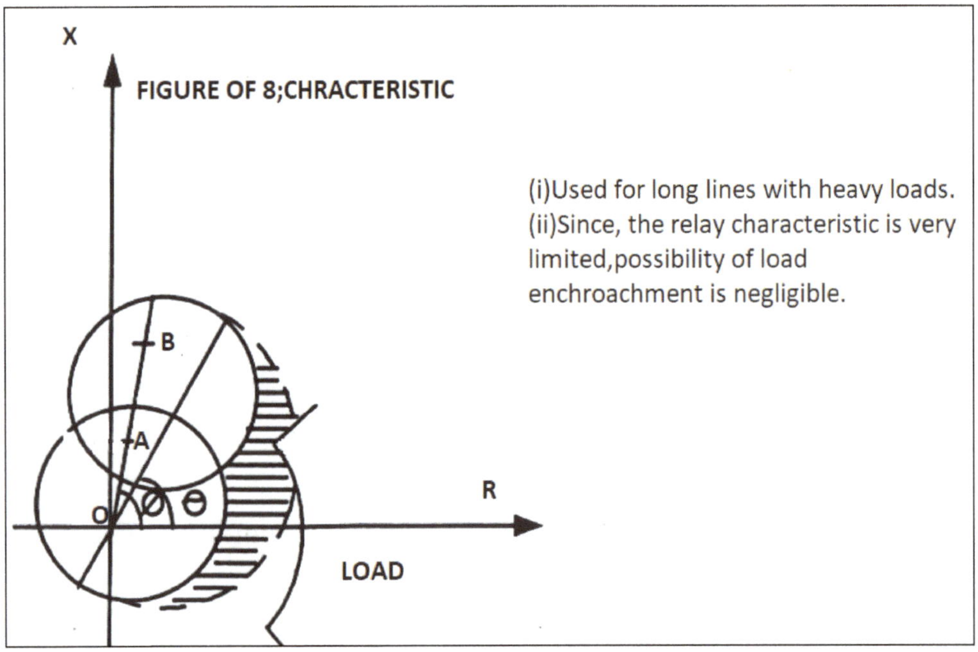

Figure 6: Figure of 8 Characteristic DPR.

6.0 Quadrilateral characteristic;

This form of polygonal impedance characteristic is shown in Figure7. The characteristic is provided with forward reach and resistive reach settings that are independently adjustable. It therefore provides better resistive coverage than any mho type characteristic for short lines.

This is especially true for earth fault impedance measurement, where the arc resistances and fault resistance to earth contribute to the highest values of fault resistance. To avoid excessive errors in the zone reach accuracy, it is common to impose a maximum resistive reach in terms of the zone impedance reach. Recommendations in this respect can usually be found in the appropriate relay manuals.

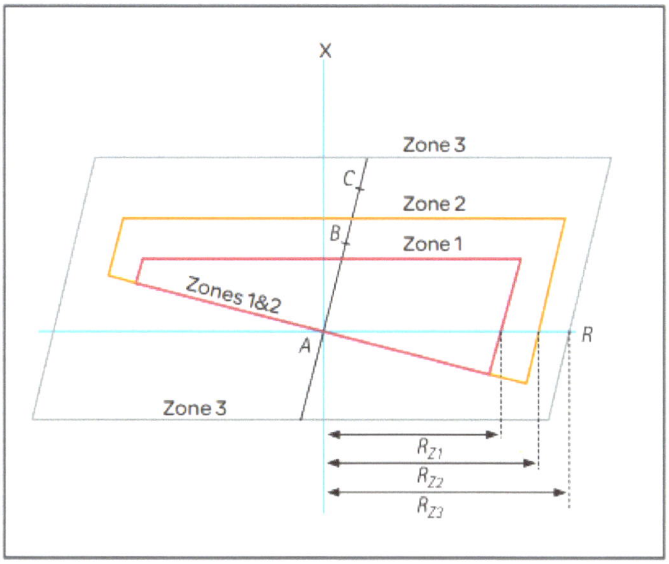

Figure 7: Quadrilateral characteristic

7.0 Basic characteristic of any DPR schemes;

- It is versatile in nature. It covers wide range of line lengths.
- High speed Operation (Typically 20 -40 ms).
- This is basically a non-unit protection, however, when used in conjunction with a signaling channel it can be converted in to Unit scheme.

8.0 Distance Relay Setting philosophy

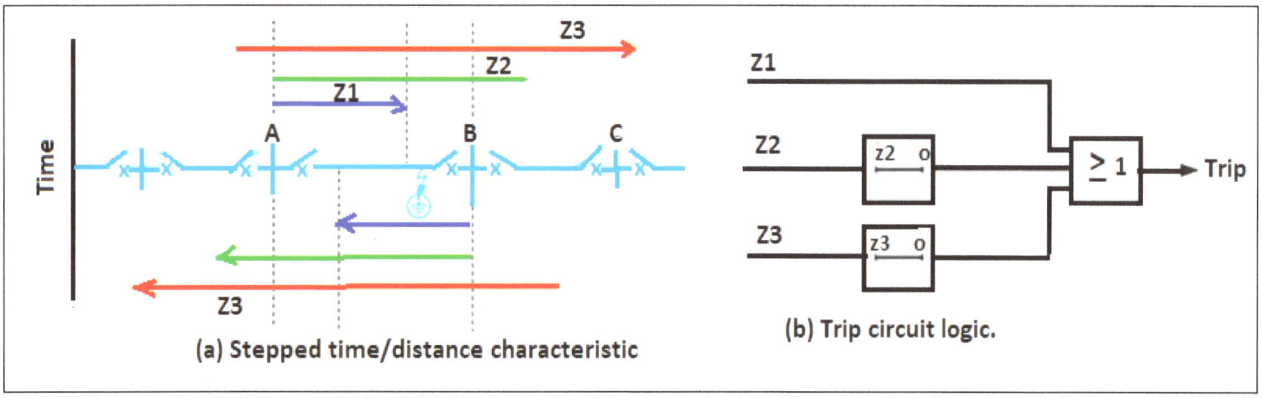

Figure 2: conventional 3 stepped time/distance characteristic

Case-1: Conventional setting of DPR.

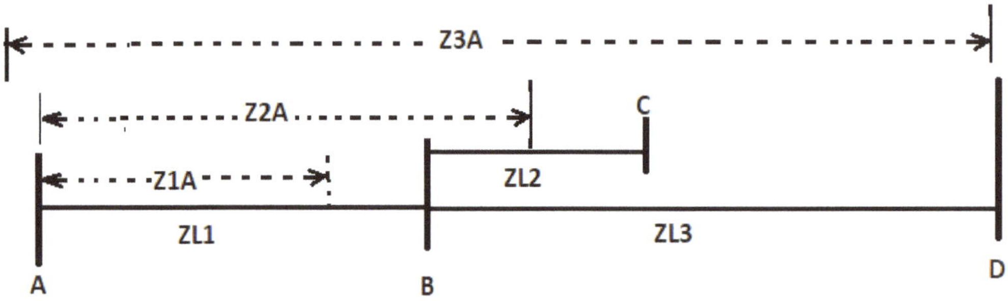

Setting;
Z1A = 80% of ZL 1. t1=instatanious
Z2A = ZL 1 + 0.5 ZL2 t2= 350ms-400ms
Z3A = 1.1(ZL 1+ZL3) t3= 1 Sec.

Case 2

Long Line Followed by a Short line.

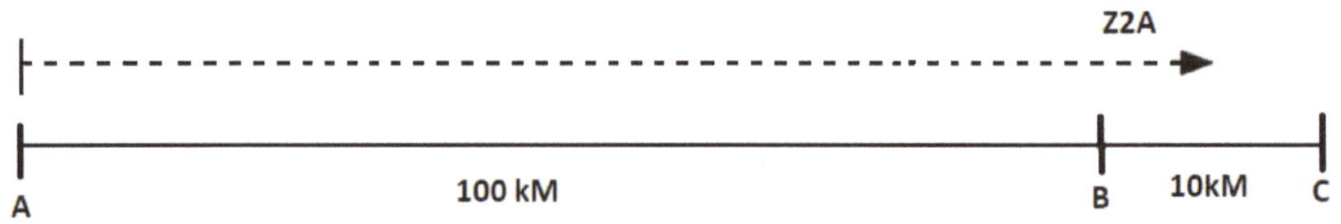

Z2 Setting based on Shortest adjoining Line will Leave very small Margin for ' Z2' to Accomodate Under reach (Typically lg A-B =100kM, Z2A=105 kM- This Leaves margin of Only 5%.

Consequently Zone 2 setting Criteria to be revised to 1.25 times the the ZL of A-B line.

Case-3 Long line followed by Short line,Over lapping Zone 2 of Long line over lapping on Zone 2 of short line.

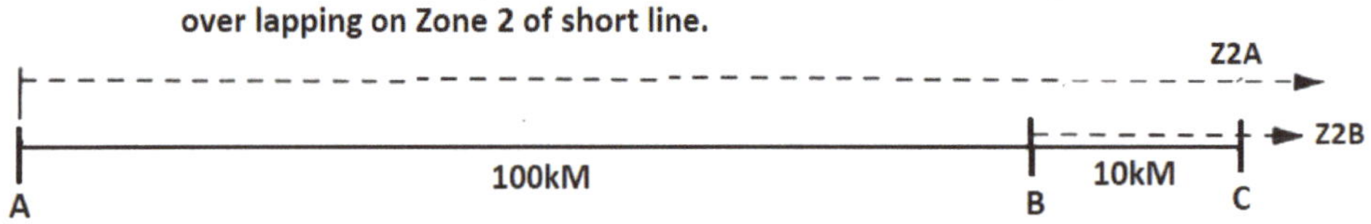

1)This will However,Make Zone 2 at 'A' Over lap with With Zone 2 reach on the Shortest Line B-C.Requiring Time Co-ordination between the Consecutive Zone2 at 'A' on A-B line and B(On B-C Line).
Note: Rquirement is first to trip Zone 2 of B-C line,thus the time co-ordination should be such that t2 for B-C line, say 250ms then the t2 from A-B line may be say 400ms.

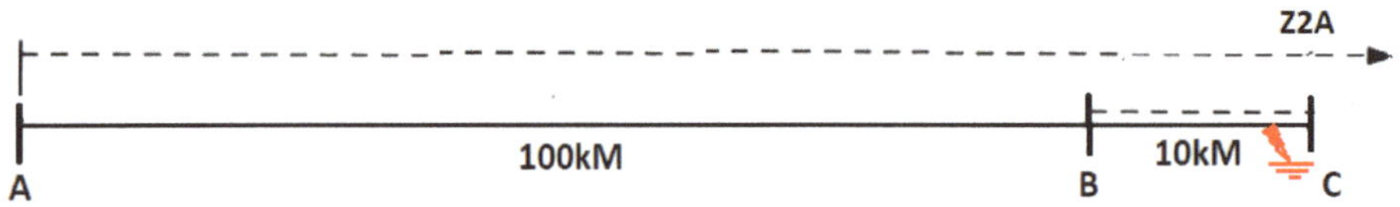

2)Since B-C line is very short such that Z2A overlaps the whole B-C line,then the best solution to have teleprotection channel on B-C line and to cover the whole B-C line with Zone 1 timimg ie instantaneous trip.As soon as it trips at B end shall simultaniously trip Breaker at C end.

Effect of Infeed on Zone 3 setting

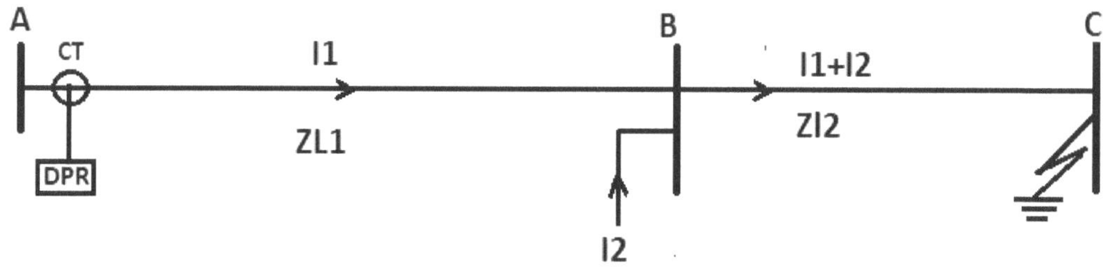

Zone 3 Setting Z3A= ZL1 +ZL2

Impedance Seen by Z3 Element at "A" considering infeed at "B"

Fault Volt.= I1.Zl1+ (I1+ I2).ZL2

Fault current =I1

Zseen = [I1.ZL1 +(I1+I2.ZL2]/(I1)

= ZL1 + [1+(I2/I1)].ZL2

= ZL! + 3ZL2 (If I2/I1= 2)

Hence Relay at 'A' will drastically under reach.

9.0 SOTF FEATURE: SOTF protection stands for "Switch On To Fault".

If any fault is persisting in the transmission line during energization of Circuit Breaker or fault happens within 20 ms from the CB closing, the SOTF protection will activate and it will trip the Circuit Breaker immediately to isolate the fault.

9.1 Why SOTF Protection?

We know that Distance Protection is responsible to isolate the fault of the transmission line. Under Zone-1 protection, the CB will trip immediately in the fault condition and it will protect up to 80% of the total length of the line. Under Zone-2 protection, the CB will trip after 300 to 500 ms in the fault condition and it will protect up to 120% of the total length of the line.

If any persisting fault detects beyond 80% of the line length, the fault will detect under Zone-2 and it will operate after a delay of 300 to 500 ms. Which is not healthy for the connecting equipment and system stability. In this situation, the SOTF protection will cover whole the line and during the persisting fault, it will immediately isolate the line.

Further, after maintenance work of transmission line, it also found many cases, the line charged from the one end whereas the line earth switch of the line not opened at the other end due to human error. In this case, also SOTF function will operate and trip the CB immediately.

9.2 Operation:

SOTF will operate by detecting the fault current and line voltage. In the persisting fault of the transmission line, the line CVT voltage will not increase after closing the CB. In that case, the SOTF function will activate as per the value of the setting of Voltage and current in the relay.

Further, the CB closing command is also used to activate the SOTF protection in the relay. After closing of CB, if any fault is detected within 500 ms, the CB will immediately trip under SOTF protection. After 500 ms, the SOTF function will automatically deactivate.

9.3 SOTF logic;

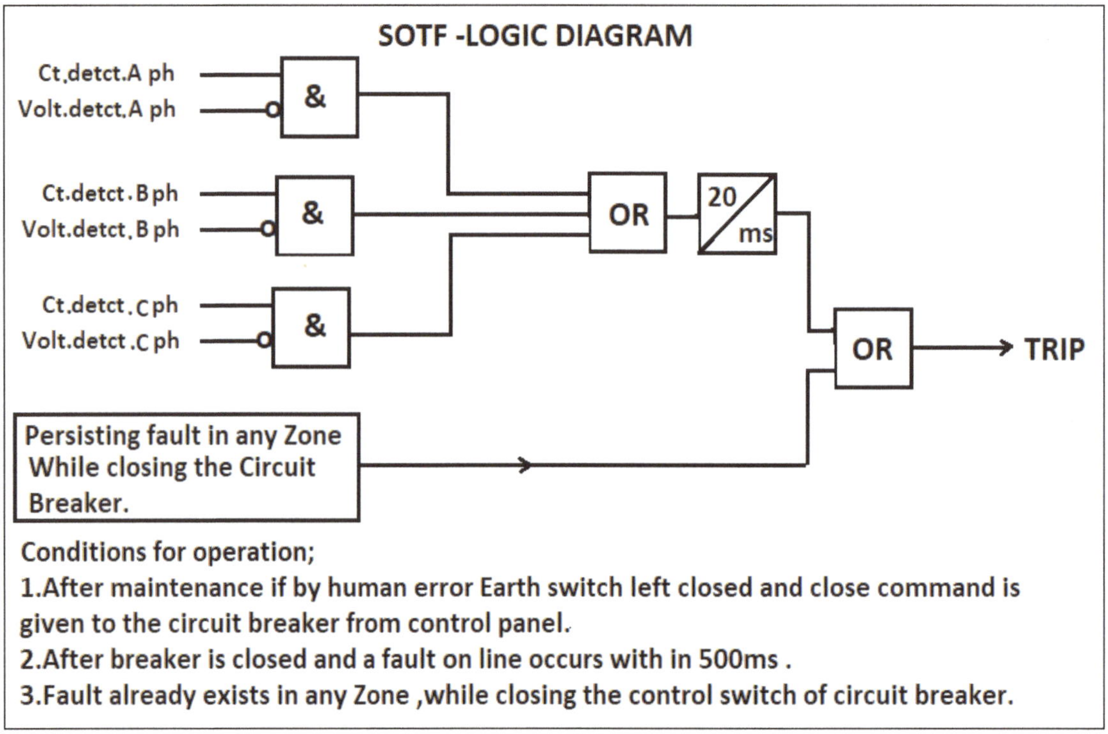

5. Phenomenon of Power-Swing & its Effect on the Distance Protection

1.0 INTRODUCTION

A balance between generated and consumed active power exists during steady-state operating conditions and is necessary for the stability of the power system. Power system disturbances cause oscillations in machine rotor angles that can result in severe power flow swings. Depending on the severity of the disturbance and the actions of power system controls, the system may remain stable or experience a large separation of generator rotor angles and eventually lose synchronism. Large power swings can cause unwanted relay operations that can further aggravate the power system disturbance and cause major power outages or blackouts. A power swing blocking (PSB) function is available in modern distance relays to prevent unwanted distance relay element operation during power swings. The out-of-step tripping (OST) function discriminates between stable and unstable power swings and initiates islanding during loss of synchronism between power system areas.

1.1. POWER SWINGS AND OUT-OF-STEP PHENOMENON;

A power swing is a system phenomenon that is observed when the phase angle of one power source starts to vary in time with respect to another source on the same network. The phenomenon generally occurs following a major disturbance, like a fault, that alters the mechanical equilibrium of one or more machines. A power swing is stable when, following a disturbance, the rotation speed of all machines returns to synchronous speed. A power swing is unstable when, following a disturbance, one or more machines do not return to synchronous speed, thereby losing synchronism with the rest of the system.

2.0. Basic Phenomenon Using the Two-Source Model

The simplest network for studying the power swing phenomenon is the two-source model, as shown in Figure 1. The left source has a phase angle advance equal to θ, and this angle will vary during a power swing. The right source represents an infinite bus, and its angle will not vary with time. As simple as it is, this elementary network can be used to model the phenomena of more complex networks.

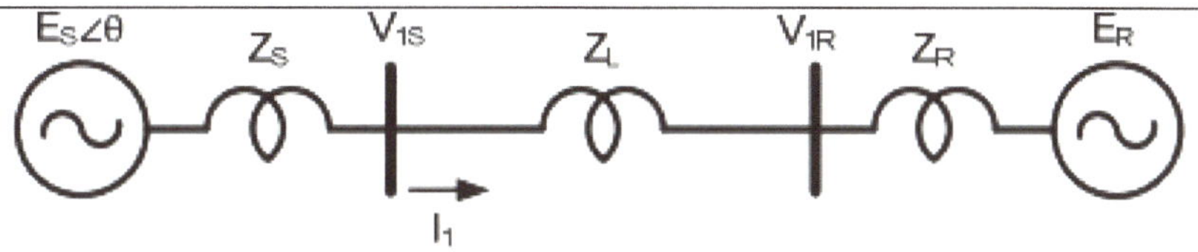

Fig. 1. Two-source equivalent elementary network.

Representation of Power Swings in the Impedance Plane

Assuming the sources have equal impedance amplitude, for

a particular phase angle θ, the location of the positive sequence impedance (Z_1) calculated at the left bus is provided by following equation

$$Z_1 = \frac{\mathbf{V1s}}{\mathbf{I1}} = Z_T \text{x} \frac{\mathbf{Es\angle\theta}}{\mathbf{Es\angle\theta - ER}} - Z_S \text{------------ } \mathbf{(1)}$$

In (1), Z_T is the total impedance, as in:

$$Z_T = Z_S + Z_L + Z_R \text{ ---------------------- (2)}$$

Assuming the two sources are equal in magnitude, the Z_1 impedance locus in the complex plane is given by (3).

$$Z_1 = \frac{ZT}{2} \text{ x} \left(1 - j\cot\frac{\theta}{2}\right) - Z_S \text{--------------(3)}$$

When angle θ varies, the locus of the Z_1 is a straight line that intersects the segment ZT orthogonally at its middle point, as shown in Figure 2.The intersection occurs when the angular difference between the two sources is 180 degrees. When generator torque angle reaches 180 degrees, the machine is said to have slipped a pole, reached an out-of-step (OOS) condition or lost Synchronism.

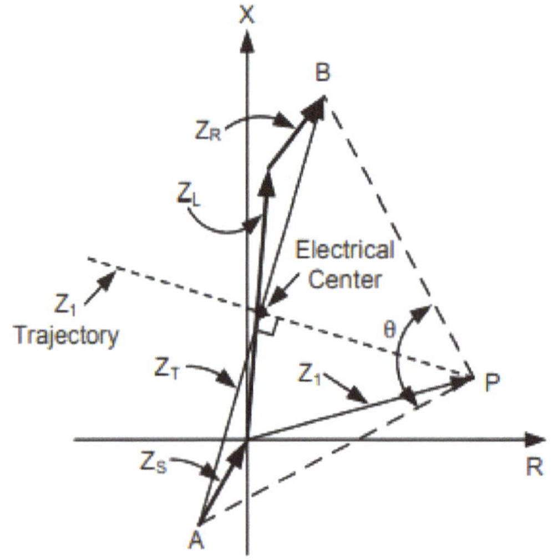

Fig. 2. Locus of the Z_1 impedance during a power swing with sources of equal magnitude.

When the two sources have unequal magnitude such that n is the ratio of ES over ER, the locus of Z1 impedance trajectory will correspond to circles shown in Figure 3.For any angle è ,the ratio of two segments joining the location of extremity of Z1 (Point P) to the total impedance extremities.

$$N = \frac{\mathbf{Es}}{\mathbf{ER}} = \frac{\mathbf{PA}}{\mathbf{PB}} \text{ ---------------------------- (4),}$$

The precise equation for the center and radius of the circles as function of the ratio n can be found in.

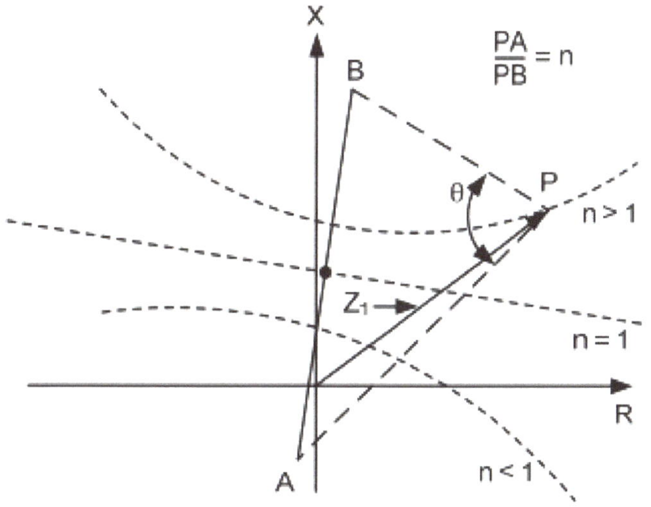

Fig. 3. Locus of the Z_1 impedance during a power swing with sources of unequal magnitude.

1. Figure 3. Locus of the Z_1 impedance during a power swing with sources of Unequal magnitude.
2. It should be noted that, in reality, a power source is a synchronous generator and is therefore not an ideal voltage source as represented in the equivalent two-source model. Furthermore, the impact of any existing automatic voltage regulator must be considered.
3. During a power swing, the ratio of two power source magnitudes is not going to remain constant. Therefore, the resulting locus of the Z_1 impedance will switch from one circle to another, depending upon the instantaneous magnitude ratio.

2.1 Rate of Change of the Positive-Sequence Impedance Starting with (1) and assuming the two sources are of equal magnitude, the time derivative of the Z_1 impedance is

Provided by(5).

$$\frac{dZ_1}{dt} = -jZ_T \bullet \frac{e^{-j\theta}}{\left(1-e^{-j\theta}\right)^2} \bullet \frac{d\theta}{dt} \qquad (5)$$

Assuming the phase angle has a linear variation with a slip frequency in radians per second given as:

$$\frac{d\theta}{dt} = \omega \qquad (6)$$

and using the identity:

$$\left|1-e^{-j\theta}\right| = 2 \bullet \sin\frac{\theta}{2} \qquad (7)$$

the rate of change of the Z1 impedance is finally expressed as:

$$\left|\frac{dZ_1}{dt}\right| = \frac{|Z_T|}{4 \bullet \sin^2 \frac{\theta}{2}} \bullet |\omega| \tag{8}$$

Equation (8) expresses the principle that the rate of change of the Z1 impedance depends upon the sources, transmission line impedances, and the slip frequency, which, in turn, depends upon the severity of the disturbance

2.2 Impact of the System Impedances on the PSB Function

To guarantee enough time to carry out blocking of the distance elements after a power swing is detected, the inner impedance of the blinder element must be placed outside the largest distance element for which blocking is required. In addition, the outer blinder impedance element should ideally be placed away from the load region to prevent PSB logic operation caused by heavy loads, thus establishing an incorrect blocking of the line mho tripping elements.

3.0. INFLUENCE OF POWER SWING ON THE DISTANCE RELAY

Power swings pose a significant challenge to the operation of protective relays, as the impedance measured during power swings can enter the relay's operating range. A power swing that is stable will eventually lead to a steady state, while a power swing that is unstable may result in a loss of synchronization.

To address these issues, protective relays employ two important schemes:

the Out-of-Step Blocking (OSB) scheme and the Out-of-Step Tripping (OST) scheme. The OSB scheme prevents protective relays from operating during stable power swings, allowing the power system to return to a stable operating state. On the other hand, the OST scheme is activated if the system loses synchronization, leading to a controlled separation of unsynchronized parts to prevent further damage.

Distance protections relays should not be activated during unstable or stable power swings. Instead, they should allow the power system to return to a stable operating state. Distance protection elements that tend to operate during steady or transient power swings should be temporarily disabled to prevent disconnection of the system at random or in locations not preselected.

In modern relays PSB function is available to differentiate between power swings and faults, blocking distance protection from operating during power swings. However, in the event of faults occurring during power swings, they need to be accurately detected and resolved with high selectivity and reliability. Serious disturbances in the power system can lead to a significant separation of rotor angles between groups of generators, potentially resulting in the loss of synchronism between these groups or neighbouring power systems.

In the case of the OST scheme, when a loss of synchronism occurs, it can use the same or different measuring elements. The operation is similar to PSB, but the expected behaviour is for the apparent impedance to pass through both the inner and outer zones (see Figure 8).

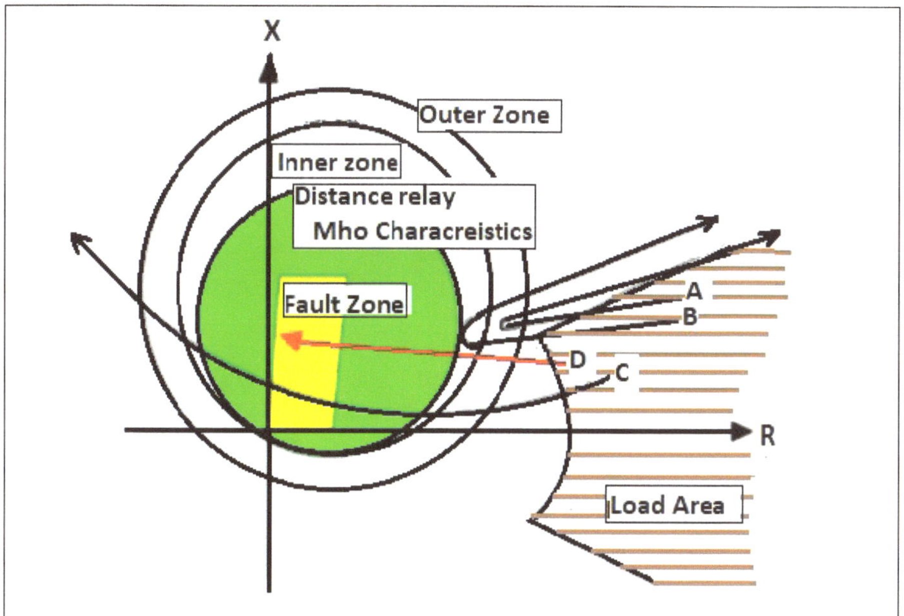

Fig.:"Mho"Type detection of Loss of Synchronism.

Where;

A: Z moves into the OOS (out-of-step) zone and slowly leaves;

B: Z moves into OOS & Trip zones and slowly leaves;

C: Z moves slowly = network becomes asynchronous;

D: Failure = Z moves rapidly into the exclusion zone.

When the criteria for power swing detection are not met, and the OST (Out of Step Tripping) scheme is selected, all relay zones are temporarily blocked to prevent premature tripping. The impedance vector "Z" is continuously monitored and when it leaves the power swing region, its "R" component is checked. If the R component retains the same sign as at the point of entry, it indicates that the power swing is in the process of stabilizing. However, if the vector passes through the Mho characteristic (trace "C" in Figure 8), it suggests a loss of synchronism.

Traditional methods are;

a. Scheme with two constraints for PSB and OST functions,
b. Single-constraint scheme for the OST function,
c. Concentric characteristic schemes

1) Continuous Impedance Calculation;

The continuous impedance calculation consists of monitoring the progression in the complex plane (Figure 10). A power swing is declared when the criteria for all three loop impedances have been fulfilled: continuity, monotony, and smoothness. Continuity verifies that the trajectory is not motionless and requires that the successive ΔR and ΔX be above a threshold. Monotony verifies that the trajectory does not change

Direction by checking that the successive ΔR and ΔX have the same signs. Finally, smoothness verifies that there are no abrupt changes in the trajectory by looking at the ratios of the successive ΔR and ΔX that must be below some threshold.

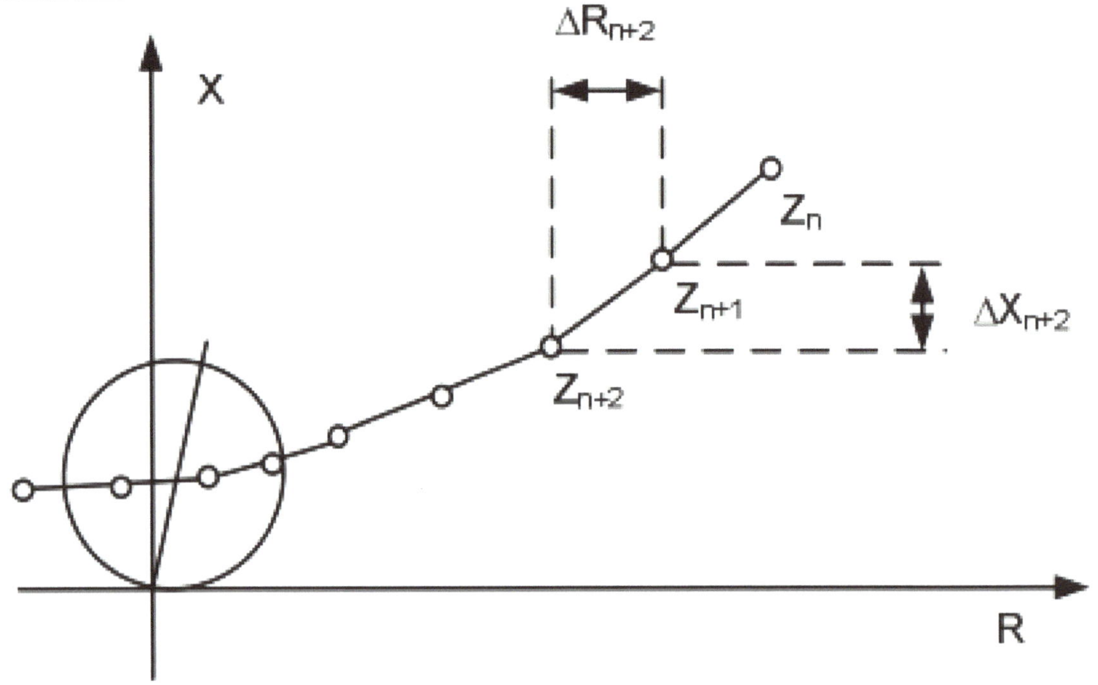

Fig. 10. Continuous impedance calculation.

The continuous impedance calculation is supplemented by a concentric characteristic to detect very slow-moving trajectories.One of the advantages of the continuous impedance calculation is that it does not require any settings and can handle slip frequencies up to 7 Hz. It does not require, therefore, any power swing studies involving complex simulations.

4.0 Impact of the System Impedances on the PSB Function

To guarantee enough time to carry out blocking of the distance elements after a power swing is detected, the inner impedance of the blinder element must be placed outside the largest distance element for which blocking is required. In addition, the outer blinder impedance element should ideally be placed away from the load region to prevent PSB logic operation caused by heavy loads, thus establishing an incorrect blocking of the line mho tripping elements.

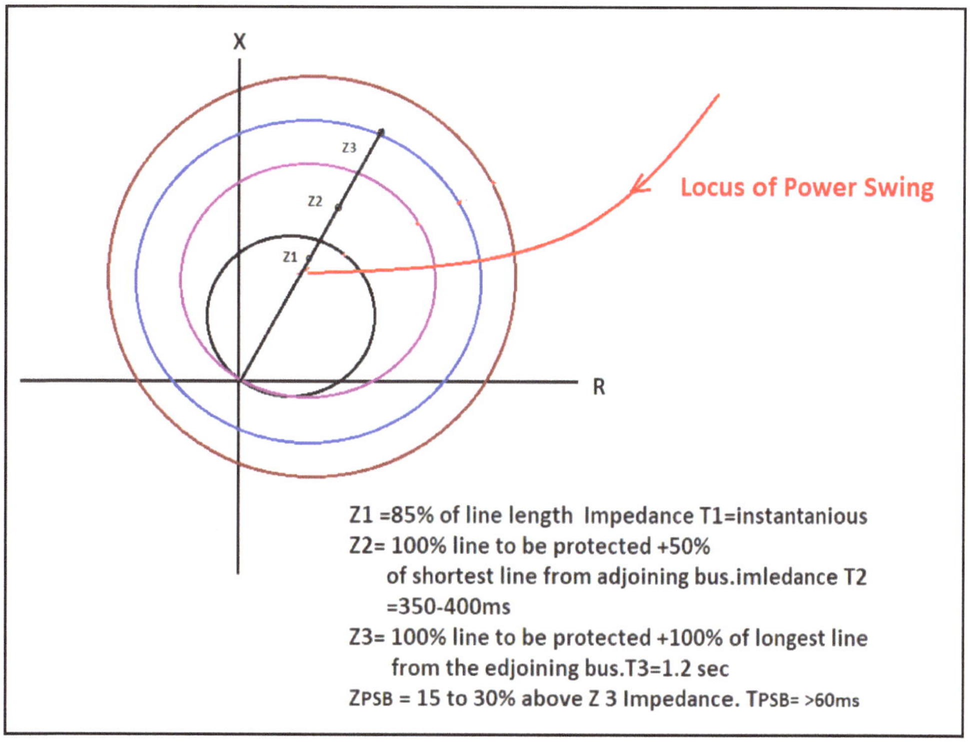

Impedance of swing Locus keeps on changing at lower rate of change.As per the the time settimg is concerned it is gererally kept at > 60ms beyond the time settimg of Zone 3 time setting.If the rate of chnge of impedance inrease, the locus of swing shall enter Zone 3 charateristics & then Zone 2 characteristic and as soon as Characteristic of Line impedance of ZL1 the trippings are blocked for a period of 2.0 Sec(till such time the system stabilizes).

In the mean while if any fault occurs the rate of change of Impedance changes at a fast pace and DPR gives a trip command to the circuit breaker accordingly.

Note: The PSB impedance is 15% to 30% > Z3 & time setting >60 ms.(ref. fig.)

As could be seen from the fig. the Z3 charaecteristic is about 10% behide the Bus and PSB(Power Swing Blocking) chateristic encompasses the Z3 charateristic by 15 to 30% > Z3 chateristic.

4.2 Conditions for blocking & tripping;

4.2.1 Blocing Phenominon;If the PSB locus enters the Z3 characteristics taking >60ms (40 to 60 ms setable) that means the pace is slower and it is identified as a Power swing and Blocks all the 3 Zones from tripping.This condition is kept for 2 secs.If during this time the system recovers to original position, the blocking of all the 3 zones is removed.

4.2.2 Tripping: When the time taken by the Power swing locus to enter in to Z3 chatercteristic in < 60ms it is identified as fault and in such cases the blocking of all the 3 Zones is removed and trippiong is effected as per the Zone where the fault lies.

In the mean while if any fault occurs the rate of change of Impedance changes at a fast pace the DPR shall give a trip command to the circuit breaker accordingly.

5.0. CONCLUSION

Power swing refers to the fluctuation in three-phase power flow that occurs when the generator angles of different synchronous machines advance or decelerate relative to each other in response to various factors.These factors can include changes in load magnitude and direction, line switching, loss of generation, faults and other disturbances within the power system.

6. Auto-Reclosing

1.0 Introduction: With the growth of Power system network, it becomes essential to maintain stability and availability of Power all the time. The whole Grid system functions in unison ie at same power frequency. Any outage due to Tripping of Inter connected EHV feeders between 2- S/S shall hamper the Stability of the system and result into the disturbance of system to the extent of even resulting into collapse of the grid system. Thus need for adopting auto-reclosing system on the EHV feeders in the grid was essentially felt.

As the name signifies the "RECLOSING OPERATION OF THE EHT FEEDER, POST TRIPPING, HAS TO BE AUTOMATIC".

This operation of auto-reclosing shall take place only and only if the system, synchronism is not lost in case of Grid system, However in case of radial feeders, no such condition exists.

2.0 Some Relevant Definitions; Before proceeding further following definitions with ref to the following fig.s should be known by the ETEs.

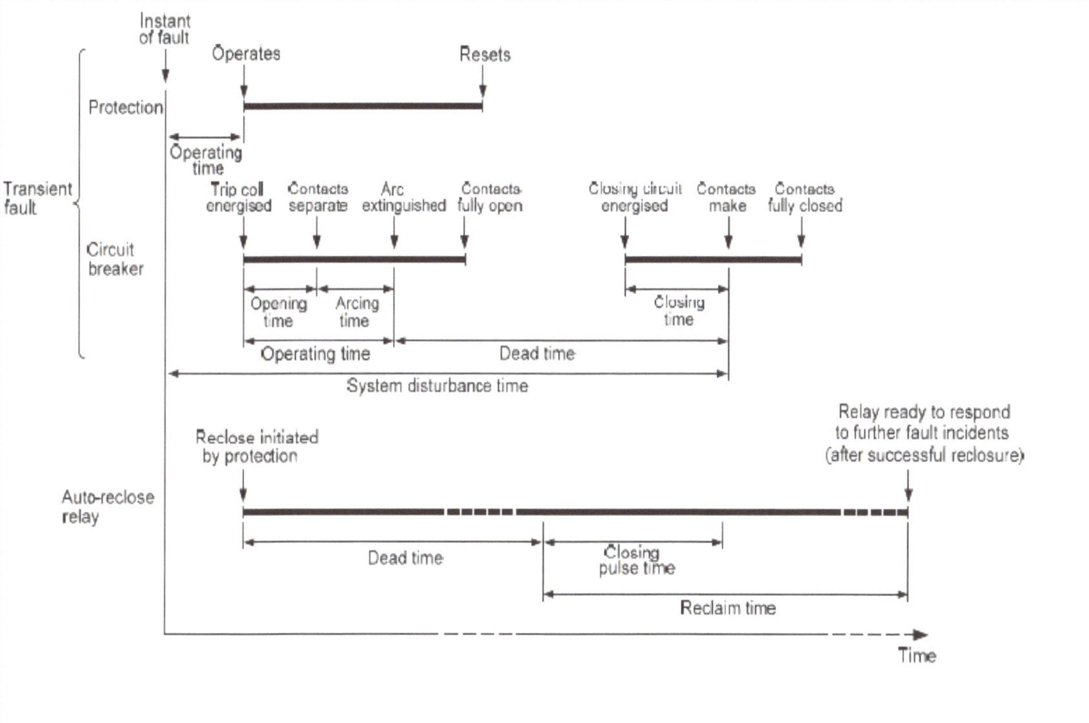

Figure 14.1: Single-shot auto-reclose scheme operation for a transient fault

Figure 1: Single shot auto-reclose scheme for a Transient fault.

As could be seen from the fig above, that there are 3- sections of horizontal lines drawn on X-axis designated for Time of operations, where as Y-axis has been shown as the instant of occurring fault.

1. First one showing the time of operation of protection.
2. Shows Operation of circuit breaker.
3. Operation of Auto-Reclose Relay.

Let us define different times and actions with reference to the above figure.

i. **Protection Operating Time**; the time from the inception of fault to closing of tripping contacts where a separate tripping relay is provided in the scheme.

1. **i. Operating time of Circuit Breaker**: The time of energizing of the trip coil until the fault arc is extinguished;

This again contains (a) opening time &(b) arcing time.

a. **Opening time:** The time between energizing of circuit breaker Trip coil and the instant of separation of the circuit breaker contacts.
b. **Arcing Time**: The time between instant of separation of the circuit breaker contacts and the instant of extinction of the fault arc.

2. **ii. Dead time Circuit (Circuit Breaker or system):** The time between the fault arc being extinguished to circuit breaker contacts re-making.(Note-Circuit Breaker contacts re-make only after the Reclosing command is received by the Circuit breaker).

   iii. De-ionizing time: The time following the extinction of an EHV/OH line fault arc necessary to ensure dispersion of ionized air so that the arc will not re-strike when the line is re-energized.

   iv. System disturbance time: The time between the inception of the fault and the circuit breaker contacts making on successful reclosing.

3. **i. Reclose initiated by Protection.**

Majority of the faults of the EHV feeder faults are of Ph to Ground nature (80% approx.).

The initiation of reclosure starts simultaneously with tripping command, the Dead time shall be up to the time the closing circuit of the breaker gets energized, then closing pulse time starts and ends when the breaker contacts fully make.

4. **Reclaim time:** The reclaim time is the time between the auto-reclose relay closing contacts to the completion of successful closing operation, then the auto–reclose relay is ready to accept further reclose command in case of subsequent incidence of fault.

5.0 Permanent Fault: The fig shown below pertains to the permanent fault condition If the fault persists or hangs-on up to the reclaim time. In such cases the logic is such that the breaker shall trip and lockout.

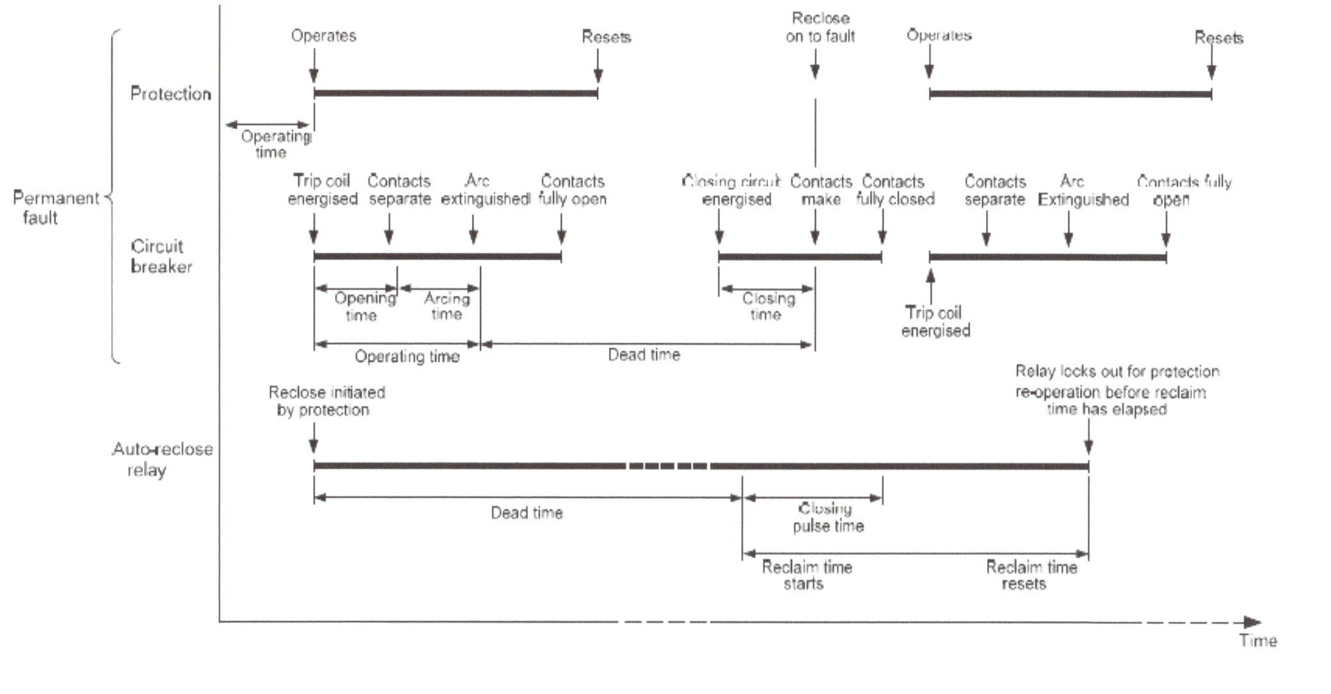

Figure 14.2: Single-shot auto-reclose scheme operation for a permanent fault

Figure 2:Opertion for a permanent fault condition.

The reclaim time is for the CB to re-coup and to return to normalcy before another

reclosure duty the circuit breaker is ready to undertake.

6.0 Condition for Auto-reclosing on EHV transmission lines. The most important conditions are maintenance of (i) System stability (ii) Synchronism. The problems involved dependent on whether the transmission system is week or strong. With a week system, loss of transmission link may lead quickly to excessive phase angle across the Circuit breaker thus preventing successful re-closer, while in a relatively strong system the rate of change of phase angle will be slow and thus delayed auto-reclose can be successfully applied.

The fig 3 shown below is an illustration between 2-power systems. Under healthy condition, the amount of synchronizing power transmitted, P, crosses the Power/ angle OAB at point X, showing that the phase displacement between the 2-systems isθo.Under fault conditions the curve OCB is applicable and the operating point changes to Y.Assuming constant power inputs to both ends of the line, there is now an accelerating power XY. As a result, the operating point moves to, an increased phase displacement θ1 between the two systems. At this point the circuit breaker trips and breaks the connection. The Phase displacement continues to increase at a rate depending up on the inertia of the two power sources.

To maintain synchronism the circuit breaker must be reclosed in a time short enough to prevent the Phase angle exceeding θ2.This angle is such that the area (2) stays greater than area (1), which is the condition for maintenance of synchronism.(refer Figure 3)

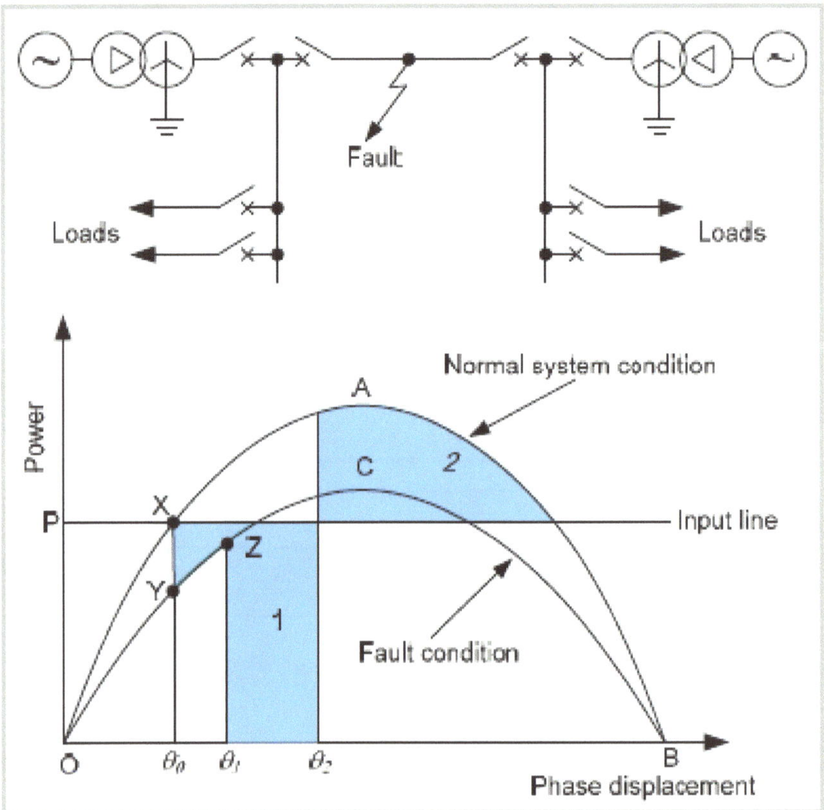

Figure 14.4: Effect of high-speed three-phase auto-reclosing on system stability for a weak system

Figure: 3

7.0 Protection Characteristics: The High speed DPR gives operating time of 40ms in conjunction with fast operating circuit breaker high speed protection reduces the duration of fault arc and thus the total system disturbance time.

It is essential to effect, the tripping of the breaker at both ends simultaneously.

8.0 De ionisation of the fault arc:

It is mportant to know the time that must be allowed for complete deionization of arc, to prevent arc restriking when voltage is reapplied.Time taken for deionisation depends on system voltage, conductor spacing fault current fault duration wind speed and capacitive coupling from adjacent conductor.The higher the votage of the transmission line longer is the time required for deionisation. Following table indicates typical values of deionization time for 3 Phase faults.

Line voltage(kV)	Minimun De-energization time(seconds)
66	0.1
110	0.15
132	0.17
220	0.28
400	0.5

If single phase tripping and auto-reclosing is used specially on long Transmission lines capacitive coupling between healthy phases and the faulty phase tends to maintain the arc and hence the de-energization time values given in the table need to be extended.

This is a particular problem in long EHV transmission lines, however where shunt compensation is applied, a neutral reactor can be used to balance the system inter-phase capacitance and thus reduce arcing time.

9.0 Single Phase Auto reclosing: As per statistics almost 80% Of faults on Transmission line are single phase to earth faults. In order to maintain the System stability it is essential to adopt auto-reclosing schemes on the Transmission Lines. When 3-Ph auto–reclosing is applied on a single transmission line connecting two sources, the tripping of all 3-phases may cause the two systems to drift apart. No interchange of synchronizing Power can take place during dead time.

If only the faulty phase is tripped, synchronizing power can still be interchanged through the healthy phases.

10.0 Three-Zone Distance Relay schemes: In MPPTCL as well as in all the constituents of WREB, 3-Zone system is adopted. We are having almost all the Numerical distance relays.

The Zone settings are as follows;

- 1st Zone: 80% of the protected line (instantaneous)
- 2nd Zone: 100% of the protected line +50% of the next shortest line or 120%of the original line whichever is more. (0.3sec)
- 3rd Zone: 100% of the line +100% of the longest line from the next station. (0.7 Sec).
- 4th Zone: reverse coverage (10%of the line towards bus).(0.5 Sec).

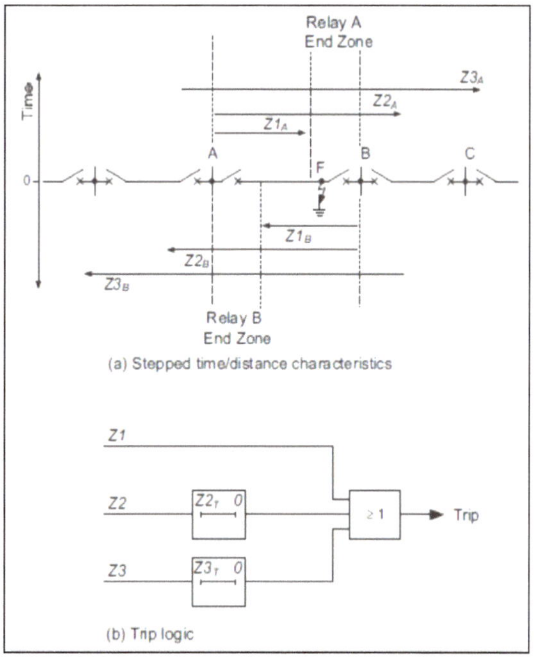

Figure 4: Typical 3 Zone Distance Protection scheme.

In most of the Transmission Utilities Auto-recloser schemes on the Transmission lines are operating in single phase mode.

To make the tripping of the poles of the faulty phase take place simultaneously a carrier signal is sent to the other end along with the tripping taking place at the sending end, the preparatory at the other end operates through the DPR starting relay and prepares the circuit to receive the command from the sending end and the tripping of the pole of the breaker takes place of the faulty phase and reclosing action thereafter takes place effecting successful auto-reclose operation.

The setting adopted for dead time is 1 Sec and the Reclaim time as 10 sec.

If the fault still hangs-on the Breaker (all the 3-poles), on both the sides shall trip and shall lock-out as this action is initiated during reclaim time.

Similar process shall take place if the fault is in the 1st zone, Ph to E fault, on the other end but the in 2nd zone-reach from this end. In such cases the carrier signal is received from the other end and successful auto-reclose operation shall take place.

11.0 Carrier Aided Auto-reclosure schemes

11.1.0 Direct Under reach- Transfer tripping scheme.;

The simplest way of reducing the fault clearance time at the terminal that clears an end zone fault in Zone 2 time is to adopt a direct transfer trip or inter-trip technique, the logic of which is shown in Figure C4.4.A contact operated by the Zone 1 relay element is arranged to send a signal to the remote relay requesting a trip. The scheme may be called a 'direct under-reach transfer tripping scheme', 'transfer trip under-reaching scheme', or 'inter-tripping under-reach distance protection scheme', as the Zone 1 relay elements do not cover the whole of the line.

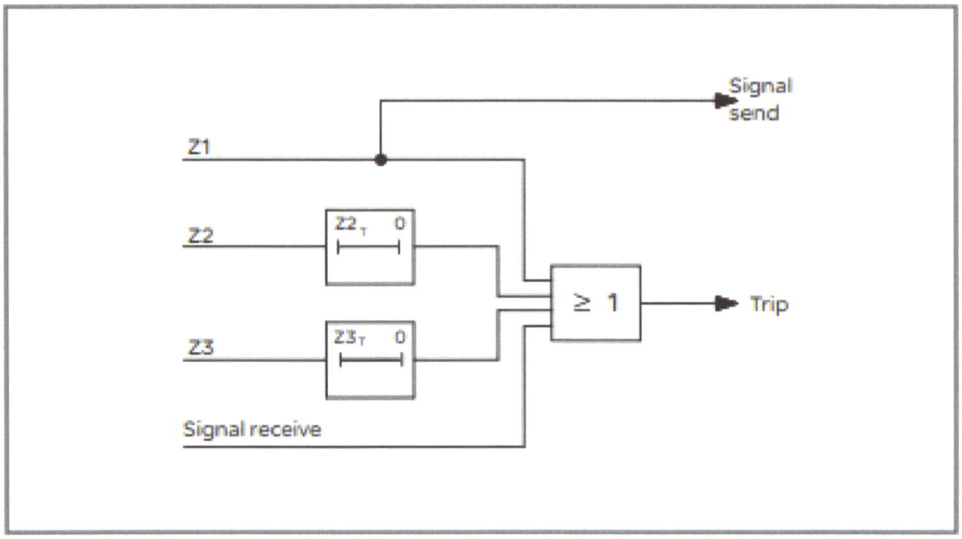

Figure 5: Direct Under reach- Transfer tripping scheme

11.2 Permissive under-reach transfer tripping (PUP) scheme;

The direct under-reach transfer tripping scheme described above is made more secure by supervising the received signal with the operation of the Zone 2 relay element before allowing an instantaneous trip, as shown in Figure C4.5. The scheme is then known as a 'permissive under-reach transfer tripping scheme' (sometimes abbreviated as PUP Z2 scheme) or

'permissive under-reach distance protection', as both relays must detect a fault before the remote end relay is permitted to trip in Zone 1 time.

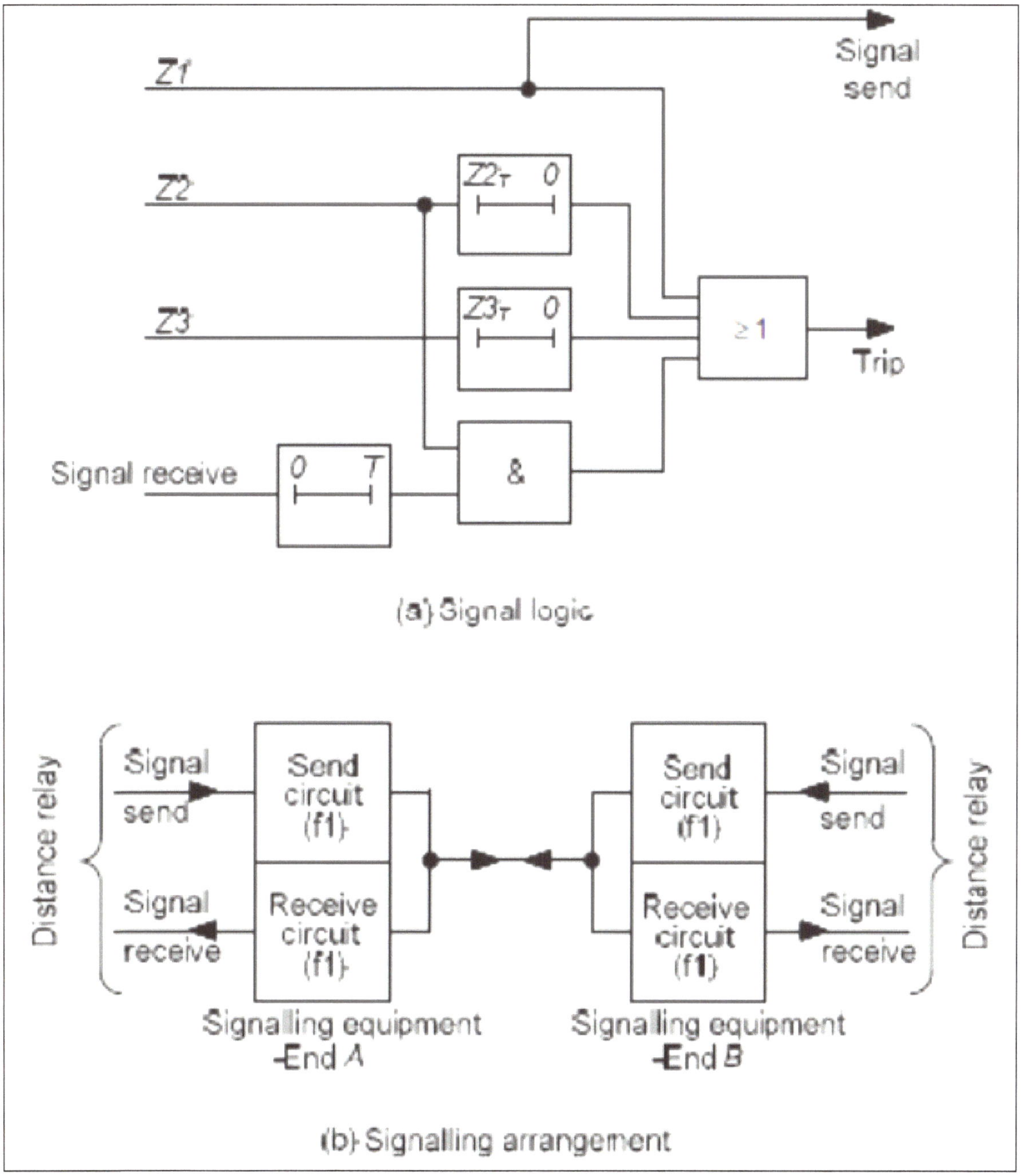

Figure 6: Permissive under-reach transfer tripping (PUP) scheme.

7. Carrier Aided Distance Relay Protection Schemes.

1.0 Carrier aided Distance Protection Schemes have many advantages.

i. For successful auto reclosing operations.
ii. Permissive Under reach scheme.
iii. Permissive over reach scheme.

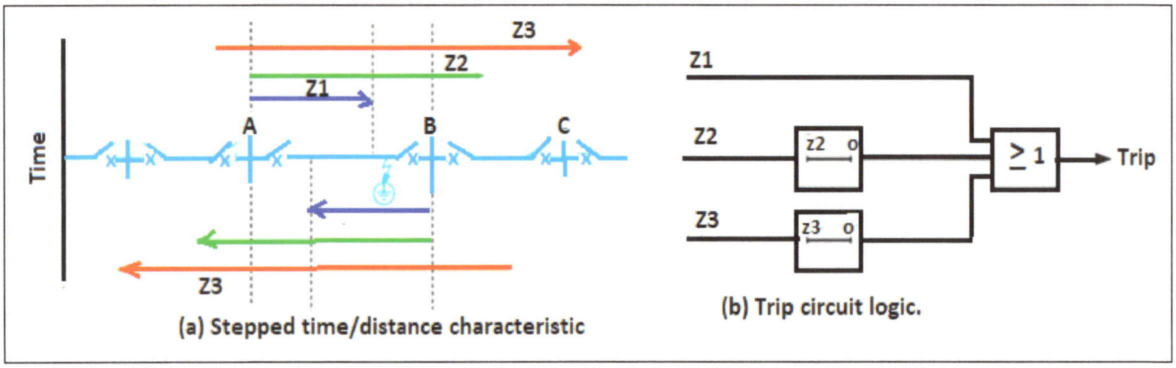

Figure 1: 3 (a) Stepped Distance relay Characteristic (b) Trip Circuit logic.

2.0 Direct under-reach transfer tripping scheme

The simplest way of reducing the fault clearance time at the terminal that clears an end zone fault in Zone 2 time is to adopt a direct transfer trip or intertrip technique, the logic of which is shown in Figure 2

A contact operated by the Zone 1 relay element is arranged to send a signal to the remote relay requesting a trip. The scheme may be called a 'direct under-reach transfer tripping scheme', 'transfer trip under-reaching scheme', or 'intertripping under-reach distance protection scheme', as the Zone 1 relay elements do not cover the whole of the line.

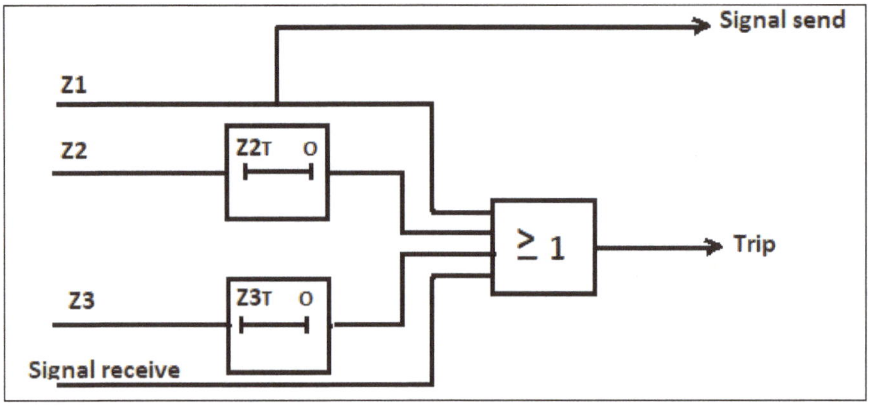

Figure 2: Logic Diagram of Direct trip through carrier signal

A fault F in the end zone at end B in Figure 1(a)results in operation of the Zone 1 relay and tripping of the circuit breaker at end B. A request to trip is also sent to the relay at end A. The receipt of a signal at A initiates tripping immediately

because the receive relay contact is connected directly to the trip relay. The disadvantage of this scheme is the possibility of undesired tripping by accidental operation or mal-operation of signaling equipment, or interference on the communications channel. As a result, it is not commonly used.

3.0 Permissive under-reach transfer tripping (PUP) scheme;

The direct under-reach transfer tripping scheme described above is made more secure by supervising the received signal with the operation of the Zone 2 relay element before allowing an instantaneous trip, as shown in Figure 3. The scheme is then known as a 'permissive under-reach transfer tripping scheme' (sometimes abbreviated as PUP Z2 scheme) or 'permissive under-reach distance protection', as both relays must detect a fault before the remote end relay is permitted to trip in Zone 1 time.

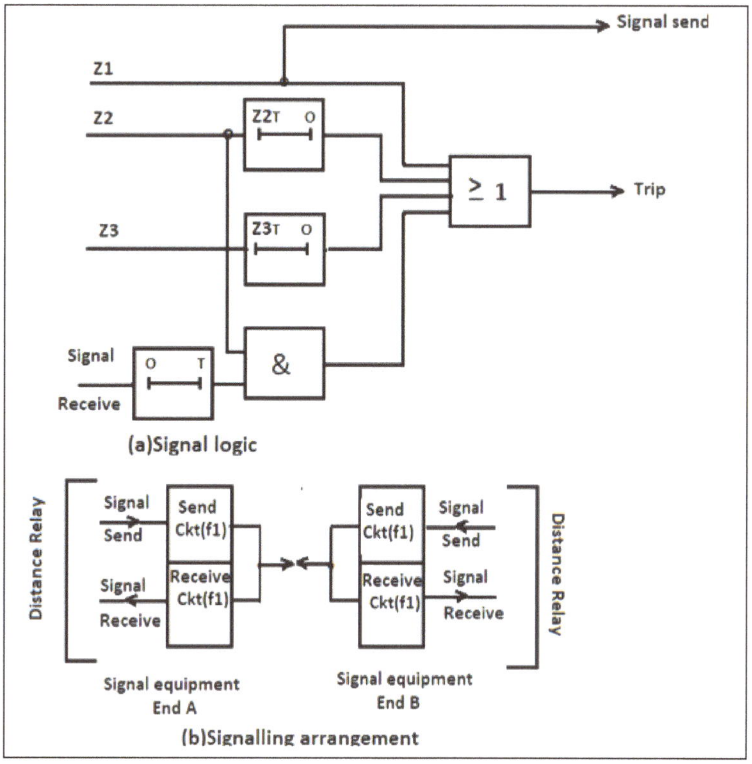

Figure 3: Permissive under-reach transfer tripping scheme.

4.0 Successful Auto reclosing Operations;

The concept is the same as Permissive under-reach Trip transfer scheme. Say for instance, as soon as the sending end circuit breaker trip in 1**st** zone (time-Instantaneous) whereas the other end DPR senses in 2nd Zone (t2-350 to 400 ms), ie if there is a source in the downstream and if there is no carrier aided Autureclosing scheme, as the fault is in 2nd zone so the fault shall be hanging on the line till t-2 time and may create a havoc in the system.To overcome such contingency Single phase carrier aided Auto-reclosing system is adopted in most of the countries including India.

Advantage of using Single Phase reclosing system;

Since, majority of faults are Single-Phase to ground, the synchronism of the system is not lost as remaining two phases are tied up in the Grid system.

Mode of operation:

i. As soon as Circuit breaker trips from either end on Single Phase to ground fault in 1st Zone and the fault lies in 2nd Zone reach from the other side.
ii. A carrier signal shall be transmitted to the other side.
iii. As a preparatory the DPR shall sense the fault in the operating direction & zone.
iv. And shall be prepared to receive the carrier signal & shall give a trip command to the faulty phase breaker pole (Breaker used is SPR type breaker(Single Pole Recloser type).

5.0 Permissive over-reach transfer tripping (POP) scheme;

In this scheme, a distance relay element set to reach beyond the remote end of the protected line is used to send an intertripping signal to the remote end. However, it is essential that the receive relay contact is monitored by a directional relay contact to ensure that tripping does not take place unless the fault is within the protected section; see Figure C4.8. The instantaneous contacts of the Zone 2 unit are arranged to send the signal, and the received signal, supervised by Zone 2 operation, is used to energize the trip circuit. The scheme is then known as a 'permissive over-reach transfer tripping scheme (abbreviated to 'POP'), or 'permissive overreach distance protection scheme.

Since the signaling channel is keyed by over-reaching Zone 2 elements, the scheme requires duplex communication channels - one frequency for each direction of signaling.

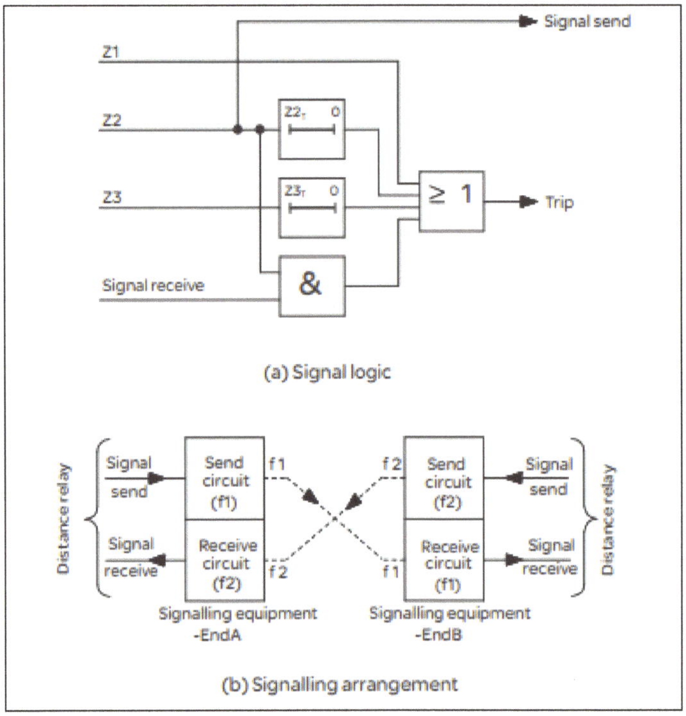

Figure 4: Logic diagram of permissive over-reach transfer tripping Scheme (abbreviated to 'POP')

8. Concept, Characteristics and Constraints of EHV CTs - A Guide to Budding Engineers

1.0 Introduction

The Current Transformers (CTs) are Instrument Transformers and are very vital equipments used on the HV, EHV and UHV level systems. These transform the Primary current (which is directly not measurable as being of HV/EHV levels) into Secondary current which is almost the replica of the primary current and can be easily used in the secondary instruments/equipments like protective relays, control panels, energy meters and for SCADA applications etc. Therefore the CTs are used as input devices producing secondary currents proportional to the primary currents within the required accuracy limits.

2.0 Theory /Principle of Current Transformers:

2.1 The Current Transformer follows the principle of maintaining balance between Primary Ampere-Turns and Secondary Ampere-Turns. Thus the following Ampere-Turn equation holds good.

$I_p \times N_p = I_s \times N_s,$

Where; I_p= Primary current,

N_p=Primary No. of turns.

I_s=Secondary current.

N_s= Secondary No. of turns.

$K_n = N_s / N_p$, turn ratio or Transformation ratio.

I_m=core magnetizing component.

I_e=Iron loss component, I_o=Exciting current. θ=the phase angle of the Current Transformer. **Note:** N_p (No of primary Turns can be 1, ie it is a bar primary CT).

2.2 Following figures depict the basic connection of CTs:

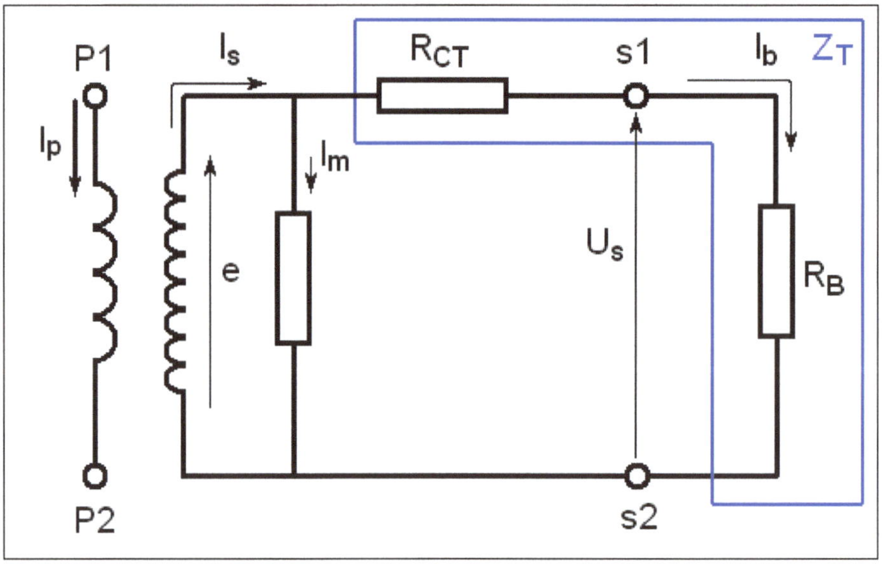

Fig:1(a)

Where

P1-P2 =Primary terminals, Ip =Primary Current,

S1-S2= Secondary terminals., Is=Secondary Current, Im=Core Magnetizing Current,

R_{CT}= Internal winding resistance of the CT, R_B=Resistance of the connected Burden.

$Z_T = R_{CT} + R_{B.}$

Note: In the above figure the Primary Turns have been shown, However in some CTs there is bar primary ie the CT has primary turn as 1 only.In such CTs the primary lead /conductor passes through the Toridal ring type core containing secondary winding.{Ref fig. 1.(b) }.

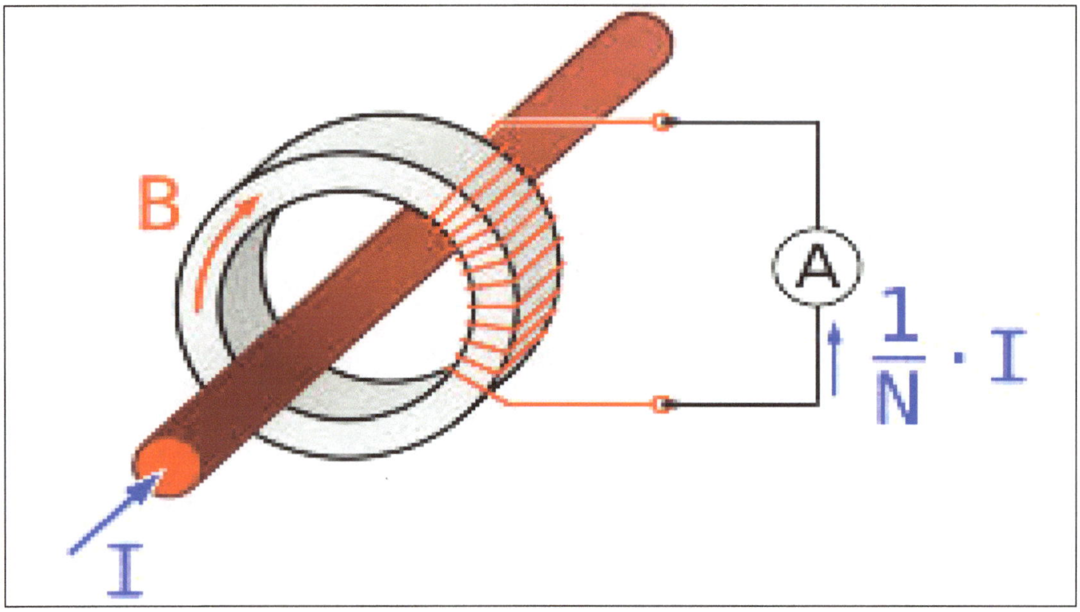

Fig:1(b)

2.3 The vector diagram of CTs is as follows;

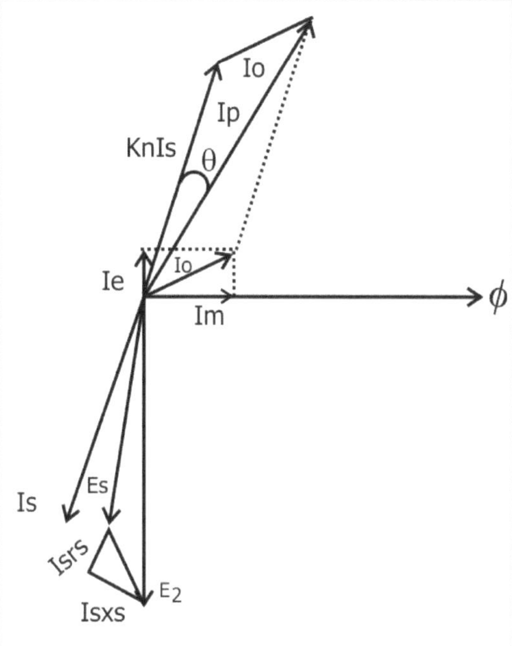

Fig : 2 Vector Diagram

2.3 CT Errors:

i. **Ratio Error**: The ratio error of the CT=$\frac{kn.Is - Ip}{Ip}$ 100. It is +ve, if K x I_s is more than I_p. that means for a given Primary Current, the Secondary current is high, however generally the ratio error is – ve.

ii. **Phase Angle of the CT** ($\angle\theta$): It is observed from the vector diagram that secondary current vector is not in line with primary current vector. The secondary current vector Is when reversed is making an angle of θ with the Pry current vector Ip. Is vector when reversed shall never lie on the Ip vector as this angle is caused due to the Magnetizing and Iron loss component of the secondary current. Thus phase angle error has to always exist.
The angle is said to be +ve by which secondary current vector, when reversed leads the Primary current vector. It is expressed in minutes.

iii. **Composite Error:** The secondary current not only contains ratio error and phase angle, but it also contains the effect of harmonics in the exciting current in the Composite Error. This is defined as the difference between the rms value of the ideal secondary current and the actual secondary current. The primary current is always assumed to be sinusoidal.

$$E_c = \frac{100}{\mathbf{Ip}}\sqrt{1/\mathbf{T}\int_0^{\mathbf{T}}(\mathbf{Kn.is} - \mathbf{ip})^2\,\mathbf{dt}}.$$

Where

K n = rated transformation ratio,

I p = rms value of the primary current, ip= instantaneous value of primary current.

is = instantaneous value of secondary current.

T = duration of one cycle.

The secondary wave shapes of Ideal secondary current, actual secondary current and the exciting current are shown in the fig. 3.

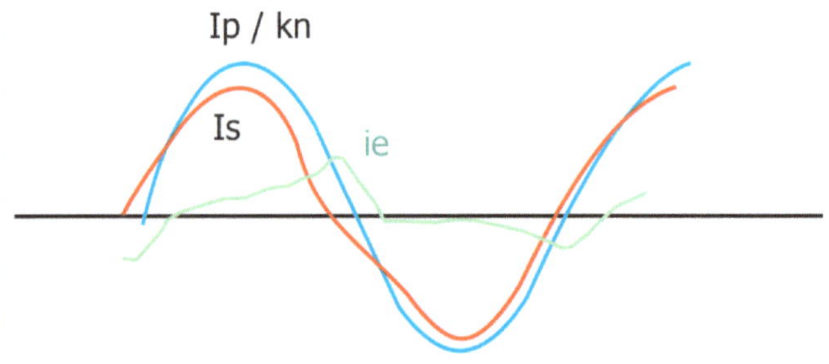

Fig : 3 Wave Shape of composite error

3.0. In addition to above, some of the relevant facts/ definitions / terminology concerning to the CTs must be known, which are mentioned in the following paragraphs;

3.1 Burden: It is the impedance of the secondary circuit in Ohms and Power factor. It is expressed in Volt ampere.

Volt Ampere should not be confused with a source which can deliver the rated current. Say 15 VA with 5A.secondary current does not mean 3voltsx5 amps. It is the square of the current flowing through the CT Sec winding resistance, resistance of 2 x the lead length and the impedance of the relays/meters connected in the circuit. For example, if the total impedance

of the instrument/relay, twice the lead resistance and the CT Secy winding resistance is say 3.0 ohms and the CT secondary rated current is say 5 amps thus the VA burden would be

I^2xZ, $(5)^2x3.0=25x3.0$, $\therefore$ VA Burden =75 AV.

3.2 Rated Burden: Output of the CT secondary to which it can feed maintaining the required accuracy stipulated in the Standards (IS: 2705 or IEC: 60044-1 etc)

3.3 Total Burden: It is the sum of the total burdens (in VA) connected to the CT secondary.

Total Burden in VA=I^2x {(resistance of 2 x length of connecting lead) + Resistance of CT Secondary winding)} + VA burden of the meters/relays connected to it.

Note: (i) In practice, in order to reduce the losses in the control cable of the CT, copper control cable of 4 sq mm is used between secondary terminals of the CTs from 220kV, 400kV and 765kV switch yards/sub-stations to the control rooms (due to very long distances) whereas other control cabling from switch yards to the control rooms is done with 2.5 sq mm copper control cables. It is for academic interest to know that resistance per Km of copper cables, of 2.5 sq mm is 7.41 Ohms/Km and that of 4.00 sq mm is 4.61 Ohms/km at 20^{0} C. (As per IS: 8130 -1984).

(ii) Resistance of Lead R_{lead} = (ρ x L)/A, here ρ (Rho) is the resistivity of copper as 1.79 x 10^{-8}Ω meter at 20°C and 2.16x10^{-8} Ω meter at 75° C.

L =2 x Actual length of the control cable lead from CT in the switch yard to the Relay. (2 is taken as it is twice lead length ie to and fro length). It should be in meters. A- Area of cross section of the conductor in Square Meter.

Example: Resistance of 50 meter copper control cable with cross section area of 4.0 Sq mm used for protection purposes is calculated below;

R_{lead}= (ρ x L)/A = (2.16x10^{-8}x 2x50)/4x10^{-6}.

Here 4mm^{2}=4x10^{-6} Sq Meters,

L= twice the lead length=2X50meters=100meters.

$$R_{lead} = \frac{2.16\text{x}10^{-6}}{4\text{x}10^{-6}} = 0.54\ \Omega$$

3.4. Dependency of secondary current on Primary current: The primary current is independent of the secondary current. The secondary current is governed by the Primary current only. Vice -versa, it is not possible.

4.0. Effect of secondary open circuiting;

Since in a CT Primary AT is balanced with Secondary AT, in a closed secondary circuit (Secondary circuit is closed through connected burden or directly) the counter flux produced by the secondary keeps the core flux below the saturation level and as the Primary current is independent of secondary current, the secondary circuit should never be allowed to get open circuited with the primary circuit carrying the current. In such situations the total primary AT is used up in magnetizing the core leading to its saturation. This results in increased EMF with voltage shooting up to a very high value depending up on the primary current which is hazardous to insulation as well as to human life.

Note: In practice, to facilitate shorting of the secondary, at the CT terminals in the C&R panels, the CT terminals are provided with shorting links and also sliding links.

5.0 Choice of secondary current:

The CTs are generally available in 1 Amp or 5 Amp secondary current. It is the choice of the user to choose between them, but there are certain constraints due to which the secondary current has to be fixed for a particular application.

The CTs are though installed in series with the equipment (Motors/Generators/Transformers/Capacitor banks/feeders etc) installed at some distance, the connections from the secondary terminals of the CTs are through leads/cables to the secondary protective relays /meters etc which are installed on the control & relay panels.

Where the primary current is high and the distance of the equipment is quite much(>30mtrs) as in the case of EHV Sub stations, the secondary current is selected as 1 Amp. Whereas when the distance of equipments is short (generally< 30mtrs) such as in small 33/11kV S/s, 3.3 kV, 6.6kV and 11kV indoor switchgears, 5 Amp secondary is generally selected.

The **VA burden** of the cable (Loss in the cable) imposed on the CTs is governed by to the resistance of the lead/cable and square of the secondary current flowing through it.

Say the lead (to & fro length) resistance is R ohms and the secondary current is of 1 amps. The VA burden imposed would be $I^2xR=1R=R$ VA, whereas in case of 5 Amp secondary current it would be $I^2xR=5^2xR=25R$ VA. ie 25 times more than with that of CT with 1 Amp secondary current.

5.1 Brief comparison of the CTs having 1 Amp and 5 amp secondary windings:

S.No	1 Amp Secondary	5 Amp Secondary
1	**Low VA burden of leads (I^2R).**	**Lead VA burden (I^2R) is 25 times more than that of 1 Amp. Secondary.**
2	**Preferred when CTs are outdoor and lead length is more(generally >30 mtrs), eg EHV Sub Stations etc.**	**Preferred where lead length is less (<30 mtrs, as per practice). eg CTs used in small 33/11kV Sub Stations, indoor switchgear cubicles, metering equipments for Billing and HT consumer metering or where primary rating of full load current is very high say for 500 MW, Generator with terminal voltage of 21 kV, the full load current would be 16166A, thus CTs used are 20,000/5.A**
3	**Development of high peak voltage when secondary is left open.**	**Development of comparatively low peak voltage when secondary gets opened.**

5.0 Reference Standards: The CTs are manufactured in INDIA confirming to Indian Standard IS: 2705(part 1 to 4) -1992 and /or to the International Electro Technical Standards, IEC: 60044-1.

6.0 Classification of CTs:

The CTs are classified basically as per their use. They are of 2 types (1) Measuring Current Transformer (2) Protective Current Transformer.

7.0 Measuring Current Transformer: This is also known as Metering CT, it should confirm to IS:27O5-2 or IEC:60044-1 The cross section area of the cores are so designed that, the core saturates at about 5 times the rated current (with low saturation current). This would facilitate safety to the instruments, while heavy fault current flows in the Primary circuit.

8.0 Before going into the details of characteristic / duty to be performed by the metering CTs, one should be aware of some definitions / terminology relevant to the equipment.

8.1 Instrument Security Factor (IFS/FS):

The ratio of instrument limits primary current to the rated primary current.

Note: In the event of system fault, the fault current flowing through the primary winding of a CT, the safety of the apparatus supplied by the CT is greatest when the value of the instrument security factor (SF) is small.

8.2 Rated Instrument Limit Primary Current (IPL):

The Value of the minimum primary current at which the composite error of the measuring CT is equal to or greater than 10%, while the secondary burden is equal to the rated burden.

Note: The composite error should be greater than 10%, in order to protect the apparatus supplied by the CT against the high current produced in the event of system fault.

9.0 Accuracy Classes;

a. **Standard Accuracy classes:** 0.1,0.2,0.5 and1.0 class-The accuracy limits are defined from 5% of current onwards, however the declared accuracy limits ie 0.1,0.2,0.5 & 1.0% respectively and 5,10,30,60 minutes respectively would be at 100% to 120% of rated current at 25% to 100% of rated burden. {Ref: IS: 2705(part-2)}
b. **Special Application Accuracy classes:** 0.2S and 0.5S are used for special applications. The accuracy limits are defined from1% of rated current, however the declared accuracy limits ie 0.2% & 0.5%and 10&30 minutes respectively would be at 20% to 120% rated current at 25% to 100% of rated burden. Precision metering of High value HT consumers and Tariff meters etc are categorized as special application metering and need 'S' class CTs. {Ref:IS:2705(part-2)}.

10.0 Limits of Accuracy for Measuring CTs; The limits of accuracies as per IS: 2705(part 2) and IEC: 60044-1 are as follows;

Table No.1.Limits of Errors of Standard Accuracy classes of CTs;

Accuracy class	± % Current(Ratio) error at percentage of Rated Current				± Phase displacement in minutes at percentage of Rated current			
	5	20	100	120	5	20	100	120
0.1	0.4	0.2	0.1	0.1	15	8	5	5
0.2	0.75	0.35	0.2	0.2	30	15	10	10
0.5	1.5	0.75	0.5	0.5	90	45	30	30
1.0	3.0	1.5	1.0	1.0	180	90	60	60

Table no. 2. Limits of Errors of Special Application Accuracy classes of CTs;

Accuracy Class	±% Current(Ratio) error at percentage of Rated Current					± Phase displacement in minutes at percentage of Rated current				
	1	5	20	100	120	1	5	20	100	120
0.2S	0.75	0.35	0.2	0.2	0.2	30	15	10	10	10
0.5S	1.5	0.75	0.5	0.5	0.5	90	45	30	30	30

Note: Almost all the Distribution Utilities follow the convention mentioned here under;

i. 0.2 and 0.2S are used for the voltages above 33kV level.

ii. 0.5 and 0.5S are used up to 33kV level.

11.0 Criteria/Methodology generally followed by the Power Utilities/Distribution Utilities while selecting CTs for HT Consumer metering and Tariff metering:

SN	HT consumer Metering	Tariff metering
1	kV level - 33kV	kV level -220kV
2	Maximum demand - 5 MVA, Current equivalent =87.5 A	Max load -150 MVA, Current equivalent = 394A
3	Distance of meters from CTs -15 meter	Distance of CTs from meter- 25 Mtrs (In EHV S/s yard)
4	CT ratio needed 100/5 A (Since distance is less than 30 mtrs., 5Amp is selected.) Note: (i) CT rating should in variably be 120% of the Max rated current. Therefore 1.2 x 87.5=105A, Thus 100/5A CT would be suitable. (ii) In order to rule out any confusion about connected ratio or also any mal practice by the HT costumer, most of the distribution Utilities are procuring single ratio CTs for HT Consumer metering application.	CT ratio needed 400/1A (though distance of metering from the CTs is less than 30 mtrs, but 1 Amp is used as the secondary current for EHV CTs for all the core is 1 Amp only. Note: (i) CT should have ratio of 800-400/1A, Connected to 400/1A (metering core) along with other cores for protection purposes etc. (ii) Over load of 120% can be taken care of by this CT (CTs can carry 120% rated current continuously with defined accuracy limits).
5	Requirement of VA; (a)Loop resistance of connecting lead of 4.0 Sq mm copper Cable=2x4.61x15/1000=0.1383 ohms. (Where as resistance of copper cable of 4 sq mm is 4.61Ohm/km.) (b)Energy meter's internal burden=4 VA. VA Burden on the CT =1.5x{(0.1383x5x5)+4}=11.186 VA Selected VA for CT is 15 VA	Requirement of VA; (a) Loop resistance of connecting lead of 4.0 Sq mm copper Cable=2x4.61x25/1000=0.2305 ohms. (b)Energy meter's internal burden=2 VA. VA Burden on the CT =1.5x{(0.2305x1x1)+2}=3.345VA Selected VA for CT is 10 VA
6	ISF<5	ISF<5
7	Class of accuracy- 0.2S	Class of accuracy- 0.2S
8	Thus the CT metering core shall have details as follows; 100/5A,0.2S,15VA,<5	Thus the CT metering core shall have details as follows; 800-400/1A,0.2S,10VA,<5

12.0 The Metering CTs are typically designated as follows;

Ratio/Accuracy class/VA Burden//ISF.

A typical example is as follows: 400-200/1A/0.2/10VA/5.

13.0 Protection CTs;

Protection CTs play very important and vital role in Power System. At the time of faults the excessive fault current is fed to the protection relays through these CTs for their operation.

Type of Class: Basically there are two types of CTs/Cores of the main CTs. (i) 5P10 and 5P20.

iii. (ii) PS or PX (IEC equivalent).

13.1 Accuracy Limit Factor (ALF): It is defined as the multiple of the rated current up to which the CT follows the accuracy limit. In other words it is the ratio of the fault current to the rated current up to which the CT shall follow the accuracy limit.

14.0 5P10 & 5P20 Class CTs: They are generally used for Over current (Instantaneous and IDMT relays) and E/F protection of Feeders, Transformers etc.

- 5 signifies the % composite error.
- P signifies as Protection core.
- 10, 20 signify the quantum of fault current in terms of multiples of the rated current or ALF(Accuracy Limit Factor) up to which the CT shall have the defined percentage of composite error i.e. ± 5%.

The CTs of 10P10 or 10P20 are also available which are used in O/C protection of the feeders and Transformers etc wherein CT composite error to the range of ± 10% is tolerable at 10, 20 times of rated current.

The Standard Accuracy Limit factors as per IS: 2705 -part 3 & IEC60044-1 are 5, 10, 15, 20 & 30.

Table 3: Limits of Errors (As per IS: 2705 (Part 3) & IEC: 60044-1;

Accuracy class	Current Error at Rated Primary Current	Phase Displacement at Rated Primary Current	Composite Error at Rated Accuracy Limit Primary Current
	Percent	Minutes	Percent
5P	±1	±60	5
10P	±3	-	10
15P	±5	-	15

15.0 Accuracy limit factor is inversely proportional to the actual burden if lower than the rated burden:

If the burden is lower than the rated burden, the accuracy limit factor shall elevate beyond the rated Accuracy Limit Factor. The formula for actual ALF;

$$ALF_{ACT} = ALF_{RTD} \text{ x } \frac{\text{Sin} + \text{Sn}(\text{Rated Burden})}{\text{Sin} + \text{Sa}(\text{Actual Burden})}.$$

Example:- Calculate actual Accuracy limit factor if secondary burden is reduced to 1/3rd, the CT details are 200/5 A, 5P10,15VA and Rct = 0.15Ω?

ALF_{Rated} = 10 (From CT data 5P10),

Sin (Burden due to internal resistance of the CT) = $(5A)^2$ x 0.15=25 x 0.15=3.75VA,

Sn (Rated burden) = 15VA (From CT data), Sa (Actual Burden)= 15÷3= 5VA,

$$\mathbf{ALF_{ACT} = ALF_{RTD} \text{ x }} \frac{\text{Sin} + \text{Sn}(\text{Rated Burden})}{\text{Sin} + \text{Sa}(\text{Actual Burden})}$$

$$ALF_{ACT} = 10x \frac{(3.75+15)}{(3.75+5)} = 10x \frac{18.75}{8.75} = 21.42$$

It is clear from the above example that the ALF increases if the connected actual burden is lower than the rated one. The CT in this condition shall maintain accuracy up to fault current of 21.42 times of rated current as against 10 times the rated current and shall not saturate till then.

16.0 Knee Point Voltage;

It is a magnetizing characteristic of a CT Keeping the Primary Terminals open and connecting AC supply(in volts) to the secondary terminals, the voltage is gradually increased observing excitation current ie (in mili amps) taken by the core. (Pl ref Fig: 4 excitation curve) Initially, as the voltage is raised, it cause the magnetizing current to increases to a certain limit beyond which when the Voltage is increased to10% it causes the magnetizing current to increase to 50%.This voltage is defined as Knee point voltage and written as Vk. Beyond this voltage the Curve flattens. (Ref. Curve Fig: 4).Knee point Voltage signifies the saturation voltage level of the CT of protection core.

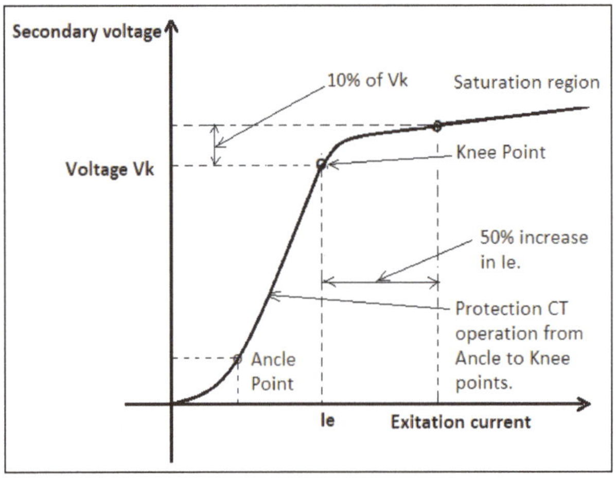

Figure 4: Knee point voltage curve.

17.0 PS Core {ref. IS: 2705 (Part 4)} [Known as Px core as per IEC: 60044-1]: Though abbreviation PS is not elaborated anywhere, however it is assumed to be "Special Protection Class" CT core. This core is used particularly where current balance is required to be maintained. The turn ratio error is limited to ± 0.25%, which helps in maintaining current balance in Differential protections & restricted E/F protection, which is the prime requirement between associated CTs, particularly during through fault conditions. The 5P/10P class CTs cannot match the characteristic as that of PS class. The core of this class is such that very high current is needed for saturation of the core.

The CT of PS class is specified in terms of;

i. Rated Knee point voltage Vk.
ii. Maximum exciting (Im) at knee- point voltage or at the specified fraction thereof.
iii. Secondary winding resistance (Rct) of the CT at 75^{0}C.

17.1 Requirement of Vk (Knee point voltages) of some of the known Differential relays;

i. **ALSTOM Make;**

a. Type DTH31/32(Static);

Vk ≥ 40xI (Rct+2Rl),

where Rct = internal resistance of CT., Rl = Lead resistance of cable from CT to Control panel(single length.)

b. **Type MBCH 12/13;**

Vk ≥ 24 x In (Rct+2Rl), In =Relay rated Current.

c. Type KBCH 120,130 and 140.

Vk ≥ 24 x In (Rct+2Rl), In =Relay rated Current.

ii. **ABB Make;**

a. Type RADSB;

Vk ≥ 30x In. (Rct+2.Rl+Rt+Zr),

In =Secondary Current corresponding to rated Primary Current.

Rct = resistance of CT secondary winding.

Rl = resistance of single lead from CT to relay.

Rt = resistance of interposing CT.

Zr = reflected burden of Relay.

iii. **EASUN REYROLLE Make;**

DUOBIAS-M (Numeric)

Vk≥ 4xIf (Rct+2Rl), where If=3ph fault current limited by transformer impedance or high-set setting whichever is greater, Rct= CT secy winding resistance, 2 Rl=twice the lead resistance from CT to relay.

iv. **AREVA Make:**

a. MICOM P633, (Digital relay):

$Vsat \geq (Rop+Ri).k.I'_{1,max.}$

Where;

Vsat =Saturation Voltage, Rop=Actual connected operating burden, Ri =Internal burden, $I'_{1.max}$ = Non-off set max primary Current converted to secondary side,

k=Over dimensioning factor or Transient factor,

= 1+ωT, wherein ω =System Angular frequency and T=Primary system time constant ie L/R of the system.

18.0 CT Saturation: The CTs are liable to saturate if the Primary current at the time of fault exceeds beyond its ALT limit. Such saturation may occur if the CTs have very low ratio and the fault current exceeds beyond 20 times of the rated current, in such cases the protective relays (O/C) become inoperative.

Please refer to the following figure 5 depicting wave shape of the secondary current at the time of saturation.

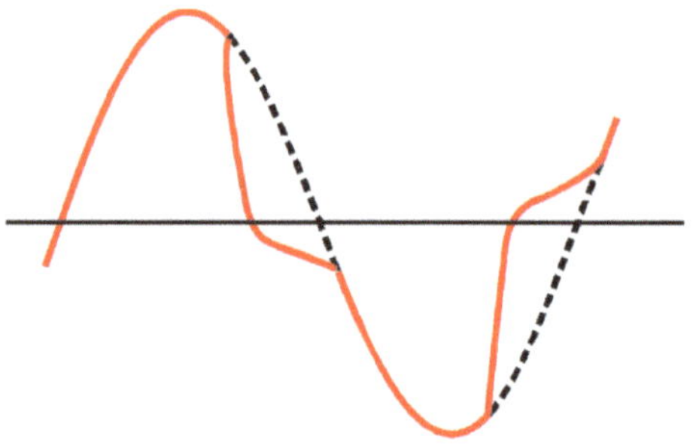

Fig. 5 : Steady-state saturation with AC current

To avoid saturation the CT should develop voltage such that $Vk \geq I_{f'} (R_b + R_{CT} + R_L)$,

I_f= secondary fault current for fault at zone-1reach. (Amps),

R_b=Relay burden in ohms.

R_{CT}=Resistance of CT winding in ohms.

R_L=Resistance of lead (2xsingle lead length) (ohms).

19.0 Transient saturation: In the transmission lines, since they predominantly contain inductance, whenever fault occurs, the fault current contains component of DC current also. In practical applications, the DC transient current is in exponential form and does not sustain but decays in the time equivalent to the time constant L / R {Ref.fig: 7(a) below}, L is the inductance in Henry and R is the resistance of the transmission line in ohms up to the location of occurrence of the fault. However in practice full line length is considered for calculation purposes.

Due to the DC transient component of the fault current, the total flux in the CT core is distorted and affects the secondary current {Ref fig: 7(b) below} and the curve of CT current at secondary will be as shown in Fig: 6. As soon as the DC Transient vanishes, the steady state condition of the CT gets restored.

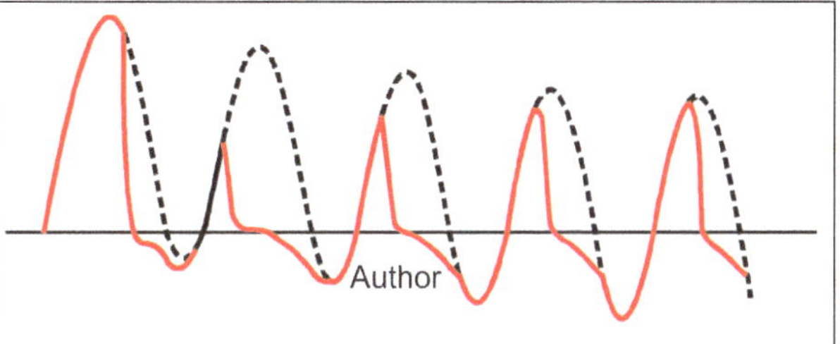

Fig. 6 : Transient saturation with offset current

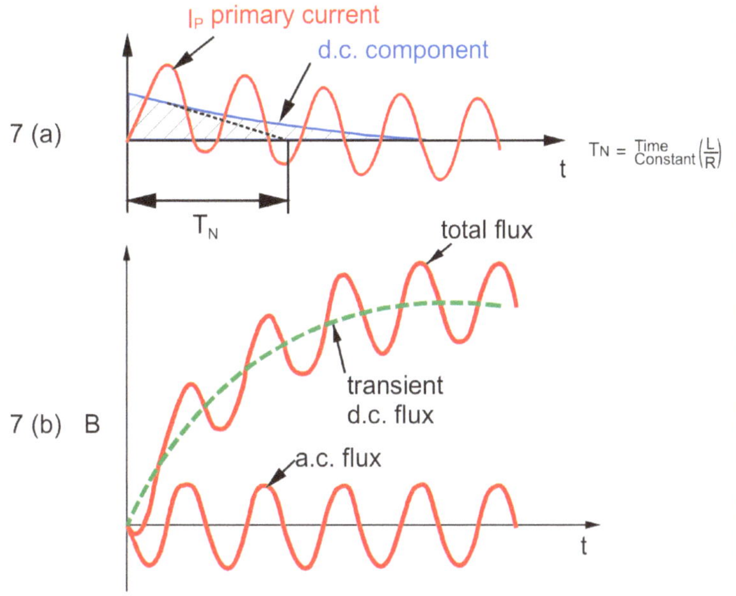

Fig. 7 (a) Primary short Circuit current with DC component.

Fig. 7 (b) Effect of DC component on CT Flux.

As has been mentioned that the DC transient persists for a period of about the Time constant T_N of the Transmission line {ref Fig: 6(a)}, care should be taken while procuring/selecting CT for faithful operation of the Distance relay. The Vk should be should be as follows.

This formula for Vk shall take care for the transient DC decay time of all kV class of Transmission lines. The CT should develop voltage such that

$Vk \geq \mathbf{I_f} (R_b + R_{CT} + R_L)$,

Where,

X =System Reactance (ohms), R= System Resistance (ohms).

I_f= secondary fault current for fault at zone-1reach.(Amps),

R_b=Relay burden in ohms.

R_{CT}=Resistance of CT winding in ohms.

R_L=resistance of lead (2xsingle length)(ohms).

The term (1+X/R) is an additional term as compared to requirement of development of voltage to avoid saturation in case of AC saturation (ref equation in para.17).

20.0 Ratio Selection: Required CT Ratios are selected from reconnection of Primary Terminals in majority of the cases, However in some cases the ratios selection is provided at secondary terminals. Further, in some cases the ratios are selected reconnecting primary terminals as well as from secondary terminals (as the case may be).

i. Selection through Primary terminal reconnection: If there are 2 ratios say 400- 200A/1-1 A.
ii. The reconnection at primary terminals is such that for lower ratio the winding shall be in series and for higher ratio, the reconnection is such that the primary current shall get a parallel path.
iii. Following sketches depict the ratio selection of ratios at primary terminals as well as through secondary terminals;
iv. For higher ratio reconnect the primary terminals to P1-C1 and C2-P2 and for lower ratio C1-C2{ref fig8(a)}
v. Each core is identified with the prefix numerical number. eg core1:1S1-1S2, core 2: 2S1-2S2, core3: 3S1- 3S2, core 4: 4S1-4S2 & core 5: 5S1-5S2 etc.
vi. Where selection of ratios are in secondary only the lower ratio shall be 1S1-1S2, 2S1-2S2 etc and the higher ratio shall be 1S1-1S3, 2S1-2S3 etc.{Ref Fig:8(b)}
vii. Figure 8(c) depicts the ratio selection through both Primary and Secondary terminals.

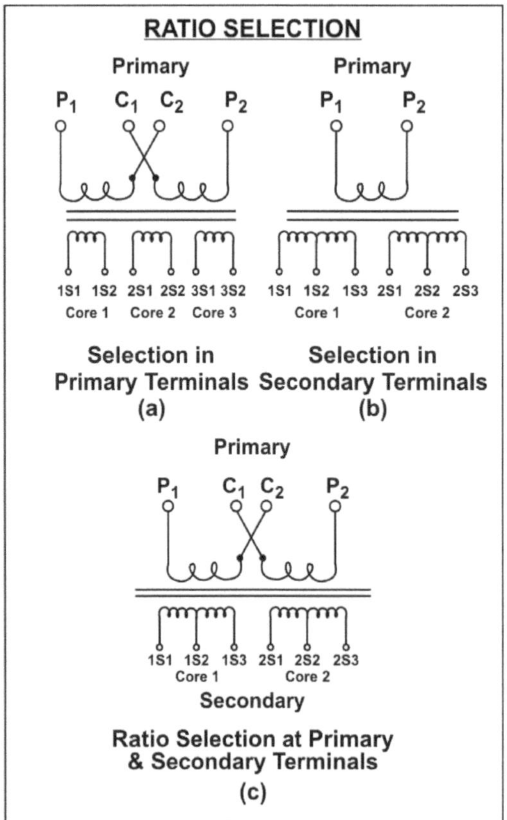

Fig. : 8

21.0 Types of CTs: There are two types of EHV class CTs,(i) Dead Tank Type and (ii) Live Tank type. They have been designated as per their design of placement of CT windings. The constructional details have not been covered this article. However, sketches/figures of Live Tank Type and Dead Tank Type CTs have been shown in fig.9(a) and fig.9(b) below for academic interest only.

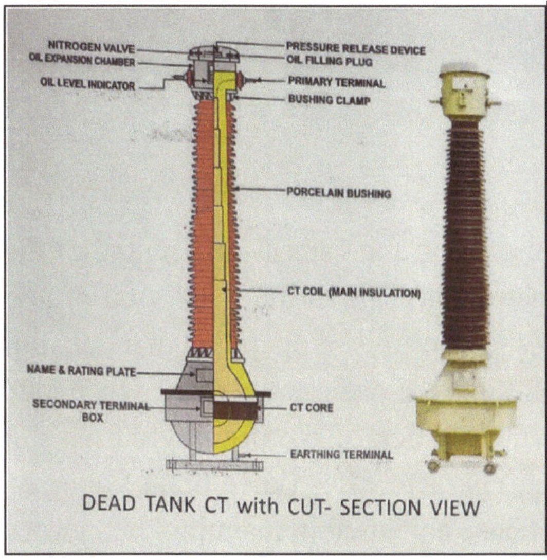

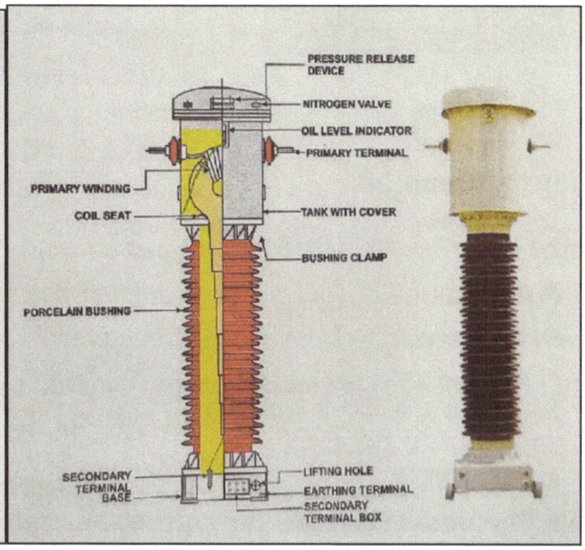

Dead Tank Type CT Live Tank Type CT

Figure 9(a) Figure 9(b)

22.0 Conclusion: The above narration is particularly of EHV class CTs. Practically all the salient practical and important features have been covered in this article which may prove to be a guide for budding Protection and Metering Engineers.

23.0 References:

1. IEC: 60044-1, and IS: 2705 (Part 1 to part 4),
2. Write up of ABB on Calculation of Accuracy Limit Factor.
3. Brochures /Instruction manuals of and Differential relays mentioned in the paper.
4. Write up on basics of Current and Voltage Transformers by Siemens AG 2007.

9. Prevention Intrusion of Zero Sequence Current (I_0) from Differential Relay Protection Circuit of Transformers with Grounded Neutral - (Part-I)

1.0 Synopsis/Preamble;

As the name signifies, the Differential protection is based on the current differential of the primary (input) and the secondary(output) currents of a transformer. This is the most vital protection of a transformer categorized as Unit protection covering a specific zone of operation of a transformer from HV CTs to LV CTs. This protection is implemented through the differential relay which identifies the current differential in the secondary circuit through the replica of the primary currents. The main constraint in this protection is to have CTs of similar characteristics (knee point voltage) and their connections should be such that the currents in each phase are equal in magnitude in normal running conditions as well as in the through-fault conditions and also elimination of the Zero sequence current (Io) from the Differential protection Secondary circuitry.

This Article covers concept of Transformer Differential - Protection, CT secondary connections vis a' vis eliminating phase-shift due to vector group difference feeding equivalent secondary currents from HV & LV using Auxiliary CTs and avoiding intrusion of Zero sequence currents (Io) in case of external Earth-Faults in star-connected neutral- earthed Transformers.

2.0 Key words: Equipment no 87, Auxiliary CTs, Io-Zero sequence current

3.0 Concept of Transformer Differential Protection: This is the most critical and vital protection of the Power Transformer. It compares the Secondary Currents of HV and LV side of the transformer in magnitude and phase. The direction of HV and LV secondary currents in the Operating Coil is such that they oppose each other in the normal operating condition. The HV and LV currents are made available to this relay through the main CTs.

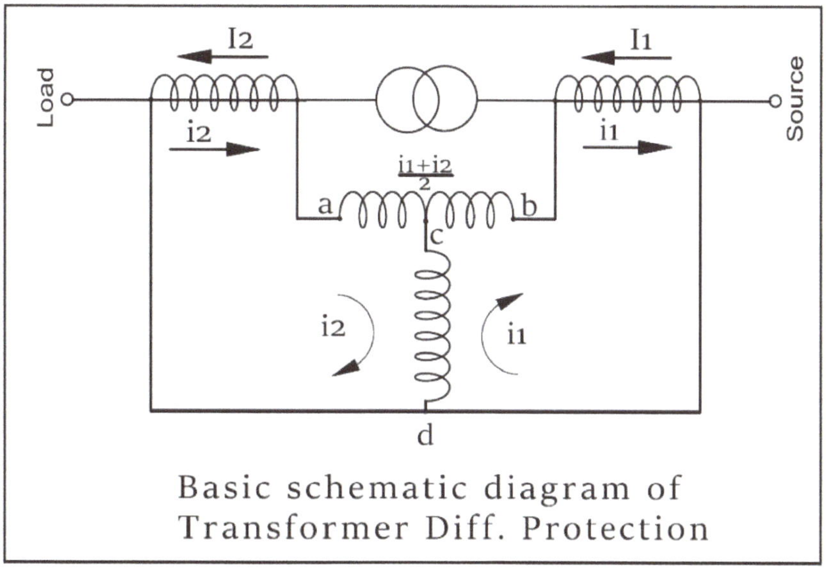

Figure 1A.Basic schematic of transformer Differential protection

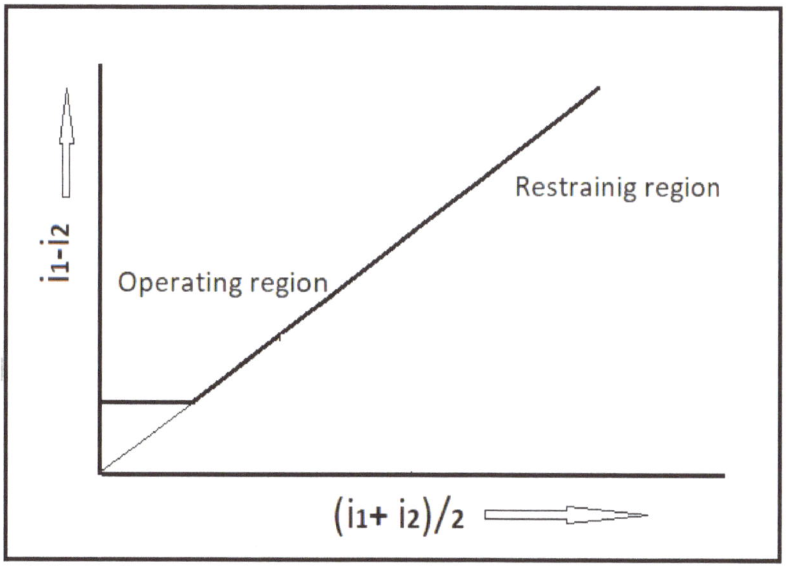

Figure 1 B; Transformer Differential Protection & Characteristic showing operating and restraining zones.

Figure 1A & B Basic theory of Transformer Differential Protection & Characteristic showing operating and restraining zones.

1.1 Analysis of the characteristic: (Refer figure 1A &B)

i. In case of feeding a load, the I1 and I2 are equal thus the relay restrains.

ii. In case of fault beyond differential Zone; In this case also the differential relay shall restrain as I1 =I2 are equal and cancel each other in the secondary current loops.

iii. In case there is a fault in differential Zone in LV side with Radial feed: the Differential Relay's operating element shall receive secondary current from HV side CT only. There will not be any Fault current from LV side and the differential relay shall operate.

iv. In case there is a fault with Power source on LV side as well: In such cases the fault will be fed from either ends ie the direction of current shall reverse in the LV side CTs and the Differential relay operating element/ coil shall get the added operating current ie it shall get I1+I2 and operate.

v. In case of fault in HV side in differential Zone similar condition to (iv) shall take place and the Differential relay shall operate.

1.2 The effect of variation in LV-HV voltages-currents due to OLTC operations and variation in characteristics of the CTs:

The effect of variation in LV-HV voltages-currents due to OLTC operations and also difference in CTs characteristics as far as ratio errors and the limit for saturation are concerned, these are also taken care of by the restraining elements/coils (in 'T' formation) on either side of the operating element/ coil through Bias settings. The characteristic of Differential relay is such that the lower portion of the characteristic depicts non-operating zone while the upper portion over the characteristic curve shows the operating-Zone. (Refer figure:1B).

1.3 Characteristic of Modern Numerical Differential relay;

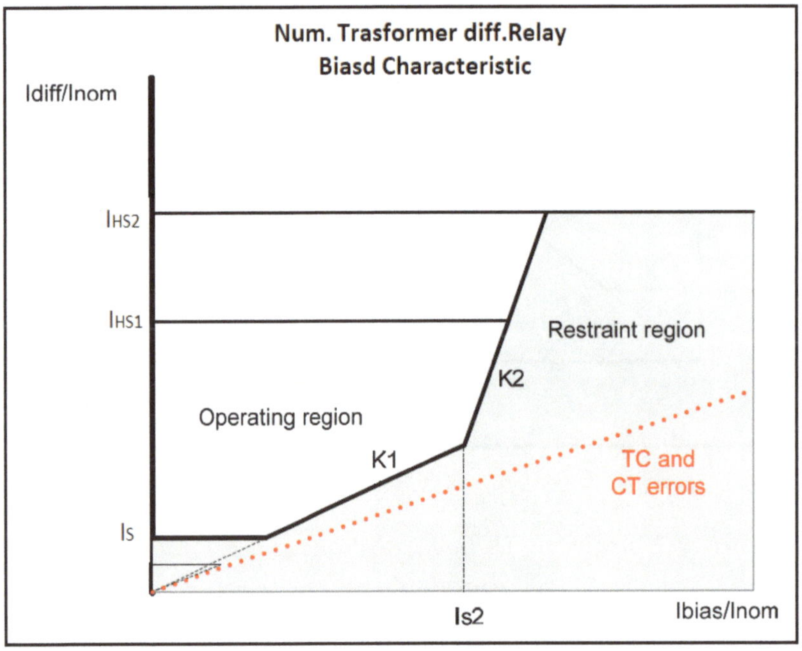

Figure 1C: Biased characteristic of modern Num.Diff relay.

Brief description of the slopes as depicted in the figure1C;

Slope 1; It is flat and depends on I_S setting.

Slope 2; Denoted by K1and it is settable. It shall take care of internal faults up to full-load current and also segregate the restraining area for taking care against Tap changer operation and the CT error. It is generally set at 30%.

Slope 3; Denoted by K2 and it is settable. It is used to ensure stability under heavy through fault conditions, which may possibly saturate any of the CTs. It generally set at 80%.

Note: The above brief description of the Biased characteristic of the Numerical Differential Relay has been given only to apprise the readers. Further, details etc. shall be described in the Part-2 of this article.

2.0 Magnetizing inrush current;

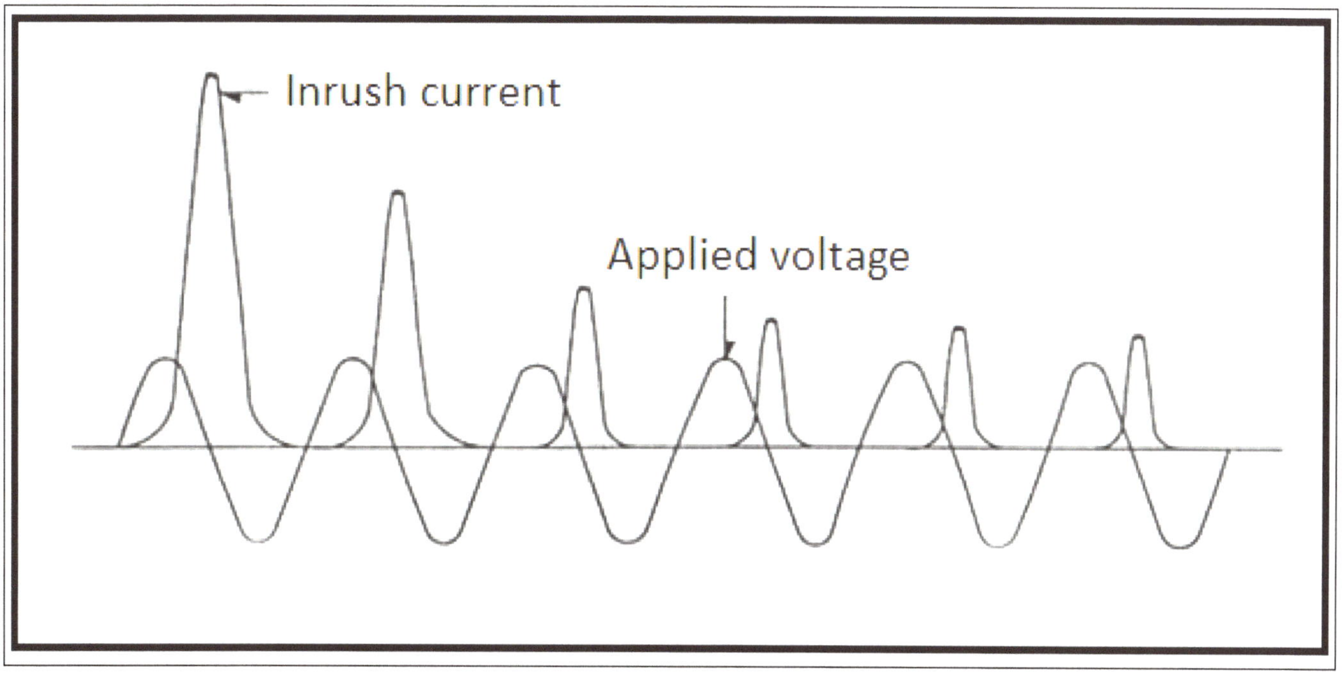

Figure 2: Graphical depiction of 2nd harmonic Inrush current.

The phenomenon of inrush-current is seen as a fault by the Differential relay when energizing the transformer from HV side keeping the LV side open(unloaded). 2nd harmonic current to the tune of 63% of the fundamental is predominately contained in the inrush-current. Thus, the Differential relays must have 2nd harmonic restraints to combat such situations.

3.0 Condition of Over Excitation (Over-fluxing): The condition of over excitation can be experienced by the Generators Transformer when unit connected generator is separated while exporting VARs, experiences over excitation due to sudden increase of HV voltage resulting into saturation due to Over excitation condition. This condition is seen as a fault by the differential relay. The 5th Harmonic component is substantially present in the magnetizing current of an over excited Transformer. To curb this phenomenon the relay should possess 5th harmonic restraint characteristics.

4.0 Basic schematic of Transformer differential Protection;

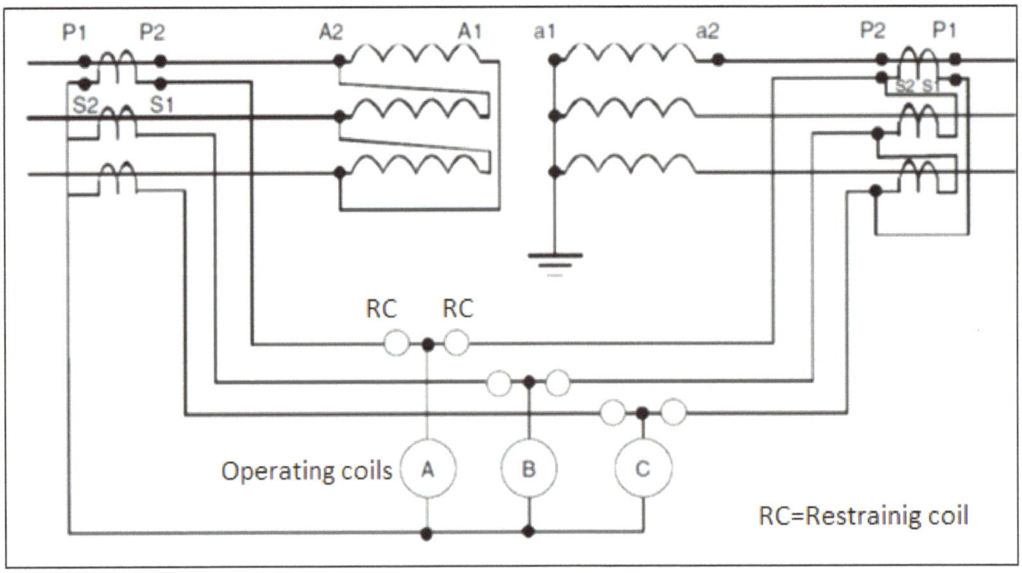

Figure3: Basic schematic diagram of differential protection of 2 winding Dyn1 transformer.

It can be seen from the Schematic diagrams;

I. **Figure 3**;
   a. Basic scheme of differential protection of for 2 winding Dyn1 transformer;
   b. the conventional differential connections are made such that the secondary circuit of CTs on the Delta side of the Transformer winding are connected in star formation and the CT's secondary in the star side of the transformers are connected in delta formation but just opposite to the vector group of the Transformer ie the CTs secondary circuit delta in the instance case. The vector group of the 2-winding transformer is Dyn1(-30°) then the CTs secondaries on Star side of the transformers would be in (+30°).
   c. This is done to match the correct phase in the differential protection relay.
   d. It may be mentioned here that on either side of the transformers, the full load current would be different.
   e. For example, in 20 MVA 132kV /33 kV Dyn1, Transformer the full load currents thereof are; on the HV side 87.5 A and on the LV side 350 A. On HV side required CTs would be of 87.5/1 A and on LV side the CTs needed would be 350/0.577A.
   f. On LV side why this odd ratio? Answer is to have equal current in the leads to be connected to the differential relay ie the current in the external wires would be $\sqrt{3}$x 0.577 ≈ 1 A.
   g. The CTs mentioned in sr. f) are to be tailor-made (got manufactured exclusively for the instance case). Therefore, for each Transformer of different MVA, voltage ratio and full load currents, the CTs have to be tailor-made.
   h. Standard rating CTs are available in the market ie for the instant case the CTs have to be; for HV side -100/1 A and for LV side-400/1 A.

i. For such main CTs, Auxiliary CTs (Interposing CTs) are required which would not only transform secondary of the CTs to required ratios but also would connect the CT external leads to correct Phases.

II. **Figure 4**; Basic scheme of differential protection of an Auto-transformer, YNa0d11.

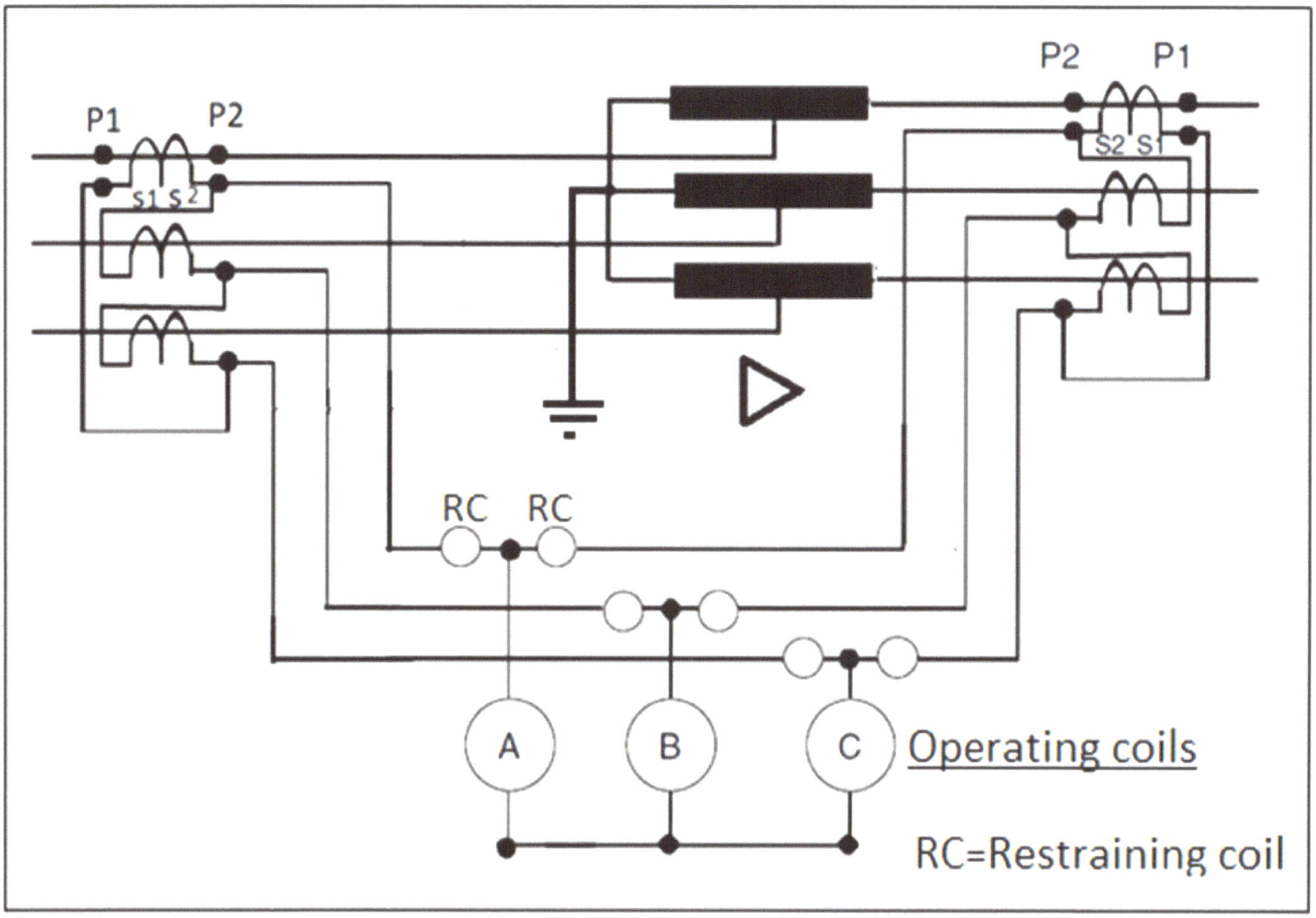

Figure 4: Basic schematic diagram of differential protection of an auto-transformer YNa0d11.

The above criteria hold good for the windings are in star formation being an auto-Transformer with a common Neutral. In this case Delta-tertiary is also there as a stabilizing winding.

Note:

a. The whole exercise is done to prevent the replica of Zero Sequence current to to enter into the differential protection Circuit and mal-operate the Differential Relay for external earth faults.

b. It may be kept in mind that the secondary return neutral wire of star connected Auxiliary CT or the Star point of all the 3 operating current coils or modules should not be grounded. As they would form a parallel path to the main CTs grounded neutral.

4.0 Practical Differential protection circuits with Auxiliary CTs (Interposing CTs):

4.1 Practical sample Connection diagram of Differential protection with auxiliary CTs of a 20 MVA, 132kV/33kV, Dyn11, Transformer: The Connection diagram for Differential protection of 20 MVA, 132kV/33kV, Dy11, Transformer. (with Full load currents HV-87.5A & LV- 350A). Ratios of CTs installed on HV side-100/1A and LV side- 400/1A respectively is shown in figure-.

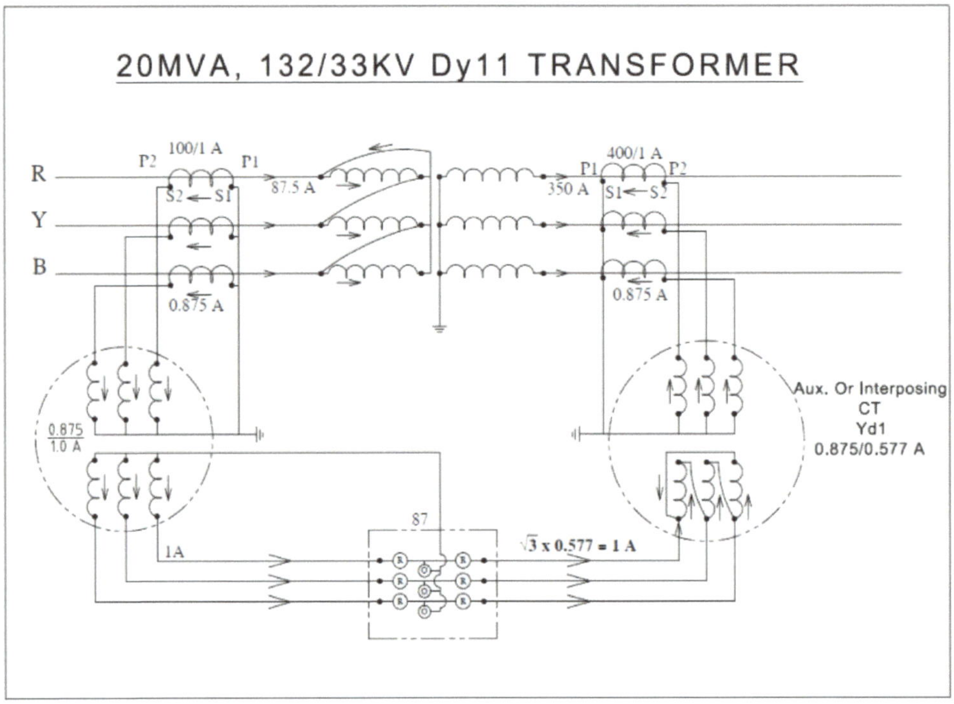

Figure-5: Differential connection diagram using Interposing CTs on 20 MVA, 132/33kV Dyn11 Transformer.

Note:

i. 'O' stands for operating element,
ii. 'R' stands for restraining element.

5.0 WHY USE INTERPOSING CTs (Auxiliary CTS)?

Explanation:

i. The Full load currents are HV-87.5 A and LV-350 Amp.
   How do you arrive at this?

$$\text{HV Full load current} = \frac{\text{MVA.1000}}{\sqrt{3}\text{kV}} = \frac{20\text{x}1000}{\sqrt{3}\text{x}132} = \frac{20000}{1.732\text{x}132} = 87.479\text{ A} \approx 87.5\text{ A}$$

Since the voltage ratio is $\frac{132}{33}$ kV= 4. Therefore, the LV-current has to be

87.5 x 4 = 350 A.

Note: Since, the primary CTs 87.5/1 A on HV side and 350/1 A on LV side have to be tailor-made as such odd ratio CTs are not in the manufacturing range of the vendors, therefore, the main primary CTs used are HV side 100/1-1-1 A and LV side 400/1-1-1 A.

ii. The Differential relay must get 1 Amp current while the Transformer is in service to its full capacity ie at 20 MVA, (with full load currents).

iii. Now, to get 1 amp at the terminals of the Differential relay, the Interposing CTs are used.

At full load, the main Transformer CTs shall carry and transform them into secondary currents as follows;

HV side $\frac{87.5}{100} = 0.875$ Amp. LV side: $\frac{350}{400} = 0.875$ Amp.

iv. It may be borne in mind that main CTs are available with 1 Amp secondary or 5 Amp. Secondary currents only. The CTs with 1 amp. secondary is used in EHV sub stations where the lead lengths of the control cables are very long, whereas 5 amp. secondary currents are generally used in the Sub-transmission sub-stations.

v. *Now, as per the prime requirement of the Differential Protection, the Zero sequence currents should be prevented to enter into the Differential relay's circuit. This may trip the Differential relay spuriously whenever any Phase to ground faults occur outside the differential zone.*

- To avoid such spurious trippings, the auxiliary CTs on the star side be connected in Delta and the Delta side be connected in star formation such connections are formed in the Secondary windings of the auxiliary CTs so as to get 1 amp at the differential relay terminals.
- We have to accordingly form the delta in the side where star formation of the Transformer winding is there. In the instant case pleases refer to the figure the auxiliary CTs on the HV side are connected in Star formation with a Ratio of $\frac{0.875}{1}$ A and on LV side the ratio should be $\frac{0.875}{0.577}$ A.
- Why, $\frac{0.875}{0.577}$ A? The reason is when the current on the external leads from delat formed Auxiliary CTs would be $\sqrt{3} \times 0.577 = 0.999 \text{ A} \approx 1$ A.

vi. General practice followed while cabling/ wiring the CTs in the switch yard, all the 3 secondaries (of R, Y, B Phases) of the CTs are wired up in star formation resulting into R-Y-B-N leads, which are brought to the control-room through control cables and are terminated on the terminal block of the Control & Relay panel of the Transformer.

vii. The Star point (Neutral point) of the secondaries of main CTs should be towards the Transformer side or it may be away from the Transformer on both HV and LV the sides. The position of P1 and P2 (the primary terminals) decide this.

In instant case P1s of the CT primary terminal are towards transformer side on HV as well as on LV side.

Thus to form neutral of secondary CT circuit, all (*)S1 are to be looped together.

Note about the CTs with reference to IS: 2705;

(*) this is the prefix for the core terminals to indicate the number of the core.
Each secondary core (winding) has 2 terminals. S1 and S2.
Say there are 3 cores in a CT then they will be numbered as 1S1-1S2, 2S1-2S2 and 3S1-3S2.
The CTs generally have 3 cores.

5.1.1 Metering Core.

i. 5P10, 20-Protection core.
ii. PS core (Special protection core).: This is used for Differential protection (as per IEC:60044-1revised to IEC 61869-1:2007 it is Px core)

6.0 In case of YNyn0 Transformer:

Figure 6 depicts the schematic diagram for Differential potection for a 132/33kV, 20 MVA YNyn0 Transformer;

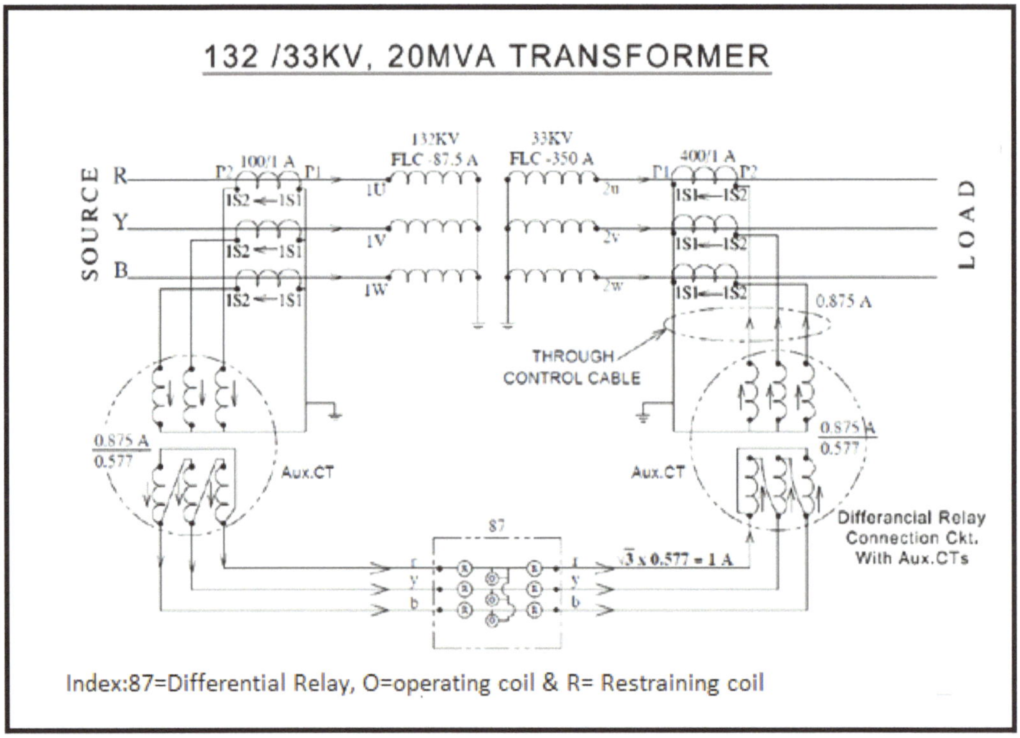

Figure 6: Differential Protection connection diagram of 20 MVA, 132kV/33kV, YNyn0, Transformer with Interposing CTs (Aux. CTs).

Note: Such EHV transformers are used in the Grid sub-stations.

- The Primary Terminals (P1 & P2) of main 132kV and 33kV CTs should be connected in such a way that P1 of CTs of HV side and that of LV side must face towards the Transformer.
- The star points of the secondary winding on both 132kV and 33kV side are formed towards the Transformer.
    - In the instant case since P1 of HV and LV side CTs are facing towards the Transformer therefore all the 1S1 are shorted together so as to form the star point ie Neutral, which in-turn becomes the secondary neutral lead. This secondary is then Earthed.
- The Auxiliary Current transformers are required as the CTs available in the market are either having 1.0 Amp or 5.0Amp secondary current.
- The reason for selection of ratio for the Auxiliary Current Transformers;
    - HV side CT ratio is 100/1A so for full load current (87.5A) flow the secondary current of main HV side CTs would be 0.875A.
    - LV side CT ratio is 400/1A so for full load current of 350 A, the main CTs secondary current would be 350/400 A=0.875A.
- Why the ratio of Aux. CTs is selected as 0.875/0.577A?
    - The answer is as line currents in delta connection would be $\sqrt{3} \times 0.577 \approx 1$Amp.
    - the Auxiliary CTs are used in delta formation for eliminating Zero sequence current in the Differential circuit.
- '87' is the device number allocated for the differential relay.

Note: Part -II of this Article shall contain;

- the relevant details about prevention methodology of Zero sequence current (I_0)in the modern Numerical differential relay.
- Salient points regarding Setting etc including inrush current restraint function.

10. A Concept of Elimination of Zero Seq. Current in Numerical Differential Relays and Other Salient Features Thereof - (Part-II)

1.0 Preamble: Unlike the electro-mechanical and static Differential relays following main features are incorporated in the numerical Differential Relays using some Algorithms making them fool-proof;

1. *there is no need for using interposing CTs (Aux.CTs) for matching the phase shifts between the HV & LV phases, current ratio correction and for prevention of Zero sequence current from entering into Differential protection circuit.*
2. *inhibiting(blocking) from tripping due to 2nd harmonic current in the inrush current using cross blocking feature, however, it has the facility to trip the transformer instantaneously in case of existence of fault in the windings at the time of switching-on of the transformer using hi-set option over riding the 2nd harmonic restrain feature.*

The author has narrated the concept of elimination of Zero-sequence current through Matrices, implementation of Cross blocking scheme to block unwanted tripping due to inrush current in particular and other salient features of numerical Differential relays.

2.0 Key words; 2nd harmonic current, Inrush current, Zero seq. current, Matrix/ Matrices, cross blocking etc.

3.0: Digital Differential Protection -measuring value acquisition and processing.

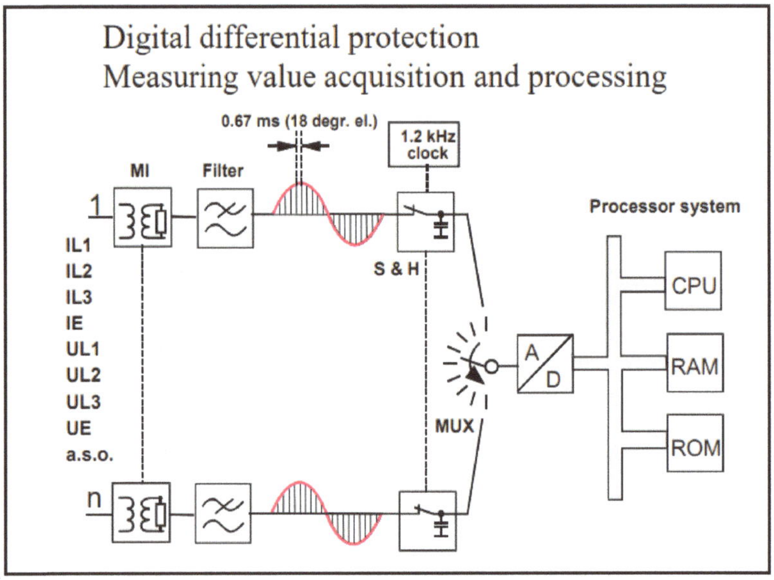

Figure 1: Digital Differential Protection working principle based on measuring value acquisition and processing.

3.1 Extraction of momentary values of fault and importance thereof; the fault current curve decides Extraction Of momentary values of the depending upon to which portion of the characteristic curve the replica of primary fault current in terms of secondary current terms lies on the curve.

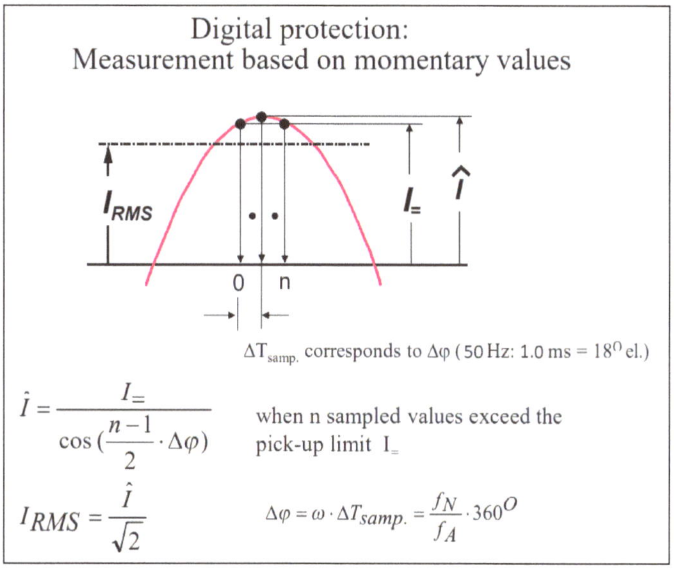

Figure2: Theory of calculating momentary values.

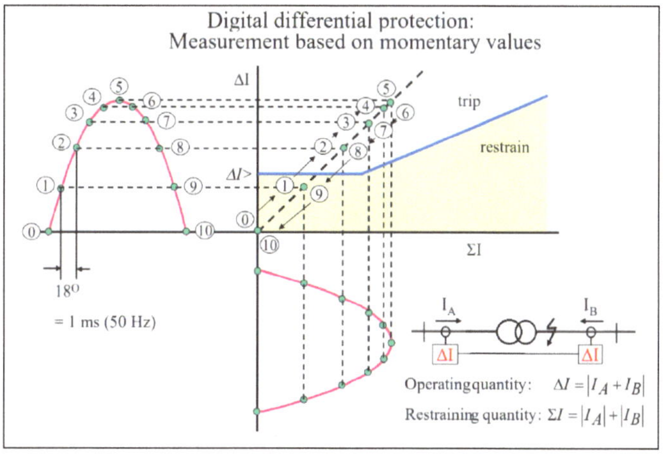

Figure 3: Graph drawn on extraction of momentary values which decide whether to trip or restrain, depends upon the zone it falls in?

4.0 Salient features of Numerical Differential Relay: Salient features are common practically in all the numerical Transformer Differential relays of different manufacturers. They use different Algorithms to incorporate relevant features and make these Relays as IED (Intelligent Electronic Device).

i. **Characteristic;** The Numerical Differential relay has 2 knee points characteristics.
ii. **Internal compensations;** Following are the internal compensations which are essential and are the back-bone of any Numerical Differential relay which varies from manufacturer to manufacturer as per the Algorithms developed by them. Following are the basic concepts mentioned here-under;

➢ Phase shift compensation,
➢ Zero sequence compensation (also known as Zero sequence Current Trap).
➢ Current compensation.

Every Numerical Differential Relay is designed for performing the above functions successfully.

4.1 Transformer Numerical Differential Relay Characteristics;

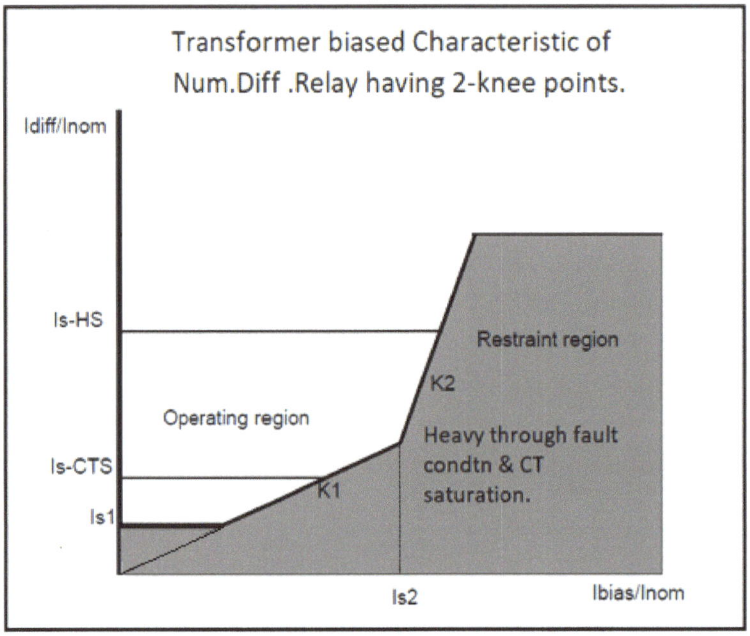

Figure 4: Transformer Num. Differential relay Characteristics

Tripping characteristic of differential has two knee point.

a. The lower slope provides sensitivity for internal faults.
b. The higher slope provides stability under following conditions;

i. Voltage variation causing variation in currents due to OLTC operation.
ii. Heavy through fault conditions.
iii. CT saturation due to transient faults.
iv. Mismatch in characteristics of CTs used in HV & LV sides.
v. Magnetizing inrush currents.
vi. High exciting currents caused by transformer overexcitation.

Note: (a) The % slope and Hi set settings are selectable.

(b) Most of the Utilities/entrepreneurs may follow the recommended settings which can be implemented without any hitch or hesitation are as follows;

i. Slope 1-20%,
ii. Slope 2-50%,
iii. Hi set-6 pu.

4.2 Phase shift Compensation;

- In the earlier generation of Differential Relays, the auxiliary CTs (Interposing CTs) were used for Amplitude, Phase correction and for eliminating replica of zero sequence current in the Differential protection circuit. However, these features are incorporated in the Num. Differential Relays of different manufacturers through different Algorithms, thus Auxiliary CTs (Interposing CTs) are not required in the present scenario.
- Phase compensation is carried out following the matrices as defined in the IEEE Std. C37 91-2021 for each vector group for Delta/Delta, Delta/star or Star/delta connections for the 3 phase transformers.
- Sample matrices for phase shift correction of Delta or Star winding for matching the respective phase secondary currents in the Numerical Differential Relays & of the
- Transformers considering the vector group of the transformers are as tabulated here under;

<table>
<tr><td>ABC, 1*30° cw</td><td>30°</td><td rowspan="2">Dyn11
&
YNd11</td><td rowspan="2">$\frac{1}{\sqrt{3}}\begin{bmatrix} 1 & -1 & 0 \\ 0 & 1 & -1 \\ -1 & 0 & 1 \end{bmatrix}$
Delta A-B (DAB)</td></tr>
<tr><td>ACB, 1*30° ccw</td><td>330°</td></tr>
<tr><td>ABC, 11*30° cw</td><td>330°</td><td rowspan="2">Dyn1
&
YNd1</td><td rowspan="2">$\frac{1}{\sqrt{3}}\begin{bmatrix} 1 & 0 & -1 \\ -1 & 1 & 0 \\ 0 & -1 & 1 \end{bmatrix}$
Delta A-C (DAC)</td></tr>
<tr><td>ABC, 11*30° ccw</td><td>30°</td></tr>
</table>

Table1: Matrices for Dyn11&Ynd11and Dyn1 &Ynd1

Note: It is observed from the rotation ie cw (clock wise) & ccw (counter clock wise); the HV vector is taken as reference and LV vector is rotated accordingly and the relevant matrix is selected.

4.3 Prevention of intrusion of zero sequence current: In case of star winding It is very essential to prevent intrusion of the replica of zero sequence current to enter into the numerical Differential Circuit/module of grounded neutral star connected winding for any out of zone Phase to ground faults failing which, there is every chance for mis- operation of relay resulting into inadvertant tripping of the Transformer.The Algorithm developed to this effect follows the equation of Zero Sequence current and the Matrix depicted here under;

$$\begin{bmatrix} I'_A \\ I'_B \\ I'_C \end{bmatrix} = \begin{bmatrix} I_A - I_0 \\ I_B - I_0 \\ I_C - I_0 \end{bmatrix}$$

Also, $I_0 = (1/3)\,(I_A + I_B + I_C)$:

$$\begin{bmatrix} I'_A \\ I'_B \\ I'_C \end{bmatrix} = \begin{bmatrix} I_A - \frac{1}{3}(I_A + I_B + I_C) \\ I_B - \frac{1}{3}(I_A + I_B + I_C) \\ I_C - \frac{1}{3}(I_A + I_B + I_C) \end{bmatrix} = \frac{1}{3}\begin{bmatrix} 2I_A - I_B - I_C \\ -I_A + 2I_B - I_C \\ -I_A - I_B + 2I_C \end{bmatrix}$$

$$\begin{bmatrix} I'_A \\ I'_B \\ I'_C \end{bmatrix} = \frac{1}{3}\begin{bmatrix} +2 & -1 & -1 \\ -1 & +2 & -1 \\ -1 & -1 & +2 \end{bmatrix}\begin{bmatrix} I_A \\ I_B \\ I_C \end{bmatrix} = \frac{1}{3}\begin{bmatrix} 2I_A - I_B - I_C \\ -I_A + 2I_B - I_C \\ -I_A - I_B + 2I_C \end{bmatrix}$$

4.3.1 Condition for verifying non-intrusion of Zero seq. current;

Addition of the matrices of HV side and the LV side is the differential current IΔ-A, IΔ-B, IΔ-C. The condition for no-intrusion of the zero – seq. current into the Differential module is fulfilled if the result of addition of these two matrices is equal to zero.

$$\begin{vmatrix} I\Delta A \\ I\Delta B \\ I\Delta C \end{vmatrix} = \begin{vmatrix} IA^{*} \\ IB^{*} \\ IC^{*} \end{vmatrix} + \begin{vmatrix} Ia^{**} \\ Ib^{**} \\ Ic^{**} \end{vmatrix} = 0$$

5.0 Numerical measured values adaption: The dictionary meaning of 'adaption' is "to change behaviour or adjustment".

a. Basic of Differential Protection is conceptualized on current comparison basis. In the Transformer Differential protection, the HV and LV side currents are compared. Prior to current comparison two adaption are required which are described in the following paragraphs b) and c)
b. On the earthed star-point winding, the Zero sequence component of the current must be eliminated or compensated for.
c. On delta winding vector group must be compensated for.

5.1 Example; Measured values adaption in 20 MVA,132/33kV, Dyn11Transformer Numerical relay for an external 'c' Phase to ground fault on the LV side, proving that protection remains stable.

The Matrix developed for Delta winding on HV side or LV side depending upon the vector group of the transformer. A few of the Matrices are shown in the table 1.

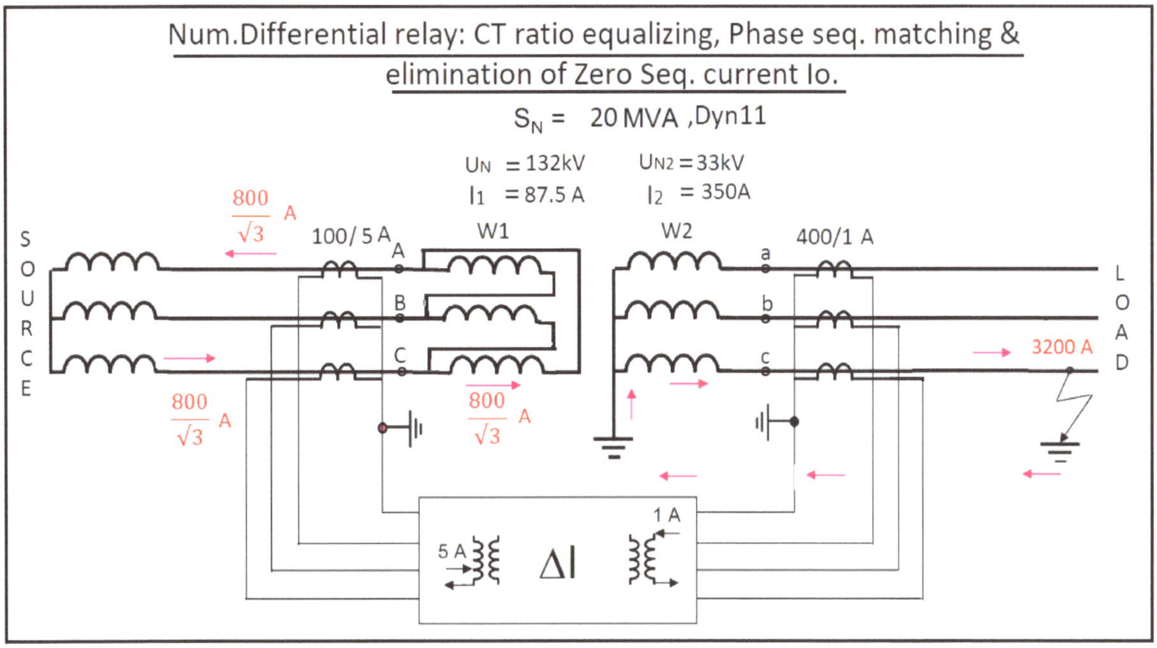

Figure 5: Circuit diagram depicting out of zone ph-ground fault on LV side of the 20 MVA Transfpormer.

132kV/33kV, Dyn11, Power Transformer.

$$I1_{F\,HV} = \frac{20,000}{\sqrt{3}\times 132} = \mathbf{87.5A}\ \mathbf{I2}_{F\,LV} = \frac{20,000}{\sqrt{3}\times 33} = \mathbf{350A}$$

$$\mathbf{If1} = \frac{\mathbf{UN2}}{\mathbf{UN1}} \times \frac{1}{\sqrt{3}} \times \mathbf{if\ 2} = \frac{33}{132} \times \frac{3200}{\sqrt{3}} = \frac{800}{\sqrt{3}}\ \mathbf{A = 462\ A}$$

$$\mathbf{J1} = \frac{5}{100} \times \frac{1}{5} \times \frac{800}{\sqrt{3}} = \frac{8}{\sqrt{3}} = \mathbf{4.62\ A.}$$

$$\mathbf{I1norm} = \frac{\textbf{Primary of the CT}}{\textbf{Full load current}}\ \mathbf{x\ J1} = \frac{100}{87.5}\ \mathbf{x}\ \frac{8}{\sqrt{3}} = \frac{8}{\sqrt{3}}\ \mathbf{x\ 1.14\ A}$$

Calculating J2 and I2 $_{\textbf{norm.}}$

$$\mathbf{J2} = \frac{\textbf{Fault current}}{\textbf{Primary of CT}} = \frac{3200}{400} = \mathbf{8\ A,}$$

$$\mathbf{i_2\ norm} = \frac{\textbf{Primary of the CT}}{\textbf{Full load current}}\ \mathbf{x\ J2,\ i_2\ norm} = \frac{400}{350}\ \mathbf{x\ 8 = 8x1.14.}$$

Note: The secondary star points of HV and LV sides should be be towards the Transformer

5.1(a) Selecting Dyn11- Matrix;

$$\begin{vmatrix} IA' \\ IB' \\ IC' \end{vmatrix} = \frac{1}{\sqrt{3}} \begin{vmatrix} 1 & -1 & 0 \\ 0 & 1 & -1 \\ -1 & 0 & 1 \end{vmatrix} \times \begin{vmatrix} IA\ norm \\ IB\ norm \\ IC\ norm \end{vmatrix} \text{--------------------(1)}$$

$$I1\ norm = \frac{8}{\sqrt{3}}\ x\ 1.14$$

$$\begin{vmatrix} IA' \\ IB' \\ IC' \end{vmatrix} = \frac{1}{\sqrt{3}} \begin{vmatrix} 1 & -1 & 0 \\ 0 & 1 & -1 \\ -1 & 0 & 1 \end{vmatrix} \times \begin{vmatrix} +\frac{8}{\sqrt{3}}\ x\ 1.14 \\ 0 \\ -\frac{8}{\sqrt{3}}\ x\ 1.14 \end{vmatrix} \text{--------------(2)}$$

$$IA' = +\frac{8}{3}\ x1.14, \text{-- (3)}$$

$$IB' = \frac{1}{\sqrt{3}}\left[+\frac{8}{\sqrt{3}}\ x\ 1.14\right] = \frac{8}{3}\ x\ 1.14 \text{-------------------- (4)}$$

$$IC' = \frac{1}{\sqrt{3}}\left[-\frac{8}{\sqrt{3}}\ x\ 1.14 + 0 + \left(-\frac{8}{\sqrt{3}}\ x\ 1.14\right)\right],$$

$$= \frac{1}{\sqrt{3}}\left[-2.\frac{8}{\sqrt{3}}\ x\ 1.14\right] = -2.\frac{8}{3}\ x\ 1.14. \text{-- (5)}$$

5.1(b) Calculation pertaining to the LV winding Neutral grounded.

The matrix shown here under shall apply for the star connected Neutral grounded transformers;

$$\begin{bmatrix} I'_A \\ I'_B \\ I'_C \end{bmatrix} = \frac{1}{3}\begin{bmatrix} +2 & -1 & -1 \\ -1 & +2 & -1 \\ -1 & -1 & +2 \end{bmatrix}\begin{bmatrix} I_A \\ I_B \\ I_C \end{bmatrix} = \frac{1}{3}\begin{bmatrix} 2I_A - I_B - I_C \\ -I_A + 2I_B - I_C \\ -I_A - I_B + 2I_C \end{bmatrix} \text{--------------- (6)}$$

calculating J2 and i2 $_{norm.}$

$$J2 = \frac{\text{Fault current}}{\text{Primary of CT}} = \frac{3200}{400} = 8\ A,$$

$$i_2\ norm = \frac{\text{Primary of the CT}}{\text{Full load current}}\ x\ J2,\ i_2\ norm = \frac{400}{350}\ x\ 8 = 8x1.14.$$

Assuming the load currents in A & B phase as ‘0’, calculations will be made Using the matrix (6);

$$\begin{vmatrix} ia' \\ ib' \\ ic' \end{vmatrix} = \frac{1}{3}\begin{vmatrix} +2 & -1 & -1 \\ -1 & +2 & -1 \\ -1 & -1 & +2 \end{vmatrix} \times \begin{vmatrix} 0 \\ 0 \\ 8x1.14 \end{vmatrix}$$ --------- (7)

ia' $=\frac{1}{3}$ (- 8 x 1.14),--(8)

ib' $=\frac{1}{3}$ (- 8 x 1.14), --(9)

ic' $=\frac{1}{3}$ (2x 8 x1.14) --- (10)

5.1(c) Condition for verifying non-intrusion of Zero seq. current;

$$\begin{vmatrix} I\Delta A' \\ I\Delta B' \\ I\Delta C' \end{vmatrix} = \begin{vmatrix} IA' \\ IB' \\ IC' \end{vmatrix} + \begin{vmatrix} ia' \\ ib' \\ ic' \end{vmatrix} = 0$$ ---------- (11)

The addition of IA' with ia', IB' with ib' and IC' with ic' respectively should result into zero as per Matrix (11)

Values of IA', IB' and IC' are calculated in the equations (4),(5) & (6)respectively and those of ia', ab' and ic' are calculated in the equations (8),(9) & (10)respectively and then added;

IA' + ia' = + $\frac{8}{3}$ x1.14, + $\frac{1}{3}$ (- 8 x 1.14)= 0 ---------- (12)

IB' + ib' =$\frac{8}{3}$ x 1.14 + $\frac{1}{3}$ (- 8 x 1.14)= 0 ----------- (13)

IC' + ic' =(-2.$\frac{8}{3}$ x 1.14) + $\frac{1}{3}$ (2x 8 x1.14) =0 ---- (14)

Since, Equations (12),(13) & (14) result to "0", thus the condition is fulfilled.

5.2 Magnitude compensation; In the earlier generation Differential relay connections Aux. CTs (interposing CTs) were used for magnitude, phase, and trapping of zero seq. currents. Unlike in the same, the magnitude compensation takes place in the numerical relay automatically. In addition to phase shift and zero-sequence compensation, magnitude compensation is also provided in

The multiplier in front of each matrix as depicted in IEEE Std, C37 91-2021 represents the scaling adjustment required to bring the currents after compensation back to 1 p.u.

In a wye matrix, there is no change in the apparent magnitude; so the multiplier is 1.

In a delta matrix, the compensated currents are √3 times higher than the phase currents; therefore, a multiplying factor of 1/√3 is used to bring the currents back to 1 p.u.

6.0 Magnetizing Inrush Currents;

1. **Energization of the transformer.** This is the typical event where magnetizing inrush currents are a concern. The excitation voltage on HV winding keeping the LV winding open-circuited to the available full voltage. The transformer core typically saturates, depending upon design, system impedance, the remanent flux in the core and the point on the voltage wave when the transformer is energized. The current needed to supply this flux may be to the tune of 8 to 30 times the rated current. Since, there is no balancing current, the Differential relay sees it as a fault.

Analysis of a typical magnitude inrush current wave shows (fundamental = 100%):

Component	DC	2nd H	3rd H	4th H	5^{Th} H	6th H	7^{Th} H
	55%	63%	26.8%	5.1%	4.1%	3.7%	2.4%

The inrush current predominately contains 2nd harmonic current to the tune of 63% of the fundamental differential current.

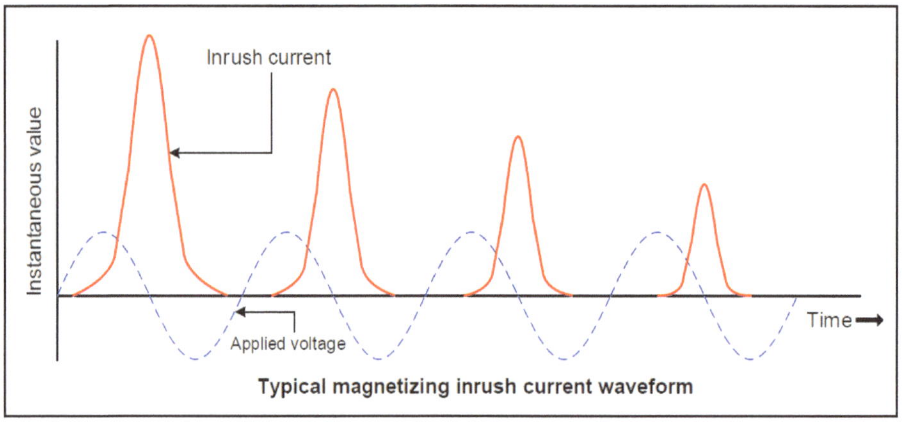

Figure 6: Typical 63% 2nd harmonic current contained in the Inrush current.

6.1 Magnetizing inrush current during fault clearing.

An external fault may significantly reduce the system voltage, and therefore reduce the excitation voltage of the transformer. When this fault is cleared, the excitation voltage returns to the normal system voltage level resulting in magnetizing inrush current. This magnetizing inrush current will be less than that of energization, as there is no remanent flux in the core. The current measured by the differential relay will be fairly linear due to the presence of load current and may result in low levels of second harmonic current.

6.2 Sympathetic inrush. Energizing a transformer on the power system can cause sympathetic magnetizing inrush currents to flow in an already energized parallel transformer. Energizing the second transformer causes a voltage drop across the resistance of the source line feeding the transformers. This voltage drop may cause a saturation of the already energized transformer in the negative direction. This saturation causes magnetizing inrush current to supply the flux. The magnitude of the magnetizing inrush current is generally not as severe as the other cases.

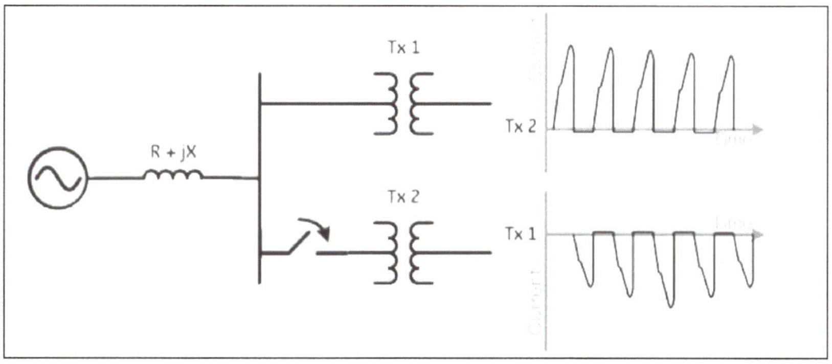

Figure 7: Sympathetic in-rush current.

7.0 % 2nd Harmonic setting; The inrush current contains predominantly very high 2nd Harmonic content where the current for an internal transformer fault typically has very low levels of 2nd Harmonic current. The simplest method of harmonic restraint uses the magnitude of the second harmonic in the differential current compared to the magnitude of the fundamental frequency component in the differential current.

$$\frac{\textbf{Idifferential}\left(\textbf{2nd harmonic}\right)}{\textbf{I differential }\left(\textbf{fundamental}\right)}$$

Tripping of the Differential element is blocked when this ratio exceeds an adjustable threshold.

Constraints:

i. If set too high, the differential element may trip during Transformer - energizing.
ii. if set too low, inrush restraint may block tripping during some internal fault event.
iii. **Sample oscillograph picture**; Sample oscillograph picture of tripping of 315 MVA 400/220/33kV transformer during energization due to inrush current at % 2nd harmonic setting provided as 12%.

Reason for tripping; since the ratio;

$$\frac{\textbf{Idifferential}\left(\textbf{2nd harmonic}\right)}{\textbf{I differential }\left(\textbf{fundamental}\right)}$$

must have been lower than 12%, therefore it must have tripped. The transformer was re-energized successfully by lowering the % 2nd harmonic setting to 10%. However, the setting of 12% was restored after energizing the Transformer.

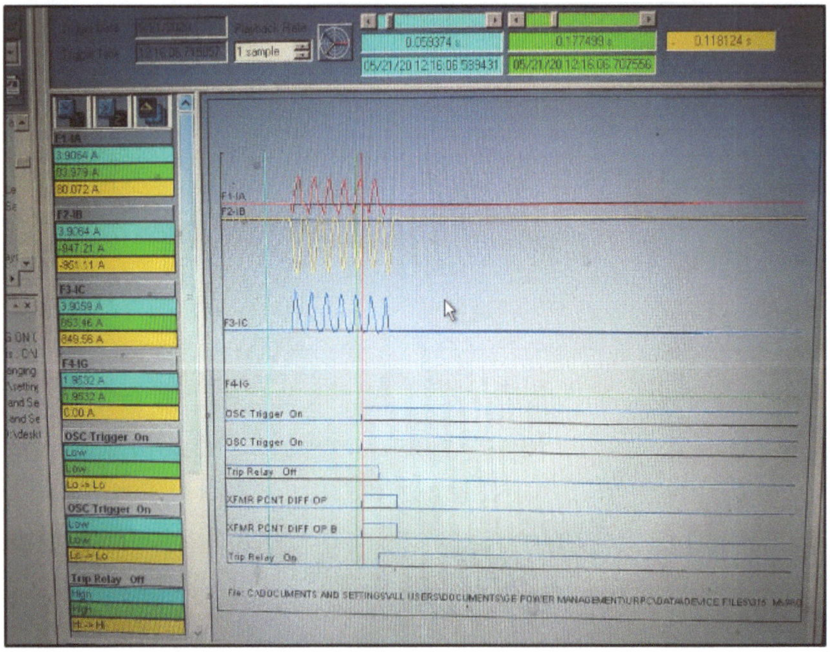

Figure 8: It may be observed that the tripping took place in B Phase (of ABC convention).

8.0 2nd harmonic blocking features;

To avoid relay mis-operation during inrush current condition, the filtered differential element uses harmonics to block/restrain the differential element.

8.1 Cross blocking features: Harmonic restraint are normally calculated individually on each phase. Typically, the operation of the Differential element on any phase operates the circuit breakers supplying the transformer. If the restraint criterion is not met on any phase, the transformer may trip, off-line during energization or due to sympathetic inrush conditions, ***However, looking to the security aspect, of the differential element against inrush currents, the restraint functions should be enabled in more than one phase.***

Cross blocking feature types

Per-phase. In per-phase mode the relay performs inrush restraint individually in each phase.

2-out-of-3. In 2-out-of-3 mode, the relay checks second harmonic level in all three phases individually. If any two phases establish a blocking condition, the remaining phase is restrained automatically.

Note: Most of the Utilities are selecting this function in order to achieve security against undesirable tripping.

Averaging. In averaging mode, the relay first calculates the average second harmonic ratio, and then applies the inrush threshold to the calculated average.

1-out-of-3. In 1-out-of-3 mode, all three phases are restrained when a blocking condition exists on any one phase. 1-out-of-3 mode typically reverts back to per-phase mode after a short time delay to allow tripping in case an internal fault occurs during energization.

8.0 Sample settings; Typical Numerical Differential relay setting philosophy of 50 MVA, 220/33kV Transformer is depicted in the figure 9.

TRANSFOREMR DIFFERENTIAL PROTECTION SETTING PARAMETER

NAME OF SUB-STATION :-	220KV STU		
DETAIL OF X-MER:-	50 MVA BBL		
TYPE OF X_MER	2 WINDING		
DETAIL OF RELAY:-	ALSTOM P643		
X-MER MVA (VA)=	50000000	VA	50 MVA
HV VOLTAGE(V) =	220000	V	
LV VOLTAGE(V)=	33000	V	
HV CT RATIO=	800	Amp	
LV CT RATIO=	1200	Amp	
TV CT RATIO=	NA		
X-MER RATIO =	6.667		
HV RATED CURRENT=	131.2198194		
LV RATED CURRENT=	874.7987963		
Secondary HV rated I=	0.164024774		
Secondary LV rated I=	0.728998997		
HV WINDING MATCH FACTOR	1/(132.21/800)	6.051	
IV WINDING MATCH FACTOR	1/(874.79/1200)	1.372	
WINDING CONFIGURATION	HV+LV/HV+LV+TV	HV+LV	
WINDING TYPE	AUTO/CONVENTIONAL	CONVENTIONAL	
HV WINDING RATTING IN MVA	50		
HV CONNECTION	YN		
HV VOLTAGE IN KV	220		
LV WINDING RATTING IN MVA	50		
LV CONNECTION	YN		
LV VOLTAGE IN KV	33		
LV VECTOR GROUP	0		
Differential relay Pick up Setting=	0.2 PU		
Differential relay Slope 1 Setting=	20%		
Differential relay Slope 2 Setting=	50%		
High set diff setting1 =	6.0 PU		
Hugh set diff setting2 =	NA		
Zero sequence filter HV	Enabled		
Zero sequence filter HV	Enabled		
2nd Harmonic Blocking	Enabled	15%	
5th Harmonic Blocking	Enabled	30%	
Cross Blocking	Enabled		
CT saturation	Enabled	1 .5 PU	

Figure 9: Typical Numerical differential relay setting of 50 MVA,220/33kV Transformer.

10.0 Other Features; Following features are generally found in the numerical Differential relays in common;

- DR (Disturbance recorder),
- Restricted Earth Fault (REF),
- Protection Overcurrent Protection (50/51),
- Directional Overcurrent Protection (67),
- Earth Fault protection,
- Distance relay protection.
- Circuit Breaker Fail (CBF) Protection,
- Overvoltage Protection, Undervoltage Protection,
- Frequency Protection, Thermal Overload Protection (49),
- Over-fluxing Protection (V/f),
- Current Transformer Supervision (CTS),
- Voltage Transformer Supervision (VTS).
- Aux D.C. failure alarm.

11.0 References:

1. Gerhard Ziegler-Numerical Differential protection Principles and application.
2. G Ziegler, siemens: Differential Protection symposium, Belo Horizonte 7 to 9 Nov. 2005.
3. Rich hunt GE Multilin, Joe Schaefer & Bob Bentert of Florida Power and Light Company: Practical Experience in Setting Transformer Differential Inrush Restraint.
4. IEEE Std C37.91-2021: IEEE guide for protecting Power Transformers.
5. Manual -Differential relays -MICOM-P64X Alstom

11. REF Protection- A Complementary Protection to the Differential Protection

1.0 Synopsis: Most vital and important Protection to protect the Transformer is the Biased Differential Protection which provides excellent protection against Ph to Ph faults and phase to earth faults for the faults within the differential zone protection covering 80%, however it is not capable of providing protection against Single Phase to Earth faults near the neutral in a star connected Transformer, with Neutral Solidly earthed(grounded) because of its feature of biasing. To address this constraint REF protection is the only answer. Thus REF Protection may be called as a Complementary Protection to the Biased Differential protection as this protection too is based on Merz-price methodology.[1]

2.0 REF Protection Concept.

2.1 REF is the acronym for Restricted Earth fault. The area of protection is restricted to whole winding from bushing turrets to neutral of the Star connected transformers with solidly earthed Neutral.

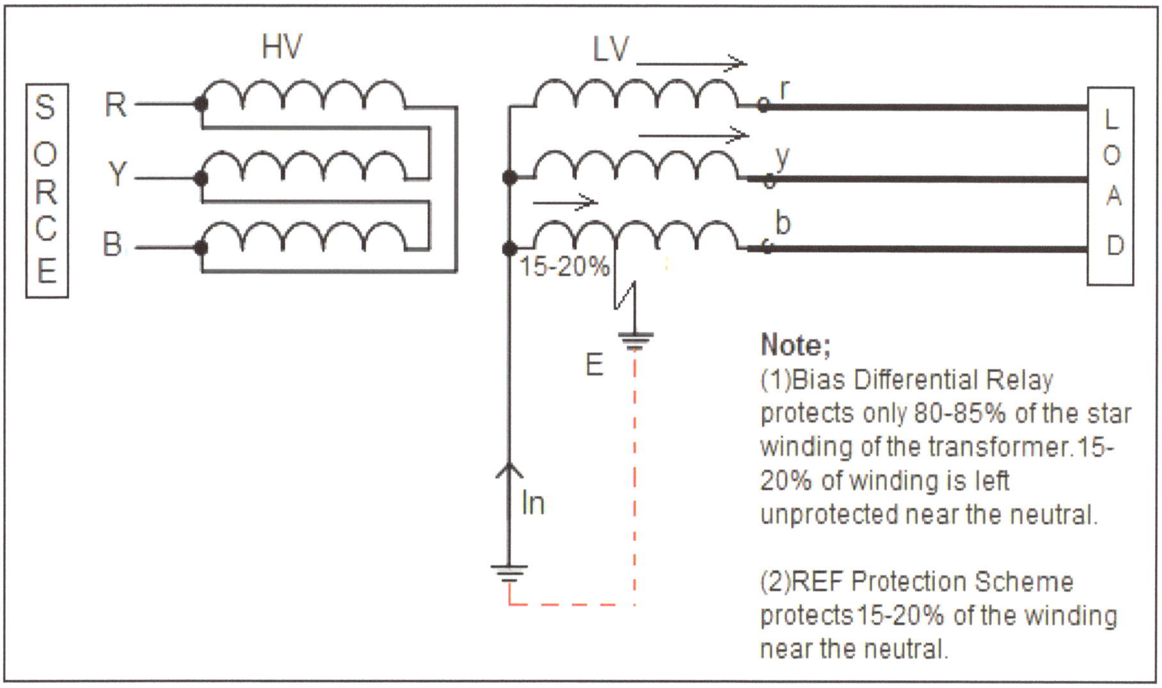

Figure 1: Depicting unprotected 15-20% of winding of Star connected transformers near the neutral against Phase to Earth Fault.

2.1.1 Type of REF Protections;

1. Low Impedance.
2. High Impedance.

3.0 Low Impedance REF Protection;

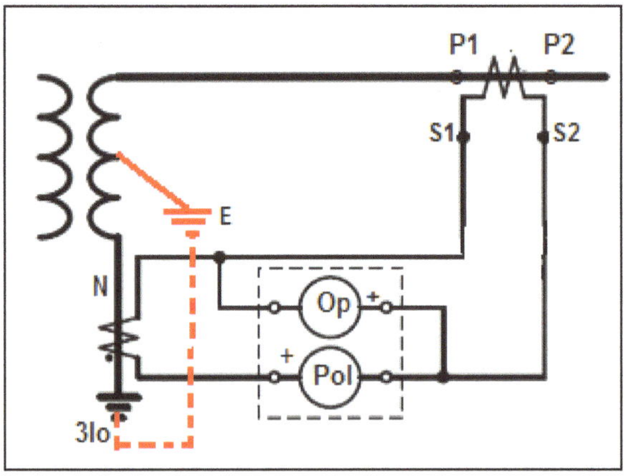

Figure 2: Schematic diagram of Low Impedance REF Protection.

Low impedance function is a percentage biases function with an additional zero sequence current directional comparison criteria felicitating excellent sensivity and stability against through fault conditions. There is no taboo about having same ratio and magnetizing characteristics of CTs in the Turret and the neutral.

Although actual Low impedance REF algorithm may differ from relay make to make, in general, Low Impedance REF protection commonly uses a directional element that compares the direction of an operating current, derived from the line/turret-end CTs, with the polarizing current obtained from the neutral CT.

The directional element compares the angle difference between polarizing current (3Io derived from line CTs) to the operating current (IN from NCT) and indicates forward (internal) fault location when angle diff is <90 degrees or reverse (external) fault location if angle diff is >90deg. The forward/ internal indication occurs if the fault is within the protected winding, between the line-end CTs and the neutral CT.[2]

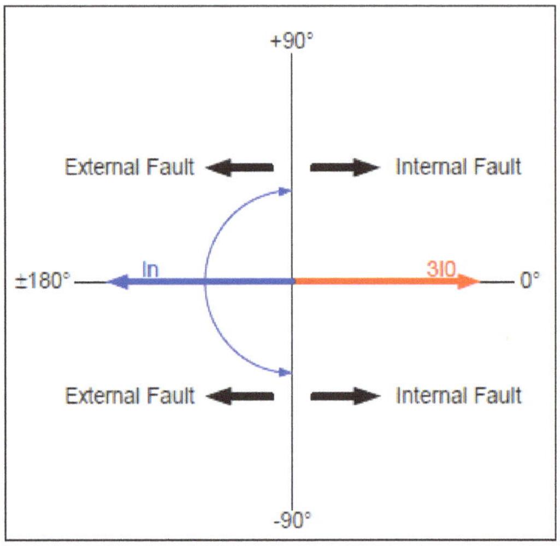

Figure 3: Graphical representation of Internal or external fault. [3]

3.1 High Impedance REF Protection;

The operating principal of the high impedance REF is to balance the sum of the residual phase currents with the output of a current transformer in the neutral. The CT connections of the scheme are show in Figure 4 where the transformer is solidly earthed significant part of the winding is protected. For external faults the secondary y sum of the measured residual current of line/turret CTs are balanced against the secondary of neutral current in opposition, causing no current to flow in the High Z REF relay thus this protection should be stable, however for internal faults near the neutral said relay shall sense the fault and operate. This protection is widely used as it has better sensitivity against CT to saturation.

3.1.1 Depiction of Out-of zone fault;

As per the High Impedance REF scheme, the secondary current of Earth fault current (3Io) entering the source through the Neutral is balanced against the residual current so obtained from all the Phase

CT(Turret CT) Secondary windings connected in parallel ie all the S1 terminals and all the S2 terminals are connected in parallel(please refer Figure 4) and are balanced against the Neutral CT through the REF relay 64R.

In case of out of zone earth fault the direction of secondary current from the Turret CT and the earth fault current(Io) entering the neutral are since opposite to each other in their respective loops cancel each other and the REF relay 64 R shall not operate.

Note: In this High Z scheme it is essential to same ratio and characteristics of Turret CTs and Neutral CT ie the knee-point voltage should be same and the CTs should be of PS Class confirming to IS: 2705(part-4) or PX class confirming to IEC 61869(Earlier IEC 600440)/IEEE Std. C57.13 class C.[4,5]

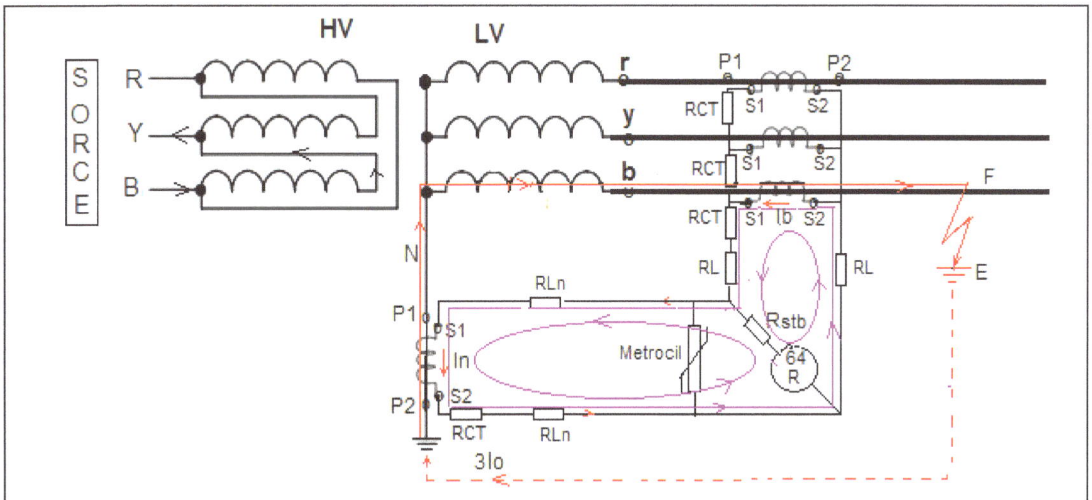

Fig:Out of zone LV Bph to Earth fault.since Ib& In cancel each other ,thus no current would flow through the REF relay arm.

Figure 4: Schematic diagram of circuit for out of zone fault in REF Protection.

3.1.2 REF protection in 20% of unprotected winding:

The % Bias Differential relay with 20% bias setting and with operating current setting of 0.2PU can cover 80-85% of the winding only. High Impedance REF protection takes care of faults between phase winding and earth against the un-protected 20% of the winding. In a star connected transformer the voltage also reduces in proportion to the remaining turns (distance) from the neutral. Thus whenever Phase to Earth fault occurs within 20% distance from the neutral, the driving voltage of earth-fault current very low. The secondary replica of this low earth fault current (3Io) is sufficient to activate the REF relay 64 R.

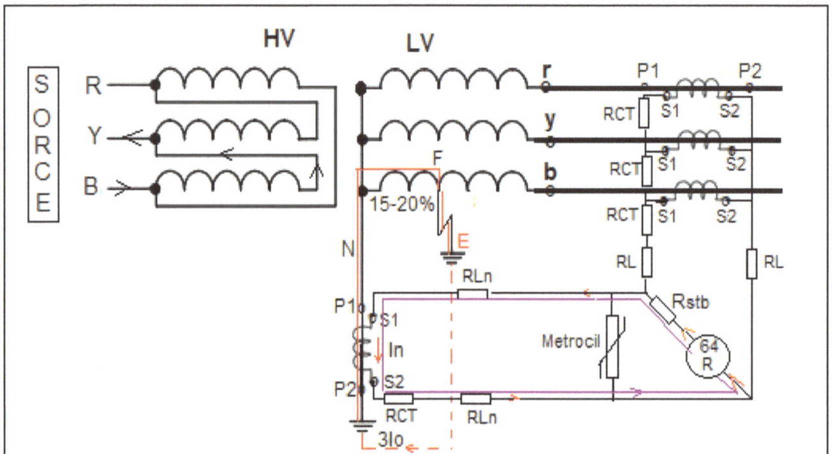

Fig: LV B phase winding earthed near the neutral actuating the REF relay 64R with the secondary replica of Zero seq.current.

Figure 5: Schematic diagram depicting High Impedance REF fault with in 15-20% of the

Star winding nearer to the neutral.

The LV B Phase to earth fault occurs at a location F of the winding in the vicinity of 15-20% of the winding. The zero sequence Earth fault current (3Io) shall follow the path shown in red coloured dotted lines and shall enter the source through the Neutral N (please refer to fig.5).The secondary replica of zero Seq. Earth fault current (3Io) is (In) and is in the opposite direction of the Primary Current (as per characteristic of a CT).This(In) shall circulate in the loop actuating the REF 64R relay and in turn give a trip command to the Circuit breaker.

A question comes to mind that why there is no contribution of the secondary currents from the turret CT? This is because the induced EMF from the HV Source winding B-Y has the tendency to rise from Neutral to the Phase in a healthy transformer winding but in this instant case it is getting earthed at point “F”, thus there is no contribution from the Turret CT of LV -B phase.

It is not out of place to mention here that the Star connected Transformer winding have graded insulation ie insulation level decreases vis’ a vis decrease in voltage too of the winding as it approaches neutral. This concept was used in the REF relay to protect the unprotected portion of the winding through the differential relay.

3.2.1 Protection of Delta winding in conjunction with zig-Zag Transformer;

The delta winding is protected in conjunction with the Zig-zag Transformer as depicted in the schematic diagram in fig.6

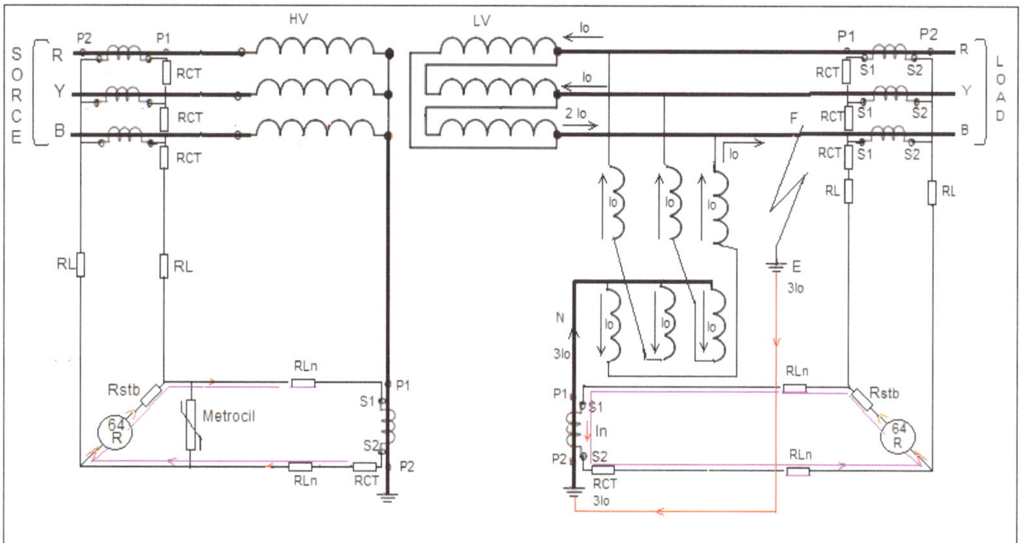

Fig.: Schematic diagram for REF protection of Delta winding including Zig Zag transformer.

Figure 6: Schematic diagram of protection of Delta winding as a whole and also Zig-zag Transformer windings.

2.2.3 REF Protection in an Auto-transformer:

The difference between an Auto transformer and 2 winding Star/Star transformer is an Auto transformer has only one Neutral and 2 winding Star/star transformer has 2 neutrals.ie one each in the HV side and LV side.

In this protection secondary replica of neutral current '3Io' is balanced against the secondary of HV turret CTs paralleled together and also secondary IV Turret CTs paralleled together against the High Z REF relay as shown in the figure 7.

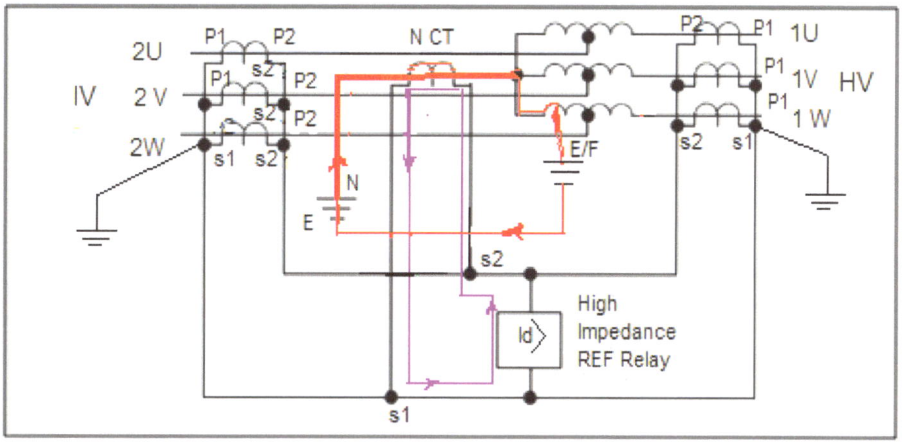

Figure 7: Schematic diagram of REF protection in Auto transformer.[6]

3.0 CT saturation and its effect on REF protection;

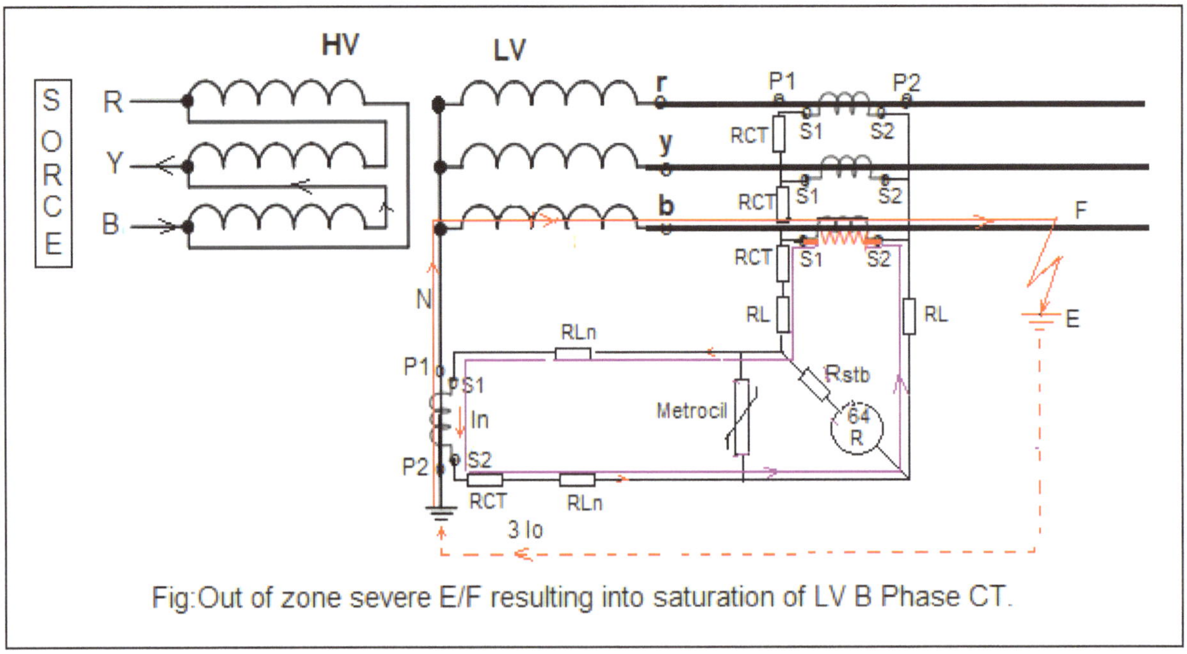

Figure 8: Schematic diagram of High Z REF for Out of Zone sever fault causing CT saturation.

Transient CT errors are caused by CT saturation due to out of zone fault of high magnitude (Figure 8). When fully saturated the CT secondary provides no current it behaves like a resistance in the secondary circuit. The Differential current in the secondary circuit is nothing but the secondary replica of the primary Zero sequence Current 3Io which enters the Transformer through the neutral. It has two paths, either through this resistance or through the REF relay. A stabilizing register is connected in series with the REF relay 64 R(a voltage operated relay) in put to ensure that the operating voltage at current setting is greater than the maximum voltage which can appear across the element/stabilizing register during maximum assigned through fault current.

3.1.1 Purpose of Stabilizing register;

The stabilizing register "Rst" is connected in series with the REF relay 64R.This is provided to offer larger resistance to the path of secondary differential current. Thus the differential current during out of Zone heavy fault current which causes CT of the faulted phase to saturate.

Sample Calculatuoion of stabilizing regiater;[7]

$$\text{Rst} = \frac{\textbf{Vs}}{\textbf{Is}}\text{, where Vs=If(RCT+2RL),}$$

$$\textbf{Say Vs} = 60\textbf{Vand Is}\left(\textbf{operating current setting}\right)\textbf{is } 0.1\textbf{A}$$

$$\textbf{Rst} = \frac{60}{0.1} = 600©$$

Note: Please see the fig.7 Stabilising registers mounted in a Relay panel is shown therein.

3.1.2 Purpose of Metrocil (Non-Linear Register);

It is Voltage limiting Non-Linear Register. It is connected in parallel with the REF Relay and the Stabilizing Register to limit the peak voltage developed across them to safeguard them and also to safeguard the connecting cable from insulation failure. The value thereof is to be limited to 3 kV safeguard against the insulation failure of the connecting leads.

Formula and sample Calculation [7]

$$\mathbf{Vp = 2\sqrt{2Vk(Vf - Vk)}},$$

$$\mathbf{Vf = If \times \frac{CT\ sec}{CT\ pry}(RCT + 2RL + Rr)\ \ where\ \ If : Fault\ current,}$$

$\mathbf{RCT : Resistance\ of\ CT\ winding, RL :: Lead\ resistance\ \&\ Rr\ : Relay\ resistance}$

$\mathbf{Vk = Knee\ point\ voltage\ of\ the\ CT}$

$$\mathbf{Vf = If \times \frac{CT\ sec}{CT\ pry}(RCT + 2RL + Rst + Rr)},$$

where,

$\mathbf{RCT : Resistance\ of\ CT\ wdg, RL : lead\ resistance, Rst\ Stabilizing\ register\ \&Rr : Relay\ resistan}$

$$\mathbf{Say\ If = 7873\ A,\ CT\ ratio\ \frac{1}{400}\ RCT = 0.5\Omega, RL = 0.98\Omega, Rst = 455\Omega\ \&\ Rr - negligible},$$

$$Vf = 7873 \times \frac{1}{400}(0.5 + 2 \times 0.98 + 455)$$

$$= 9004\mathbf{V}$$

$$\mathbf{Vp = 2\sqrt{2Vk(Vf - Vk)}},$$

$$= 2\sqrt{2 \times 200(9004 - 200)}$$

$$= 3753\ \mathbf{V}$$

Note:

- Since this value is above >3000V therefore Non -linear resistance is required.
- Vk value should be actual value as per name plate details of The CT.
- Note: Please see the fig.9 Metrocils are mounted in a Relay panel is shown therein.

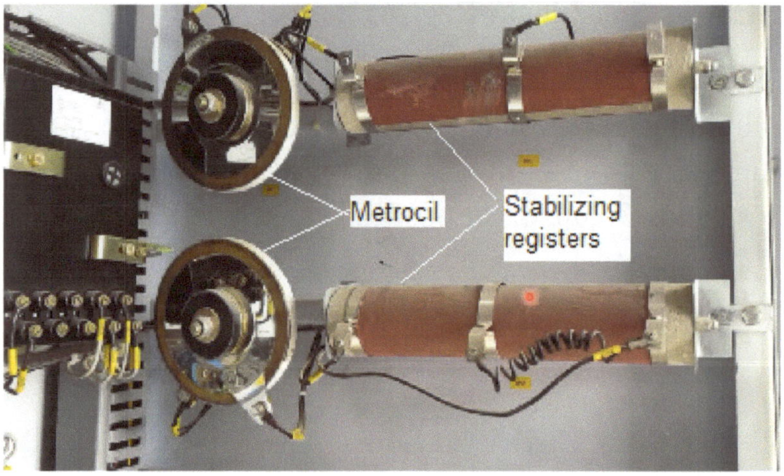

in Figure 9: Metrosils & Stabilizing registers mounted a Relay panel.

4.0 True calculation of a 132/33kV, 50 MVA Transformer;

1. **132/33 KV 50 MVA Two Winding Transformer with HV CTR-400/1, LV CTR-1200/1 and % Z= 10.10**

i. For High Impedance REF protection we have to set the value of Variable Resistance or stabilising resistance (Rs) on the basis of current setting on which we want to operate the REF.

ii. **Stabilising Resistance Rs** $= \frac{\mathbf{Vs}}{\mathbf{Is}}$, here Vs is Voltage across REF element and Is is current setting which we put for REF.

iii. **Voltage across Element;** $\mathrm{Vs} = \mathrm{If} \times \frac{\mathrm{CT\ sec}}{\mathrm{CTpri}}(\mathrm{Rct} + \mathrm{Rw})$,

Where **'If'** is the maximum through fault current and Rct & Rw are CT core and lead resistance.

iv. **Rct & Rw** are either calculated theoretically as per the core resistance mention at CT secondary name plate and lead cable or directly measures on the CT terminal at C&R panel where the REF leads are connected

v. **Through Fault current If**

$$\mathrm{If} = \frac{\mathrm{Iratd}}{\mathrm{Z\ pu}}$$

vi. Now Calculations for **HV REF** are as follows: -

a. We kept Current Pick setting for REF. Is = 0.1 Amp.

b. **Rct + Rw** measured directly at C&R panel on Terminal where REF Cable is Connected (This is the combination HV Turret CT and HV Neutral Turret CT) which comes to 3.3 Ohm.

c. Through Fault current **If** = (HV Rated current)/ (% Impedance).

$$\mathbf{Iratd} = \frac{\mathbf{MVA} \times 1000}{\sqrt{3} \times \mathbf{KV}} = \frac{50 \times 1000}{1.732 \times 132} = 218.7\mathbf{A}$$

$$\mathbf{If} = \frac{\mathbf{Iratd}}{\mathbf{Zpu}} = \frac{218.7}{0.1010} = 2165.34 \ \mathbf{Amp}$$

d. **Voltage across Element**

HV CT prim = 400 Amp and CT sec = 1 Amp

$$\mathbf{Vs = If \times \frac{CT\ sec}{CTpri}(Rct + Rw)},$$

$$= 2165.34 \times \left(\frac{1}{400}\right) \times 3.3$$

$$= 17.86\ V$$

Therefore Stabilising Resistance **Rst;**

$$\frac{\mathbf{Vs}}{\mathbf{Is}} = \frac{17.86}{0.1}$$

= **178.6 Ohm**

e. **Hence we have to set The Stabilising resistor at 178.6 ohm for HV REF Protection with 0.1 Amp Current pick up**.

vii. **Now Calculations for LV REF are as follows: -**

a. We kept Current Pick setting for REF. Is = 0.1 Amp.

b. Rct + Rw measured directly at C&R panel on Terminal where REF Cable is Connected (This is the combination HV Turret CT and HV Neutral Turret CT) which comes as 3.3 Ohm.

c. **Through Fault current** $\mathbf{If = \frac{LV\ Iratd}{Zpu}}$

$$= \frac{50 \times 1000}{\sqrt{3} \times 33} \div 10.10\%$$

d. $= 8661.37$ **Amp Voltage across Element**

HV CT prim = 1200 Amp and CT sec = 1 Amp

$$\mathbf{Vs = If \times \frac{CT\ sec}{CTpri}(Rct + Rw)}$$

$$= 8661.37 \times \left(\frac{1}{1200}\right) \times 3.3$$

$$= 23.8\ V$$

e. **Therefore Stabilising Resistance Rst = Vs/Is**

$$\frac{\mathbf{Vs}}{\mathbf{Is}} \frac{1}{2} \frac{4.052\mathbf{x}1.732}{220\mathbf{x}1000} \frac{\mathbf{Ich}}{\mathbf{V}} \frac{15x0.132x10^6}{\sqrt{3}\ x220x10^3} = \frac{\mathbf{Line\,length\,x\,Line\,chrging\,MVARx}10^6}{\sqrt{3}\mathbf{x\,kV\,rating\,of\,line\,x}10^3} = \frac{23.8}{0.1}$$

= 238.18 Ohm

f. **Hence we have to set The Stabilising resistor at 238.18 ohm for LV REF Protection with 0.1 Amp Current pick up.**

5.0 Comparisons between Low Z and High Z REF protection.

Sr.No	LOW IMP. REF	HIGH IMP. REF
1	The input impedance of the low-impedance REF is very low.	The input impedance of the high-impedance REF is very high.
2	Low-impedance REF protection has low inherent stability against CT saturation for external faults.	High-impedance REF protection has high inherent stability against CT saturation for external faults.
3	The operating current of the low-impedance REF protection is not realized by CT connection. The relay measures all four CTs necessary to realize the element.	The operating current of the High-impedance REF protection is realized by CT connection
4	Ratios of the CTs may not be same	The ratios, knee point voltage and the winding resistance of the Turret CTs and the neutral CT should be same.

6.0 References;

1. Mohammad Gholami, Ehsan Hajipour & Mehdi Vakikian of Sharif University of technology, Tehran, Iran-High Impedance REF Protection: Fundamental concepts, Design and performance. International Conference on Electrical Engineering in Tehran Iran-2016.
2. In consultation with Er. Sthitaprajnyan Sharma, Application manager, Schweitzer Engineering Laboratories Inc.
3. Meyer Kao and Garry Cobet(Tennessee Valley Authority):Considerations and Experiences in implementing ground Differential protection for Transformer protection at TVA.
4. IS: 2705(part4).
5. IEC 61869.1(previously IEC 60044.1).
6. Alstom's Network Protection and Automation Guide.
7. Alstom's Manual of Micom 642
8. electricaltm.word press.com *under* REF PROTECTION *Posted on* March 27, 2013 *by*

7.0 K.K.Murty & Pranay Joshi

12. Line Differential Protection – Concept & Sample Calculations.

Preamble:

Line differential Protection can be applied on overhead lines and cables. It is a very reliable unit protection. All faults on the line, between the line CTs, are cleared instantaneously. It has ability to detect high resistive earth faults It is a preferred protection scheme for underground cable and also for short lines where Distance protection would have limitations.

Due to its factor of reliability, it is used as an additional fault detecting relay for short line protection by most of the State Transmission Utilities as well as the Central Transmission Utility. It is also used as fault detecting protection in case of Hybrid line configuration comprising of Overhead lines and underground cables.

Line Differential protection is usually applied as Primary protection as it is instantaneous and not coordinated with any other Protection. However, in the modern line Differential protection. IED may also have Distance protection and over current protection that is coordinated as back up protection.

Key words: *Line Differential Relay, Line Capacitances & Fiber-Optic Communication link.*

1.0 Concept cum features of this protection;

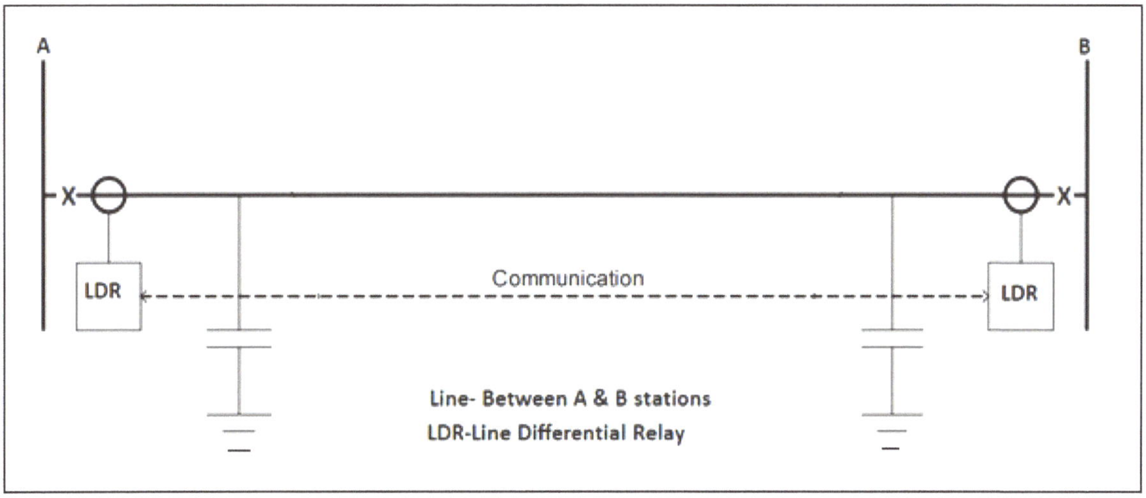

Figure 1: Schematic of line on which Line Differential Protection is shown implemented.

1.1 Advantages of this protection[1];

- Line differential protection provides a sensitive and selective protection.
- No need for time co-ordination as line differential protection only operates for faults with in the protected zone ie Line only thus, it is also called Unit protection.

- Protection operation time can be less than 20ms.
- No voltage measurement is needed (unlike Directional O/C Relay protection and DPR protection).

Requirement/challenges;

- Communication media is always required (optical Fiber or pilot wire).
- There are challenges related to use of multiplex channels or public switched network.
- Repeaters are required beyond a certain distance.
- There is no inherent back up protection unless a multi-function protection IED is used.

1.1 Features of this protection;

- This is current differential protection.
- Two redundant communication channels are usually provided that covers all the three terminals.
- Since this is a unit protection, there is no concept of Local and remote.
- Each terminal is designated by its Terminal designation.
- If time synchronization for measurement is done by a terminal, that terminal can be described as Master terminal. Each terminal can also have independent GPS time synch provision.
- Derivation of the differential and the bias currents are derived using e local and remote current measurements taken at the same point in time, if time is not synchronized, this would cause errors.
- Delays involved in communicating measurements between terminals calls for compensation. This process is called "time alignment".

  - **Differential current (IDiff)** = vector summation of all currents entering the protected zone.
  - **Bias current (IBias)** = half the scalar sum of the currents entering the protected zone.

1.2 Special features;

i. In case of failure of optical fiber communication between the local and remote Line Differential relays then the protection of line is automatically switches over to protect the line with Distance protection features if, such IED is provided in this Protection..
ii. With fiber optic-communication there is no length limitation on Transmission lines for Line Differential Protection. Relay to relay dedicated communication channel can be used up to 200 to 250 km. Above this length fiber optic repeaters are required. In modern numericaAl differential relays. Errors due to communication channel propagation delay as well as errors due to channel asymmetry can be compensated in modern Line Differential protection. [3]
iii. As most numerical Line differential Protection relays also have built in Distance protection in which auto-reclosure can be configured for single phase to ground faults.[3]

1.2. Bias Characteristic Graph:

The differential and bias currents are compared against a tripping criterion which is defined by a dual-slope characteristic in fig.2.shows the tripping criteria for protection of a three-terminal feeder, but the principle is similar for a two-terminal feeder also wherein instead of three currents only two currents are considered ie I1 & I2 only.

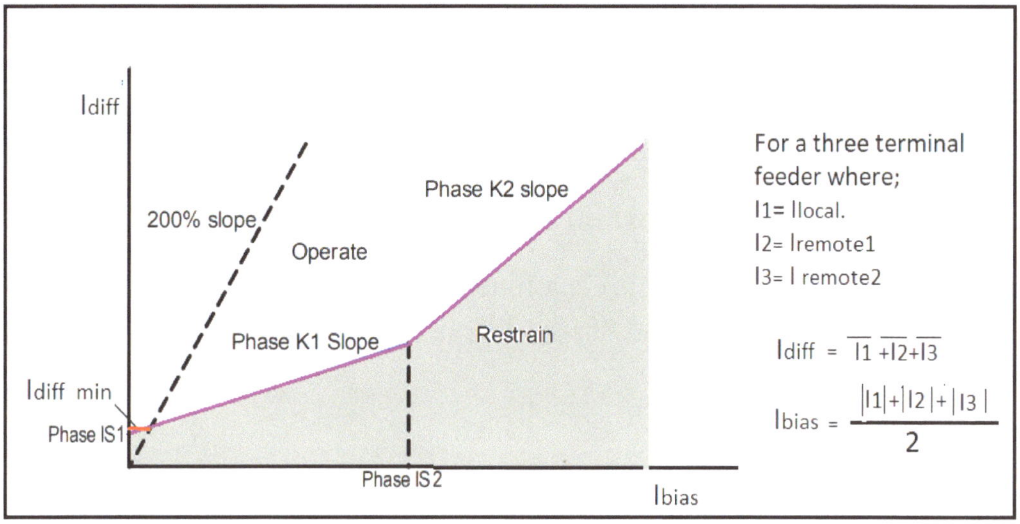

Figure 2: Bias Characteristic graph.[2]

2.0 PHASE CURRENT DIFFERENTIAL TRIPPING CRITERIA;[2]

The phase current differential characteristic is defined by four settings:

- **Phase Is1:**The basic differential current setting which determines the minimum pick-up level of the protection.
- **Phase k1:**The lower percentage bias setting used when the bias current is below the Phase Is2 setting. This provides stability for small CT mismatches, while ensuring good sensitivity to resistive faults under heavy load conditions.
- **Phase Is2:**A bias current threshold setting, above which the higher percentage bias setting Phase k2 is used.
- **Phase k2**:The higher percentage bias setting used to improve protection stability under heavy through fault current conditions.

2.1 Operating & restraining phenomenon [2];

Please refer Figure 2 Bias Characteristic graph:

- ➢ If the differential current falls above the graph. (Figure 2:), **relay shall trip**.
- ➢ If it is below the graph, the **relay shall restrain**.

As when the phase-differential protection issues a trip command, it sends a per-phase ‘differential inter trip’ signal to the remote terminals to enable remote breaker to trip. This is to ensure that all ends of the protected line trip, even for marginal fault conditions.

By default, the delay characteristic is set to definite time with a zero second delay, resulting in instantaneous tripping, however it can be programed for either definite time or on inverse time for grading with other protections. grading or coordination with other protection. A small definite time delay can be inherent to overcome transient phenomenon.

Since differential and bias current calculations are made on a per-phase basis, the values vary on a per-phase basis.

Note; for optimum stability, the highest value of the three-phase bias currents is used to restrain all three phases. The worst condition for protection to remain stable is saturation of CT at one end.

2.2 Philosophy for minimum Trip current Level [2];

The minimum trip current level (Idiff min) is a function of Phase Is1 and Phase k1 as shown in the following calculation (note: Phase Is1 and Phase k1 have been shortened to just Is1 and k1 for clarity).

$$Idiff = k1(Ibias) + Is1 \ldots$$

$$Idiff = (k1)(0.5Idiff) + Is1 \ldots$$

$$Is1 = Idiff - k1.0.5Idiff = Idiff.(1 - 0.5k1) \ldots$$

$$Idiff = Is1 / (1 - 0.5k1)$$

If, k1 has a value of 30% for example, this means that the minimum trip current is 1.176 Is1.

Where, Is1 (basic differential current setting which determines the minimum pick-up level of the protection.)

2.3 Charging current compensation;

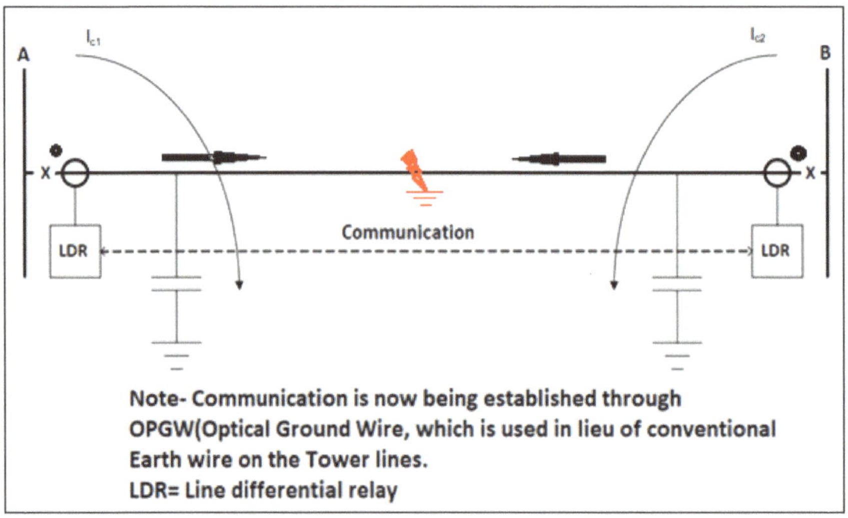

Figure 3: Typical schematic diagram of Line Differential Protection, showing Line capacitances & location of fault.[2]

There are capacitances between the line - phase conductors and between phase conductors and earth. These capacitances give rise to line charging currents which are seen by the Line differential protection Relay as "false" differential currents.

The magnitude of the charging current is dependent of the line capacitance and the system voltage. To overcome this constraint, the charging current compensation is made available in Line differential protection. When enabled, this compensation will measure the fundamental frequency differential current under steady state undisturbed conditions and then subtract it, making the resulting differential current zero (or close to zero).

Note: All small pre-fault differential currents are subtracted, no matter what may be their origin. As the differential protection is phase segregated, this action is made separately for each phase.

The charging current is generated because there is a voltage applied over the line capacitances as seen in the fig.3. If an external short circuit with negligible fault resistance occurs close to the line, the voltage in the fault location will be approximately zero. Consequently, zero voltage will also be applied over part of the line capacitance, which in turn will decrease the charging current compared to the pre-fault value.

3.0 Calculation for charging current;

Say for 220kV line of 15km line.

Line charging MVR/km for 220kV line = 0.132 MVAR / km,

Formula for Line charging current; $\dfrac{\textbf{Line length x Line chrging MVARx}10^6}{\sqrt{3}\textbf{x kV rating of line x } 10^3}$

Therefore, line charging current for say, 15 km line length =

$$\frac{15x0.132x10^6}{\sqrt{3}\ x\ 220x\ 10^3} = 5.196 \text{ Amps}$$

3.1 True calculation of charging current for 11.7 km 220kV Line;

$$\text{Line charging current} = \frac{11.7 \text{ x } 0.132\text{x}10^6}{\sqrt{3}\text{x } 220\text{x}10^3} = 4.052 \text{ A}$$

$$\textbf{Susceptance (B)} = \omega C = \frac{\textbf{Ich}}{\textbf{V}}$$

$$\textbf{Susceptance} = \frac{4.052 \textbf{ x } 1.732}{220\textbf{x } 1000} = 0.03190\text{x}10^{-3} \text{ Sec.}$$

Since, this calculated Susceptance value being very low, therefore considering the default value of 0.2 sec., while Charging current calculations

4.0 Operation as Distance protection Relay;

In every numerical line difference relay, there is a feature, if optical fiber link fails, then it switches – over to Distance Protection Relay, in which local CVTs & CTs connected to Line Difference Protection are used in Distance Protection.

The settings thereof are similar to any other Distance protection for any feeder ie. Z1, Z2 & Z3, impedances.

5.0 Constraints in respect of usage of Line Differential Relay [3];

With fiber-optic communication there is no length limitation on Transmission lines for Line Differential Protection. Relay to relay dedicated communication channel can be used up to 200 to 250 km. above this length fiber optic repeaters are required. In modern numerical differential relays, Errors due to communication channel propagation delay as well as errors due to channel asymmetry can be compensated in modern Line Differential protection. [3] Can be compensated by appropriate settings and communication algorithm.

Note:

- It can be used in any network including weak in-feed condition can be used for any type of network earthing. However, this protection may not be suitable for high impedance earth**ing.**
- The general consensus in our country is to use Line Differential Relays preferably for Short lines only.
- If underground cable is used as transmission line protected through Line differential relay, auto - reclosure is not used.

6.0 Settings adopted for 11.7km 220kV line;

Differential current setting;

i. CT ratio 800/1Amp..
ii. CVT ratio 220kV/110V
iii. Characteristic slope:- 20%
iv. Operating current:- 0.15 A
v. Charging current:- not considered as it is << 0.2A default value (refer Cl.3.1)
vi. **Differential Current calculation:**
   a. *Please refer Cl.2 [calculation is reproduced for Easy reference;*
      - *Phase Is1: The basic differential current setting which determines the minimum pick-up level of the protection.*
      - *Phase k1: The lower percentage bias setting used when the bias current is below the Phase Is2 setting. This provides stability for small CT mismatches, while ensuring good sensitivity to resistive faults under heavy load conditions.]*

   b. *minimum Trip current ie operating; (please refer Cl.2.2)*

   Idiff = phase Is1/ (1 - 0.5.phase k1).where, Phase K1 is % slope.

vii. **Operating current;**
(Differential current calculation adopting Characteristic slope of 20%);
Idiff = phase Is1/ (1 - 0.5.phase k1).
Idiff= 0.15/1- 0.5 x 0.20) = 0.1666 Amp

viii. **Restating current;**

$\frac{1}{2}$ of Operating current;

= 0.1666/2=0.0833 A

Note:

i. The settings are opted for 11.7 km Line & set on the Line Differential Relay type.Ic.1-545.001 with distance relay which runs into large number of pages; 76pages (pdf A4 copies).

ii. **Relevant settings adopted on the Line differential relay:**

Screen shot of relevant setting (enclosed in red font rectangle) of Line Diff. IC 1.545.001 is reproduced hereunder; As could be seen therefrom phase Is2=2.00A & Phase k2 = 50%.(pl ref.fig.5)

Please ref. the characteristic curve here-under (fig.4)

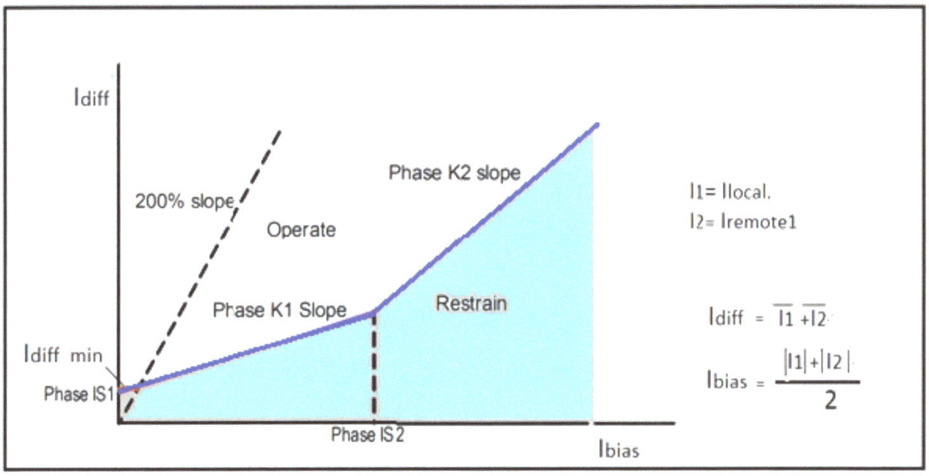

Figure 4: characteristic graph for 1-remote terminal.

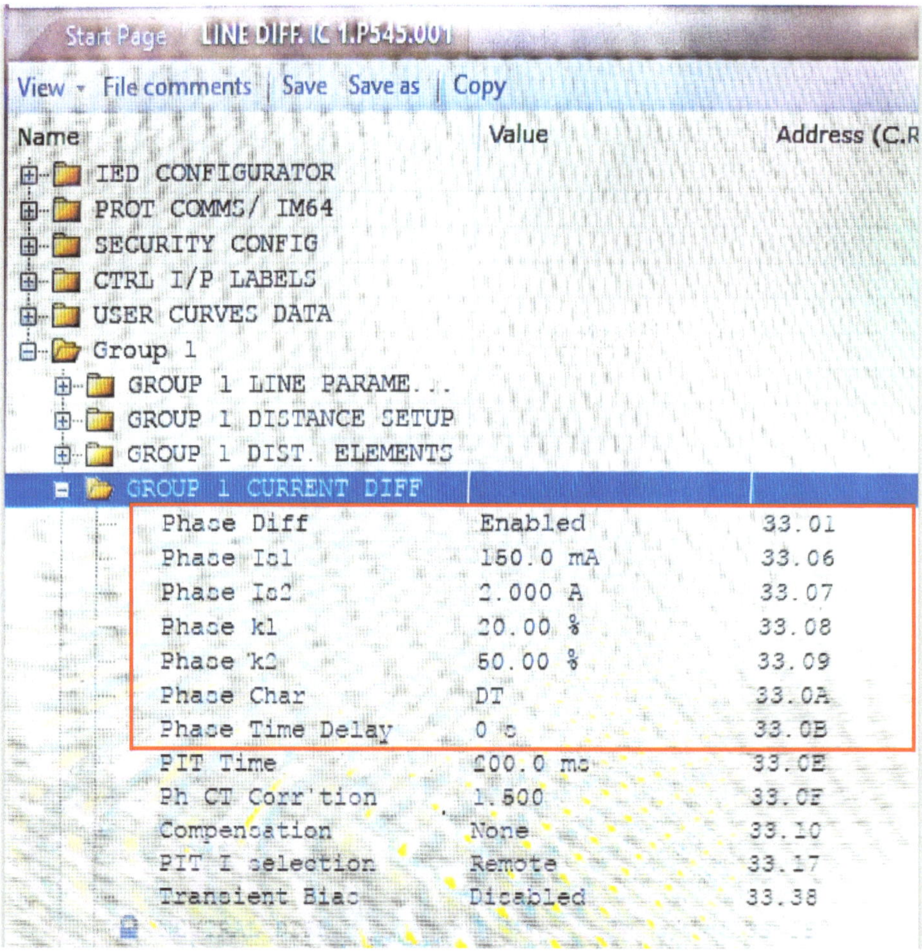

Figure 5:pdf copy of actual relevant settings provided on the Line differential relay for the said 11.7 kV Line.

Note:

i. Kindly refer Cl.6.0.as regards settings adopted.

ii. Kindly refer fig.5. enclosed in red rectangle are the real settings adopted in the Line Differential Relay.

iii. For clarity the settings adopted are reproduced – hereunder; (table 6).

Start page -Line Diff. IC1. P 545.001		
View	**File comments/save save as**	
Name		
⊞ 📁	**IED Configurator**	
⊞ 📁	**PROT COMMS / IM 64**	
⊞ 📁	**SECURITY CONFIG.**	
⊞ 📁	**CTRL I/P LABLES**	

USER CVURVE DATA		
GROUP 1		
	GROUP 1	LINE PARAMETETR
	GROUP 1	DISTANCE SETUP
	GROUP1	DISTANCE ELEMENTS
GROUP 1 CURRENT DIFF		
Phase Diff	**Enabled**	**33.01**
Phase IS1	**150.0mA**	**33.06**
Phase IS 2	**2.000 A**	**33.07**
Phase k 1	**20.00%**	**33.08**
Phase k 2	**50.00%**	**33.09**
Phase Char	**DT**	**33.0A**
Phase Time Delay	**0s**	**33.0B**
PIT Time	**200.0 ms**	**33.0E**
Ph CT correction	**1.500**	**33.0F**
Compensation	**None**	**33.10**
PIT I Selection	**Remote**	**33.17**
Transient Bias	**Disabled**	**33.38**

Table 6.

7.0 Conclusion;

1. This protection is preferably used for short lines by most of the STUs and CTU in addition to DPR.
2. It has an advantage of built in Single phase to ground fault auto-reclose function.
3. If the transmission line is through Underground cable, Auto reclose function should be disabled

8.0 References;

1. ABB's Line deferential Protraction manual RED-615.
2. MiCOM P54x (P543, P544, P545 & P546) Current Differential Protection Relay-mannual.
3. In consultation with Er.Shri Sanjay Patki, former Vice-president-Corporate Engineering (Protection Expert), Tata Power, Mumbai and a distinguished member in CIGRE, Paris.

13. Digital Low-Impedance Bus Differential Protection

Basic Operating concept of Bus-bar Differential Protection: The CT secondary, phase-wise, of all incoming and outgoing feeders and Transformers are connected in parallel and the differential relay is connected across the CT Secondary bus wires. For external faults, the summated in-flow of currents from healthy circuits is balanced by the current flow from the faulty circuit.pl. ref. figures A & B below;

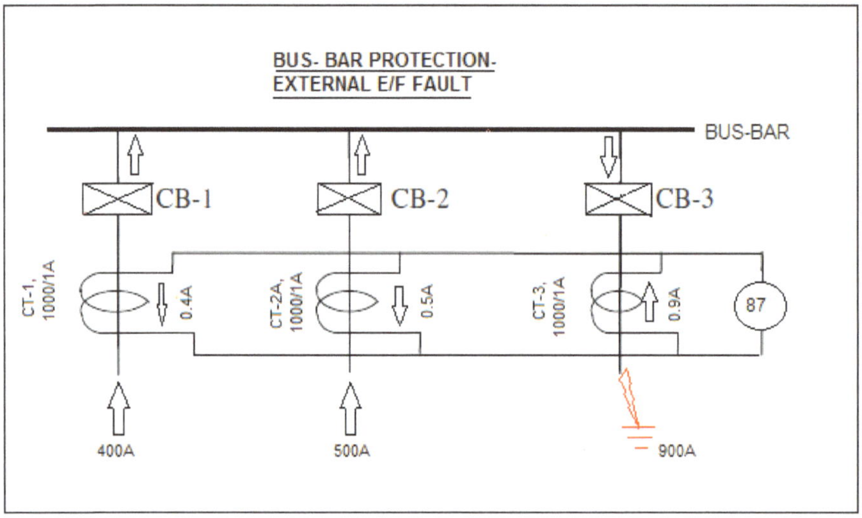

Figure-A

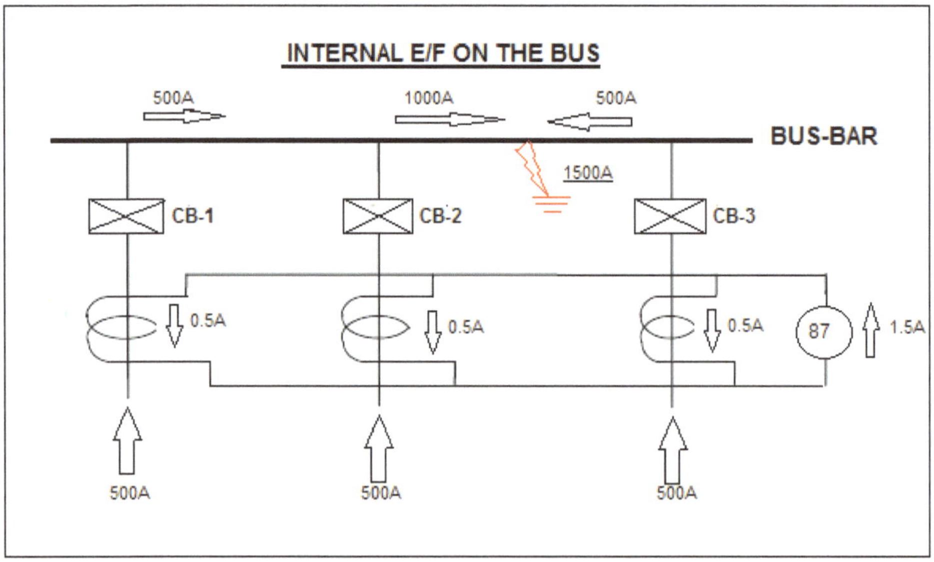

Figure B

Current Distribution: The figures. above depicts differential current status for external and internal faults.

Introduction: Keeping in view the concept of Bus bar Protection system narrated above, modern numerical Bus bar protection relays have been developed.

Protection of power system busbars is one of the most critical relaying applications. Busbars are areas in power systems where fault current levels could be very high. In spite of that, some of the circuits connected to the bus may have their Current Transformers (CTs) insufficiently rated.

This creates a danger of significant CT saturation and jeopardizes security of the busbar.

A false tripping of a transmission bus-bar may drastically change system topology and may hamper the power system stability. Hence, the requirement of a maximum security of bus-bar protection is very essential.

1.0 Bus configuration;

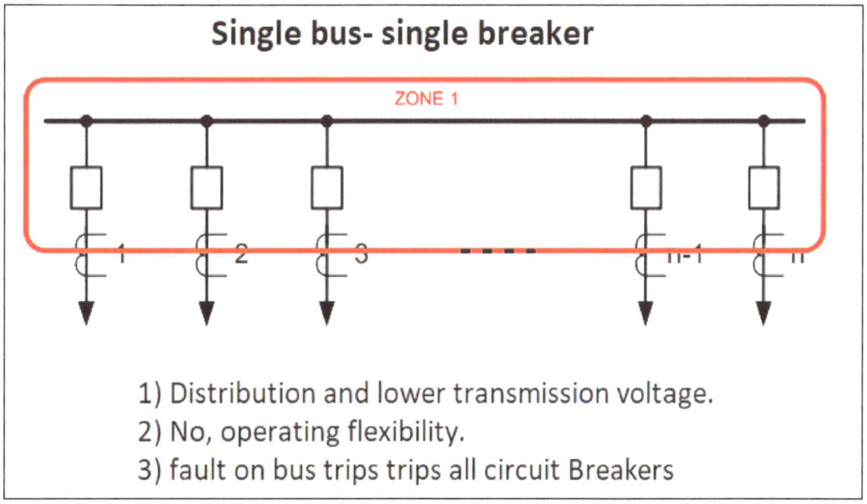

Figure 1

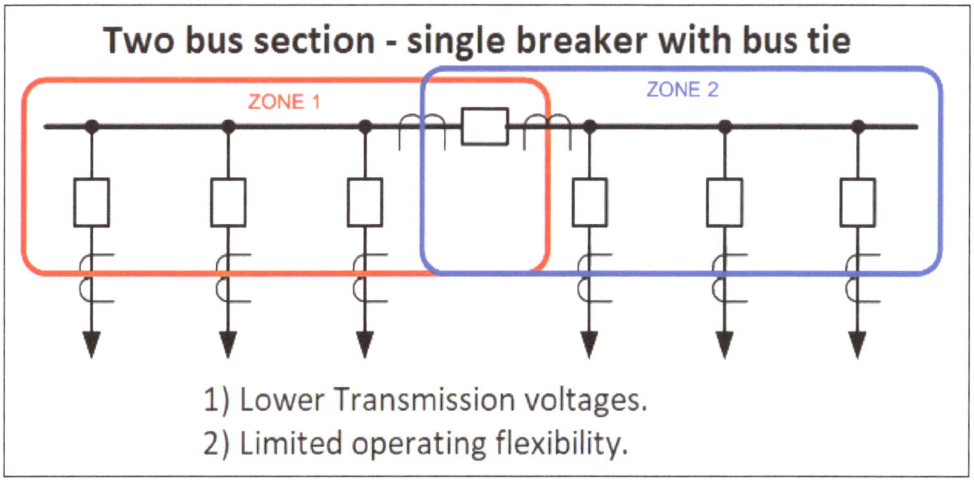

Figure 2

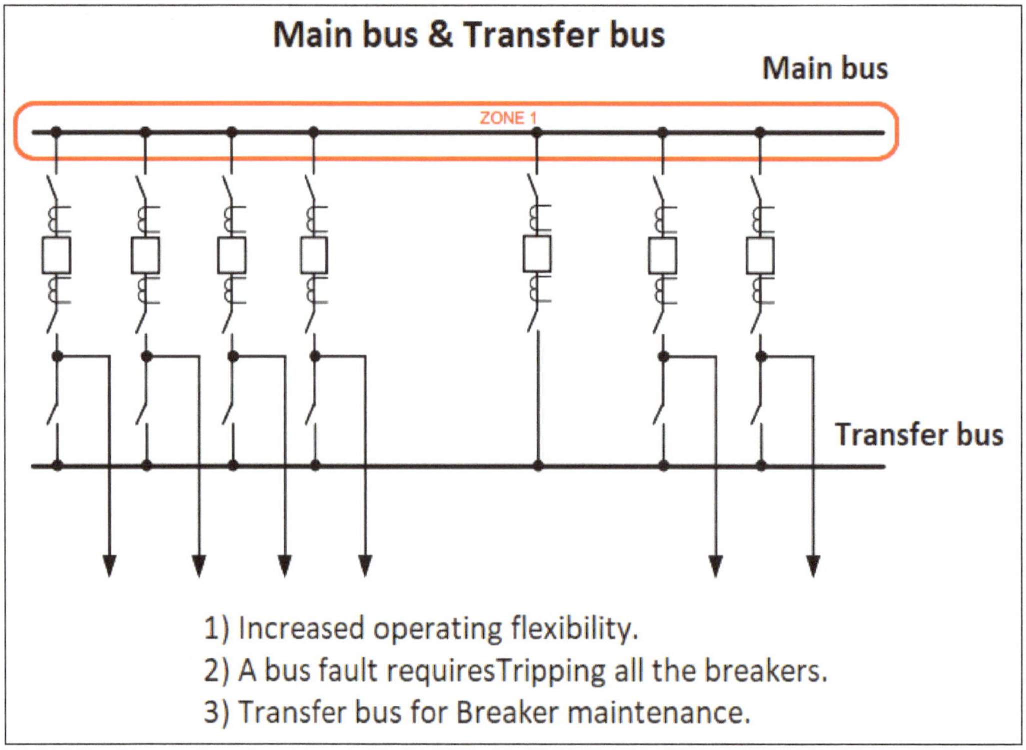

Figure 3

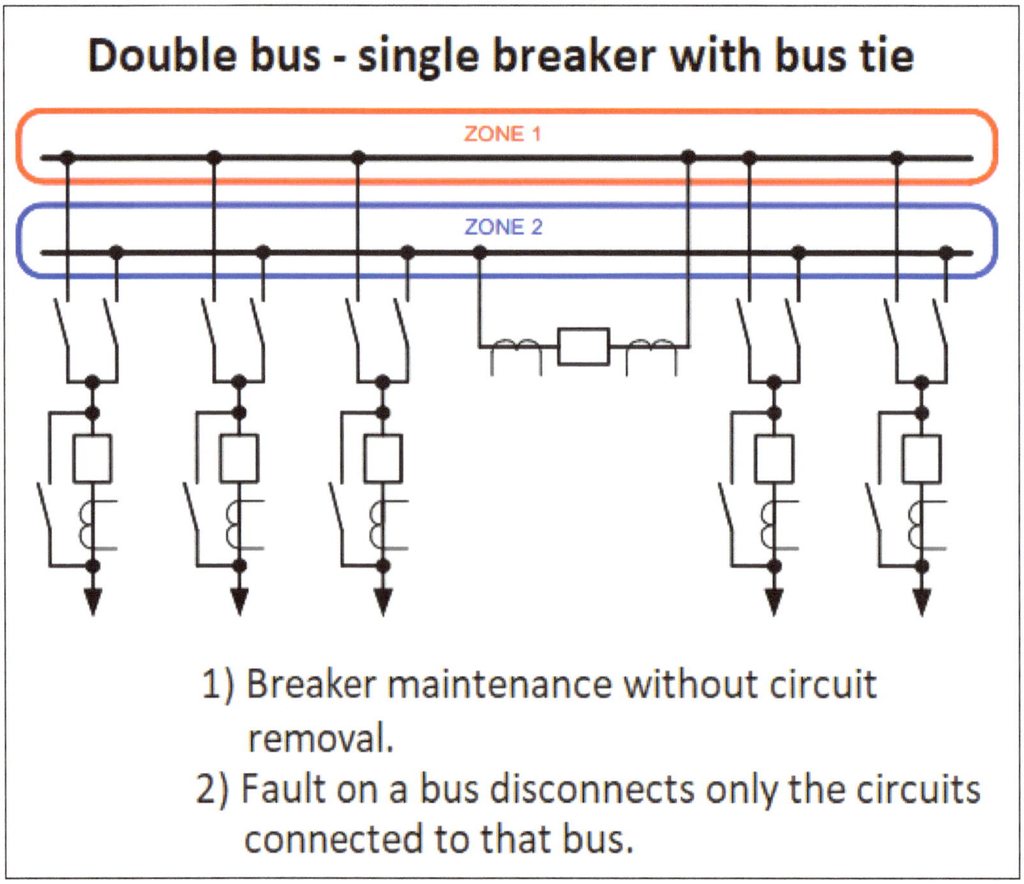

Figure 4

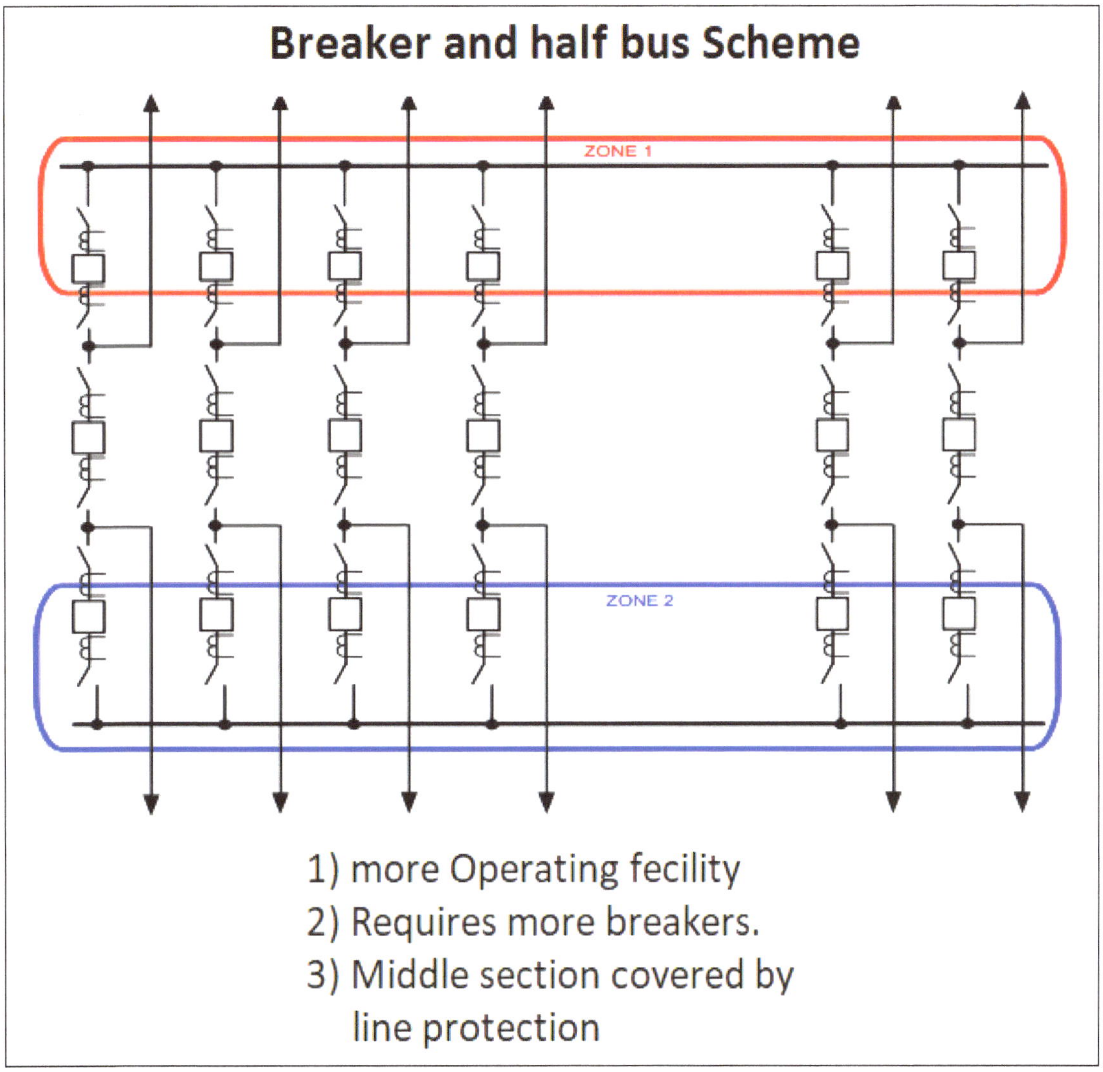

Figure 5

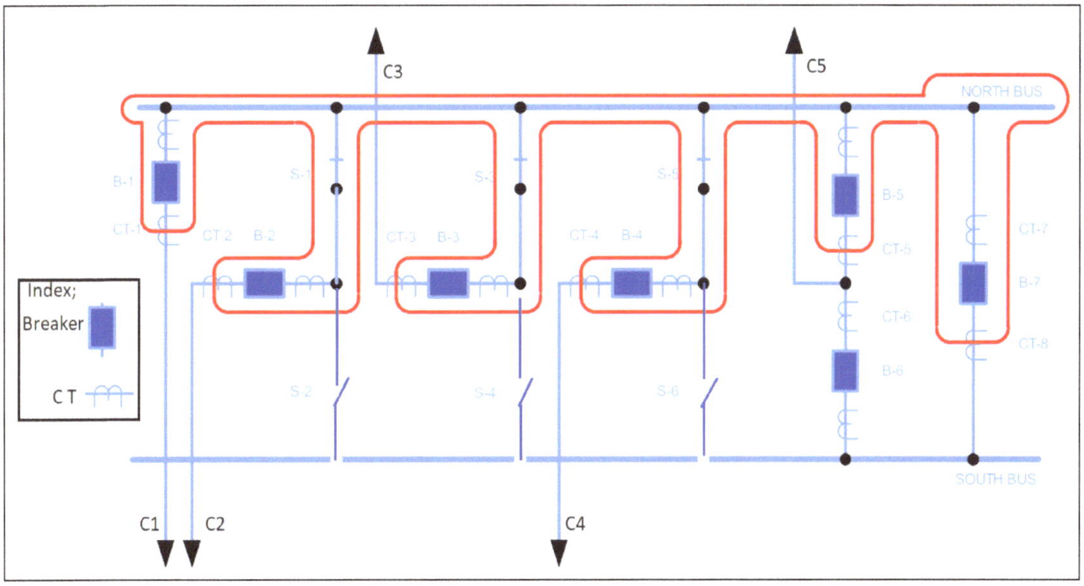

Figure 6

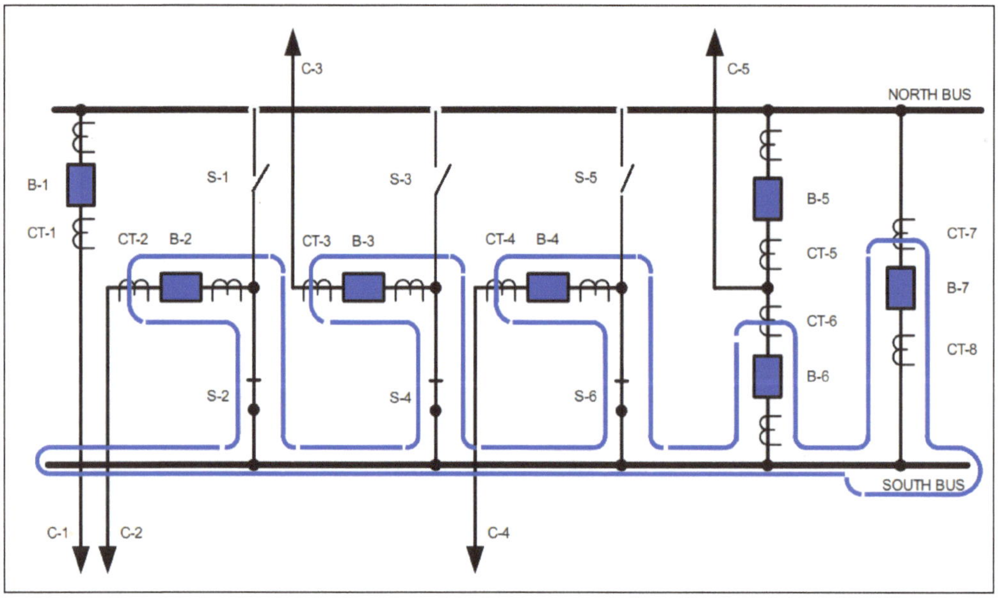

Figure 7

2.0 Bus-bar Protection Techniques;

combined with economic factors yield a number of solutions for busbar protection.

2.1 Percentage Differential;

Percent differential relays create a restraining signal in addition to the differential signal and apply a percent (restrained) characteristic. The choices of the restraining signal include "sum", "average" and "maximum" of the bus currents.

The choices of the characteristic include typically single-slope and double-slope characteristics. This low-impedance approach does not require dedicated CTs, can tolerate substantial CT saturation and provides.

This low-impedance approach does not require dedicated CTs, can tolerate

Substantial CT saturation and provides for high-speed tripping.

This principle became really attractive with the advent of microprocessor-based relays because of the following:

- Advanced algorithms supplement the percent differential protection function making the relay very secure.
- Protection of re-configurable bus-bars becomes easier as the dynamic bus replica (bus image) can be accomplished without switching current secondary circuits.
- Integrated Breaker Fail (BF) function can provide optimal tripping strategy depending on the actual configuration of a busbar.
- Distributed architectures are proposed that place Data Acquisition Units (DAU) in bays and replace current wires by fiber optic communications.

Microprocessor-based low-impedance bus-bar relays are developed in one of the two architectures:

2.2 **Distributed busbar protection** uses DAUs installed in each bay to sample and pre-process the signals and provide trip rated output contacts (Figure 8). It uses a separate Central Unit (CU) for gathering and processing all the information and fiber-optic communications between the CU and DAUs to deliver the data. Sampling synchronization and/or time-stamping mechanisms are required. This solution brings advantages of reduced wiring and increased compuational power allowing for additional functions such as back-up OC protection or BF per ciruit.

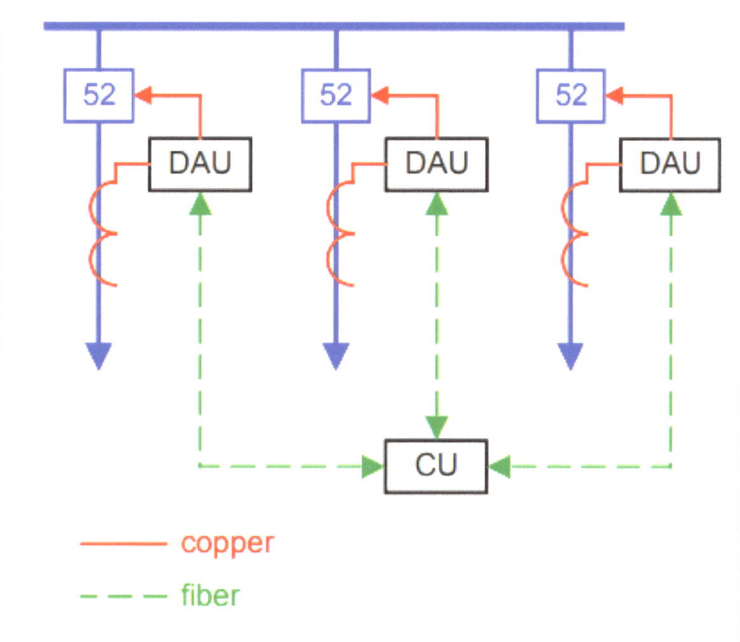

Fig. Distribured Bus bar protection

Figure 8

2.3 Centralized busbar protection requires wiring all the signals to a central location, where a single relay does the entire processing (Figure 8). The wiring cannot be reduced and the calculations cannot be distributed between a number of DAUs imposing more computational demand for the central unit. On the other hand, this architecture is perceived as more reliable and suits better retrofit applications.

Algorithms for low-impedance relays are aimed at;

a. Improving the main differential algorithm by providing better filtering, faster response, better restraining technique, robust switch-off transient blocking, etc.
b. Incorporating a saturation detection mechanism that would recognize CT saturation on external faults in a fast and reliable manner.
c. Applying a second protection principle such as phase directional (phase comparison) for better security.

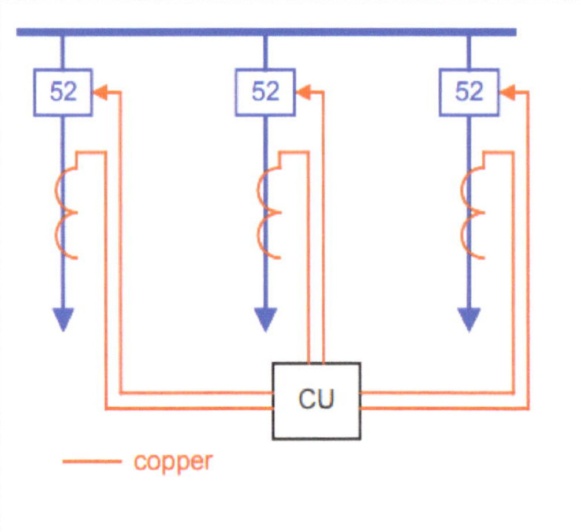

Fig. Cerntralized Bus-bar Protection

Figure 9

3.0 Characteristic of Bus-bar Differential relay;

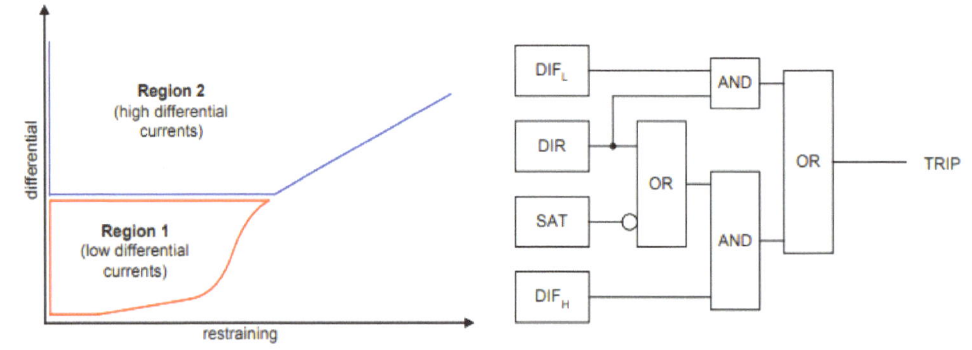

a) Two regions of the differential characteristic. b)Adaptive trip logic.

Figure 10.

The differential protection function uses a double-slope doublebreakpoint characteristic. In order to enhance the security, the operating region of the characteristic is divided into two areas (Figure 6) having diverse operating modes

The bottom portion of the characteristic applies to comparatively low differential currents and has been introduced to deal with CT saturation on low-current external faults. Certain distant external faults may cause CT saturation due to extremely long time constants of the d.c. components or due to multiple autoreclosure shots. The saturation, however, is difficult to detect in such cases. Additional security is permanently applied to this region without regard to the saturation detector

The top region includes the remaining portion of the differential characteristic and applies to comparatively high differential currents. If, during an external fault, the spurious differential current is high enough so that the differential ñ restraining current trajectory enters the top region, then such CT saturation is guaranteed to be detected by the saturation detector.

The differential current is produced as a sum of the phasors of the input currents of a differential bus zone taking into account the connection status of the currents, i.e. applying the dynamic bus replica of the protected bus zone. The CT ratio matching is performed before forming the differential and restraining currents.

3.1 Differential Principle

3.1.1 Differential and Restraining Currents;

a. **The differential current** is produced as a sum of the phasors of the input currents of a differential bus zone taking into account the connection status of the currents, i.e. applying the dynamic bus replica of the protected bus zone. The CT ratio **matching is** performed before forming the differential and restraining currents.
b. **The restraining current** is produced as a maximum of the magnitudes of the phasors of the bus zone input currents taking into account the connection status of the currents

4.0 Differential Characteristic;

The relay uses a double-slope double-breakpoint operating characteristic shown in Figure 11.

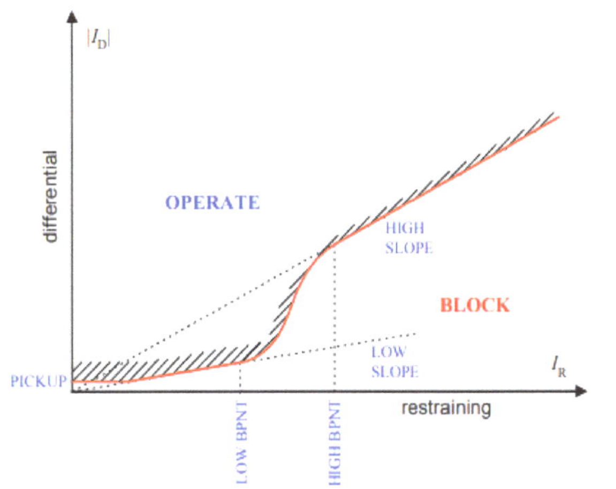

Fig. Percentage characteristic & its setting points.

Figure 11

The **PICKUP** setting is provided to cope with spurious differential signals when the bus carries a light load and there is not any effective restraining signal.

The first **breakpoint** (LOW BPNT) is provided to specify the limit of guaranteed linear operation of the CTs.

This value defines the upper limit for the application of the first, lower slope (LOW SLOPE). The second breakpoint (HIGH BPNT) is provided to specify the limits of operation of the CTs with substantial saturation. This point defines the lower limit for the application of the second slope (HIGH SLOPE).

5.0 Effect of CT saturation;

a. The CT saturation condition is declared by the saturation detector when the magnitude of the restraining signal becomes larger than the higher breakpoint (HIGH BPNT) and at the same time the differential current is below the first slope (LOW SLOPE). The said condition is of a transient nature and requires a seal-in.
b. The saturation detector takes care that of the fact that any CT operates correctly for a short period of time even under very large primary currents that would subsequently cause a very deep saturation.
c. As a result of that, in the case of an external fault, the differential current stays very low during the initial period of linear operation of the CTs while the restraining signal develops rapidly. The restraining signal, however, yields by at least a few milliseconds.
d. During internal faults, both the differential and restraining currents develop simultaneously. This creates characteristic patterns for the differential - restraining trajectory as depicted below;

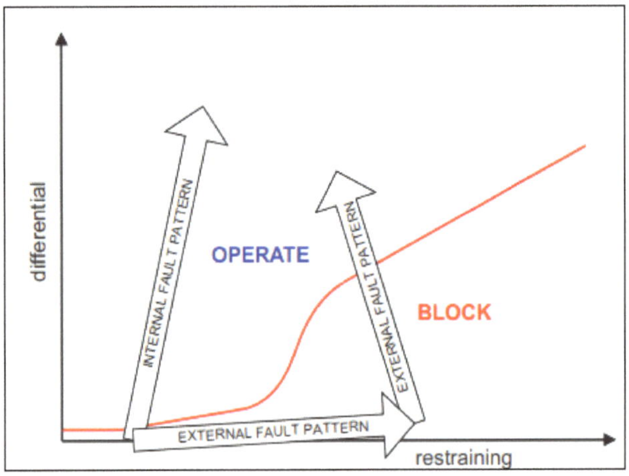

Figure 12

The saturation detector is capable of detecting saturation occurring in approximately 2 ms into a fault. It is worth emphasizing that the saturation detector, although having no dedicated settings, uses the main differential characteristic for proper operation. This must be kept in mind when setting the characteristic as its parameters must retain their original meaning.

14. En-masse operation of Buchholz relays due to Earth quake- An experience

EPISODE: En-masse tripping of EHV Transformers (35-40 nos) on BUCHHOLZ indication on the 22nd May, 1997 at 04.22 hrs in Jabalpur and neighboring area affecting large no of villages and Urban area in habitants when a strong jolt of earth quake took place of the magnitude of around 6.0 MMs (Moment magnitude scale-rector scale).resulting in to whole area in and around Jabalpur plunged in to darkness.

[**Note**: Major damage occurred to structures in the Jabalpur District, Mandla District, Seoni District and Chhindwara District in Madhya Pradesh. Jabalpur and Mandla were the worst affected districts. A total of 887 villages were affected. Approximately 8,546 houses collapsed and nearly 52,690 houses were partially damaged. Some aftershock activity occurred, but they did not cause any additional damage. Longitudinal ground cracks were observed in some locations of the affected area.]

Restoration of Supply:

At that time I was an Executive Engineer (S/S & Transmission line maintenance).I had reached the 220kV S/s Jbp within 10 mins time. I was the only available Engineer concerning the S/Ss. On that ill-fated day all other Senior Engineers were out of station on either WREB meetings, Meetings at Bhopal & other private engagements etc. The Executive Engineer (Testing) got stuck behind the debris of his damaged house due to earth-quake. The whole responsibility of restoration of supply lied on me as none of the Sr. Engineers responsible for decision making could be contacted on P&T phones as the main Telephone exchange of Jabalpur had also got affected.

Constraints:

1. I took indications of tripping of all the EHV transformers affected from the S/ss of Jabalpur and neighboring places, analyzed the same and reached to the final conclusion that "As tripping of all the Transformers had taken place on Buchhloz- trip indication not accompanying with any other indication as Differential or PRD/PRV and cancellation of Buchholz Trip indication in case of each and every affected transformer confirmed that It was due to the seismic effect of the earth-quake only.
2. I had to take my own decision on my field experience putting my career at stake, against the protocol and ethics of testing, to charge and restore the supply as early as possible in the interest of Public at large. I instructed all the Operators of the affected S/S on PLCC to charge their transformers following their protocol /Operational practices in vogue in their S/s. If all the transformers were first tested and then charged, perhaps, it would have taken about 3-5 days to restore the Power supply pooling testing teams from other Testing Divisions.

3. Such episode must have been experienced by the MSEB Engineers too when Latur had also suffered Jolts of Earth quake in 1993, which was even more severe than the Jabalpur earth quake intensity (perhaps 6.4 MMs).

Such incidences of Earth-quake are very- very rare may be once in a blue-moon or so, however, it is to confirm that the tripping of the Transformers on Buchholz Relay operation is due to the Seismic effect of the Earth Quake only. Thus, you are all requested to ponder and come to some conclusion in this regard as to, whether to have some protection against earth quake seismic conditions or not, weighing the essentiality vis' a vis the cost factor involved to innovate and implement such protection.

Opinion; In my view following arrangement may be innovated by the Relay Manufacturers in agreement with the Transformer Manufacturers;

"The Relay/ module should be designed on the principle of measurement of noise of the Transformer humming through acoustic noise level meter and it should be embedded in the body of the transformer itself which should operate sensing the pronounced noise and jolts due to seismic effect of the earth-quake much earlier that the operation of the conventional Buchholz relay such that its command in-turn actuates the master trip relay/Master trip module earlier than the conventional Buchholz relay. Appropriate logic may be thought over and designed by the Relay manufacturers."

ON BACK COVER PAGE

Compendium of Articles on Substations & Relay Protections for budding and Practicing Engineers of the Transmission Utilitied.

Contains- 45+years of rich experience of the author as a Practicing Engineer.

EHV Substations: Bus configuration, innovovations-Synchronizing without synchroscope and use of Transferbus, Substation equipment like; CVTs, Circuit Breakers, Isolators & LAs & Brief Introduction of GIS and Earthing system successfully adopted in hiilly site, etc.

EHV Transformers:Transformer & Reactor – Over view, KYT(Know Your Transformer).

On-line reconditioning of old Transformers. EHV Condenser Bushings. Concept of SFRA, In-house retrofitting of 420kV Bushing, Extension of 220kV Turret and repair of Tap-changer of 160MVA, Transformer.RLA of Transformers & methodology to enhance life thereof.

Relays & Protections Schemes; Concepts & description of various. Relays & Protection schemes including auto-reclosing etc.Usage of Control Switching Device in Switching in Shunt capacitor banks, Operation of Buchholz relays due to Earth Quake: (Seismic condition); En-masse tripping of 35 nos. of EHV Transformers on Buchholz Indication and successfully energizing them.

15. Breaker Failure Protection (50BF) in conjunction with "GOOSE communication system" in IEC 61850 Protocol.

By: K.K.Murty, Former-C.E. (T&C) &

Rajesh Shrivastava, Former-C.E. (T&C)

MPPTCL, Jabalpur.

1.0 Synopsis: With the growing complexities in any EHV Grid-system network containing 765kV, 400kV, 220 kV systems with associated Transformers/ICTs and Bus configurations, failure of Circuit Breaker and non-operation of its associated conventional Breaker failure protection wired up with hard copper wires may lead to system disturbance and affect the life expectancy of out-door equipments/ may lead to equipment failure as well. Keeping this in view the IEC TC 57 committee has felt its importance and has incorporated GOOSE Communication in the IEC 61850. Modern IEDs available in the market are IEC 61850 compliant. The signal transmission time in GOOSE communication is < 4ms (extremely fast).This feature has been utilised in Breaker Fail protection (50BF) schemes in the EHV switch-yards. This shall not only safeguard the Power system from disturbances but also prevent failure of outdoor equipment /hamper life expectancy thereof.

The authors narrates salient points of IEC61850 Protocol in the Automation of EHV Sub-stations and the core topic on GOOSE communication feature of IEC61850 Protocol in Circuit Breaker Failure Protection scheme

2.0 Key words: SAS (Substation Automation System), IEC 61850, GOOSE (Generic Object Oriented Sub station Event)messages, Interoperability, ISO/OSI (International standards Organisation/ Open System Interconnection)

3.0 IEC 61850 Structure;

Prior to utilising IEC: 61850 protocol, other protocols like Modbus, DNP3 and IEC 60870-5-101 protocol were used.

The IEC Committee TC 57 had developed IEC 61850 and had published it as" COMMUNICATION NETWORKS AND SYSTEMS FOR POWER UTILITY AUTOMATION." Sub-station automation is based on this IEC 61850 Protocol.

The important feature of this protocol is Time-Synchronisation and Inter–opera ability.

Before proceeding to the core subject of GOOSE Communication feature in IEC 61850 protocol for substation automation system, it is pertinent to appraise the readers with the different parts of the IEC 61850 standard which have been specifically designed for communication needs for Power Utility Automation.

Structure of the IEC 61850 Standard.

Part	Title
1	Introduction and overview
2	Glossary
3	General requirement
4	System and project management
5	Communication requirements for functions Device Models.
6	Configuration description Language for Communication in electrical Sub Station related to IEDs
7	Basic communication structure for Substation and Feeder equipment
7.1	Principle and models.
7.2	Abstract communication interface(ACSI)
7.3	Common Data Classes(CDC)
7.4	Compatible Logic node classes and data classes.
8	Specific communication service
8.1	Mapping to MMS(ISO/IEC9506-part-1 & Part -2 and ISO/IEC 8802-3
9	Specific Communication Service Mapping(SCSM)
9.1	Sampled Values over Special Unidirectional Multi-drop Point to point Link
9. 2	Sampled Value over ISO/IEC 8802-3
10	Conformance testing

Table 1:Parts of IEC 61850

Though Part 3, 4 and 5 define the general and specific functional requirement for communication in Substation automation, other parts are equally important, which are the building-blocks of the Substation automation system.

Part 7.2: defines Abstract services,

Part 7.4: abstraction of Data object referred to as logic nodes.

Part 8.1: mapping of abstract data object and services onto Manufacturing messaging specification (MMS).

Part 9.1: Mapping of sample measured values (Unidirectional Point-to-Point and bidirectional Multi point accordingly) on to an Ethernet data frame.

Part 9.2: Defines in other words “Process bus”

Part 6: Defines XML (Extensible Mark up language) based substation configuration language (SCL). It allows formal description of relationship between the substation automation system and the substation (Switch yard)

At the application level, the switch yard topology and relation of switch yard Structure to the SAS functions (Logical nodes) configured on the IEDs can be described. Each device is provided with a SCl file which shall describe configuration of it.

IEC 61850 has not only taken over the drive of communication technology suitable for of power Utilities, but it has also adopted the best possible protocols and configurations for high functionality and reliable data transmission.

In a nut-shell, the salient features and advantages of IEC 61850 are enumerated here under:

- Inter-operability by various manufacturer's IEDs as an integrated system.
- High data transfer amoung IEDs considdering peer to peer communication model instead of master – slave communication model in previewed protocol.
- Data definition based on advanced object oriented model which contains whole data specification instead of single–oriented model with each data definition by numeric address.
- Supporting functiomality of devices to provide better Communication.
- Communication extendibility and data integrity.
- Support of sampled value exchange.
- File transfer for disturbance recording Communication services to connect primary equipment such as Instrument transducers to relays
- Increasing reliability by proer bus toology.
- Offers a complete set of specifications covering all communication issues inside a Sub-Station.

4.0 Substation Automation methodology - brief description;

Substation automation consists of 3-level stages and their functions which are as follows;

1. **Process level functions**; It is the lowest level where the Substation outdoor equipment Circuit breakers, CTs etc are installed. The relevant information are received by the sensors and actuators, and to send them to the bay level devices,
   The other major task of the process level function is to receive the control command from the bay level device and execute it at appropriate switch level.

2. **Bay level functions**: Acquires the data from the bay and mainly acts on primary (Power circuit) equipment of the bay. It is the middle level where IEDs (Intelligent Electronic Devices) which are multifunction electronic devices they contain protection and metering etc. They are physically connected through hard wires. The basic process of transferring data is through binary and anologue input or output informations such as voltage and current transformer outputs and trip controls from the protective relays.The MU (Merging unit) is the device to collect the instantaneous values of current and voltages from CT & PT, sample the same and send them to protection and control unit
3. **Station Level Fuctions :** They are of two types;

   i. Process related functionas act on the data from multiple bays or sub station level data base.these functions are used to convey the control commands for for primary equipment Circuit breaker and collect the sub station datalike voltage, current, pf, etc from the bay level devices (please refer figure 1).Each by includes ond primary equipment Like transformer and feeders.

ii. Interface related functions enable interactive interface of sub staion automation syatem to the local operator HMI (Human Machine Interface), to a remote control center TCI (Tele Control Interface) or to a Remote monitoring Center for monitoring, TMI (Tele Monitoring Interface).

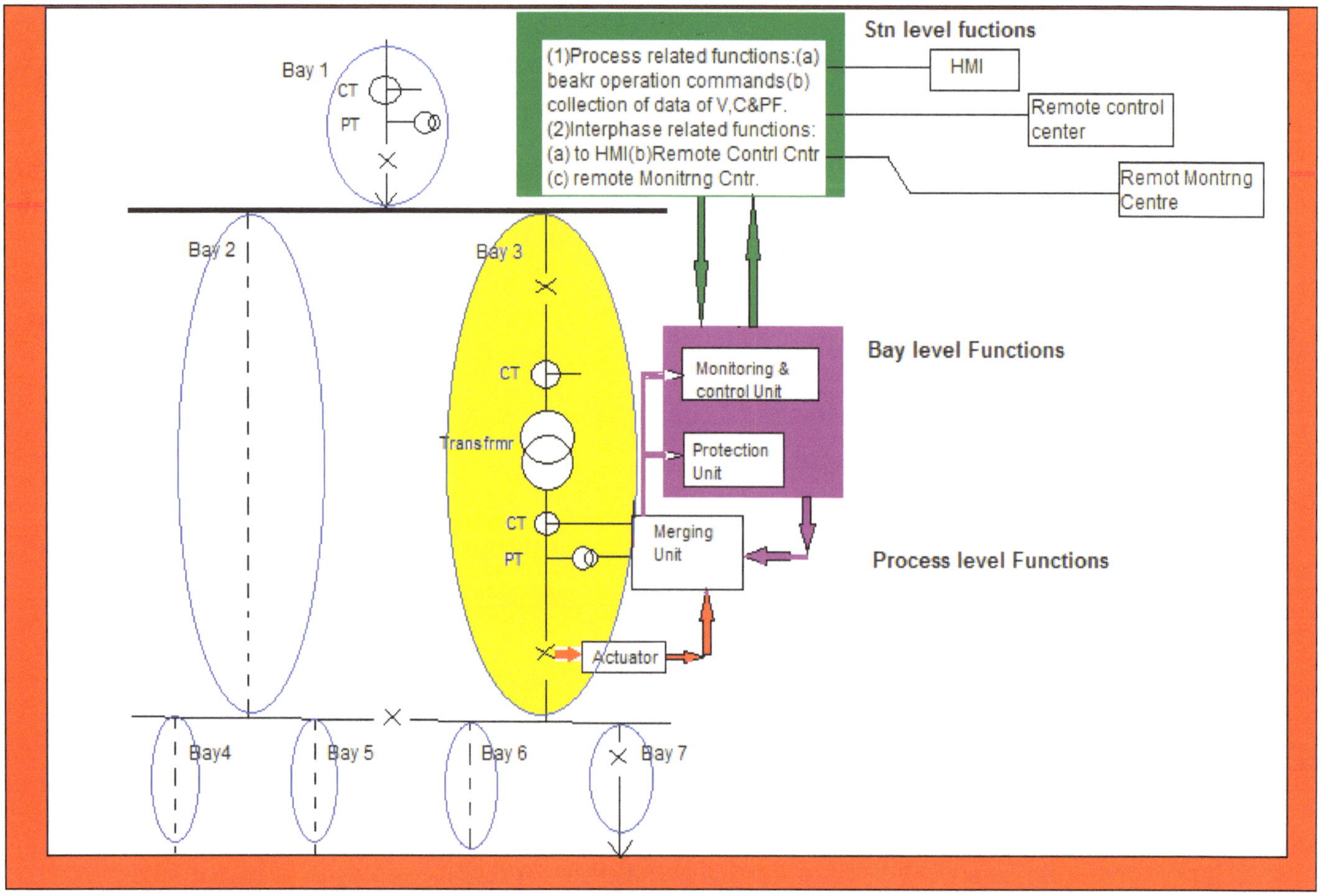

Figure. 1: Schematic of sub-station automation system with IEC 61850 Protocol

4.0 Conceptual Sub Station Automation topology:

Figure No.2 depicts the Sub station automation topology of two bays.

The process bus is established through standard local area net work(LAN)..This Switch filters and fowards data packet between LAN segments.

The sequence of activities/operations from Staion Bus level to Process level is as follows;

1. **Station Level**; The station Bus is created by installing a multiport Ethernet Switch.This switch is a device in computer network that filters and forwards data packets between LAN segments.

   Generally Router, Human Machine Interface(HMI) and Engineering Controls are connected with Ethernal switch at station level.
   Router fecilitates data communication between substation and Master Control Center(MCC).

2. **Bay Control Level:** If Station level node has to communicate with Process level node, it will send message to Bay lenel node through Ethernet switch. Ethernet switch will send the message to appropriate node at bay level.This bay level node executes its function and forwards the message to process level node through merging Unit. merging Unit acts like a switch and provides appropriate path to messages.
   GOOSE Message is also performed in the bay level to communicate with Peer to peer IEDs of different bays, particularly in case of Bus bar protection and Breaker fail protection.

3. **Process level** :The Merging Unit acts like a switch and provides appropriate path to messages. Eventually, the function will be performed by the process level devices.

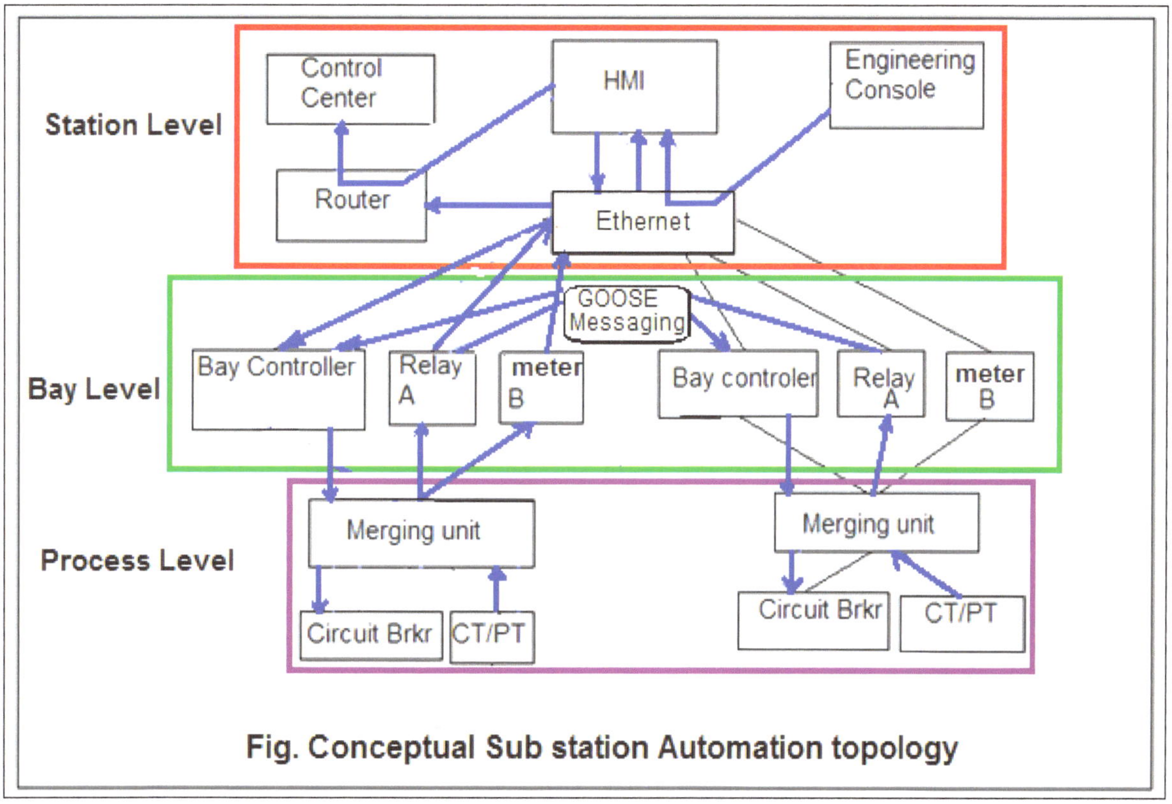

Figure 2: Conceptual Sub Station Automation topology

5.0 Communication system Structure: IEC 61850 basically has state of art communication structure. Following figure (Fig. 3) depicts graphical representation of the Communication structure.

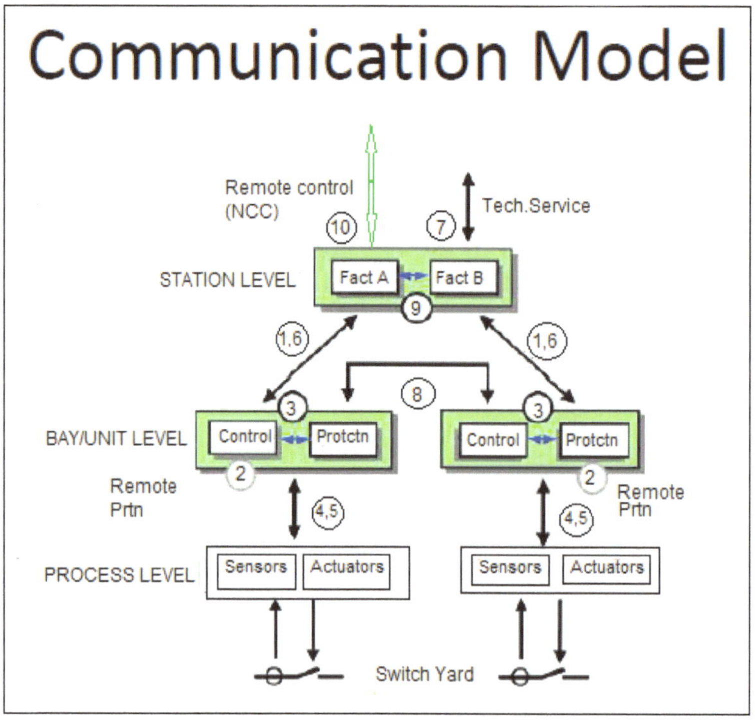

Figure 3: basic communication structure.

Methodology of data exchange as per IEC 61850	**Nomenclature of Nos. depicting data Exchange**
Protection data exchange between Bay and Station level	1.
Protection data exchange between Bay level and remote protection	2
Data exchange within bay level	3
CT& PT instantaneous data exchange between process & bay levels	4
Control-data exchange between process and process & bay level	5
Control data exchange between bay and station level	6
Data exchange between substation and remote engineer's workplace	7
Direct data exchange between the bays especially for fast functions like interlocking	8
Data exchange with in station level	9
Control data exchange between substation devices(devices and a remote control centre	10

Table:2

6.0 Inter-operability: Sub Station Automation Systems based on DNP3 & IEC 60870-5-101 protocols had some limitations of peer to peer communication calling for protocol converters for peer to peer Communication between IEDs of different manufacturers.

This feature in IEC 61850 Protocol has enabled the Sub Station Automation operators to overcome the the excercise of using costly protocol converters for IEDs manufacturered by different vendors, for establishing Interoperability between them. They can be integrated in one Sub station infra-structure, this is done by defining the station bus.

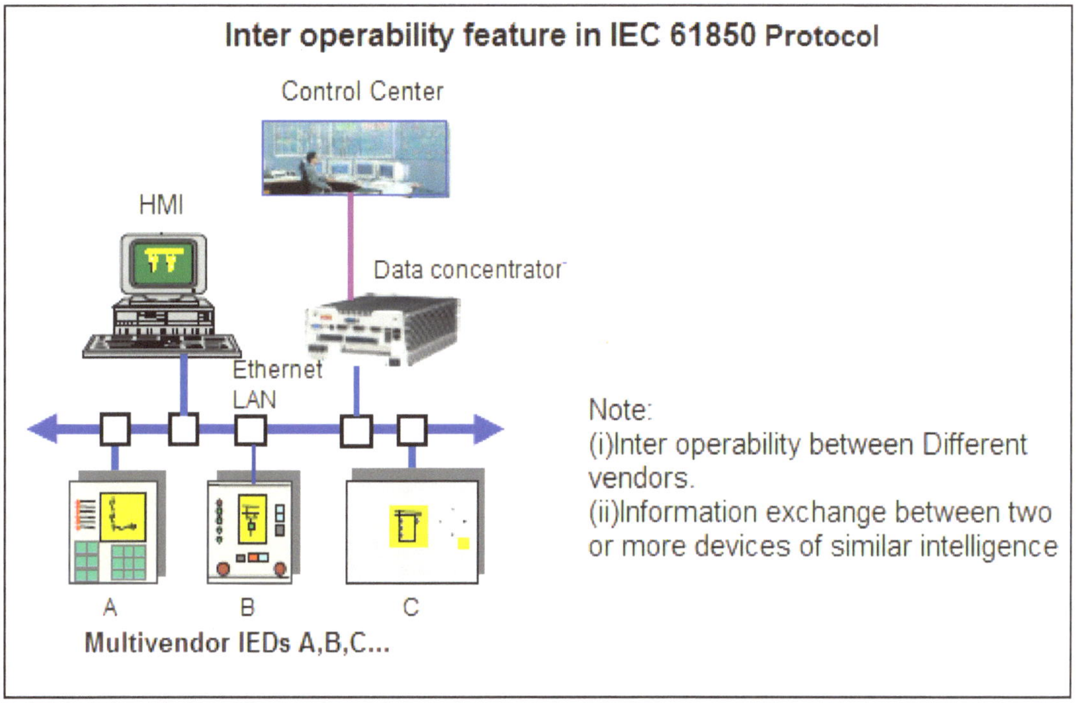

Figure 4: Interoperability feature in IEC 61850 Protocol.

7.0 Horizontal Communication: This is defined as peer-to- peer comunication between IEDs of different manufacturers or IEDs of same manufacturer.

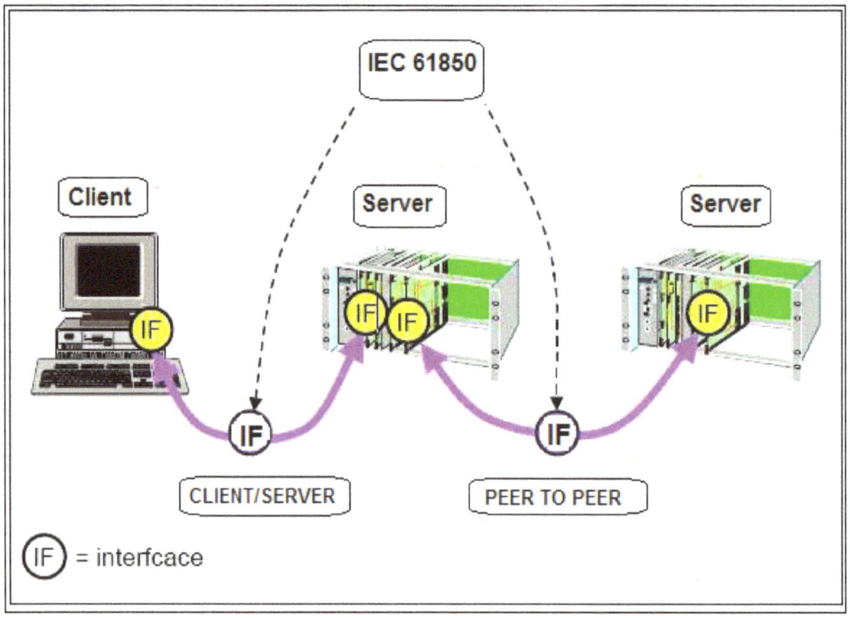

Figure 5: Peer to Peer Communication

This kind of Communication is very important specially at the time of critical function that must be highly reliable. They are definend in IEC 61850 as follows;

- GOOSE (Generic Object Oriented Sub station Event).
- GSSE (Generic Sub station Status Event).

8.0 Time Synchronization;

Time Synchronization is a very essential feature of Automation in a substation. Events of operations have to be published in chronological order of time. All the IEDs have to be Time synchronised. Range of accuracy as defined in IEC 6180 in milliseconds in LAN that should be NTP(Net work time protocol)and SNTP(Simple Net work Time Protocol) based. Radio signals and GPS (Global Position System) can be used also to implement Time synchronisation.

SNTP is able to Provide accuracy only about 1-millisecond which is not enough for raw data sample values and for merging units. As stipulated in IEEE 1588 Std. the requirement for such functions is less than 1millisecond. The solution is to use PTP (Precision Time Control) which has been developed off late having accuracy level of less than 1- micro-second.

9.0 GOOSE & GSSE Communication:

Features of GOOSE & GSSE Messages;

- GOOSE(Generic Object Oriented Sub station Event) messages:- These include many data type analogue, binary and integer values.
- GSSE(Generic Sub station Status Event). messages:- limited to support only fixed structure of binary event status data.

GOOSE and GSSE are high-speed-time-critical messaging system with high priority transmission capability to the final desitination in < 4ms. These protocols can be considered as multicast publisher/ subscriber model such that, those published GOOSE messages can be received by various IEDs which have been configuted to subscribe them.

10.0 Prioritization of GOOSE messages: Ethernet Traffic may slow down the GOOSE message, however using managed Ethernet switch GOOSE Message prioritization and segregation can be done at Ethernet network.

10.1 Etherne Switch Priority tagging

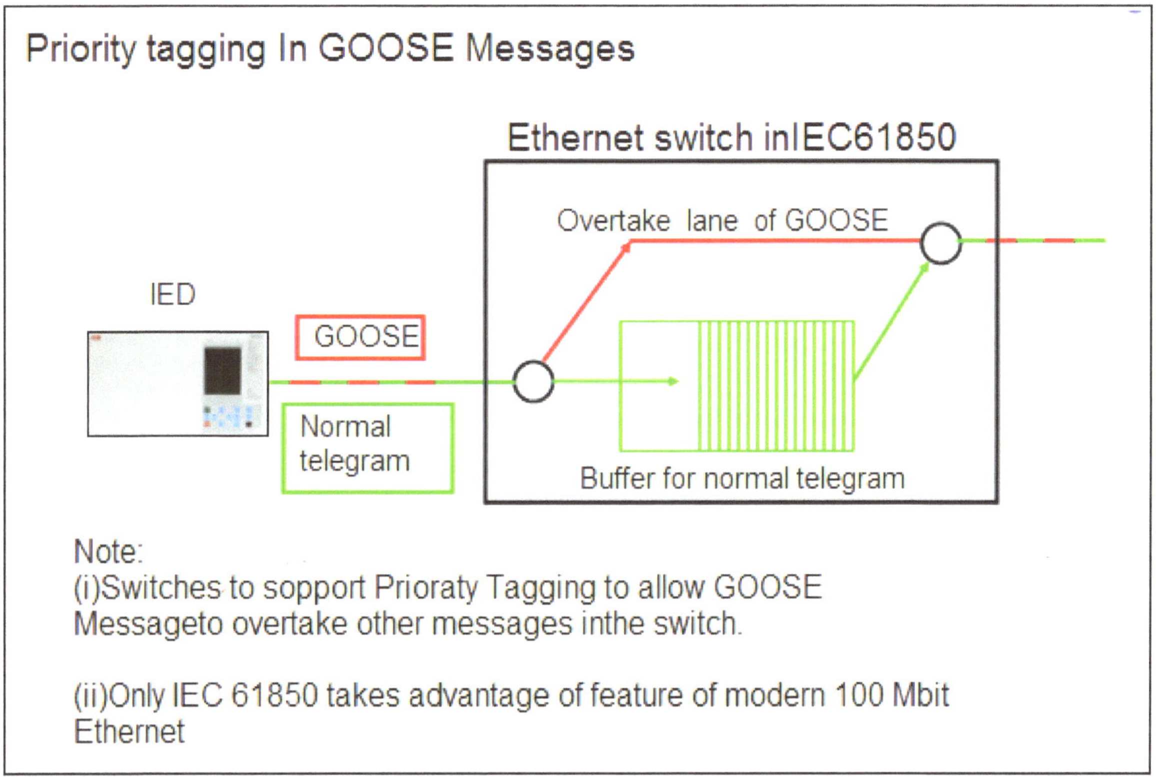

Figure 6: Priority tagging in GOOSE messages.

11.0 ISO/OSI (International Organisation of Standardization/Open System Interconnection) **Stack layers**

The ISO-OSI model is 7- layer architecture. It defines 7- layers or levels in a complete communication system.

It is worth mentioning that GOOSE Messages are sent in level 2 of ISO/OSI, TCP/IP stack, as a result GOOSE Messages can not be transmitted through Routers and Fire walls, hence out siders shall have difficulty in accessing or sending any threat to sub station automation Ethernet LAN.Therefore GOOSE and Ethernet LAN are secured for Sub Station Automation

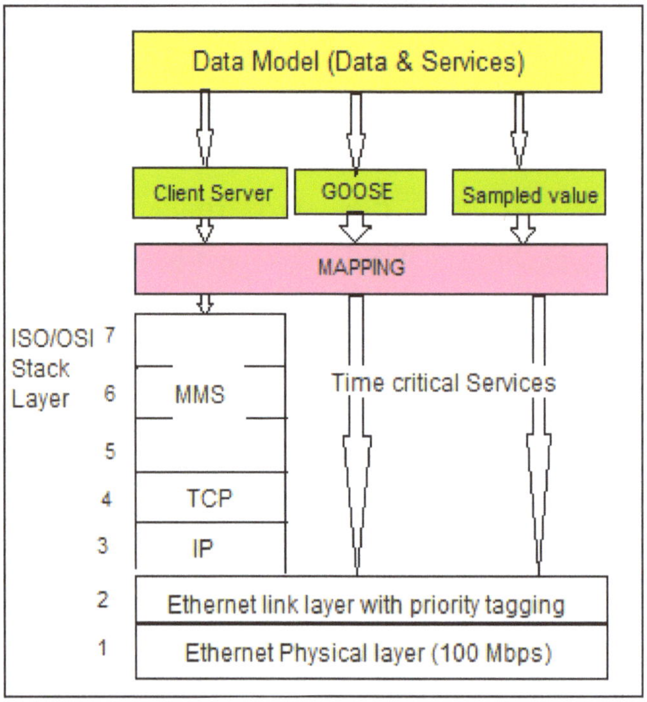

Figure 7: ISO/OSI Stack layers

The Main benifits for GOOSE in the Sub station automation are as follows;

- Fast signal transfer between Sub staion IEDs.
- Reduced Copper wiring for protection and control schemes.
- GOOSE messages are supervised signals transfer which ensures high system availability.
- Flexibility and expandability for Protection and Control schemes

12.0 Implementation of IEC 61850 protocol;

12.1 Constraints of of earlier SAS Protocol used Prior to IEC 61850 protocol;

a. Prior to IEC 61850 protocol, Sub Station Automation System was based on DNP3 & IEC 60870-5-101 protocol which had some limitations of peer to peer communication calling for protocol converters for peer to peer Communication between IEDs and hard wired connections betwwen each IEDs as depicted in a specimen SLD of a Sub-Station here under;

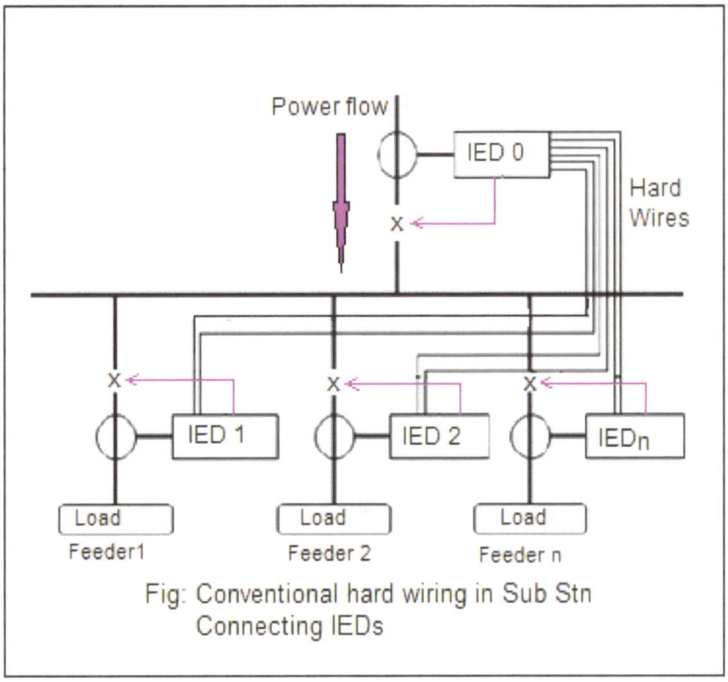

Figure 8: IED to IED connection through hard wiring(Copper).

b. With the implementation of IEC 61850 and GOOSE as a feature thereof, all such cumbersome panel to panel and IED to IED hard wiring would be eliminated and so also the protocol converters.

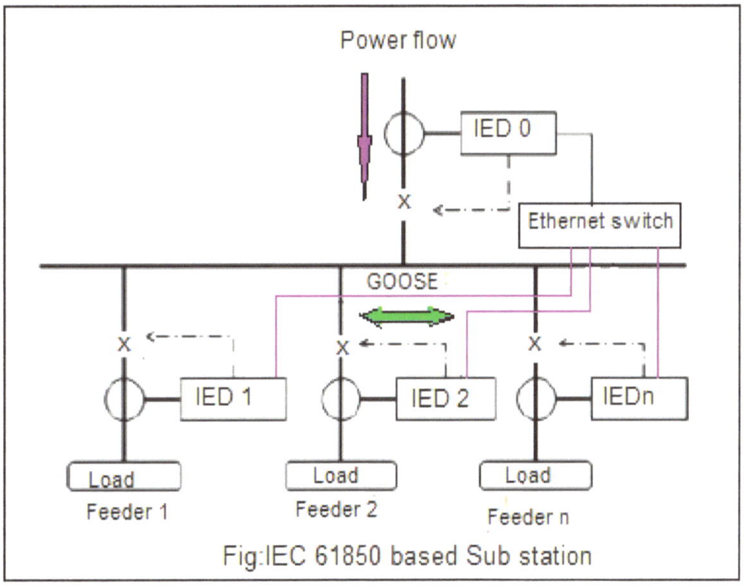

Figure 9: Connection between IEDs with implementation of IEC 61850 protocol(without hard wires)

13.0 Breaker Failure Protection- Concept:

In order to safeguard the power system against Cascade trippings/ System disturbance consequent to Local circuit breaker failure to open and clear the fault. Breaker failure protection is implemented. This protection is also known as LBB protection (Local breaker Backup protection).

13.1 Reasons for Breaker not opening on faults;

- Non availability of adequate Aux DC supply to trip coil.
- Loose connections to trip-coil.
- Trip coil assembly mechanically damaged.
- Defective breaker auxiliary switch.
- Trip latch of breaker not Properly aligned.
- Leakage of gas.

13.2 Breaker failure Protection logic;

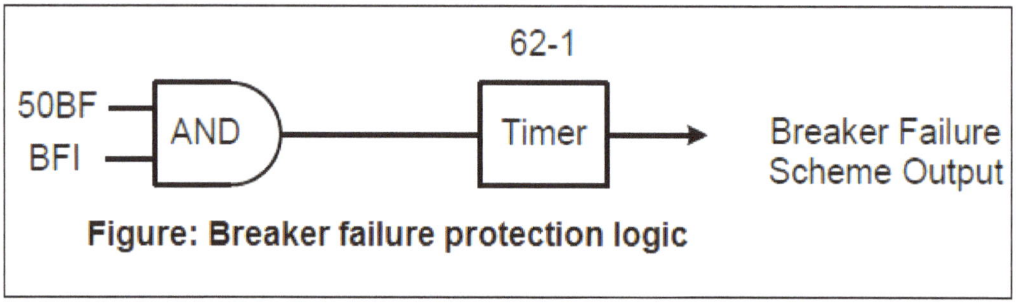

Figure 10: Breaker failure logic.

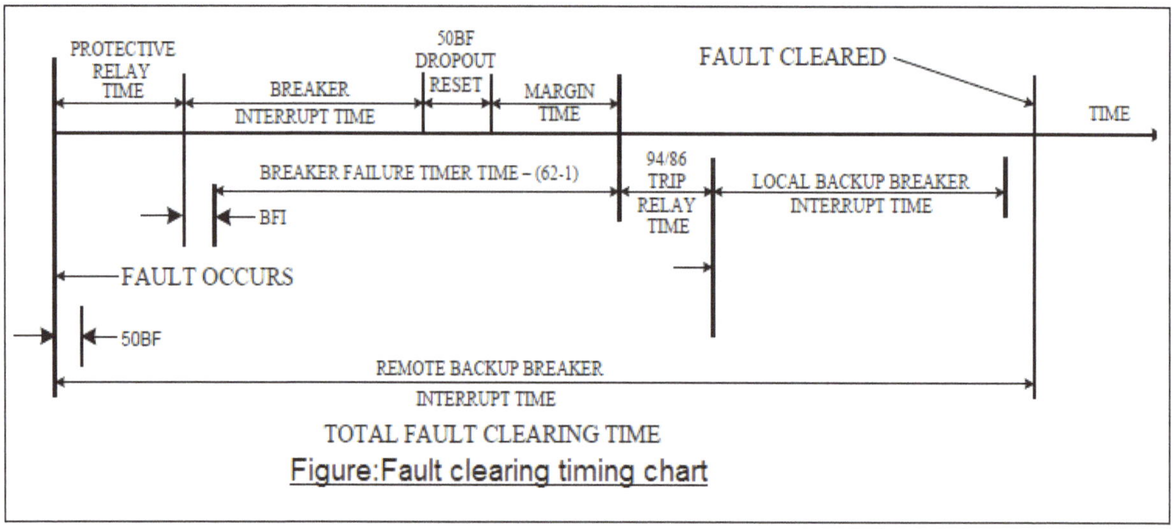

Figure 11: Fault clearing time chart

13.3 Sequence of Operations: Non-opening of local breaker on account of the reasons ennumerated in cl13.1, the sequence of operation is as follows (Ref. figure 10);

i. 50 BF is an Over-current sensor(O/C Relay) will pick-up simultaneously along with the operation of protective relay.

ii. BFI (Breaker fail innitiator circuit), It will get energized soon as the the trip signal is transmitted to the Breaker.

iii. 50BF and BFI together shall initiate the Breaker failure Timer (62-1) and shall startcounting the time.

iv. The time setting is given a margin of time above the Breaker interruption time+ 50BFdrop our time.

v. 50BF will drop out in case there is no breaker failure as soon as the Breaker Opens.

vi. With the dropping-off of 50BF the 62-1timer shall reset.

vii. If the Braker fails to open, after the set-time of

Breaker failure timer (62-1)which is >(breaker tripping time + 50BF drop out time + Margin time).

viii. The trip relay 86 shall get energized and shall give tripping command to all the adjacent breakers.

ix. The backup relay time of remote bus is co-ordinated such that it is more than total time from initiation of fault to clearing of fault through Breaker fail protection.

14.0 Breaker Failure Protection(50BF)in conjunction with GOOSE communication:

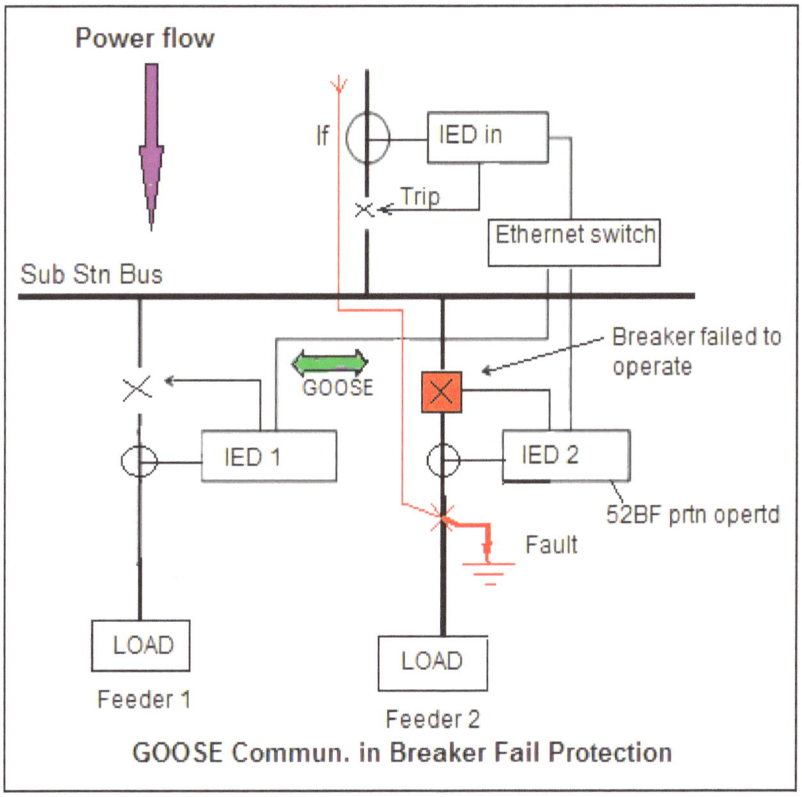

Figure 12: BFP through GOOSE communication in a S/s having single infeed and radial feders.

Breaker failure protection is also known as 50BF as per ANSI / IEEE protection code and is an essential Protection to minimize the Power system disturbance and failure of/ strain on other equiment during a fault condition in which circuit breaker fails to open after receiving a protection trip command from IED.

If the Circuit breaker fails to trip within a predefined period of time, IED sends a trip command to the up-stream breaker to clear the fault.Here fault clearance time is the essence of protction scheme.Prior to introduction of IEC 61850 containing GOOSE system for communication of command from one IED to another, Copper wiring was the only choice earlier.

In the hard-wired system the signal has to pass through out-put contact of sending IED and then through input contact of receiving IEDs.Using GOOSE message it is possible to reduce fault clearnce time in the breaker failure protection scheme and over all relay co-ordination time.

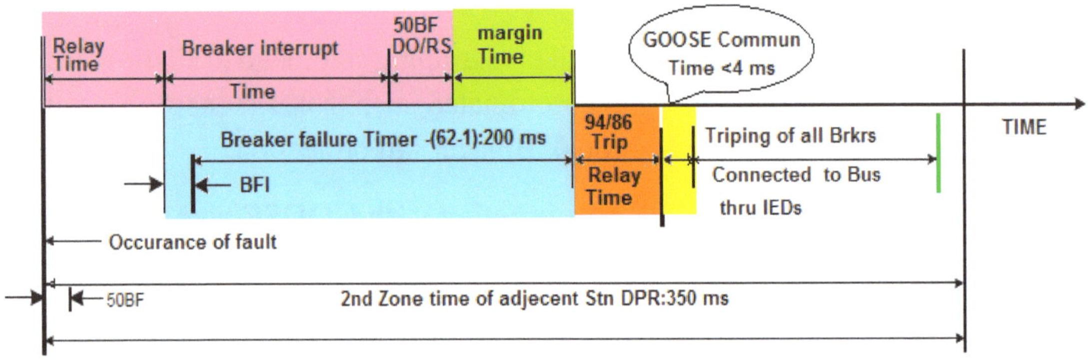

Figure: Time chart of clearing the fault in Breaker Failure Protection in conjunction with GOOSE communication System

Figure 13: Time chart indicating Breaker failure Protection in conjunction with GOOSE Communication.

Circuit breaker failure Protection is achieved by monitoring the time delay between the trip signal and actual opening of the Breaker.If the Circuit breaker fails to trip in predefined time, ie fault current has not been interrupted within a set time, the breaker failure protection 50BF is initiated.This is realised by sending a trip signal to incoming Breaker. It is depicted in the sample case as shown above.(fig. :12) &(fig 13)

Note:

a. The LBB or 50BFprotection timer is given a time of 200ms.
b. GOOSE Communication time is< 4ms.

15.0 Breaker Failure Protection In EHV S/Ss;

In case of Grid system where feeding sources are on either side of the Sub-stations with 2 bus(2 main bus + 1 Aux bus) sysyem or 1&1/2 breaker schemes, the Circuit breakers of all the feeders connected onto that particular bus wherein the the ill-fated feeder 's Circuit breaker Failure protection operated, are sent trip signal through GOOSE communication to individual IEDs of all the feeder to trip.After all the circuit breakers trip, the fault is isolated.

15.1.1 **Case (1) :**The figure No.14 depicts a 2- Main bus and 1- Auxilliary Bus system in an EHV Sub station in a Grid system having feeding source of power on either side of this Sub station.

Case (1) Please refer to the fig. no 14 in which the system a has been shown as all the feeders are on main bus 1.In the normal case, the load of feeder A is supplied

from this particular Sub station. Operation of the Breaker Failure Protection on feeder no.D has innitiated due to non-operation of the circuit breaker of feder D despite having fault on this feeder. The IED of feeder D gives trip command through GOOSE communication to the IEDs of the breakers of feeder A, feeder B and feeder C which are connected to the same Bus as feeder D, to trip their breakers immediately in the stipulated time period of <4ms to 4 ms.

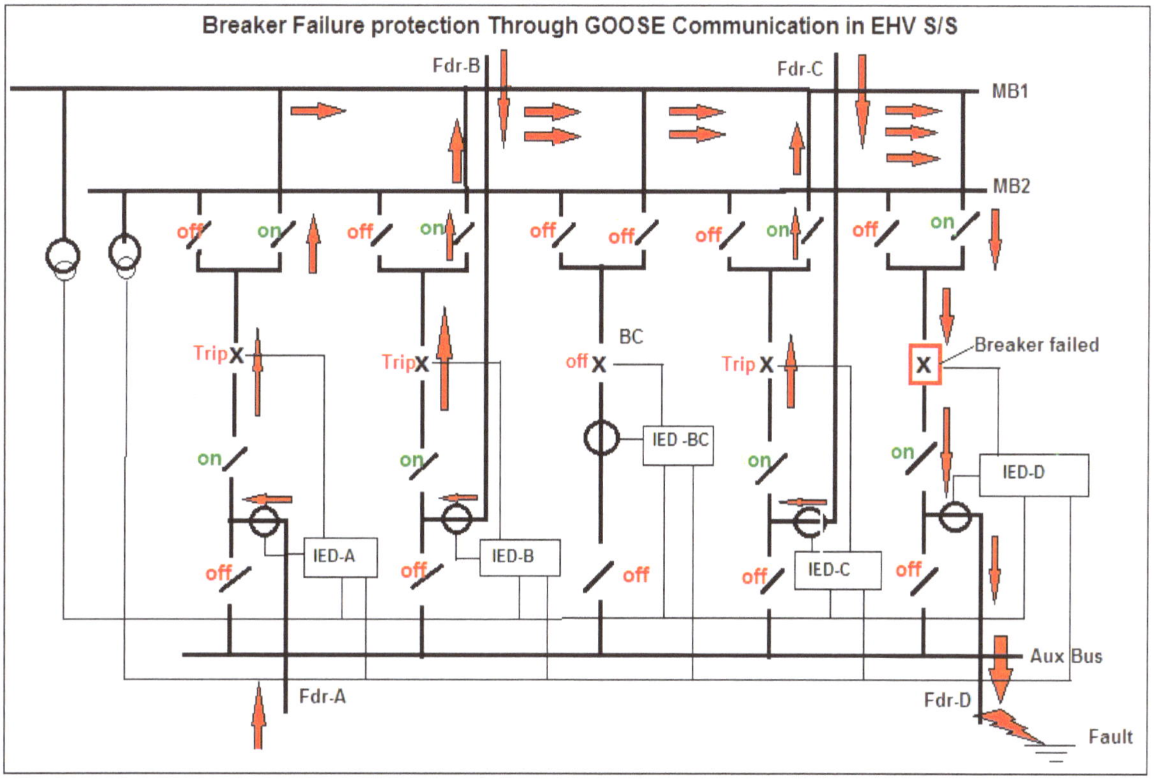

Figure 14: BFP of an EHV S/S all the feeders are fed from MB1

15.1.2 **Case (2)**: When some of the feeders are fed from bus No1and remainig feeders(in the instant case only -1 feeder) are being fed from Bus No 2 via bus-coupler.(Figure: 15) In such cases all the feeder breakers connected on Bus No -1 shall receive Trip command through GOOSE message communication sysyem and also the Bus coupler circuit breaker. In minimum possible time of < 4ms to 4ms.

Thus the feeders connected on to the bus No 2- shall not be disconnected and shall keep on functioning as usual.

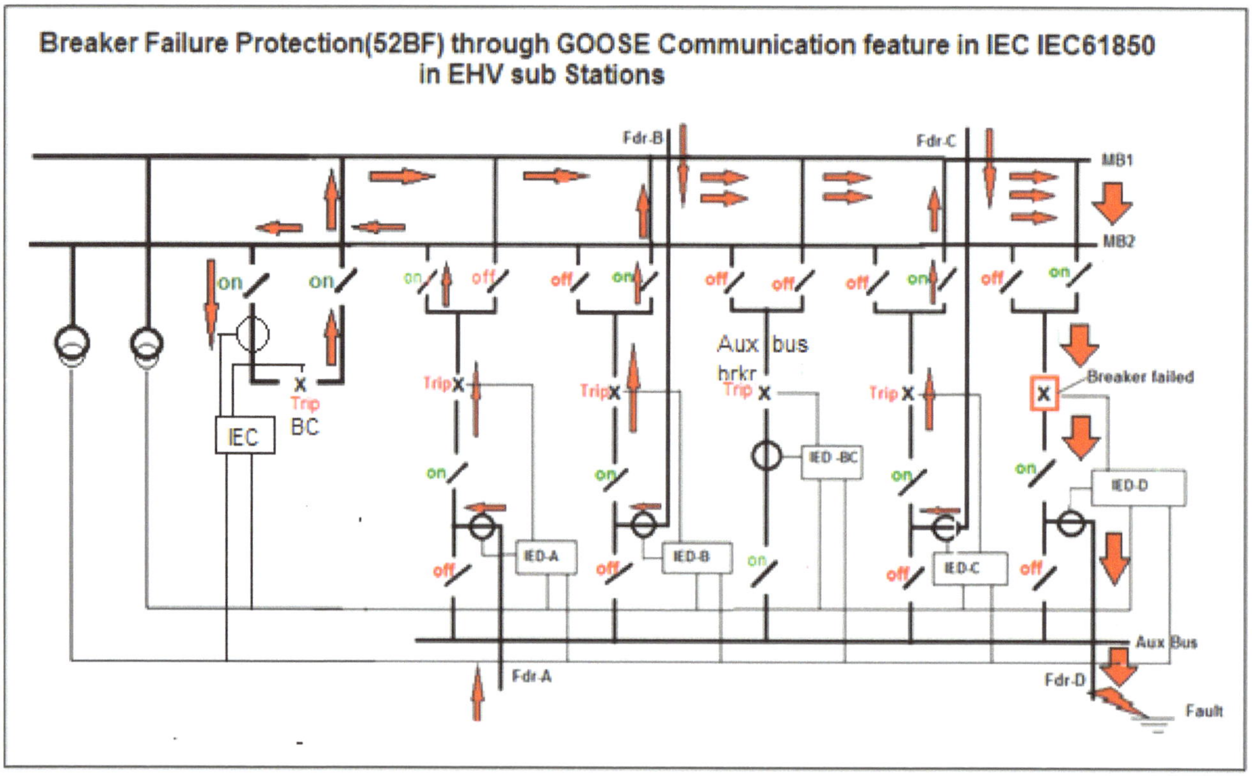

Figure 15: BFP of an EHV Sub station having some feeders on MB1 and one feeder on MB2

By using GOOSE communication the basic advantage is, it reduces time of operation as low as 4 ms and also gets-rid-of cumbersome hard-copper wirings.

16.0 Conclusion:

1. Substation automation using IEC 61850 protocol containig GOOSE messeging feature is very advantageous for Breaker Failure Protection, Early clearance of fault would, not only save the sysyem from disturbance but also shall prevent the outdoor equipment like Transformers CTs, PTs etc from failure and/or shall not hamper the life expectancy thereof.
2. In India too, all the STUs, CTUs and other private Transmission units may adopt Sub station Automation system in their EHV S/Ss above 220 kV in particular.

17.0 References:

i. ABB- Substation automation solution with IEC 61850.
ii. Overview of IEC61850 and benefits paper by R.E.Mackiewicz, member IEEE.
iii. Industrial communications Hand Book-Chapter 7, IEC 61850.
iv. Saeed Roostee; Substation automation using IEC61850.
v. R.C.Gupta Member IEEE: Substation Automation using IEC 61850 Standard.
vi. IEEE PSRCC K 23 WG headed by Roger Whittaker as Chairman -- Summary of revision, IEEE C 37.119-2016, Guide For Breaker Failure Protection of Power Circuit Breakers.

18.0 The Author:

Er.K.K.Murty holds a Bachelor's degree (Hons) in Electrical Engineering obtained from the University of Jabalpur in the year 1968. He was a former Chief Engineer and Head of Department (Testing & Commun.) in M.P. Power Transmission Co. Ltd. Jabalpur (India)., he was a member of the panel of Expert Professionals at the Central Power Research Institute (CPRI), Bangalore, from 2008 to 2012. Previous to this, he worked as an Advisor (Testing) at SOUTHCO, a DISCOM in the State of Orissa, a metering consultant to M. P. Electricity Regulatory Commission and a Course Director for the Graduate Electrical Engineering Trainees at the Training Institute of M.P. Power Transmission Co. Ltd., Jabalpur. He has published many technical aricles in the national and international journals and presented techical papers at various national and international conferences pertaining to the Power Transformers and other eqipments of power sector. Mr. Murty is a member of India's Society of Power Engineers (MSPE), a Fellow of Institution of Engineers, India (FIE) a Chartered Engineer (CE) and a Member CIGRE' India. He had been awarded a plaque by the Institution of Engineers(India), Kolkata, in Oct. 2015, in recognition of his eminance and contribution to the profession of Electrical Engineering at the National level.

Contact details (i) email address: kkmurty05@yahoo.com,

(ii) mob nos +91 9425154250, +91 7828149226

END

www.ingramcontent.com/pod-product-compliance
Lightning Source LLC
LaVergne TN
LVHW070942210926
848482LV00002B/74

* 9 7 9 8 8 9 4 7 5 0 2 0 0 *